T0140330

Cognitive Science and Technology

Series editor

David M.W. Powers, Adelaide, Australia

More information about this series at http://www.springer.com/series/11554

The seadragons were intrigued by calculus and flocked to the teacher

James K. Peterson

Calculus for Cognitive Scientists

Higher Order Models and Their Analysis

Springer

James K. Peterson
Department of Mathematical Sciences
Clemson University
Clemson, SC
USA

ISSN 2195-3988 ISSN 2195-3996 (electronic)
Cognitive Science and Technology
ISBN 978-981-13-5720-6 ISBN 978-981-287-877-9 (eBook)
DOI 10.1007/978-981-287-877-9

Printed on acid-free paper

This Springer imprint is published by SpringerNature
The registered company is Springer Science+Business Media Singapore Pte Ltd.

I dedicate this work to the biology students who have learned this material over the last 10 semesters, the practicing biologists, the immunologists, the cognitive scientists, and the computer scientists who have helped an outsider think much better and to my family who have listened to my ideas in the living room and over dinner for many years. I hope that this text helps inspire everyone who works in science to consider mathematics and computer science as indispensable tools in their own work in the biological sciences.

Acknowledgments

We would like to thank all the students who have used the various iterations of these notes as they have evolved from handwritten to the fourth fully typed version here. We particularly appreciate your interest as this course is required and uses mathematics; a combination that causes fear in many biological science majors. We have been pleased by the enthusiasm you have brought to this interesting combination of ideas from many disciplines. Finally, we gratefully acknowledge the support of Hap Wheeler in the Department of Biological Sciences during the years 2006 to 2014 for believing that this material would be useful to biology students.

For this new text on a follow-up course to the first course on calculus for cognitive scientists, we would like to thank all of the students from Spring 2006 to Fall 2014 for their comments and patience with the inevitable typographical errors, mistakes in the way we explained topics, and organizational flaws as we have taught second semester of calculus ideas to them. This new text starts assuming you know something the equivalent of a first semester course in calculus and particularly know about exponential and logarithm functions, first-order models and the MATLAB tools needed to solve the models numerically. In addition, you need to know a fair bit of a start into calculus for functions of more than one variable and the ideas of approximation to functions of one and two variables. These are not really standard topics in just one course in calculus, which is why our first volume was written to provide coverage of all those things. In addition, all of the mathematics subserve ideas from biological models so that everything is wrapped together in a pleasing package!

With that background given, in this text, we add new material on linear and nonlinear systems models and more biological models. We also cover a useful way of solving what are called linear partial differential equations using the technique named *Separation of Variables*. To make sense of all this, we naturally have to dip into mathematical waters at appropriate points and we are not shy about that! But rest assured, everything we do is carefully planned because it is of great use to you in your attempts to forge an alliance between cognitive science, mathematics, and computation.

Contents

Part IV Interesting Models

List of Figures

List of Tables

List of Code Examples

Abstract

This book tries to show how mathematics, computer science, and biology can be usefully and pleasurably intertwined. The first volume (J. Peterson, *Calculus for Cognitive Scientists: Derivatives, Integration and Modeling* (Springer, Singapore, 2015 in press)) discussed the necessary one- and two-variable calculus tools as well as first-order ODE models. In this volume, we explicitly focus on two-variable ODE models both linear and nonlinear and learn both theoretical and computational tools using MATLAB to help us understand their solutions. We also go over carefully on how to solve cable models using separation of variables and Fourier series. And we must always caution you to be careful to make sure the use of mathematics gives you insight. These cautionary words about the modeling of the physics of stars from 1938 should be taken to heart:

> Technical journals are filled with elaborate papers on conditions in the interiors of model gaseous spheres, but these discussions have, for the most part, the character of exercises in mathematical physics rather than astronomical investigations, and it is difficult to judge the degree of resemblance between the models and actual stars. Differential equations are like servants in livery: it is honourable to be able to command them, but they are "yes" men, loyally giving support and amplification to the ideas entrusted to them by their master—Paul W. Merrill, The Nature of Variable Stars, 1938, quoted in Arthur I. Miller Empire of the Stars: Obsession, Friendship, and Betrayal in the Quest for Black Holes, 2005.

The relevance of this quotation to our pursuits is clear. It is easy to develop sophisticated mathematical models that abstract from biological complexity something we can then analyze with mathematics or computational tools in an attempt to gain insight. But as Merrill says,

> it is difficult to judge the degree of resemblance between the models and actual [biology]

where we have taken the liberty to replace *physics* with our domain here of *biology*. We should never forget the last line

> Differential equations are like servants in livery: it is honourable to be able to command them, but they are "yes" men, loyally giving support and amplification to the ideas entrusted to them by their master.

We must always take our modeling results and go back to the scientists to make sure they retain relevance.

History

Based On:
Notes On MTHSC 108 for Biologists developed during the
Spring 2006 and Fall 2006,
Spring 2007 and Fall 2007 courses
The first edition of this text was used in
Spring 2008 and Fall 2008,
The course was then relabeled at MTHSC 111
and the text was used in
Spring 2009 and Fall 2009 courses
The second edition of this text was used in
Spring 2010 and Fall 2010 courses
The third edition was used in the
Spring 2011 and Fall 2011,
Spring 2012 and Summer Session I courses
The fourth edition was used in
Fall 2012, Spring 2013, and Fall 2013 courses
The fifth edition was used in the Spring 2014 course
Also, we have used material from notes on
Partial Differential Equation Models
which has been taught to small numbers of students since 2008.

Part I
Introduction

Chapter 1
Introductory Remarks

In this course, we will try to introduce beginning cognitive scientists to more of the kinds of mathematics and mathematical reasoning that will be useful to them if they continue to live, grow and learn within this area. In our twenty first century world, there is a tight integration between the areas of mathematics, computer science and science. Now a traditional Calculus course for the engineering and physical sciences consists of a four semester sequence as shown in Table 1.1.

Unfortunately, this sequence of courses is heavily slanted towards the needs of the physical sciences. For example, many of our examples come from physics, chemistry and so forth. As long as everyone in the class has the common language to understand the examples, the examples are a wonderful tool for adding insight. However, the typical students who are interested in cognitive science often find the language of physics and engineering to be outside their comfort zone. Hence, the examples lose their explanatory power. Our first course starts a different way of teaching this material and this text will continue that journey. Our philosophy, as usual, is that all parts of the course must be integrated, so we don't want to use mathematics, science or computer approaches for their own intrinsic value. Experts in these separate fields must work hard to avoid this. This is the time to be generalists and always look for connective approaches. Also, models are carefully chosen to illustrate the basic idea that we know far too much detail about virtually any biologically based system we can think of. Hence, we must learn to throw away information in the search of the appropriate abstraction. The resulting ideas can then be phrased in terms of mathematics and simulated or solved with computer based tools. However, the results are not useful, and must be discarded and the model changed, if the predictions and illuminating insights we gain from the model are incorrect. We must always remember that throwing away information allows for the possibility of mistakes. This is a hard lesson to learn, but important. Note that models from population biology, genetics, cognitive dysfunction, regulatory gene circuits and many others are good examples to work with. All require massive amounts of abstraction and data pruning to get anywhere, but the illumination payoffs are potentially quite large.

J.K. Peterson, *Calculus for Cognitive Scientists*,
Cognitive Science and Technology, DOI 10.1007/978-981-287-877-9_1

Table 1.1 Typical engineering calculus sequence

Course	Topics
MTHSC 106	Simple limit ideas
	Functions and continuity
	Differentiation and applications
	Simple integration and applications
MTHSC 108	More integration and applications
	Sequences and series
MTHSC 206	Polar coordinates
	Space curves
	Double and triple integrals
	Vectors and functions of more than one variable
	Partial derivatives
	2D and 3D applications
MTHSC 208	First order ordinary differential equations (ODEs)
	Linear second order differential equations
	Complex numbers and linear independence of functions
	Systems of linear differential equations
	Matrices, eigenvalues and eigenvectors
	Qualitative analysis of linear systems
	Laplace transform techniques

1.1 A Roadmap to the Text

In this course, we introduce enough relevant mathematics and the beginnings of useful computational tools so that you can begin to understand a fair bit about Biological Modeling. We present a selection of nonlinear biological models and slowly build you to the point where you can begin to have a feel for the model building process. We start our model discussion with the classical Predator–Prey model in Chap. 10. We try to talk about it as completely as possible and we use it as a vehicle to show how graphical analysis coupled with careful mathematical reasoning can give us great insight. We discuss completely the theory of the original Predator–Prey model in Sect. 10.1 and its qualitative analysis in Sect. 10.5. We then introduce the use of computational tools to solve the Predator–Prey model using MatLab in Sect. 10.11. While this model is very successful at modeling biology, the addition of self-interaction terms is not. The self-interaction models are analyzed in Chap. 11 and computational tools are discussed in Sect. 11.8.

In Chap. 12, we show you a simple infectious disease model. The nullclines for this model are developed in Sect. 12.1 and our reasoning why only trajectories that start with positive initial conditions are biologically relevant are explained in Sect. 12.2.

The infectious versus susceptible curve is then derived in Sect. 12.3. We finish this Chapter with a long discussion of how we use a bit of mathematical wizardry to develop a way to estimate the value of ρ in these disease models by using data gathered on the value of R'. This analysis in Sect. 12.6, while complicated, is well worth your effort to peruse!

In Chap. 13, we show you a simple model of colon cancer which while linear is made more complicated by its higher dimensionality—there are 6 variables of interest now and graphical analysis is of less help. We try hard to show you how we can use this model to get insight as to when point mutations or chromosomal instability are the dominant pathway to cancer.

In Chap. 15, we go over a simple model of insulin detection using second order models which have complex roots. We use the phase shifted form to try to detect insulin from certain types of data.

1.2 Code

The code for much of this text is in the directory **ODE** in our code folder which you can download from **Biological Information Processing** (http://www.ces.clemson.edu/~petersj/CognitiveModels.html). These code samples can then be downloaded as the zipped tar ball **CognitiveCode.tar.gz** and unpacked where you wish. If you have access to MatLab, just add this folder with its sub folders to your MatLab path. If you don't have such access, download and install **Octave** on your laptop. Now Octave is more of a command line tool, so the process of adding paths is a bit more tedious. When we start up an Octave session, we use the following trick. We write up our paths in a file we call **MyPath.m**. For us, this code looks like this

Listing 1.1: How to add paths to octave

```
function MyPath()
%
s1 = '/home/petersj/MatLabFiles/BioInfo/:';
s2 = '/home/petersj/MatLabFiles/BioInfo/GSO:';
s3 = '/home/petersj/MatLabFiles/BioInfo/HH:';
s4 = '/home/petersj/MatLabFiles/BioInfo/Integration:';
s5 = '/home/petersj/MatLabFiles/BioInfo/Interpolation:';
s6 = '/home/petersj/MatLabFiles/BioInfo/LinearAlgebra:';
s7 = '/home/petersj/MatLabFiles/BioInfo/Nernst:';
s8 = '/home/petersj/MatLabFiles/BioInfo/ODE:';
s9 = '/home/petersj/MatLabFiles/BioInfo/RootsOpt:';
s10 = '/home/petersj/MatLabFiles/BioInfo/Letters:';
s11 = '/home/petersj/MatLabFiles/BioInfo/Graphs:';
s12 = '/home/petersj/MatLabFiles/BioInfo/PDE:';
s13 = '/home/petersj/MatLabFiles/BioInfo/FDPDE:';
s14 = '/home/petersj/MatLabFiles/BioInfo/3DCode';
s = [s1,s2,s3,s4,s5,s6,s7,s8,s9,s12];
addpath(s);
end
```

The paths we want to add are setup as strings, here called **s1** etc., and to use this, we start up Octave like so. We copy **MyPath.m** into our working directory and then do this

Listing 1.2: Set paths in octave

```
octave>> MyPath();
```

We agree it is not as nice as working in MatLab, but it is free! You still have to think a bit about how to do the paths. For example, in Peterson (2015c), we develop two different ways to handle graphs in MatLab. The first is in the directory **GraphsGlobal** and the second is in the directory **Graphs**. They are not to be used together. So if we wanted to use the setup of **Graphs** and nothing else, we would edit the **MyPath.m** file to set **s = [s11];** only. If we wanted to use the **GraphsGlobal** code, we would edit **MyPath.m** so that **s11 = '/home/petersj/MatLabFiles/BioInfo/GraphsGlobal:';** and then set **s = [s11];**. Note the directories in the **MyPath.m** are ours: the main directory is **'/home/petersj/MatLabFiles/BioInfo/** and of course, you will have to edit this file to put your directory information in there instead of ours.

All the code will work fine with **Octave**. So pull up a chair, grab a cup of coffee or tea and let's get started.

1.3 Final Thoughts

As we said in Peterson (2015a), we want you to continue to grow in a multidisciplinary way and so we think you will definitely need to learn more. Remember, every time you try to figure something out in science, you will find there is a lot of stuff you don't know and you have to go learn new tricks. That's ok and you shouldn't be afraid of it. You will probably find you need more mathematics, statistics and so forth in your work, so don't forget to read more as there is always something interesting over the horizon you are not prepared for yet. But the thing is, every time you figure something out, you get better at figuring out the next thing! Also, we have written several more companion texts for you to consider on your journey. The next on is **Calculus for Cognitive Scientists: Partial Differential Equation Models**, Peterson (2015b) and then there is the fourth volume on bioinformation processing, **BioInformation Processing: A Primer On Computational Cognitive Science**, Peterson (2015c) which starts you on building interesting neural systems, which this material will prepare you for. Enjoy your journeys!

References

J. Peterson, *Calculus for Cognitive Scientists: Derivatives, Integration and Modeling*, Springer Series on Cognitive Science and Technology (Springer Science+Business Media Singapore Pte Ltd, Singapore, 2015a in press)

J. Peterson, *Calculus for Cognitive Scientists: Partial Differential Equation Models*, Springer Series on Cognitive Science and Technology (Springer Science+Business Media Singapore Pte Ltd, Singapore, 2015b in press)

J. Peterson, *BioInformation Processing: A Primer On Computational Cognitive Science*, Springer Series on Cognitive Science and Technology (Springer Science+Business Media Singapore Pte Ltd, Singapore, 2015c in press)

Part II
Review

Chapter 2
Linear Algebra

We need to use both *vector* and *matrix* ideas in this course. This was covered already in the first text (Peterson 2015), so we will assume you can review that material before you start into this chapter. Here we will introduce some new ideas as well as tools in MatLab we can use to solve what are called linear algebra problems; i.e. systems of equations. Let's begin by looking at inner products more closely.

2.1 The Inner Product of Two Column Vectors

We can also define the *inner product* of two vectors. If $\boldsymbol{V}$ and $\boldsymbol{W}$ are two column vectors of size $n \times 1$, then the product $\boldsymbol{V}^T \boldsymbol{W}$ is a matrix of size 1×1 which we identify with a real number. We see if

$$\boldsymbol{V} = \begin{bmatrix} V_1 \\ V_2 \\ V_3 \\ \vdots \\ V_n \end{bmatrix} \text{ and } \boldsymbol{W} = \begin{bmatrix} W_1 \\ W_2 \\ W_3 \\ \vdots \\ W_n \end{bmatrix}$$

then we define the 1×1 matrix

$$\boldsymbol{V}^T \boldsymbol{W} = \boldsymbol{W}^T \boldsymbol{V} = \langle \boldsymbol{V}, \boldsymbol{W} \rangle \left[V_1 W_1 + V_2 W_2 + V_3 W_3 + \cdots + V_n W_n \right]$$

and we identify this one by one matrix with the real number

$$V_1 W_1 + V_2 W_2 + V_3 W_3 + \cdots + V_n W_n$$

This product is so important, it is given a special name: it is the **inner product** of the two vectors $\boldsymbol{V}$ and $\boldsymbol{W}$. Let's make this formal with Definition 2.1.1.

J.K. Peterson, *Calculus for Cognitive Scientists*,
Cognitive Science and Technology, DOI 10.1007/978-981-287-877-9_2

Definition 2.1.1 (*The Inner Product Of Two Vectors*)
If $\boldsymbol{V}$ and $\boldsymbol{W}$ are two column vectors of size $n \times 1$, the inner product of these vectors is denoted by $< \boldsymbol{V}, \boldsymbol{W} >$ which is defined as the matrix product $\boldsymbol{V}^T \boldsymbol{W}$ which is equivalent to the $\boldsymbol{W}^T \boldsymbol{V}$ and we interpret this 1×1 matrix product as the real number

$$V_1 W_1 + V_2 W_2 + V_3 W_3 + \cdots + V_n W_n$$

where V_i are the components of $\boldsymbol{V}$ and W_i are the components of $\boldsymbol{W}$.

2.1.1 Homework

Exercise 2.1.1 *Find the dot product of the vectors* $\boldsymbol{V}$ *and* $\boldsymbol{W}$ *given by*

$$\boldsymbol{V} = \begin{bmatrix} 6 \\ 1 \end{bmatrix} \text{ and } \boldsymbol{W} = \begin{bmatrix} 7 \\ 2 \end{bmatrix}.$$

Exercise 2.1.2 *Find the dot product of the vectors* $\boldsymbol{V}$ *and* $\boldsymbol{W}$ *given by*

$$\boldsymbol{V} = \begin{bmatrix} -6 \\ -8 \end{bmatrix} \text{ and } \boldsymbol{W} = \begin{bmatrix} 2 \\ 6 \end{bmatrix}.$$

Exercise 2.1.3 *Find the dot product of the vectors* $\boldsymbol{V}$ *and* $\boldsymbol{W}$ *given by*

$$\boldsymbol{V} = \begin{bmatrix} 10 \\ -4 \end{bmatrix} \text{ and } \boldsymbol{W} = \begin{bmatrix} 2 \\ 80 \end{bmatrix}.$$

We add, subtract and scalar multiply vectors and matrices as usual. We also suggest you review how to do matrix–vector multiplications. Multiplication of matrices is more complex as we discussed in the volume (Peterson 2015). Let's go through it again a bit more abstractly. Recall the dot product of two vectors $\boldsymbol{V}$ and $\boldsymbol{V}\boldsymbol{W}$ is defined to be

$$\langle \boldsymbol{V}, \boldsymbol{W} \rangle = \sum_{i=1}^{n} V_i W_i$$

where n is the number of components in the vectors. Using this we can define the multiplication of the matrix $\boldsymbol{A}$ of size $n \times p$ with the matrix $\boldsymbol{B}$ of size $p \times m$ as follows.

$$\begin{bmatrix} \text{Row 1 of A} \\ \text{Row 2 of A} \\ \vdots \\ \text{Row n of A} \end{bmatrix} \begin{bmatrix} \text{Column 1 of B} \mid \cdots \mid \text{Column n of B} \end{bmatrix}$$

$$= \begin{bmatrix} <\text{Row 1 of A, Column 1 of B}> \ldots <\text{Row 1 of A, Column n of B}> \\ <\text{Row 2 of A, Column 1 of B}> \ldots <\text{Row 2 of A, Column n of B}> \\ \vdots \quad\quad \vdots \quad\quad \vdots \\ <\text{Row n of A, Column 1 of B}> \ldots <\text{Row n of A, Column n of B}> \end{bmatrix}$$

We can write this more succinctly with we let $\boldsymbol{A}_i$ denote the rows of $\boldsymbol{A}$ and $\boldsymbol{B}^i$ be the columns of $\boldsymbol{B}$. Note the use of subscripts for the rows and superscripts for the columns. Then, we can rewrite the matrix multiplication algorithm more compactly as

$$\begin{bmatrix} \boldsymbol{A}_1 \\ \boldsymbol{A}_2 \\ \vdots \\ \boldsymbol{A}_n \end{bmatrix} \begin{bmatrix} \boldsymbol{B}^1 \mid \cdots \mid \boldsymbol{B}^n \end{bmatrix} = \begin{bmatrix} \langle \boldsymbol{A}_1, \boldsymbol{B}^1 \rangle & \ldots & \langle \boldsymbol{A}_1, \boldsymbol{B}^n \rangle \\ \langle \boldsymbol{A}_2, \boldsymbol{B}^1 \rangle & \ldots & \langle \boldsymbol{A}_2, \boldsymbol{B}^n \rangle \\ \vdots & \vdots & \vdots \\ \langle \boldsymbol{A}_n, \boldsymbol{B}^1 \rangle & \ldots & \langle \boldsymbol{A}_n, \boldsymbol{B}^1 \rangle \end{bmatrix}$$

Thus, the entry in row i and column j of the matrix product $\boldsymbol{AB}$ is

$$\boldsymbol{AB}_{ij} = \langle \boldsymbol{A}_i, \boldsymbol{B}^j \rangle.$$

Comment 2.1.1 *If* $\mathbf{A}$ *is a matrix of any size and* $\mathbf{0}$ *is the appropriate zero matrix of the same size, then both* $\mathbf{0} + \mathbf{A}$ *and* $\mathbf{A} + \mathbf{0}$ *are nicely defined operations and the result is just* $\mathbf{A}$.

Comment 2.1.2 *Matrix multiplication is not communicative: i.e. for square matrices* $\mathbf{A}$ *and* $\mathbf{B}$*, the matrix product* $\mathbf{A}\ \mathbf{B}$ *is not necessarily the same as the product* $\mathbf{B}\ \mathbf{A}$.

2.2 Interpreting the Inner Product

What could this number $< \boldsymbol{V}, \boldsymbol{W} >$ possibly mean? To figure this out, we have to do some algebra. Let's specialize to nonzero column vectors with only 2 components. Let

$$\boldsymbol{V} = \begin{bmatrix} a \\ c \end{bmatrix} \text{ and } \boldsymbol{W} = \begin{bmatrix} b \\ d \end{bmatrix}$$

Since these vectors are not zero, only one of the terms in (a, c) and in (b, d) can be zero because otherwise both components would be zero and we are assuming these vectors are not the zero vector. We will use this fact in a bit. Now here $< \boldsymbol{V}, \boldsymbol{W} > = ab + cd$. So

$$(ab + cd)^2 = a^2b^2 + 2abcd + c^2d^2$$

$$\begin{aligned} || \boldsymbol{V} ||^2 \, || \boldsymbol{W} ||^2 &= \left(a^2 + c^2\right) \, \left(b^2 + d^2\right) \\ &= a^2b^2 + a^2d^2 + c^2b^2 + c^2d^2 \end{aligned}$$

Thus,

$$\begin{aligned} || \boldsymbol{V} ||^2 \, || \boldsymbol{W} ||^2 - (< \boldsymbol{V}, \boldsymbol{W} >)^2 &= a^2b^2 + a^2d^2 + c^2b^2 + c^2d^2 - a^2b^2 - 2abcd - c^2d^2 \\ &= a^2d^2 - 2abcd + c^2b^2 \\ &= (ad - bc)^2 \,. \end{aligned}$$

Now, this does look complicated, doesn't it? But this last term is something squared and so it must be non-negative! Hence, taking square roots, we have shown that

$$|< \boldsymbol{V}, \boldsymbol{W} >| \leq || \boldsymbol{V} || \, || \boldsymbol{W} ||$$

Note, since a real number is always less than or equal to it absolute value, we can also say

$$< \boldsymbol{V}, \boldsymbol{W} > \leq || \boldsymbol{V} || \, || \boldsymbol{W} ||$$

And we can say more. If it turned out that the term $(ad - bc)^2$ was zero, then $ad - bc = 0$. There are then a few cases to look at.

1. If all the terms a, b, c and d are not zero, then we can write $ad = bc$ implies $a/c = b/d$. We know the vector $\boldsymbol{V}$ can be interpreted as the line segment starting at $(0, 0)$ on the line with equation $y = (a/c)x$. Similarly, the vector $\boldsymbol{W}$ can be interpreted as the line segment connecting $(0, 0)$ and (b, d) on the line $y = (b/d)x$. Since $a/c = b/d$, these lines are the same. So both points (a, c) and (b, d) line on the same line. Thus, we see these vectors lay on top of each other or point directly opposite each other in the $x - y$ plane; i.e. the angle between these vectors is 0 or π radians (that is $0°$ or $180°$).
2. If $a = 0$, then bc must be 0 also. Since we know the vector $\boldsymbol{V}$ is not the zero vector, we can't have $c = 0$ also. Thus, b must be zero. This tells us $\boldsymbol{V}$ has components $(0, c)$ for some non zero c and $\boldsymbol{V}$ has components $(0, d)$ for some non zero d. These components also determine two lines like in the case above which either point in the same direction or opposite one another. Hence, again, the angle between the lines determined by these vectors is either 0 or π radians.
3. We can argue just like the case above if $d = 0$. We would find the angle between the lines determined by the vectors is either 0 or π radians.

We can summarize our results as a Theorem which is called the Cauchy–Schwartz Theorem for two dimensional vectors.

Theorem 2.2.1 (Cauchy Schwartz Theorem For Two Dimensional Vectors)
If $\boldsymbol{V}$ *and* $\boldsymbol{W}$ *are two dimensional column vectors with components* (a, c) *and* (b, d) *respectively, then it is always true that*

$$< \boldsymbol{V}, \boldsymbol{W} > \leq |< \boldsymbol{V}, \boldsymbol{W} >| \leq || \boldsymbol{V} || \, || \boldsymbol{W} ||$$

Moreover,

$$|< \boldsymbol{V}, \boldsymbol{W} >| = || \boldsymbol{V} || \, || \boldsymbol{W} ||$$

if and only the quantity $ad - bc = 0$*. Further, this quantity is equal to* 0 *if and only if the angle between the line segments determined by the vectors* $\boldsymbol{V}$ *and* $\boldsymbol{W}$ *is* $0°$ *or* $180°$.

Here is yet another way to look at this: assume there is a non zero value of t so that the equation below is true.

$$\boldsymbol{V} + t\,\boldsymbol{W} = \begin{bmatrix} a \\ c \end{bmatrix} + t \begin{bmatrix} b \\ d \end{bmatrix} = \begin{bmatrix} 0 \\ 0 \end{bmatrix}$$

This implies

$$\begin{bmatrix} a \\ c \end{bmatrix} = -t \begin{bmatrix} b \\ d \end{bmatrix}$$

Since these two vectors are equal, their components must match. Thus, we must have

$$\begin{aligned} a &= -t\, b \\ c &= -t\, d \end{aligned}$$

Thus,

$$a\, d = (-t\, b)\, \frac{c}{-t} = b\, c$$

and we are back to $ad - bc = 0$! Hence, another way of saying that the vectors $\boldsymbol{V}$ and $\boldsymbol{W}$ are either $0°$ or $180°$ apart is to say that as vectors they are multiples of one another! Such vectors are called **collinear** vectors to save writing. In general, we say two n dimensional vectors are collinear if there is a nonzero constant t so that $\boldsymbol{V} = t\, \boldsymbol{W}$ although, of course, we can't really figure out a way to visualize these vectors!

Now, the scaled vectors $\boldsymbol{E} = \frac{\boldsymbol{V}}{||\boldsymbol{V}||}$ and $\boldsymbol{F} = \frac{\boldsymbol{W}}{||\boldsymbol{W}||}$ have magnitudes of 1. Their components are $(a/ \, || \boldsymbol{V} ||, c/ \, || \boldsymbol{V} ||$ and $(b/ \, || \boldsymbol{W} ||, d/ \, || \boldsymbol{W} ||$. These points lie on a circle of radius 1 centered at the origin. Let θ_1 be the angle $\boldsymbol{E}$ makes with the positive x-axis. Then, since the hypotenuse distance that defines the $\cos(\theta_1)$ and $\sin(\theta_1)$ is 1, we must have

$$\cos(\theta_1) = \frac{a}{\| \boldsymbol{V} \|}$$
$$\sin(\theta_1) = \frac{c}{\| \boldsymbol{V} \|}$$

We can do the same thing for the angle θ_2 that $\boldsymbol{F}$ makes with the positive x axis to see

$$\cos(\theta_2) = \frac{b}{\| \boldsymbol{W} \|}$$
$$\sin(\theta_2) = \frac{d}{\| \boldsymbol{W} \|}$$

The angle between vectors $\boldsymbol{V}$ and $\boldsymbol{W}$ is the same as between vectors $\boldsymbol{E}$ and $\boldsymbol{F}$. Call this angle θ. Then $\theta = \theta_1 - \theta_2$ and using the formula for the cos of the difference of angles

$$\begin{aligned}
\cos(\theta) &= \cos(\theta_1 - \theta_2) \\
&= \cos(\theta_1)\ \cos(\theta_2) + \sin(\theta_1)\ \sin(\theta_2) \\
&= \frac{a}{\| \boldsymbol{V} \|} \frac{b}{\| \boldsymbol{W} \|} + \frac{c}{\| \boldsymbol{V} \|} \frac{d}{\| \boldsymbol{W} \|} \\
&= \frac{ab + cd}{\| \boldsymbol{V} \| \| \boldsymbol{W} \|} \\
&= \frac{< \boldsymbol{V}, \boldsymbol{W} >}{\| \boldsymbol{V} \| \| \boldsymbol{W} \|}
\end{aligned}$$

Hence, the ratio $< \boldsymbol{V}, \boldsymbol{W} > /(\| \boldsymbol{V} \| \| \boldsymbol{W} \|)$ is the same as $\cos(\theta)$! So we can use this simple calculation to find the angle between a pair two dimensional vectors.

The more general proof of the Cauchy Schwartz Theorem for n dimensional vectors is a journey you can take in another mathematics class! We will state it though so we can use it later if we need it.

Theorem 2.2.2 (Cauchy Schwartz Theorem For n Dimensional Vectors)
If $\boldsymbol{V}$ and $\boldsymbol{W}$ are n dimensional column vectors with components $(V_1, \ldots, V_n)$ and $(W_1, \ldots, W_n)$ respectively, then it is always true that

$$< \boldsymbol{V}, \boldsymbol{W} > \ \leq\ | < \boldsymbol{V}, \boldsymbol{W} > | \leq \| \boldsymbol{V} \| \| \boldsymbol{W} \|$$

Moreover,

$$|< \boldsymbol{V}, \boldsymbol{W} >| = \| \boldsymbol{V} \| \| \boldsymbol{W} \|$$

if and only the vector $\boldsymbol{V}$ is a non zero multiple of the vector $\boldsymbol{W}$.

Theorem 2.2.2 then tells us that if the vectors $\boldsymbol{V}$ and $\boldsymbol{W}$ are not zero, then

$$-1 \leq \frac{< \boldsymbol{V}, \boldsymbol{W} >}{|| \boldsymbol{V} || \, || \boldsymbol{W} ||} \leq 1$$

and by analogy to what works for two dimensional vectors, we can use this ratio to define the cos of the angle between two n dimensional vectors even though we can't see them at all! We do this in Definition 2.2.1.

Definition 2.2.1 (*The Angle Between n Dimensional Vectors*)
If $\boldsymbol{V}$ and $\boldsymbol{W}$ are two non zero n dimensional column vectors with components $(V_1, \ldots, V_n)$ and $(W_1, \ldots, W_n)$ respectively, the angle θ between these vectors is defined by

$$\cos(\theta) = \frac{< \boldsymbol{V}, \boldsymbol{W} >}{|| \boldsymbol{V} || \, || \boldsymbol{W} ||}$$

Moreover, the angle between the vectors is $0°$ if $< \boldsymbol{V}, \boldsymbol{W} > = 1$ and is $180°$ if $< \boldsymbol{V}, \boldsymbol{W} > = -1$.

2.2.1 Examples

Example 2.2.1 Find the angle between the vectors $\boldsymbol{V}$ and $\boldsymbol{W}$ given by

$$\boldsymbol{V} = \begin{bmatrix} -6 \\ 13 \end{bmatrix} \text{ and } \boldsymbol{W} = \begin{bmatrix} -8 \\ 1 \end{bmatrix}.$$

Solution *Compute the inner product* $< \boldsymbol{V}, \boldsymbol{W} > = (-6)(-8) + (13)(1) = 61$. *Next, find the magnitudes of these vectors:* $|| \boldsymbol{V} || = \sqrt{(-6)^2 + (13)^2} = \sqrt{205}$ *and* $|| \boldsymbol{W} || = \sqrt{(-8)^2 + (1)^2} = \sqrt{65}$. *Then, if* θ *is the angle between the vectors, we know*

$$\begin{aligned} \cos(\theta) &= \frac{< \boldsymbol{V}, \boldsymbol{W} >}{|| \boldsymbol{V} || \, || \boldsymbol{W} ||} = \frac{61}{\sqrt{205}\,\sqrt{65}} \\ &= 0.5284 \end{aligned}$$

Hence, since $\boldsymbol{V}$ *is in quadrant 2 and* $\boldsymbol{W}$ *is in quadrant 2 as well, we expect the angle between them should be between 0° and 90°. Your calculator should return* $\cos^{-1}(0.5284) = 58.10°$ *or* 1.0141 *rad. You should graph these vectors and see this visually too.*

Example 2.2.2 Find the angle between the vectors $\boldsymbol{V}$ and $\boldsymbol{W}$ given by

$$\boldsymbol{V} = \begin{bmatrix} -6 \\ -13 \end{bmatrix} \text{ and } \boldsymbol{W} = \begin{bmatrix} 8 \\ 1 \end{bmatrix}.$$

Solution *Compute the inner product* $< \boldsymbol{V}, \boldsymbol{W} >= (-6)(8) + (-13)(1) = -61$. *Next, find the magnitudes of these vectors:* $\| \boldsymbol{V} \| = \sqrt{(-6)^2 + (-13)^2} = \sqrt{205}$ *and* $\| \boldsymbol{W} \| = \sqrt{(8)^2 + (1)^2} = \sqrt{65}$. *Then, if* θ *is the angle between the vectors, we know*

$$\cos(\theta) = \frac{< \boldsymbol{V}, \boldsymbol{W} >}{\| \boldsymbol{V} \| \| \boldsymbol{W} \|} = \frac{-61}{\sqrt{205}\sqrt{65}}$$
$$= -0.5284$$

Hence, since $\boldsymbol{V}$ *is in quadrant 3 and* $\boldsymbol{W}$ *is in quadrant 1, we expect the angle between them should be larger than* $90°$. *Your calculator should return* $\cos^{-1}(-0.5284) = 121.90°$ *or* 2.1275 *rad. You should graph these vectors and see this visually too.*

Example 2.2.3 Find the angle between the vectors $\boldsymbol{V}$ and $\boldsymbol{W}$ given by

$$\boldsymbol{V} = \begin{bmatrix} 6 \\ -13 \end{bmatrix} \text{ and } \boldsymbol{W} = \begin{bmatrix} 8 \\ 1 \end{bmatrix}.$$

Solution *Compute the inner product* $< \boldsymbol{V}, \boldsymbol{W} >= (6)(8) + (-13)(1) = 35$. *Next, find the magnitudes of these vectors:* $\| \boldsymbol{V} \| = \sqrt{(6)^2 + (-13)^2} = \sqrt{205}$ *and* $\| \boldsymbol{W} \| = \sqrt{(8)^2 + (1)^2} = \sqrt{65}$. *Then, if* θ *is the angle between the vectors, we know*

$$\cos(\theta) = \frac{< \boldsymbol{V}, \boldsymbol{W} >}{\| \boldsymbol{V} \| \| \boldsymbol{W} \|} = \frac{35}{\sqrt{205}\sqrt{65}}$$
$$= 0.3032$$

Hence, since $\boldsymbol{V}$ *is in quadrant 4 and* $\boldsymbol{W}$ *is in quadrant 1, we expect the angle between them should be between* $0°$ *and* $180°$. *Your calculator should return* $\cos^{-1}(0.3032) = 72.35°$ *or* 1.2627 *rad. You should graph these vectors and see this visually too.*

2.2.2 Homework

Exercise 2.2.1 *Find the angle between the vectors* $\boldsymbol{V}$ *and* $\boldsymbol{W}$ *given by*

$$\boldsymbol{V} = \begin{bmatrix} 5 \\ 4 \end{bmatrix} \text{ and } \boldsymbol{W} = \begin{bmatrix} 7 \\ 2 \end{bmatrix}.$$

Exercise 2.2.2 *Find the angle between the vectors* $\boldsymbol{V}$ *and* $\boldsymbol{W}$ *given by*

$$\boldsymbol{V} = \begin{bmatrix} -6 \\ -8 \end{bmatrix} \text{ and } \boldsymbol{W} = \begin{bmatrix} 9 \\ 8 \end{bmatrix}.$$

Exercise 2.2.3 *Find the angle between the vectors* $\boldsymbol{V}$ *and* $\boldsymbol{W}$ *given by*

$$\boldsymbol{V} = \begin{bmatrix} 10 \\ -4 \end{bmatrix} \text{ and } \boldsymbol{W} = \begin{bmatrix} 2 \\ 8 \end{bmatrix}.$$

Exercise 2.2.4 *Find the angle between the vectors* $\boldsymbol{V}$ *and* $\boldsymbol{W}$ *given by*

$$\boldsymbol{V} = \begin{bmatrix} 6 \\ 1 \end{bmatrix} \text{ and } \boldsymbol{W} = \begin{bmatrix} 7 \\ 2 \end{bmatrix}.$$

Exercise 2.2.5 *Find the angle between the vectors* $\boldsymbol{V}$ *and* $\boldsymbol{W}$ *given by*

$$\boldsymbol{V} = \begin{bmatrix} 3 \\ -5 \end{bmatrix} \text{ and } \boldsymbol{W} = \begin{bmatrix} 2 \\ -3 \end{bmatrix}.$$

Exercise 2.2.6 *Find the angle between the vectors* $\boldsymbol{V}$ *and* $\boldsymbol{W}$ *given by*

$$\boldsymbol{V} = \begin{bmatrix} 1 \\ -4 \end{bmatrix} \text{ and } \boldsymbol{W} = \begin{bmatrix} -2 \\ -3 \end{bmatrix}.$$

2.3 Determinants of 2 × 2 Matrices

Since the number $ad - bc$ is so important is all of our discussions about the relationship between the two dimensional vectors $\boldsymbol{V}$ and $\boldsymbol{W}$ with components (a, c) and (b, d) respectively, we will define this number to be the **determinant** of the matrix $\boldsymbol{A}$ formed by using $\boldsymbol{V}$ for column 1 and $\boldsymbol{W}$ for column 2 of $\boldsymbol{A}$. That is

$$\boldsymbol{A} = \begin{bmatrix} a & b \\ c & d \end{bmatrix} = [\boldsymbol{V}\ \boldsymbol{W}] = \left[\boxed{\begin{bmatrix} a \\ c \end{bmatrix}} \boxed{\begin{bmatrix} b \\ d \end{bmatrix}} \right]$$

We then formally define the **determinant** of the 2×2 matrix $\boldsymbol{A}$ by Definition 2.3.1.

Definition 2.3.1 (*The Determinant Of A* 2×2 *Matrix*)
Given the 2×2 matrix $\boldsymbol{A}$ defined by

$$\boldsymbol{A} = \begin{bmatrix} a & b \\ c & d \end{bmatrix},$$

the determinant of $\boldsymbol{A}$ is the number $ad - bc$. We denote the determinant by $det(\boldsymbol{A})$ or $\mid \boldsymbol{A} \mid$.

Comment 2.3.1 *It is also common to denote the determinant by*

$$det\boldsymbol{A} = \begin{vmatrix} a & b \\ c & d \end{vmatrix}.$$

Also, note that if we looked at the transpose of $\boldsymbol{A}$, we would find

$$\boldsymbol{A}^T = \begin{bmatrix} a & c \\ b & d \end{bmatrix} = [\boldsymbol{Y}\ \boldsymbol{Z}] = \left[\boxed{\begin{bmatrix} a \\ b \end{bmatrix}} \boxed{\begin{bmatrix} c \\ d \end{bmatrix}} \right].$$

Notice that $det\left(\boldsymbol{A}^T\right)$ is $(a)(d) - (b)(c)$ also. Hence, if $det\left(\boldsymbol{A}^T\right)$ is zero, it means that $\boldsymbol{Y}$ and $\boldsymbol{Z}$ are collinear. Hence, it the $det\,(\boldsymbol{A})$ is zero, both the vectors determined by the rows of $\boldsymbol{A}$ and the columns of $\boldsymbol{A}$ are collinear. Let's summarize what we know about this new thing called the **determinant** of $\boldsymbol{A}$.

1. If $\mid \boldsymbol{A} \mid$ is 0, then the vectors determined by the columns of $\boldsymbol{A}$ are collinear. This also means that the vectors determined by the columns are multiples of one another. Also, the vectors determined by the columns of $\boldsymbol{A}^T$ are also collinear.
2. If $\mid \boldsymbol{A} \mid$ is not 0, then the vectors determined by the columns of $\boldsymbol{A}$ are not collinear which means these vectors point in different directions. Another way of saying this is that these vectors are **not** multiples of one another. The same is true for the columns of the transpose of $\boldsymbol{A}$.

2.3.1 Worked Out Problems

Example 2.3.1 Compute the determinant of

$$\boldsymbol{A} = \begin{bmatrix} 16.0 & 8.0 \\ -6.0 & -5.0 \end{bmatrix}$$

Solution $\mid \boldsymbol{A} \mid = (16)(-5) - (8)(-6) = -32.$

Example 2.3.2 Compute the determinant of

$$\boldsymbol{A} = \begin{bmatrix} -2.0 & 3.0 \\ 6.0 & -9.0 \end{bmatrix}$$

Solution $\mid \boldsymbol{A} \mid = (-2)(-9) - (3)(6) = 0.$

Example 2.3.3 Determine if the vectors $\boldsymbol{V}$ and $\boldsymbol{W}$ given by

$$\boldsymbol{V} = \begin{bmatrix} 4 \\ 5 \end{bmatrix} \text{ and } \boldsymbol{W} = \begin{bmatrix} -2 \\ 3 \end{bmatrix}$$

are collinear.

Solution *Form the matrix* $\boldsymbol{A}$ *using these vectors as the columns. This gives*

$$\boldsymbol{A} = \begin{bmatrix} 4 & -2 \\ 5 & 3 \end{bmatrix}$$

The calculate $\mid \boldsymbol{A} \mid = (4)(3) - (-2)(5)$. *Since this value is* 22 *which is not zero, these vectors are not collinear.*

Example 2.3.4 Determine if the vectors $\boldsymbol{V}$ and $\boldsymbol{W}$ given by

$$\boldsymbol{V} = \begin{bmatrix} -6 \\ 4 \end{bmatrix} \text{ and } \boldsymbol{W} = \begin{bmatrix} 3 \\ -2 \end{bmatrix}$$

are collinear.

Solution *Form the matrix* **A** *using these vectors as the columns. This gives*

$$\boldsymbol{A} = \begin{bmatrix} -6 & 3 \\ 4 & -2 \end{bmatrix}$$

The calculate $\mid \boldsymbol{A} \mid = (-6)(-2) - (3)(4)$. *Since this value is* 0, *these vectors are collinear. You should graph them in the* $x - y$ *plane to see this visually.*

2.3.2 *Homework*

Exercise 2.3.1 *Compute the determinant of*

$$\begin{bmatrix} 2.0 & -3.0 \\ 6.0 & 5.0 \end{bmatrix}$$

Exercise 2.3.2 *Compute the determinant of*

$$\begin{bmatrix} 12.0 & -1.0 \\ 4.0 & 2.0 \end{bmatrix}$$

2.4 Systems of Two Linear Equations

We can use all of this material to understand simple two linear equations in two unknowns x and y. Consider the problem

$$2\,x + 4\,y = 7 \tag{2.1}$$

$$3\,x + 4\,y = -8 \tag{2.2}$$

Now consider the equation below written in terms of vectors:

$$x \begin{bmatrix} 2 \\ 3 \end{bmatrix} + y \begin{bmatrix} 4 \\ 4 \end{bmatrix} = \begin{bmatrix} 7 \\ -8 \end{bmatrix}$$

Using the standard ways of multiplying vectors by scalars and adding vectors, we see the above can be rewritten as

$$\begin{bmatrix} 2x \\ 3x \end{bmatrix} + \begin{bmatrix} 4y \\ 4y \end{bmatrix} = \begin{bmatrix} 7 \\ -8 \end{bmatrix}$$

or

$$\begin{bmatrix} 2x + 4y \\ 3x + 4y \end{bmatrix} = \begin{bmatrix} 7 \\ -8 \end{bmatrix}$$

This last vector equation is clearly the same as the original Eqs. 2.1 and 2.2:

$$2x + 4y = 7$$
$$3x + 4y = -8$$

Further, in this example, letting

$$\boldsymbol{V} = \begin{bmatrix} 2 \\ 3 \end{bmatrix}, \quad \boldsymbol{W} = \begin{bmatrix} 4 \\ 4 \end{bmatrix}, \quad \text{and } \boldsymbol{D} = \begin{bmatrix} 7 \\ -8 \end{bmatrix}$$

we see Eqs. 2.1 and 2.2 are equivalent to the vector equation

$$x\, \boldsymbol{V} + y\, \boldsymbol{W} = \boldsymbol{D}.$$

We can also write the system Eqs. 2.1 and 2.2 in an equivalent matrix–vector form. Recall the original system which is written below:

$$2\,x + 4\,y = 7$$
$$3\,x + 4\,y = -8$$

We have already identified this system is equivalent to the vector equation

$$x\, \boldsymbol{V} + y\, \boldsymbol{W} = \boldsymbol{D}$$

where

$$\boldsymbol{V} = \begin{bmatrix} 2 \\ 3 \end{bmatrix}, \quad \boldsymbol{W} = \begin{bmatrix} 4 \\ 4 \end{bmatrix} \text{ and } \boldsymbol{D} = \begin{bmatrix} 7 \\ -8 \end{bmatrix}$$

Now use $\boldsymbol{V}$ and $\boldsymbol{W}$ as column one and column two of the matrix $\boldsymbol{A}$

$$\boldsymbol{A} = \begin{bmatrix} \boldsymbol{V} \ \boldsymbol{W} \end{bmatrix} = \begin{bmatrix} 2 & 4 \\ 3 & 4 \end{bmatrix}$$

Then, the original system can be written in the matrix–vector form

$$\begin{bmatrix} 2 & 4 \\ 3 & 4 \end{bmatrix} \begin{bmatrix} x \\ y \end{bmatrix} = \begin{bmatrix} 7 \\ -8 \end{bmatrix}$$

We call the matrix $\boldsymbol{A}$ the **coefficient** matrix of the system given by Eqs. 2.1 and 2.2.

Now we can introduce a new type of notation. Think of the column vector

$$\begin{bmatrix} x \\ y \end{bmatrix}$$

as being a **vector** variable. We will use a bold font and a capital letter for this and set

$$\boldsymbol{X} = \begin{bmatrix} x \\ y \end{bmatrix}$$

Then, the original system can be written as

$$\boldsymbol{A}\,\boldsymbol{X} = \boldsymbol{D}.$$

We typically refer to the vector $\boldsymbol{D}$ as the **data** vector associated with the system given by Eqs. 2.1 and 2.2.

2.4.1 Worked Out Examples

Example 2.4.1 Consider the system of equations

$$\begin{aligned} 1\,x \ + \ 2\,y &= 9 \\ -5\,x \ + \ 12\,y &= -1 \end{aligned}$$

Find the matrix vector equation form of this system.

Solution *Define* $\boldsymbol{V}$, $\boldsymbol{W}$ *and* $\boldsymbol{D}$ *as follows:*

$$\boldsymbol{V} = \begin{bmatrix} 1 \\ -5 \end{bmatrix}, \quad \boldsymbol{W} = \begin{bmatrix} 2 \\ 12 \end{bmatrix} \quad \text{and } \boldsymbol{D} = \begin{bmatrix} 9 \\ -1 \end{bmatrix}$$

Then define the matrix $\boldsymbol{A}$ *using* $\boldsymbol{V}$ *and* $\boldsymbol{W}$ *as its columns:*

$$\boldsymbol{A} = \begin{bmatrix} 1 & 2 \\ -5 & 12 \end{bmatrix}$$

and the system is equivalent to

$$\boldsymbol{A}\,\boldsymbol{X} = \boldsymbol{D}.$$

Example 2.4.2 Consider the system of equations

$$\begin{aligned} 7\,x + 5\,y &= 2 \\ -3\,x + -4\,y &= 1 \end{aligned}$$

Find the matrix vector equation form of this system.

Solution *Define* $\boldsymbol{V}$, $\boldsymbol{W}$ *and* $\boldsymbol{D}$ *as follows:*

$$\boldsymbol{V} = \begin{bmatrix} 7 \\ -3 \end{bmatrix}, \quad \boldsymbol{W} = \begin{bmatrix} 5 \\ -4 \end{bmatrix} \quad \text{and } \boldsymbol{D} = \begin{bmatrix} 2 \\ 1 \end{bmatrix}$$

Then define the matrix $\boldsymbol{A}$ *using* $\boldsymbol{V}$ *and* $\boldsymbol{W}$ *as its columns:*

$$\boldsymbol{A} = \begin{bmatrix} 7 & 5 \\ -3 & -4 \end{bmatrix}$$

and the system is equivalent to

$$\boldsymbol{A}\,\boldsymbol{X} = \boldsymbol{D}.$$

2.4.2 Homework

Exercise 2.4.1 *Consider the system of equations*

$$\begin{aligned} 1\,\alpha + 2\,\beta &= 3 \\ 4\,\alpha + 5\,\beta &= 6 \end{aligned}$$

Find the matrix vector equation form of this system.

Exercise 2.4.2 *Consider the system of equations*

$$
\begin{aligned}
-1\,w \;+\; 3\,z &= 21 \\
6\,w \;+\; 7\,z &= 12
\end{aligned}
$$

Find the matrix vector equation form of this system.

Exercise 2.4.3 *Consider the system of equations*

$$
\begin{aligned}
-7\,u \;+\; 14\,v &= 8 \\
25\,u \;+\; -2\,v &= 8
\end{aligned}
$$

Find the matrix vector equation form of this system.

2.5 Solving Two Linear Equations in Two Unknowns

We now know how to write the system of two linear equations in two unknowns given by Eqs. 2.3 and 2.4

$$a\,x + b\,y \;=\; D_1 \tag{2.3}$$

$$c\,x + d\,y \;=\; D_2 \tag{2.4}$$

in an equivalent matrix–vector form. This system is equivalent to the vector equation

$$x\,\boldsymbol{V} + y\,\boldsymbol{W} = \boldsymbol{D}$$

where

$$\boldsymbol{V} = \begin{bmatrix} a \\ c \end{bmatrix}, \quad \boldsymbol{W} = \begin{bmatrix} b \\ d \end{bmatrix} \text{ and } \boldsymbol{D} = \begin{bmatrix} D_1 \\ D_2 \end{bmatrix}$$

Finally, using $\boldsymbol{V}$ and $\boldsymbol{W}$ as column one and column two of the matrix $\boldsymbol{A}$

$$\boldsymbol{A} = \begin{bmatrix} \boldsymbol{V} \; \boldsymbol{W} \end{bmatrix} = \begin{bmatrix} a & b \\ c & d \end{bmatrix}$$

Then, the original system was written in vector and matrix–vector form as

$$x\,\boldsymbol{V} + y\,\boldsymbol{W} = \begin{bmatrix} a & b \\ c & d \end{bmatrix} \begin{bmatrix} x \\ y \end{bmatrix} = \begin{bmatrix} D_1 \\ D_2 \end{bmatrix}$$

Now, we can solve this system very easily as follows. We have already discussed the inner product of two vectors. So we could compute the inner product of both sides of $x\,\boldsymbol{V} + y\,\boldsymbol{W} = \boldsymbol{D}$ with any vector $\boldsymbol{U}$ we want and get

$$< \boldsymbol{U}, x\,\boldsymbol{V} + y\,\boldsymbol{W} > = < \boldsymbol{U}, \boldsymbol{D} >$$

We can simplify the left hand side to get

$$x < \boldsymbol{U}, \boldsymbol{V} > + y\ < \boldsymbol{U}, \boldsymbol{W} > = < \boldsymbol{U}, \boldsymbol{D} >$$

Since this is true for any vector $\boldsymbol{U}$, let's try to find useful ones! Any vector $\boldsymbol{U}$ that satisfies $< \boldsymbol{U}, \boldsymbol{W} > = 0$ would be great as then the y would drop out and we could solve for x. The angle between such vector $\boldsymbol{U}$ and $\boldsymbol{W}$ would then be 90° or 270°. We will call such vectors **orthogonal** as the lines associated with the vectors are perpendicular.

We can easily find such a vector. Since $\boldsymbol{W}$ defines a line through the origin with slope d/b, from our usual algebra and pre-calculus courses, we know the line through the origin which is perpendicular to it has negative reciprocal slope: i.e. $-b/d$. A line with the slope $-b/d$ corresponds with a vector with components $(d, -b)$. The usual symbol for *perpendicularity* is $\perp$ so we will label our vector orthogonal to $\boldsymbol{W}$ as $\boldsymbol{W}^{\perp}$. We see that

$$\boldsymbol{W}^{\perp} = \begin{bmatrix} d \\ -b \end{bmatrix}$$

and as expected

$$< \boldsymbol{W}^{\perp}, \boldsymbol{W} > = (d)(b) + (-b)(d) = 0$$

Thus, we have

$$\begin{aligned} < \boldsymbol{W}^{\perp}, \boldsymbol{D} > &= x < \boldsymbol{W}^{\perp}, \boldsymbol{V} > + \ y\ < \boldsymbol{W}^{\perp}, \boldsymbol{W} > \\ &= x < \boldsymbol{W}^{\perp}, \boldsymbol{V} > \end{aligned}$$

This looks complicated, but it can be written in terms of things we understand. Let's actually calculate the inner products. We find

$$< \boldsymbol{W}^{\perp}, \boldsymbol{V} > = (d)(a) + (-b)(c) = det\ (\boldsymbol{A})$$

and

$$< \boldsymbol{W}^{\perp}, \boldsymbol{D} > = (d)(D_1) + (-b)(D_2) = det \begin{bmatrix} D_1 & b \\ D_2 & d \end{bmatrix}.$$

Hence, by taking the inner product of both sides with $\boldsymbol{W}^{\perp}$, we find the y term drops out and we have

$$x\ det \begin{bmatrix} a & b \\ c & d \end{bmatrix} = det \begin{bmatrix} D_1 & b \\ D_2 & d \end{bmatrix}$$

Thus, if $det\ (\boldsymbol{A})$ is not zero, we can solve for x to get

$$x = \frac{det \begin{bmatrix} D_1 & b \\ D_2 & d \end{bmatrix}}{det \begin{bmatrix} a & b \\ c & d \end{bmatrix}} = \frac{det \begin{bmatrix} \boldsymbol{D} & \boldsymbol{W} \end{bmatrix}}{det \begin{bmatrix} \boldsymbol{V} & \boldsymbol{W} \end{bmatrix}}$$

We can do a similar thing to find out what the variable y is by taking the inner product of both sides of of $x\ \boldsymbol{V} + y\ \boldsymbol{W} = \boldsymbol{D}$ with the vector $\boldsymbol{V}^{\perp}$ and get

$$x\ < \boldsymbol{V}^{\perp}, \boldsymbol{V} > + y\ < \boldsymbol{V}^{\perp}, \boldsymbol{W} > = < \boldsymbol{V}^{\perp}, \boldsymbol{D} >$$

where

$$\boldsymbol{V}^{\perp} = \begin{bmatrix} c \\ -a \end{bmatrix}$$

and as expected

$$< \boldsymbol{V}^{\perp}, \boldsymbol{V} > = (c)(a) + (-a)(c) = 0$$

Going through the same steps as before, we would find that if $det\ (\boldsymbol{A})$ is non zero, we could solve for y to get

$$y = \frac{det \begin{bmatrix} a & D_1 \\ c & D_2 \end{bmatrix}}{det \begin{bmatrix} a & b \\ c & d \end{bmatrix}} = \frac{det \begin{bmatrix} \boldsymbol{V} & \boldsymbol{D} \end{bmatrix}}{det \begin{bmatrix} \boldsymbol{V} & \boldsymbol{W} \end{bmatrix}}$$

Let's summarize:

1. Given any system of two linear equations in two unknowns, there is a coefficient matrix $\boldsymbol{A}$ with first column $\boldsymbol{V}$ and second column $\boldsymbol{W}$ that is associated with it. Further, the right hand side of the system defines a data vector $\boldsymbol{D}$.
2. If $det\ (\boldsymbol{A})$ is not zero, we can solve for the unknowns x and y as follows:

$$x = \frac{det\ ([\boldsymbol{D}\ \boldsymbol{W}])}{det\ ([\boldsymbol{V}\ \boldsymbol{W}])}$$
$$y = \frac{det\ ([\boldsymbol{V}\ \boldsymbol{D}])}{det\ ([\boldsymbol{V}\ \boldsymbol{W}])}$$

This method of solution is known as **Cramer's Rule**.

3. This system of two linear equations in two unknowns is associated with two column vectors $\boldsymbol{V}$ and $\boldsymbol{W}$. You can see there is a **unique** solution if and only if $\mid \boldsymbol{A} \mid$ is not zero. This is the same as saying there is a unique solution if and only if the vectors are not *collinear*.

We can state this as Theorem 2.5.1.

Theorem 2.5.1 (Cramer's Rule)
Consider the system of equations

$$\begin{aligned} a\,x + b\,y &= D_1 \\ c\,x + d\,y &= D_2. \end{aligned}$$

Define the vectors

$$\boldsymbol{V} = \begin{bmatrix} a \\ c \end{bmatrix}, \quad \boldsymbol{W} = \begin{bmatrix} b \\ d \end{bmatrix}, \textit{ and } \boldsymbol{D} = \begin{bmatrix} D_1 \\ D_2 \end{bmatrix}$$

Also, define the matrix $\boldsymbol{A}$ *by*

$$\boldsymbol{A} = \left[\boldsymbol{V}\ \boldsymbol{W}\right]$$

Then, if $det(\boldsymbol{A}) \neq 0$, *the unique solution to this system of equations is given by*

$$x = \frac{det\ ([\boldsymbol{D}\ \boldsymbol{W}])}{det\ (\boldsymbol{A})}$$
$$y = \frac{det\ ([\boldsymbol{V}\ \boldsymbol{D}])}{det\ (\boldsymbol{A})}$$

2.5.1 Worked Out Examples

Example 2.5.1 Solve the system

$$\begin{aligned} -2\,x\ +\ 4\,y &= 6 \\ 8\,x\ +\ -1\,y &= 2 \end{aligned}$$

using Cramer's Rule.

Solution *Solve*

$$\begin{aligned} -2\,x \,+\, 4\,y &= 6 \\ 8\,x \,+\, -1\,y &= 2 \end{aligned}$$

using Cramer's Rule. We have

$$\boldsymbol{V} = \begin{bmatrix} -2 \\ 8 \end{bmatrix}, \quad \boldsymbol{W} = \begin{bmatrix} 4 \\ -1 \end{bmatrix} \quad \textit{and } \boldsymbol{D} = \begin{bmatrix} 6 \\ 2 \end{bmatrix}$$

$$\begin{aligned} x &= \frac{det\ ([\boldsymbol{D}\ \boldsymbol{W}])}{det\ ([\boldsymbol{V}\ \boldsymbol{W}])} \\ &= \frac{det\left(\begin{bmatrix} 6 & 4 \\ 2 & -1 \end{bmatrix}\right)}{det\left(\begin{bmatrix} -2 & 4 \\ 8 & -1 \end{bmatrix}\right)} \\ &= \frac{-30}{-30} = 1, \\ y &= \frac{det\ ([\boldsymbol{V}\ \boldsymbol{D}])}{det\ ([\boldsymbol{V}\ \boldsymbol{W}])} \\ &= \frac{det\left(\begin{bmatrix} -2 & 6 \\ 8 & 2 \end{bmatrix}\right)}{det\left(\begin{bmatrix} -2 & 4 \\ 8 & -1 \end{bmatrix}\right)} \\ &= \frac{-52}{-30} = -\frac{26}{15} \end{aligned}$$

Example 2.5.2 Solve the system

$$\begin{aligned} -5\,x \,+\, 1\,y &= 8 \\ 9\,x \,+\, -10\,y &= 2 \end{aligned}$$

using Cramer's Rule.

Solution *We have*

$$\boldsymbol{V} = \begin{bmatrix} -5 \\ 9 \end{bmatrix}, \quad \boldsymbol{W} = \begin{bmatrix} 1 \\ -10 \end{bmatrix} \quad \textit{and } \boldsymbol{D} = \begin{bmatrix} 9 \\ -2 \end{bmatrix}$$

$$
\begin{aligned}
x &= \frac{det\ ([\boldsymbol{D}\ \boldsymbol{W}])}{det\ ([\boldsymbol{V}\ \boldsymbol{W}])} \\
&= \frac{det\left(\begin{bmatrix} 8 & 1 \\ 2 & -10 \end{bmatrix}\right)}{det\left(\begin{bmatrix} -5 & 1 \\ 9 & -10 \end{bmatrix}\right)} \\
&= \frac{-82}{41} = -\frac{-82}{41}. \\
y &= \frac{det\ ([\boldsymbol{V}\ \boldsymbol{D}])}{det\ ([\boldsymbol{V}\ \boldsymbol{W}])} \\
&= \frac{det\left(\begin{bmatrix} -5 & 8 \\ 9 & 2 \end{bmatrix}\right)}{det\left(\begin{bmatrix} -5 & 1 \\ 9 & -10 \end{bmatrix}\right)} \\
&= \frac{-82}{41}
\end{aligned}
$$

2.5.2 *Homework*

Exercise 2.5.1 *Solve the system*

$$
\begin{aligned}
-3\,x + 4\,y &= 6 \\
8\,x + 9\,y &= -1
\end{aligned}
$$

using Cramer's Rule.

Exercise 2.5.2 *Solve the system*

$$
\begin{aligned}
2\,x + 3\,y &= 6 \\
-4\,x + 0\,y &= 8
\end{aligned}
$$

using Cramer's Rule.

Exercise 2.5.3 *Solve the system*

$$
\begin{aligned}
18\,x + 1\,y &= 1 \\
-9\,x + 3\,y &= 17
\end{aligned}
$$

using Cramer's Rule.

Exercise 2.5.4 *Solve the system*

$$\begin{aligned} -7\,x \,+\, 6\,y &= -4 \\ 8\,x \,+\, 1\,y &= 1 \end{aligned}$$

using Cramer's Rule.

Exercise 2.5.5 *Solve the system*

$$\begin{aligned} -90\,x \,+\, 1\,y &= 1 \\ 80\,x \,+\, -1\,y &= 1 \end{aligned}$$

using Cramer's Rule.

2.6 Consistent and Inconsistent Systems

So what happens if $det\,(\boldsymbol{A}) = 0$? By the remark above, we know that the vectors $\boldsymbol{V}$ and $\boldsymbol{W}$ are collinear. We also know from our discussions in Sect. 2.3 that the columns of $\boldsymbol{A}^T$ are collinear. Hence, there is a non zero constant r so that

$$\begin{bmatrix} a \\ b \end{bmatrix} = r \begin{bmatrix} c \\ d \end{bmatrix}$$

Thus, $a = r\,c$ and $b = r\,d$ and the original system can be written as

$$\begin{aligned} r\,c\,x + r\,d\,y &= D_1 \\ c\,x + d\,y &= D_2 \end{aligned}$$

or

$$\begin{aligned} r\,(c\,x + d\,y) &= D_1 \\ c\,x + d\,y &= D_2 \end{aligned}$$

You can see we do not really have two equations in two unknowns since the top equation on the left hand side is just a multiple of the left hand side of the bottom equation. This can only make sense if $D_1/r = D_2$ or $D_1 = r\, D_2$. We can conclude that the relationship between the components of $\boldsymbol{D}$ must be just right! Hence, we have the system

$$\begin{aligned} c\,x + d\,y &= D_1/r \\ c\,x + d\,y &= D_2 \end{aligned}$$

Now subtract the top equation from the bottom equation. You find

$$0\,x + 0\,y = 0 = D_2 - D_1/r$$

This equation only makes sense if when you subtract the top from the bottom equation, the new right hand side is 0! We call such systems **consistent** if the right hand side becomes 0 and *inconsistent* if not zero. So we have a great test for *inconsistency*. We scale the top or bottom equation just right to make them identical and subtract the two equations. If we get $0 = \alpha$ for a nonzero α, the system is *inconsistent.*

Here is an example. Consider the system

$$\begin{aligned} 2\,x + 3\,y &= 8 \\ 4\,x + 6\,y &= 9 \end{aligned}$$

Here, the column vectors of $\boldsymbol{A}^T$ are

$$\boldsymbol{Y} = \begin{bmatrix} 2 \\ 3 \end{bmatrix} \text{ and } \boldsymbol{Z} = \begin{bmatrix} 4 \\ 6 \end{bmatrix}$$

We see $\boldsymbol{Z} = 2\,\boldsymbol{Y}$ and we have the system

$$\begin{aligned} 2\,x + 3\,y &= 8 = D_1 \\ 2\,(2\,x + 3\,y) &= 9 = D_2 \end{aligned}$$

This system would be *consistent* if the bottom equation was exactly two times the top equation. For this to happen, we need $D_2 = 2\,D_1$; i.e., we need $9 = 2 \times 8$ which is impossible. So these equations are *inconsistent.* As mentioned earlier, an even better way to see these equations are inconsistent is to subtract the top equation from the bottom equation to get

$$0x + 0\,y = 0 = 1$$

which again is not possible. Remember, *consistent* equations when $det(\boldsymbol{A}) = 0$ would have the some multiple of **top − bottom** equation = zero.

Another way to look at this situation is to note that the column vectors, $\boldsymbol{V}$ and $\boldsymbol{W}$, of $\boldsymbol{A}$ are collinear. Hence, there is another non zero scalar s so that $\boldsymbol{V} = s\,\boldsymbol{W}$. We can then rewrite the usual vector form of our system as

$$\begin{aligned} \boldsymbol{D} &= x\,\boldsymbol{V} + y\,\boldsymbol{W} \\ &= x\,s\boldsymbol{W} + y\,\boldsymbol{W} \end{aligned}$$

This says that the data vector $\boldsymbol{D} = (xs + y)\,\boldsymbol{W}$. Hence, if there is a solution x and y, it will only happen in $\boldsymbol{D}$ is a multiple of $\boldsymbol{W}$. This says $\boldsymbol{D}$ is collinear with $\boldsymbol{W}$ which in turn is collinear with $\boldsymbol{V}$. Going back to our sample

$$\begin{aligned} 2\,x + 3\,y &= 8 \\ 4\,x + 6\,y &= 9. \end{aligned}$$

We see $\boldsymbol{D}$ with components $(8, 9)$ is not a multiple of $\boldsymbol{V}$ with components $(2, 4)$ or $\boldsymbol{W}$ with components $(3, 6)$. Thus, the system must be inconsistent.

2.6.1 Worked Out Examples

Example 2.6.1 Consider the system

$$\begin{aligned} 4\,x + 5\,y &= 11 \\ -8\,x - 10\,y &= -22 \end{aligned}$$

Determine if this system is consistent or inconsistent.

Solution *We see immediately that the determinant of the coefficient matrix* $\mathbf{A}$ *is zero. So the question of consistency is reasonable to ask. Here, the column vectors of* $\mathbf{A}^T$ *are*

$$\boldsymbol{Y} = \begin{bmatrix} 4 \\ 5 \end{bmatrix} \quad \text{and} \quad \mathbf{Z} = \begin{bmatrix} -8 \\ -10 \end{bmatrix}$$

We see $\mathbf{Z} = -2\,\boldsymbol{Y}$ *and we have the system*

$$\begin{aligned} 4\,x + 5\,y &= 11 = D_1 \\ -2\,(4\,x + 5\,y) &= -22 = D_2 \end{aligned}$$

This system would be consistent *if the bottom equation was exactly minus two times the top equation. For this to happen, we need* $D_2 = -2\, D_1$; *i.e., we need* $-22 = -2 \times 11$ *which is true. So these equations are* consistent.

Example 2.6.2 Consider the system

$$\begin{aligned} 6\,x + 8\,y &= 14 \\ 18\,x + 24\,y &= 48 \end{aligned}$$

Determine if this system is consistent or inconsistent.

Solution *We see immediately that the determinant of the coefficient matrix* $\mathbf{A}$ *is zero. So again, the question of consistency is reasonable to ask. Here, the column vectors of* $\mathbf{A}^T$ *are*

$$\boldsymbol{Y} = \begin{bmatrix} 6 \\ 8 \end{bmatrix} \text{ and } \boldsymbol{Z} = \begin{bmatrix} 18 \\ 24 \end{bmatrix}$$

We see $\boldsymbol{Z} = 3\,\boldsymbol{Y}$ *and we have the system*

$$\begin{aligned} 6\,x + 8\,y &= 14 = D_1 \\ 3\,(6\,x + 8\,y) &= 48 = D_2 \end{aligned}$$

This system would be consistent *if the bottom equation was exactly three times the top equation. For this to happen, we need* $D_2 = 3\, D_1$; *i.e., we need* $-48 = 3 \times 14$ *which is not true. So these equations are* inconsistent.

2.6.2 Homework

Exercise 2.6.1 *Consider the system*

$$\begin{aligned} 2\,x + 5\,y &= 1 \\ 8\,x + 20\,y &= 4 \end{aligned}$$

Determine if this system is consistent or inconsistent.

Exercise 2.6.2 *Consider the system*

$$\begin{aligned} 60\,x + 80\,y &= 120 \\ 6\,x + 8\,y &= 13 \end{aligned}$$

Determine if this system is consistent or inconsistent.

Exercise 2.6.3 *Consider the system*

$$\begin{aligned} -2\,x + 7\,y &= 10 \\ 20\,x - 70\,y &= 4 \end{aligned}$$

Determine if this system is consistent or inconsistent.

Exercise 2.6.4 *Consider the system*

$$\begin{aligned} x + y &= 1 \\ 2\,x + 2\,y &= 3 \end{aligned}$$

Determine if this system is consistent or inconsistent.

Exercise 2.6.5 *Consider the system*

$$\begin{aligned} -11\,x - 3\,y &= -2 \\ 33\,x + 9\,y &= 6 \end{aligned}$$

Determine if this system is consistent or inconsistent.

2.7 Specializing to Zero Data

If the system we want to solve has zero data, then we must solve a system of equations like

$$\begin{aligned} a\,x + b\,y &= 0 \\ c\,x + d\,y &= 0. \end{aligned}$$

Define the vectors $\boldsymbol{V}$ and $\boldsymbol{W}$ as usual. Note $\boldsymbol{D}$ is now the zero vector

$$\boldsymbol{V} = \begin{bmatrix} a \\ c \end{bmatrix}, \quad \boldsymbol{W} = \begin{bmatrix} b \\ d \end{bmatrix}, \quad \text{and} \quad \boldsymbol{D} = \begin{bmatrix} 0 \\ 0 \end{bmatrix}$$

Also, define the matrix $\boldsymbol{A}$ by

$$\boldsymbol{A} = \begin{bmatrix} \boldsymbol{V} \; \boldsymbol{W} \end{bmatrix}$$

Then, if $det(\boldsymbol{A}) \neq 0$, the unique solution to this system of equations is given by

$$\begin{aligned} x &= \frac{det\left(\begin{bmatrix} 0 & b \\ 0 & d \end{bmatrix}\right)}{det\ (\boldsymbol{A})} \\ &= \frac{0}{ad - bc} = 0 \end{aligned}$$

$$\begin{aligned} y &= \frac{det\left(\begin{bmatrix} a & 0 \\ c & 0 \end{bmatrix}\right)}{det\ (\boldsymbol{A})} \\ &= \frac{0}{ad - bc} = 0 \end{aligned}$$

Hence, the unique solution to a system of the form $\boldsymbol{A}\ \boldsymbol{X} = \boldsymbol{0}$ is $x = 0$ and $y = 0$. But what happens if the determinant of $\boldsymbol{A}$ is zero? In this case, we know the column vectors of $\boldsymbol{A}^T$ are collinear: i.e.

$$\boldsymbol{Y} = \begin{bmatrix} a \\ b \end{bmatrix} \quad \text{and} \quad \boldsymbol{Z} = \begin{bmatrix} c \\ d \end{bmatrix},$$

are collinear and so there is a non zero constant r so that $a = rc$ and $b = rd$. This gives the system

$$\begin{aligned} r\ (c\,x + b\,y) &= 0 \\ c\,x + d\,y &= 0. \end{aligned}$$

Now if you multiply the bottom equation by r and subtract from the top equation, you get 0. This tells us the system is *consistent*. The original system of two equations is thus only one equation. We can choose to use either the original top or bottom equation to solve. Say we choose the original top equation. Then we need to find x and y choices so that

$$a\,x + b\,y = 0$$

There are in finitely many solutions here! It is easiest to see how to solve this kind of problem using some examples.

2.7.1 Worked Out Examples

Example 2.7.1 Find all solutions to the consistent system

$$-2\,x + 7\,y = 0$$
$$20\,x - 70\,y = 0$$

Solution *First, note the determinant of the coefficient matrix of the system is zero. Also, since the bottom equation is -10 times the top equation, we see the system is also consistent. We solve using the top equation:*

$$-2\,x + 7\,y = 0$$

Thus,

$$7\,y = 2\,x$$
$$y = (2/7)\,x$$

We see a solution vector of the form

$$\boldsymbol{X} = \begin{bmatrix} x \\ y \end{bmatrix} = \begin{bmatrix} x \\ (2/7)\,x \end{bmatrix} = x\begin{bmatrix} 1 \\ (2/7) \end{bmatrix}$$

will always work. There is a lot of ambiguity here as the multiplier x is completely arbitrary. For example,if we let $x = 7\,c$ for an arbitrary c, then solving for y, we find $y = 2\,c$. We can then rewrite the solution vector as

$$c\begin{bmatrix} 2 \\ 7 \end{bmatrix}$$

in terms of the arbitrary multiplier c. It does not really matter what form we pick, however we often try to pick a form which has integers as components.

Example 2.7.2 Find all solutions to the consistent system

$$4\,x + 5\,y = 0$$
$$8\,x + 10\,y = 0$$

Solution *First, note the determinant of the coefficient matrix of the system is zero. Also, since the bottom equation is 2 times the top equation, we see the system is also consistent. We solve using the bottom equation this time:*

$$8\,x + 10\,y = 0$$

Thus,

$$\begin{aligned} 10\,y &= -8\,x \\ y &= -(4/5)\,x \end{aligned}$$

We see a solution vector of the form

$$\boldsymbol{X} = \begin{bmatrix} x \\ y \end{bmatrix} = \begin{bmatrix} x \\ -(4/5)\,x \end{bmatrix} = x \begin{bmatrix} 1 \\ -(4/5) \end{bmatrix}$$

will always work. Again, there is a lot of ambiguity here as the multiplier x is completely arbitrary. For example,if we let $x = 10\,d$ for an arbitrary d, then solving for y, we find $y = -8\,d$. We can then rewrite the solution vector as

$$d \begin{bmatrix} 10 \\ -8 \end{bmatrix}$$

in terms of the arbitrary multiplier d. Again, it is important to note that it does not really matter what form we pick, however we often try to pick a form which has integers as components.

2.7.2 Homework

Exercise 2.7.1 *Find all solutions to the consistent system*

$$\begin{aligned} x + 3\,y &= 0 \\ 6\,x + 18\,y &= 0 \end{aligned}$$

Exercise 2.7.2 *Find all solutions to the consistent system*

$$\begin{aligned} -3\,x + 4\,y &= 0 \\ 9\,x - 12\,y &= 0 \end{aligned}$$

Exercise 2.7.3 *Find all solutions to the consistent system*

$$\begin{aligned} 2\,x + 7\,y &= 0 \\ 1\,x + (3/2)\,y &= 0 \end{aligned}$$

Exercise 2.7.4 *Find all solutions to the consistent system*

$$\begin{aligned} -10\,x + 5\,y &= 0 \\ 20\,x - 10\,y &= 0 \end{aligned}$$

Exercise 2.7.5 *Find all solutions to the consistent system*

$$-12\,x + 5\,y = 0$$
$$4\,x - (5/3)\,y = 0$$

2.8 Matrix Inverses

If a matrix $\boldsymbol{A}$ has a non zero determinant, we know the system

$$\boldsymbol{A}\begin{bmatrix} x \\ y \end{bmatrix} = \begin{bmatrix} D_1 \\ D_2 \end{bmatrix}$$

has a unique solution for each right hand side vector

$$\boldsymbol{D} = \begin{bmatrix} D_1 \\ D_2 \end{bmatrix}.$$

If we could find another matrix $\boldsymbol{B}$ which satisfied

$$\boldsymbol{B}\,\boldsymbol{A} = \boldsymbol{A}\,\boldsymbol{B} = \boldsymbol{I}$$

we could multiply both sides of our system by $\boldsymbol{B}$ to find

$$\boldsymbol{B}\,\boldsymbol{A}\begin{bmatrix} x \\ y \end{bmatrix} = \boldsymbol{B}\begin{bmatrix} D_1 \\ D_2 \end{bmatrix}$$
$$\boldsymbol{I}\begin{bmatrix} x \\ y \end{bmatrix} = \boldsymbol{B}\begin{bmatrix} D_1 \\ D_2 \end{bmatrix}$$
$$\begin{bmatrix} x \\ y \end{bmatrix} = \boldsymbol{B}\begin{bmatrix} D_1 \\ D_2 \end{bmatrix}$$

which is the solution to our system! The matrix $\boldsymbol{B}$ is, of course, special and clearly plays the role of an inverse for the matrix $\boldsymbol{A}$. When such a matrix $\boldsymbol{B}$ exists, it is called the inverse of $\boldsymbol{A}$ and is denoted by $\boldsymbol{A}^{-1}$.

Definition 2.8.1 (*The Inverse of the matrix* $\boldsymbol{A}$)
If there is a matrix $\boldsymbol{B}$ of the same size as the square matrix $\boldsymbol{A}$, $\boldsymbol{B}$ is said to be inverse of $\boldsymbol{A}$ is

$$\boldsymbol{B}\,\boldsymbol{A} = \boldsymbol{A}\,\boldsymbol{B} = \boldsymbol{I}$$

In this case, we denote the inverse of $\boldsymbol{A}$ by $\boldsymbol{A}^{-1}$.

We can show that the inverse of $\boldsymbol{A}$ exists if and only if $det(\boldsymbol{A}) \neq 0$. In general, it is very hard to find the inverse of a matrix, but in the case of a 2×2 matrix, it is very easy.

Definition 2.8.2 (*The Inverse of the* 2×2 *matrix* $\boldsymbol{A}$)
Let

$$\boldsymbol{A} = \begin{bmatrix} a & b \\ c & d \end{bmatrix}$$

and assume $det(\boldsymbol{A}) \neq 0$. Then, the inverse of $\boldsymbol{A}$ is given by

$$\boldsymbol{A}^{-1} = \frac{1}{det(\boldsymbol{A})} \begin{bmatrix} d & -b \\ -c & a \end{bmatrix}$$

2.8.1 Worked Out Examples

Example 2.8.1 For

$$\boldsymbol{A} = \begin{bmatrix} 6 & 2 \\ 3 & 4 \end{bmatrix}$$

find $\boldsymbol{A}^{-1}$.

Solution *Since* $det(\boldsymbol{A}) = 18$, *we see*

$$\boldsymbol{A}^{-1} = \frac{1}{18} \begin{bmatrix} 4 & -2 \\ -3 & 6 \end{bmatrix}$$

Example 2.8.2 For the system

$$\begin{bmatrix} -2 & 4 \\ 3 & 5 \end{bmatrix} \begin{bmatrix} x \\ y \end{bmatrix} = \begin{bmatrix} 8 \\ 7 \end{bmatrix}$$

find the unique solution.

Solution *The coefficient matrix here is*

$$\boldsymbol{A} = \begin{bmatrix} -2 & 4 \\ 3 & 5 \end{bmatrix}$$

Since $det(\boldsymbol{A}) = -22$, *we see*

$$\boldsymbol{A}^{-1} = \frac{-1}{22} \begin{bmatrix} 5 & -4 \\ -3 & -2 \end{bmatrix}$$

and hence,

$$\begin{aligned}\begin{bmatrix} x \\ y \end{bmatrix} &= A^{-1} \begin{bmatrix} 8 \\ 7 \end{bmatrix} \\ &= \frac{-1}{22} \begin{bmatrix} 40 - 28 \\ -24 - 14 \end{bmatrix} = \begin{bmatrix} \frac{-12}{22} \\ \frac{38}{22} \end{bmatrix}\end{aligned}$$

2.8.2 Homework

Exercise 2.8.1 *For*

$$A = \begin{bmatrix} 6 & 8 \\ 3 & 4 \end{bmatrix}$$

find A^{-1} *if it exists.*

Exercise 2.8.2 *For*

$$A = \begin{bmatrix} 6 & 8 \\ 3 & 5 \end{bmatrix}$$

find A^{-1} *if it exists.*

Exercise 2.8.3 *For*

$$A = \begin{bmatrix} -3 & 2 \\ 3 & 5 \end{bmatrix}$$

find A^{-1} *if it exists.*

Exercise 2.8.4 *For the system*

$$\begin{bmatrix} 4 & 3 \\ 11 & 2 \end{bmatrix} \begin{bmatrix} x \\ y \end{bmatrix} = \begin{bmatrix} -5 \\ 3 \end{bmatrix}$$

find the unique solution if it exists.

Exercise 2.8.5 *For the system*

$$\begin{bmatrix} -1 & 2 \\ 1 & 2 \end{bmatrix} \begin{bmatrix} x \\ y \end{bmatrix} = \begin{bmatrix} 10 \\ 30 \end{bmatrix}$$

find the unique solution if it exists.

Exercise 2.8.6 *For the system*

$$\begin{bmatrix} 40 & 30 \\ 16 & 5 \end{bmatrix} \begin{bmatrix} x \\ y \end{bmatrix} = \begin{bmatrix} -5 \\ 10 \end{bmatrix}$$

find the unique solution if it exists.

2.9 Computational Linear Algebra

Let's look at how we can use MatLab/Octave to solve the general linear system of equations

$$\boldsymbol{A}\,\boldsymbol{x} = \boldsymbol{b}$$

where $\boldsymbol{A}$ is a $n \times n$ matrix, $\boldsymbol{x}$ is a column vector with n rows whose components are the unknowns we wish to solve for and $\boldsymbol{b}$ is the data vector.

2.9.1 A Simple Lower Triangular System

We will start by writing a function to solve a special system of equations; we begin with a lower triangular matrix system $Lx = b$. For example, if the system we wanted to solve was

$$\begin{bmatrix} 1 & -2 & 3 \\ 0 & 4 & 1 \\ 0 & 0 & 6 \end{bmatrix} \begin{bmatrix} x \\ y \\ z \end{bmatrix} = \begin{bmatrix} 9 \\ 8 \\ 2 \end{bmatrix}$$

this is easily solve by starting at the last equation and working backwards. This is called *backsolving*. Here, we have

$$\begin{aligned} z &= \frac{2}{6} \\ 4y &= 8 - z = 8 - \frac{1}{3} = \frac{23}{3} \\ y &= \frac{23}{12} \\ x &= 9 + 2y - 3z = 9 + \frac{23}{6} - 1 = \frac{71}{6} \end{aligned}$$

It is easy to write code to do this in MatLab as we do below.

2.9.2 A Lower Triangular Solver

Here is a simple function to solve such a system.

Listing 2.1: Lower Triangular Solver

```
function x = LTriSol(L,b)
%
% L is n x n Lower Triangular Matrix
% b is nx1 data vector
% Obtain x by forward substitution
%
n = length(b);
x = zeros(n,1);
for j=1:n-1
  x(j) = b(j)/L(j,j);
  b(j+1:n) = b(j+1:n) - x(j)*L(j+1:n,j);
end
x(n) = b(n)/L(n,n);
end
```

To use this function, we would enter the following commands at the Matlab prompt. For now, we are assuming that you are running Matlab in a local directory which contains your Matlab code **LTriSol.m**. So we fire up Matlab and enter these commands:

Listing 2.2: Sample Solution with LTriSol

```
A = [2 0 0; 1 5 0; 7 9 8]
A =
      2     0     0
      1     5     0
      7     9     8
b = [6; 2; 5]
b =
      6
      2
      5
x = LTriSol(A,b)
x =
      3.0000
     -0.2000
     -1.7750
```

which solves the system as we wanted.

2.9.3 An Upper Triangular Solver

Here is a simple function to solve a similar system where this time A is upper triangular. The code is essentially the same although the solution process starts at the *top* and sweeps *down*.

Listing 2.3: Upper Triangular Solver

```
function x = UTriSol(U,b)
%
% U is nxn nonsingular Upper Triangular matrix
% b is nx1 data vector
% x is solved by back substitution
%
n = length(b);
x = zeros(n,1);
for j = n:-1:2
  x(j) = b(j)/U(j,j);
  b(1:j-1) = b(1:j-1) - x(j)*U(1:j-1,j);
end
x(1) = b(1)/U(1,1);
end
```

As usual, to use this function, we would enter the following commands at the Matlab prompt. We are still assuming that you are running Matlab in a local directory and that your Matlab code **UTriSol.m** is also in this directory.

So we enter these commands in Matlab.

Listing 2.4: Sample Solution with UTriSol

```
C = [7 9 8; 0 1 5; 0 0 2]
C =
        7     9     8
        0     1     5
        0     0     2
b = [6; 2; 5]
b =
        6
        2
        5
x = UTriSol(C,b)
x =
   11.5000
  -10.5000
    2.5000
```

which again solves the system as we wanted.

2.9.4 The LU Decomposition of A Without Pivoting

It is possible to take a general matrix $\boldsymbol{A}$ and rewrite it as the product of a lower triangular matrix $\boldsymbol{L}$ and an upper triangular matrix $\boldsymbol{U}$. Here is a simple function to solve a system using the LU decomposition of A. First, it finds the LU decomposition and then it uses the lower triangular and upper triangular solvers we wrote earlier. To do this, we add and subtract multiples of rows together to remake the original matrix A into an upper triangular matrix. Let's do a simple example. Let the matrix A be given by

$$A = \begin{bmatrix} 8 & 2 & 3 \\ -4 & 3 & 2 \\ 7 & 8 & 9 \end{bmatrix}$$

Start in the row 1 and column 1 position in A. The entry there is the *pivoting element*. Divide the entries below it by the 8 and store them in the rest of column 1. This gives the new matrix A^*

$$A^* = \begin{bmatrix} 8 & 2 & 3 \\ -\frac{4}{8} & 3 & 2 \\ \frac{7}{8} & 8 & 9 \end{bmatrix}$$

If we took the original row 1 and multiplied it by the $-\frac{4}{8}$ and subtracted it from row 2, we would have the new second row

$$\begin{bmatrix} 0 & 4 & 3.5 \end{bmatrix}$$

If we took the $\frac{7}{8}$, multiplied the original row 1 by it and subtracted it from the original row 3, we would have

$$\begin{bmatrix} 0 & \frac{50}{8} & \frac{51}{8} \end{bmatrix}$$

With these operations done, we have the matrix A^* taking the form

$$A^* = \begin{bmatrix} 8 & 2 & 3 \\ -\frac{4}{8} & 4 & 3.5 \\ \frac{7}{8} & \frac{25}{4} & \frac{51}{8} \end{bmatrix}$$

The *multipliers* in the lower part of column 1 are important to what we are doing, so we are saving them in the parts of column 1 we have made zero. In MatLab, what we have just done could be written like this

Listing 2.5: Storing multipliers

```
% this is a 3x3 matrix
n = 3
% store multipliers in the rest of column 1
A(2:n,1) = A(2:n,1)/A(1,1);
% compute the new the 2x2 block which
% removes column 1 and row 1
A(2:n,2:n) = A(2:n,2:n) - A(2:n,1)*A(1,2:n);
```

The code above does what we just did by hand. Now do the same thing again, but start in the column 2 and row 2 position in the new matrix A^*. The new *pivoting element* is 4, so below it in column 2, we divide the rest of the elements of column 2 by 4 and store the results. This gives

$$A^* = \begin{bmatrix} 8 & 2 & 3 \\ -\frac{4}{8} & 4 & 3.5 \\ \frac{7}{8} & \frac{25}{16} & \frac{51}{8} \end{bmatrix}$$

We are not done. We now calculate the multiplier $\frac{25}{16}$ times the part of this row 2 past the pivot position and subtract it from the rest of row 3. We actually have a 0 then in both column 1 and column 2 of row 3 now. So, the calculations give

$$\begin{bmatrix} 0 & 0 & \frac{29}{32} \end{bmatrix}$$

although the row we store in A^* is

$$\begin{bmatrix} \frac{7}{8} & \frac{25}{16} & \frac{29}{32} \end{bmatrix}$$

We are now done. We have converted A into the form

$$A^* = \begin{bmatrix} 8 & 2 & 3 \\ -\frac{4}{8} & 4 & 3.5 \\ \frac{7}{8} & \frac{25}{16} & \frac{29}{32} \end{bmatrix}$$

Let this final matrix be called B. We can extract the lower triangular part of B using the MatLab command **`tril(B,-1)`** and the lower triangular matrix L formed from A is then made by adding a main diagonal of 1's to this. The upper triangular part of A is then U which we find by using **`triu(B)`**. In code this is

Listing 2.6: Extracting Lower and Upper Parts of a matrix

```
L = eye(n,n) + tril(A,-1);
U = triu(A);
```

In our example, we find

$$L = \begin{bmatrix} 1 & 0 & 0 \\ -\frac{4}{8} & 1 & 0 \\ \frac{7}{8} & \frac{25}{16} & 1 \end{bmatrix} \quad U = \begin{bmatrix} 8 & 2 & 3 \\ 0 & 4 & \frac{7}{2} \\ 0 & 0 & \frac{29}{32} \end{bmatrix}$$

The full code is listed below.

Listing 2.7: LU Decomposition of A Without Pivoting

```
function [L,U] = GE(A)
%
% A is nxn matrix
% L is nxn lower triangular
% U is nxn upper triangular
%
% We compute the LU decomposition of A using
% Gaussian Elimination
%

[n,n] = size(A);
for k=1:n-1
  % find multiplier
  A(k+1:n,k) = A(k+1:n,k)/A(k,k);
  % zero out column
  A(k+1:n,k+1:n) = A(k+1:n,k+1:n) - A(k+1:n,k)*A(k,k+1:n);
end
L = eye(n,n) + tril(A,-1);
U = triu(A);
end
```

Now in MatLab, to see it work, we enter these commands:

Listing 2.8: Solution using LU Decomposition

```
A = [17 24 1 8 15; 23 5 7 14 16; 4 6 13 20 22;...
          10 12 19 21 3;11 18 25 2 9]
A =
      17      24       1       8      15
      23       5       7      14      16
       4       6      13      20      22
      10      12      19      21       3
      11      18      25       2       9
[L,U] = GE(A);
L
L =
      1.0000          0           0           0           0
      1.3529     1.0000           0           0           0
      0.2353    -0.0128      1.0000           0           0
      0.5882     0.0771      1.4003      1.0000           0
      0.6471    -0.0899      1.9366      4.0578      1.0000
U
U =

     17.0000    24.0000      1.0000      8.0000     15.0000
           0   -27.4706      5.6471      3.1765     -4.2941
           0          0     12.8373     18.1585     18.4154
           0          0           0     -9.3786    -31.2802
           0          0           0           0     90.1734
b = [1; 3; 5; 7; 9]
b =
       1
       3
       5
       7
       9
y = LTriSol(L,b)
y =
      1.0000
```

```
35          1.6471
            4.7859
           -0.4170
            0.9249
     x = UTriSol(U,y)
40   x =

            0.0103
            0.0103
            0.3436
45          0.0103
            0.0103
     c = A*x
     c =
            1.0000
50          3.0000
            5.0000
            7.0000
            9.0000
```

which solves the system as we wanted.

2.9.5 The LU Decomposition of A with Pivoting

Here is a simple function to solve a system using the LU decomposition of A with what is called pivoting. This means we find the largest absolute value entry in the column we are trying to zero out and perform row interchanges to bring that one to the pivot position. The MatLab code changes a bit; see if you can see what we are doing and why! Note that this pivoting step is needed if a pivot element in a column k, row k position is very small. Using it as a divisor would then cause a lot of numerical problems because we would be multiplying by very large numbers.

Listing 2.9: LU Decomposition of A With Pivoting

```
   function [L,U,piv] = GePiv(A);
 2 %
   % A is nxn matrix
   % L is nxn lower triangular matrix
   % U is nxn upper triangular matrix
   % piv is a nx1 integer vector to hold variable order
 7 % permutations
   %
   [n,n] = size(A);
   piv = 1:n;
   for k=1:n-1
12   [maxc,r] = max(abs(A(k:n,k)));
     q = r+k-1;
     piv([k q]) = piv([q k]);
     A([k q],:) = A([q k],:);
     if A(k,k) ~=0
17     A(k+1:n,k) = A(k+1:n,k)/A(k,k);
       A(k+1:n,k+1:n) = A(k+1:n,k+1:n) - A(k+1:n,k)*A(k,k+1:n);
     end
   end
   L = eye(n,n) + tril(A,-1);
22 U = triu(A);
   end
```

We use this code to solve a system as follows:

Listing 2.10: Solving a System with pivoting

```
A = [17 24  1  8 15; 23  5  7 14 16; 4 6 13 20 22; ...
         10 12 19 21  3; 11 18 25  2  9]
A =
      17     24      1      8     15
      23      5      7     14     16
       4      6     13     20     22
      10     12     19     21      3
      11     18     25      2      9
b = [1; 3; 5; 7; 9]
b =
       1
       3
       5
       7
       9
[L,U,piv] = GePiv(A);
L
L =
      1.0000          0          0          0          0
      0.7391     1.0000          0          0          0
      0.4783     0.7687     1.0000          0          0
      0.1739     0.2527     0.5164     1.0000          0
      0.4348     0.4839     0.7231     0.9231     1.0000
U
U =
     23.0000     5.0000     7.0000    14.0000    16.0000
           0    20.3043    -4.1739    -2.3478     3.1739
           0          0    24.8608    -2.8908    -1.0921
           0          0          0    19.6512    18.9793
           0          0          0          0   -22.2222
piv
piv =

       2      1      5      3      4
y = LTriSol(L,b(piv));
y
y =
      3.0000
     -1.2174
      8.5011
      0.3962
     -0.2279
x = UTriSol(U,y);
x
x =
      0.0103
      0.0103
      0.3436
      0.0103
      0.0103
c = A*x
c =
      1.0000
      3.0000
      5.0000
      7.0000
      9.0000
```

which solves the system as we wanted.

2.10 Eigenvalues and Eigenvectors

Another important aspect of matrices is called the *eigenvalues* and *eigenvectors* of a matrix. We will motivate this in the context of 2×2 matrices of real numbers, and then note it can also be done for the general square $n \times n$ matrix. Consider the general 2×2 matrix $\boldsymbol{A}$ given by

$$\boldsymbol{A} = \begin{bmatrix} a & b \\ c & d \end{bmatrix}$$

Is it possible to find a non zero vector $\boldsymbol{v}$ and a number r so that

$$\boldsymbol{A}\,\boldsymbol{v} = r\,\boldsymbol{v}? \tag{2.5}$$

There are many ways to interpret what such a number and vector pair means, but for the moment, we will concentrate on finding such a pair $(r, \boldsymbol{v})$. Now, if this was true, we could rewrite the equation as

$$r\,\boldsymbol{v} - \boldsymbol{A}\,\boldsymbol{v} = \boldsymbol{0} \tag{2.6}$$

where $\boldsymbol{0}$ denotes the vector of all zeros

$$\boldsymbol{0} = \begin{bmatrix} 0 \\ 0 \end{bmatrix}.$$

Next, recall that the two by two identity matrix $\boldsymbol{I}$ is given by

$$\boldsymbol{I} = \begin{bmatrix} 1 & 0 \\ 0 & 1 \end{bmatrix}$$

and it acts like multiplying by one with numbers; i.e. $\boldsymbol{I}\,\boldsymbol{v} = \boldsymbol{v}$ for any vector $\boldsymbol{v}$. Thus, instead of saying $r\,\boldsymbol{v}$, we could say $r\,\boldsymbol{I}\,\boldsymbol{v}$. We can therefore write Eq. 2.6 as

$$r\,\boldsymbol{I}\,\boldsymbol{v} - \boldsymbol{A}\,\boldsymbol{v} = \boldsymbol{0} \tag{2.7}$$

We know that we can factor the vector $\boldsymbol{v}$ out of the left hand side and rewrite again as Eq. 2.8.

$$\left(r\,\boldsymbol{I} - \boldsymbol{A}\right)\,\boldsymbol{v} = \boldsymbol{0} \tag{2.8}$$

Now recall that we want the vector $\boldsymbol{v}$ to be non zero. Note, in solving this system, there are two possibilities:

(i): the determinant of $\boldsymbol{B}$ is non zero which implies the only solution is $\boldsymbol{v} = \boldsymbol{0}$.
(ii): the determinant of $\boldsymbol{B}$ is zero which implies the there are infinitely many solutions for $\boldsymbol{v}$ all of the form a constant c times some non zero vector $\boldsymbol{E}$.

Here the matrix $\boldsymbol{B} = r\boldsymbol{I} - \boldsymbol{A}$. Hence, if we want a non zero solution $\boldsymbol{v}$, we must look for the values of r that force $det(r\boldsymbol{I} - \boldsymbol{A}) = 0$. Thus, we want

$$\begin{aligned} 0 &= det(r\boldsymbol{I} - \boldsymbol{A}) \\ &= det \begin{bmatrix} r - a & -b \\ -c & r - d \end{bmatrix} \\ &= (r - a)\,(r - d) - b\,c \\ &= r^2 - (a + d)\,r + ad - bc. \end{aligned}$$

This important quadratic equation in the variable r determines what values of r will allow us to find non zero vectors $\boldsymbol{v}$ so that $\boldsymbol{A}\,\boldsymbol{v} = r\,\boldsymbol{v}$. Note that although we started out in our minds thinking that r would be a real number, what we have done above shows us that it is possible that r could be complex.

Definition 2.10.1 (*The Eigenvalues and Eigenvectors of a 2 by 2 Matrix*)
Let $\boldsymbol{A}$ be the 2×2 matrix

$$\boldsymbol{A} = \begin{bmatrix} a & b \\ c & d \end{bmatrix}.$$

Then an eigenvalue r of the matrix $\boldsymbol{A}$ is a solution to the quadratic equation defined by

$$det(r\boldsymbol{I} - \boldsymbol{A}) = 0.$$

Any non zero vector that satisfies the equation

$$\boldsymbol{A}\,\boldsymbol{v} = r\,\boldsymbol{v}$$

for the eigenvalue r is then called an eigenvector associated with the eigenvalue r for the matrix $\boldsymbol{A}$.

Comment 2.10.1 *Since this is a quadratic equation, there are always two roots which take the forms below:*

(i): the roots r_1 and r_2 are real and distinct,
(ii): the roots are repeated $r_1 = r_2 = c$ for some real number c,
(iii): the roots are complex conjugate pairs; i.e. there are real numbers α and β so that $r_1 = \alpha + \beta\,\boldsymbol{i}$ and $r_2 = \alpha - \beta\,\boldsymbol{i}$.

Let's look at some examples:

Example 2.10.1 Find the eigenvalues and eigenvectors of the matrix

$$A = \begin{bmatrix} -3 & 4 \\ -1 & 2 \end{bmatrix}$$

Solution *The characteristic equation is*

$$det\left(r\begin{bmatrix} 1 & 0 \\ 0 & 1 \end{bmatrix} - \begin{bmatrix} -3 & 4 \\ -1 & 2 \end{bmatrix}\right) = 0$$

or

$$\begin{aligned} 0 &= det\left(\begin{bmatrix} r+3 & -4 \\ 1 & r-2 \end{bmatrix}\right) \\ &= (r+3)(r-2)+4 \\ &= r^2+r-2 \\ &= (r+2)(r-1) \end{aligned}$$

Hence, the roots, or **eigenvalues**, *of the characteristic equation are* $r_1 = -2$ *and* $r_2 = 1$. *Next, we find the* **eigenvectors** *associated with these eigenvalues.*

1. *For eigenvalue* $r_1 = -2$, *substitute the value of this eigenvalue into*

$$\begin{bmatrix} r+3 & -4 \\ 1 & r-2 \end{bmatrix}$$

This gives

$$\begin{bmatrix} 1 & -4 \\ 1 & -4 \end{bmatrix}$$

The two rows of this matrix should be multiples of one another. If not, we made a mistake and we have to go back and find it. Our rows are indeed multiples, so pick one row to solve for the eigenvector. We need to solve

$$\begin{bmatrix} 1 & -4 \\ 1 & -4 \end{bmatrix}\begin{bmatrix} v_1 \\ v_2 \end{bmatrix} = \begin{bmatrix} 0 \\ 0 \end{bmatrix}$$

Picking the top row, we get

$$\begin{aligned} v_1 - 4\,v_2 &= 0 \\ v_2 &= \frac{1}{4}v_1 \end{aligned}$$

Letting $v_1 = A$*, we find the solutions have the form*

$$\begin{bmatrix} v_1 \\ v_2 \end{bmatrix} = A \begin{bmatrix} 1 \\ \frac{1}{4} \end{bmatrix}$$

The vector

$$\begin{bmatrix} 1 \\ 1/4 \end{bmatrix}$$

is our choice for an eigenvector corresponding to eigenvalue $r_1 = -2$.

2. *For eigenvalue* $r_2 = 1$*, substitute the value of this eigenvalue into*

$$\begin{bmatrix} r+3 & -4 \\ 1 & r-2 \end{bmatrix}$$

This gives

$$\begin{bmatrix} 4 & -4 \\ 1 & -1 \end{bmatrix}$$

Again, the two rows of this matrix should be multiples of one another. If not, we made a mistake and we have to go back and find it. Our rows are indeed multiples, so pick one row to solve for the eigenvector. We need to solve

$$\begin{bmatrix} 4 & -4 \\ 1 & -1 \end{bmatrix} \begin{bmatrix} v_1 \\ v_2 \end{bmatrix} = \begin{bmatrix} 0 \\ 0 \end{bmatrix}$$

Picking the bottom row, we get

$$\begin{aligned} v_1 - v_2 &= 0 \\ v_2 &= v_1 \end{aligned}$$

Letting $v_1 = B$*, we find the solutions have the form*

$$\begin{bmatrix} v_1 \\ v_2 \end{bmatrix} = B \begin{bmatrix} 1 \\ 1 \end{bmatrix}$$

The vector

$$\begin{bmatrix} 1 \\ 1 \end{bmatrix}$$

is our choice for an eigenvector corresponding to eigenvalue $r_2 = 1$.

Example 2.10.2 Find the eigenvalues and eigenvectors of the matrix

$$\boldsymbol{A} = \begin{bmatrix} 4 & 9 \\ -1 & -6 \end{bmatrix}$$

Solution *The characteristic equation is*

$$det\left(r\begin{bmatrix} 1 & 0 \\ 0 & 1 \end{bmatrix} - \begin{bmatrix} 4 & 9 \\ -1 & -6 \end{bmatrix}\right) = 0$$

or

$$\begin{aligned} 0 &= det\left(\begin{bmatrix} r-4 & -9 \\ 1 & r+6 \end{bmatrix}\right) \\ &= (r-4)(r+6)+9 \\ &= r^2 + 2r - 15 \\ &= (r+5)(r-3) \end{aligned}$$

Hence, the roots, or **eigenvalues**, *of the characteristic equation are* $r_1 = -5$ *and* $r_2 = 3$. *Next, we find the* **eigenvectors** *associated with these eigenvalues.*

1. For eigenvalue $r_1 = -5$, *substitute the value of this eigenvalue into*

$$\begin{bmatrix} r-4 & -9 \\ 1 & r+6 \end{bmatrix}$$

This gives

$$\begin{bmatrix} -9 & -9 \\ 1 & 1 \end{bmatrix}$$

The two rows of this matrix should be multiples of one another. If not, we made a mistake and we have to go back and find it. Our rows are indeed multiples, so pick one row to solve for the eigenvector. We need to solve

$$\begin{bmatrix} -9 & -9 \\ 1 & 1 \end{bmatrix}\begin{bmatrix} v_1 \\ v_2 \end{bmatrix} = \begin{bmatrix} 0 \\ 0 \end{bmatrix}$$

Picking the bottom row, we get

$$\begin{aligned} v_1 + v_2 &= 0 \\ v_2 &= -\, v_1 \end{aligned}$$

Letting $v_1 = A$, *we find the solutions have the form*

$$\begin{bmatrix} v_1 \\ v_2 \end{bmatrix} = A \begin{bmatrix} 1 \\ -1 \end{bmatrix}$$

The vector

$$\begin{bmatrix} 1 \\ -1 \end{bmatrix}$$

is our choice for an eigenvector corresponding to eigenvalue $r_1 = -5$.

2. *For eigenvalue* $r_2 = 3$, *substitute the value of this eigenvalue into*

$$\begin{bmatrix} r-4 & -9 \\ 1 & r+6 \end{bmatrix}$$

This gives

$$\begin{bmatrix} -1 & -9 \\ 1 & 9 \end{bmatrix}$$

Again, the two rows of this matrix should be multiples of one another. If not, we made a mistake and we have to go back and find it. Our rows are indeed multiples, so pick one row to solve for the eigenvector. We need to solve

$$\begin{bmatrix} -1 & -9 \\ 1 & 9 \end{bmatrix} \begin{bmatrix} v_1 \\ v_2 \end{bmatrix} = \begin{bmatrix} 0 \\ 0 \end{bmatrix}$$

Picking the bottom row, we get

$$\begin{aligned} v_1 + 9\, v_2 &= 0 \\ v_2 &= \frac{-1}{9}\, v_1 \end{aligned}$$

Letting $v_1 = B$, *we find the solutions have the form*

$$\begin{bmatrix} v_1 \\ v_2 \end{bmatrix} = B \begin{bmatrix} 1 \\ \frac{-1}{9} \end{bmatrix}$$

The vector

$$\begin{bmatrix} 1 \\ \frac{-1}{9} \end{bmatrix}$$

is our choice for an eigenvector corresponding to eigenvalue $r_2 = 3$.

2.10.1 Homework

Exercise 2.10.1 *Find the eigenvalues and eigenvectors of the matrix*

$$\boldsymbol{A} = \begin{bmatrix} 6 & 3 \\ -11 & -8 \end{bmatrix}$$

Exercise 2.10.2 *Find the eigenvalues and eigenvectors of the matrix*

$$\boldsymbol{A} = \begin{bmatrix} 2 & 1 \\ -4 & -3 \end{bmatrix}$$

Exercise 2.10.3 *Find the eigenvalues and eigenvectors of the matrix*

$$\boldsymbol{A} = \begin{bmatrix} -2 & -1 \\ 8 & 7 \end{bmatrix}$$

Exercise 2.10.4 *Find the eigenvalues and eigenvectors of the matrix*

$$\boldsymbol{A} = \begin{bmatrix} -6 & -3 \\ 4 & 1 \end{bmatrix}$$

Exercise 2.10.5 *Find the eigenvalues and eigenvectors of the matrix*

$$\boldsymbol{A} = \begin{bmatrix} -4 & -2 \\ 13 & 11 \end{bmatrix}$$

2.10.2 The General Case

For a general $n \times n$ matrix $\boldsymbol{A}$, we have the following:

Definition 2.10.2 (*The Eigenvalues and Eigenvectors of a n by n Matrix*)
Let $\boldsymbol{A}$ be the $n \times n$ matrix.

$$\boldsymbol{A} = \begin{bmatrix} A_{11} & A_{12} & \cdots & A_{n1} \\ A_{21} & A_{22} & \cdots & A_{n2} \\ \vdots & \vdots & \vdots & \vdots \\ A_{n1} & A_{n2} & \cdots & A_{nn} \end{bmatrix}.$$

Then an eigenvalue r of the matrix $\boldsymbol{A}$ is a solution to the polynomial defined by

$$det(r\boldsymbol{I} - \boldsymbol{A}) = 0.$$

Any non zero vector that satisfies the equation

$$\boldsymbol{A}\ \boldsymbol{v} = r\ \boldsymbol{v}$$

for the eigenvalue r is then called an eigenvector associated with the eigenvalue r for the matrix $\boldsymbol{A}$.

Comment 2.10.2 *Since this is a polynomial equation, there are always n roots some of which are real numbers which are distinct, some might be repeated and some might be complex conjugate pairs (and they can be repeated also!). An example will help. Suppose we started with a* 5×5 *matrix. Then, the roots could be*

1. *All the roots are real and distinct; for example,* 1, 2, 3, 4 *and* 5.
2. *Two roots are the same and three roots are distinct; for examples,* 1, 1, 3, 4 *and* 5.
3. *Three roots are the same and two roots are distinct; for examples,* 1, 1, 1, 4 *and* 5.
4. *Four roots are the same and one roots is distinct from that; for examples,* 1, 1, 1, 1 *and* 5.
5. *Five roots are the same; for examples,* 1, 1, 1, 1 *and* 1.
6. *Two pairs of roots are the same and one roots is different from them; for examples,* 1, 1, 3, 3 *and* 5.
7. *One triple root and one pair of real roots; for examples,* 1, 1, 1, 3 *and* 3.
8. *One triple root and one complex conjugate pair of roots; for examples,* 1, 1, 1, $3 + 4i$ *and* $3 - 4i$.
9. *One double root and one complex conjugate pair of roots and one different real root; for examples,* 1, 1, 2, $3 + 4i$ *and* $3 - 4i$.
10. *Two complex conjugate pair of roots and one different real root; for examples,* -2, $1 + 6i$, $1 - 6i$, $3 + 4i$ *and* $3 - 4i$.

2.10.3 The MatLab Approach

We will now discuss certain ways to compute eigenvalues and eigenvectors for a square matrix in MatLab. For a given $\boldsymbol{A}$, we can compute its eigenvalues as follows:

Listing 2.11: Eigenvalues in Matlab

```
A = [1 2 3; 4 5 6; 7 8 -1]
A =
        1       2       3
        4       5       6
        7       8      -1
E = eig(A)
E =
   -0.3954
   11.8161
   -6.4206
```

So we have found the eigenvalues of this small 3×3 matrix. To get the eigenvectors, we do this:

Listing 2.12: Eigenvectors in Matlab

```
[V,D] = eig(A)
V =
    0.7530   -0.3054   -0.2580
   -0.6525   -0.7238   -0.3770
    0.0847   -0.6187    0.8896
D =
   -0.3954         0         0
         0   11.8161         0
         0         0   -6.4206
```

Note the eigenvalues are not returned in ranked order. The eigenvalue/eigenvector pairs are thus

$$\lambda_1 = -0.3954$$
$$V_1 = \begin{bmatrix} 0.7530 \\ -0.6525 \\ 0.0847 \end{bmatrix}$$

$$\lambda_2 = 11.8161$$
$$V_2 = \begin{bmatrix} -0.3054 \\ -0.7238 \\ -0.6187 \end{bmatrix}$$

$$\lambda_3 = -6.4206$$
$$V_3 = \begin{bmatrix} -0.2580 \\ -0.3770 \\ 0.8896 \end{bmatrix}$$

Now let's try a nice 5×5 array that is symmetric:

Listing 2.13: Eigenvalues and Eigenvectors Example

```
B = [1 2 3 4 5;
     2 5 6 7 9;
     3 6 1 2 3;
     4 7 2 8 9;
     5 9 3 9 6]
B =
     1     2     3     4     5
     2     5     6     7     9
     3     6     1     2     3
     4     7     2     8     9
     5     9     3     9     6
[W,Z] = eig(B)
W =

   0.8757    0.0181   -0.0389    0.4023    0.2637
  -0.4289   -0.4216   -0.0846    0.6134    0.5049
   0.1804   -0.6752    0.4567   -0.4866    0.2571
  -0.1283    0.5964    0.5736   -0.0489    0.5445
   0.0163    0.1019   -0.6736   -0.4720    0.5594
Z =
   0.1454         0         0         0         0
        0    2.4465         0         0         0
        0         0   -2.2795         0         0
        0         0         0   -5.9321         0
        0         0         0         0   26.6197
```

It is possible to show that the eigenvalues of a symmetric matrix will be real and eigenvectors corresponding to distinct eigenvalues will be 90° apart. Such vectors are called **orthogonal** and recall this means their inner product is 0. Let's check it out. The eigenvectors of our matrix are the columns of W above. So their dot product should be 0!

Listing 2.14: Checking orthogonality

```
C = dot(W(1:5,1),W(1:5,2))
C =
    1.3336e-16
```

Well, the dot product is not actually 0 because we are dealing with floating point numbers here, but as you can see it is close to machine zero (the smallest number our computer chip can *detect*). Welcome to the world of computing!

Reference

J. Peterson, *Calculus for Cognitive Scientists: Derivatives, Integration and Modeling*, Springer Series on Cognitive Science and Technology (Springer Science + Business Media Singapore Pte Ltd., Singapore, 2015 in press)

Chapter 3
Numerical Methods Order One ODEs

Now that you are taking this course on **More Calculus for Cognitive Scientists**, we note that in the previous course, you were introduced to continuity, derivatives, integrals and models using derivatives. You were also taught about functions of two variables and partial derivatives along with more interesting models. You were also introduced to how to solve models using Euler's method. We use these ideas a lot, so there is a much value in reviewing this material. So let's dive into it again. When we try to solve systems like

$$\frac{dy}{dt} = f(t, y) \tag{3.1}$$

$$y(t_0) = y0 \tag{3.2}$$

where f is continuous in the variables t and y, and y_0 is some value the solution is to have at the time point t_0, we will quickly find that it is very hard in general to do this by *hand*. So it is time to begin looking at how the MatLab environment can help us. We will use MatLab to solve these differential equations with what are called *numerical* methods. First, let's discuss how to approximate functions in general.

3.1 Taylor Polynomials

We can approximate a function at a point using polynomials of various degrees. We can first find the constant that best approximates a function f at a point p. This is called the **zeroth order Taylor polynomial** and the equation we get is

$$f(x) = f(p) + E_0(x, p)$$

J.K. Peterson, *Calculus for Cognitive Scientists*,
Cognitive Science and Technology, DOI 10.1007/978-981-287-877-9_3

where $E_0(x, p)$ is the **error**. On the other hand, we could try to find the best straight line that does the job. We would find

$$f(x) = f(p) + f'(p)(x - p) + E_1(x, p)$$

where $E_1(x, p)$ is the **error** now. This straight line is the **first order Taylor polynomial** but we know it also as the **tangent line**. We can continue finding polynomials of higher and higher degree and their respective errors. In this class, our interests stop with the quadratic case. We would find

$$f(x) = f(p) + f'(p)(x - p) + f''(p)(x - p)^2 + E_2(x, p)$$

where $E_2(x, p)$ is the **error**. This is called the **second order Taylor polynomial** or **quadratic approximation**. Now let's dig into the theory behind this so that we can better understand the error terms.

3.1.1 Fundamental Tools

Let's consider a function which is defined locally at the point p. This means there is at least an interval (a, b) containing p where f is defined. Of course, this interval could be the whole x axis!. Let's also assume f' exists locally at p in this same interval. Now pick any x is the interval $[p, b)$ (we can also pick a point in the left hand interval $(a, p]$ but we will leave that discussion to you!). From Calculus I, recall Rolle's Theorem and the Mean Value Theorem. These are usually discussed in Calculus I, but we really prove them carefully in course called Mathematical Analysis (but that is another story).

Theorem 3.1.1 (Rolle's Theorem)
Let $f : [a, b] \to \Re$ be a function defined on the interval $[a, b]$ which is continuous on the closed interval $[a, b]$ and is at least differentiable on the open interval (a, b). If $f(a) = f(b)$, then there is at least one point c, between a and b, so that $f'(c) = 0$.

and

Theorem 3.1.2 (The Mean Value Theorem)
Let $f : [a, b] \to \Re$ be a function defined on the interval $[a, b]$ which is continuous on the closed interval $[a, b]$ and is at least differentiable on the open interval (a, b). Then there is at least one point c, between a and b, so that

$$\frac{f(b) - f(a)}{b - a} = f'(c).$$

3.1.2 The Zeroth Order Taylor Polynomial

Our function f on the interval $[p, x]$ satisfies all the requirements of the Mean Value Theorem. So we know there is a point c_x with $p < c_x < x$ so that

$$\frac{f(x) - f(p)}{x - p} = f'(c_x).$$

This can be written as

$$f(x) = f(p) + f'(c_x)(p - a).$$

Let the constant $f(p)$, a polynomial of degree 0, be denoted by

$$P_0(p, x) = f(p).$$

We'll call this the 0th order Taylor Polynomial for f at the point p. Next, let the 0th order error term be defined by

$$E_0(p, x) = f(x) - P_0(p, x) = f(x) - f(p).$$

The error or remainder term is clearly the difference or discrepancy between the actual function value at x and the 0th order Taylor Polynomial. Since $f(x) - f(p) = f'(c_x)(x - p)$, we can write all we have above as

$$E_0(p, x) = f'(c_x)(x - p), \quad \text{some } c_x \text{, with } p < c_x < x.$$

We can interpret what we have done by saying $f(p)$ is the best choice of 0th order polynomial or constant to approximate $f(x)$ near p. Of course, for most functions, this is a horrible approximation! So the next step is to find the best straight line that approximates f near p. Let's try our usual tangent line to f at p. We summarize this result as a theorem.

Theorem 3.1.3 (Zeroth Order Taylor Polynomial)
Let $f : [a, b] \rightarrow \Re$ be continuous on $[a, b]$ and be at least differentiable on (a, b). Then for each p in $[a, b]$, there is at least one point c, between p and x, so that $f(x) = f(p) + f'(c)(x - p)$. The constant $f(p)$ is called the **zeroth order Taylor Polynomial** *for f at p and we denote it by $P_0(x; p)$. The point p is called the* **base point**. *Note we are approximating $f(x)$ by the constant $f(p)$ and the error we make is $E_0(x, p) = f'(c)(x - p)$.*

3.1.2.1 Examples

Example 3.1.1 If $f(t) = t^3$, by the theorem above, we know on the interval $[1, 3]$ that at 1 $f(t) = f(1) + f'(c)(t - 1)$ where c is some point between 1 and t. Thus, $t^3 = 1 + (3c^2)(t - 1)$ for some $1 < c < t$. So here the **zeroth order Taylor Polynomial** is $P_0(t, 1) = 1$ and the **error** is $E_0(t, 1) = (3c^2)(t - 1)$.

Example 3.1.2 If $f(t) = e^{-1.2t}$, by the theorem above, we know that at 0 $f(t) = f(0) + f'(c)(t - 0)$ where c is some point between 0 and t. Thus, $e^{-1.2t} = 1 + (-1.2)e^{-1.2c}(t-0)$ for some $0 < c < t$ or $e^{-1.2t} = 1 - 1.2e^{-1.2c}t$. So here the **zeroth order Taylor Polynomial** is $P_0(t, 1) = 1$ and the **error** is $E_0(t, 0) = -1.2e^{-1.2c}t$.

Example 3.1.3 If $f(t) = e^{-0.00231t}$, by the theorem above, we know that at 0 $f(t) = f(0) + f'(c)(t - 0)$ where c is some point between 0 and t. Thus, $e^{-0.00231t} = 1 + (-0.00231)e^{-0.00231c}(t - 0)$ for some $0 < c < t$ or $e^{-0.00231t} = 1 - 0.00231e^{-0.00231c}t$. So here the **zeroth order Taylor Polynomial** is $P_0(t, 1) = 1$ and the **error** is $E_0(t, 0) = -0.00231e^{-0.00231c}t$.

3.1.3 The First Order Taylor Polynomial

If a function f is differentiable, from Calculus I, we know we can approximate its value at the point p by its tangent line, T. We have

$$f(x) = f(p) + f'(p)\,(x - p) + E_T(x, p), \tag{3.3}$$

where the tangent line T is the function

$$T(x) = f(p) + f'(p)\,(x - p)$$

and the term $E_1(x, p)$ represents the error between the true function value $f(x)$ and the tangent line value $T(x)$. That is

$$E_1(x, p) = f(x) - T(x).$$

Another way to look at this is that the tangent line is the best straight line or *linear* approximation to f at the point p. We all know from our first calculus course how these pictures look. If the function f is curved near p, then the tangent line is not a very good approximation to f at p unless x is very close to p. Now, let's assume f is actually two times differentiable on the local interval (a, b) also. Define the constant M by

$$M = \frac{f(x) - f(p) - f'(p)(x - p)}{(x - p)^2}.$$

In this discussion, this really is a constant value because we have fixed our value of x and p already. We can rewrite this equation as

$$f(x) = f(p) + f'(p)(x - p) + M(x - p)^2.$$

Now let's define the function g on an interval I containing p by

$$g(t) = f(t) - f(p) - f'(p)\,(t - p) - M(t - p)^2.$$

Then,

$$\begin{aligned} g'(t) &= f'(t) - f'(p) - 2M(t - p) \\ g''(t) &= f''(t) - 2M. \end{aligned}$$

Then,

$$\begin{aligned} g(x) &= f(x) - f(p) - f'(p)\,(x - p) - M(x - p)^2 \\ &= f(x) - f(p) - f'(p)\,(x - p) - f(x) + f(p) + f'(p)(x - p) \\ &= 0 \end{aligned}$$

and

$$g(p) = f(p) - f(p) - f'(p)\,(p - p) - M(p - p)^2 = 0.$$

We thus know $g(x) = g(p) = 0$. Also, from the Mean Value Theorem, there is a point c_x^0 between p and x so that

$$\frac{g(x) - g(p)}{p - x} = g'(c_x^0).$$

Since the numerator is $g(x) - g(p)$, we now know $g'(c_x^0) = 0$. But we also have

$$g'(p) = f'(p) - f'(p) - 2M(p - p) = 0.$$

Next, we can apply Rolle's Theorem to the function g'. This tells us there is a point c_x^1 between p and c_x^0 so that $g''(c_x^1) = 0$. Thus,

$$0 = g''(c_x^1) = f''(c_x^1) - 2M.$$

and simplifying, we have

$$M = \frac{1}{2}\, f''(c_x^1).$$

Remembering what the value of M was, gives us our final result

$$\frac{f(x) - f(p) - f'(p)(x-p)}{(x-p)^2} = \frac{1}{2} f''(c_x^1), \text{ some } c_x^1 \text{ with } p < c_x^1 < c_x^0.$$

which can be rewritten as

$$f(x) = f(p) + f'(p)(x-p) + \frac{1}{2} f''(c_x^1)(x-p)^2, \text{ some } c_x^1 \text{ with } p < c_x^1 < c_x^0.$$

We define the 1st order Taylor polynomial, $P_1(x, p)$ and 1st order error, $E_1(x, p)$ by

$$\begin{aligned} P_1(x, p) &= f(p) + f'(p)(x-p) \\ E_1(x, p) &= f(x) - P_1(x, p) = f(x) - f(p) - f'(p)(x-p) \\ &= \frac{1}{2} f''(c_x^1)(x-p)^2. \end{aligned}$$

Thus, we have shown, $E_1(x, p)$ satisfies

$$E_1(x, p) = f''(c_x^1) \frac{(x-p)^2}{2} \tag{3.4}$$

where c_x^1 is some point between x and p. Note the usual Tangent line is the same as the first order Taylor Polynomial, $P_1(f, p, x)$ and we have a nice representation of our error. We can state this as our next theorem:

Theorem 3.1.4 (First Order Taylor Polynomial)
Let $f : [a, b] \rightarrow \Re$ be continuous on $[a, b]$ and be at least twice differentiable on (a, b). For a given p in $[a, b]$, for each x, there is at least one point c, between p and x, so that $f(x) = f(p) + f'(p)(x-p) + (1/2) f''(c)(x-p)^2$. The $f(p) + f'(p)(x-p)$ is called the **first order Taylor Polynomial** *for f at p. and we denote it by $P_1(x; p)$. The point p is again called the* **base point**. *Note we are approximating $f(x)$ by the linear function $f(p) + f'(p)(x-p)$ and the error we make is $E_1(x, p) = (1/2) f''(c)(x-p)$.*

3.1.3.1 Example

Let's do some examples to help this sink in.

Problem One: Let's find the tangent line approximations for a simple exponential decay function.

Example 3.1.4 For $f(t) = e^{-1.2t}$ on the interval $[0, 5]$ find the tangent line approximation, the error and maximum the error can be on the interval.

Solution *Using base point* 0, *we have at any* t

$$\begin{aligned} f(t) &= f(0) + f'(0)(t-0) + (1/2)f''(c)(t-0)^2 \\ &= 1 + (-1.2)(t-0) + (1/2)(-1.2)^2 e^{-1.2c}(t-0)^2 \\ &= 1 - 1.2t + (1/2)(1.2)^2 e^{-1.2c} t^2. \end{aligned}$$

where c *is some point between* 0 *and* t. *Hence,* c *is between* 0 *and* 5 *also. The* **first order Taylor Polynomial** *is* $P_1(t, 0) = 1 - 1.2t$ *which is also the tangent line to* $e^{-1.2t}$ *at* 0. *The* **error** *is* $(1/2)(-1.2)^2 e^{-1.2c} t^2$.
Now let **AE(t)** *denote* **absolute value of the actual error at t** *and* **ME** *be* **maximum absolute error on the interval.** *The largest the error can be on* $[0, 5]$ *is when* $f''(c)$ *is the biggest it can be on the interval. Here,*

$$\mathbf{AE(t)} = (1/2)(1.2)^2 e^{-1.2c} t^2 \leq (1/2)(1.2)^2 \times 1 \times (5)^2 = (1/2)1.44 \times 25 = \mathbf{ME}.$$

Problem Two: Let's find the tangent line approximations for a simple exponential decay function again but let's do it a bit more generally.

Example 3.1.5 If $f(t) = e^{-\beta t}$, for $\beta = 1.2 \times 10^{-5}$, find the tangent line approximation, the error and the maximum error on $[0, 5]$.

Solution *At any* t

$$\begin{aligned} f(t) &= f(0) + f'(0)(t-0) + \frac{1}{2} f''(c)(t-0)^2 \\ &= 1 + (-\beta)(t-0) + \frac{1}{2}(-\beta)^2 e^{-\beta c}(t-0)^2 \\ &= 1 - \beta t + \frac{1}{2}\beta^2 e^{-\beta c} t^2. \end{aligned}$$

where c *is some point between* 0 *and* t *which means* c *is between* 0 *and* 5. *The* **first order Taylor Polynomial** *is* $P_1(t, 0) = 1 - \beta t$ *which is also the tangent line to* $e^{-\beta t}$ *at* 0. *The* **error** *is* $\frac{1}{2}\beta^2 e^{-\beta c} t^2$. *The largest the error can be on* $[0, 5]$ *is when* $f''(c)$ *is the biggest it can be on the interval. Here,*

$$\begin{aligned} \mathbf{AE(t)} &= |(1/2)(1.2 \times 10^{-5})^2 e^{-1.2 \times 10^{-5} c}\, t^2| \leq (1/2)(1.2 \times 10^{-5})^2 (1)(5)^2 \\ &= (1/2)1.44 \times 10^{-10}(25) = \mathbf{ME} \end{aligned}$$

3.1.4 Quadratic Approximations

We could also ask what quadratic function Q fits f best near p. Let the quadratic function Q be defined by

$$Q(x) = f(p) + f'(p)\,(x - p) + f''(p)\,\frac{(x-p)^2}{2}. \tag{3.5}$$

The new error is called $E_Q(x, p)$ and is given by

$$E_Q(x, p) = f(x) - Q(x).$$

If f is three times differentiable, we can argue like we did in the tangent line approximation (using the Mean Value Theorem and Rolle's theorem on an appropriately defined function g) to show there is a new point c_x^2 between p and c_x^1 with

$$E_Q(x, p) = f'''(c_x^2)\,\frac{(x-p)^3}{6} \tag{3.6}$$

So if f looks like a quadratic locally near p, then Q and f match nicely and the error is pretty small. On the other hand, if f is not quadratic at all near p, the error will be large. We then define the second order Taylor polynomial, $P_2(f, p, x)$ and second order error, $E_2(x, p) = E_Q(x, p)$ by

$$\begin{aligned}
P_2(x, p) &= f(p) + f'(p)(x-p) + \frac{1}{2} f''(p)(x-p)^2 \\
E_2(x, p) &= f(x) - P_2(x, p) = f(x) - f(p) - f'(p)(x-p) - \frac{1}{2}\, f''(p)(x-p)^2 \\
&= \frac{1}{6} f'''(c_x^2)\,(x-p)^3.
\end{aligned}$$

Theorem 3.1.5 (Second Order Taylor Polynomial)
Let $f : [a, b] \rightarrow \Re$ be continuous on $[a, b]$ and be at least three times differentiable on (a, b). Given p in $[a, b]$, for each x, there is at least one point c, between p and x, so that $f(x) = f(p) + f'(p)(x-p) + (1/2) f''(p)(x-p)^2 + (1/6) f'''(c)(x-p)^3$. The quadratic $f(p) + f'(p)(x-p) + (1/2) f''(p)(x-p)^2$ is called the **second order Taylor Polynomial** *for f at p and we denote it by $P_2(x, p)$. The point p is again called the* **base point**. *Note we are approximating $f(x)$ by the quadratic $f(p) + f'(p)(x-p) + (1/2) f''(p)(x-p)^2$ and the error we make is $E_2(x, p) = (1/6) f'''(c)(x-p)$.*

3.1.4.1 Examples

Let's work out some problems involving quadratic approximations.

Example 3.1.6 If $f(t) = e^{-\beta t}$, for $\beta = 1.2 \times 10^{-5}$, find the second order approximation, the error and the maximum error on $[0, 5]$.

Solution *For each t in the interval* $[0, 5]$, *then there is some* $0 < c << t < 5$ *so that*

$$\begin{aligned} f(t) &= f(0) + f'(0)(t-0) + (1/2)f''(0)(t-0)^2 \\ &\quad + (1/6)f'''(c)(t-0)^3 \\ &= 1 + (-\beta)(t-0) + \frac{1}{2}(-\beta)^2(t-0)^2 + \frac{1}{6}(-\beta)^3 e^{-\beta c}(t-0)^3 \\ &= 1 - \beta t + (1/2)\beta^2 - (1/6)\beta^3 e^{-\beta c} t^3. \end{aligned}$$

The **second order Taylor Polynomial** *is* $p_2(t, 0) = 1 - \beta t + (1/2)\beta^2\, t^2$ *which is also called the quadratic approximation to* $e^{-\beta t}$ *at* 0. *The* **error** *is* $-\frac{1}{6}\beta^3 e^{-\beta c} t^3$. *The error is largest on* $[0, 5]$ *when* $f'''(c)$ *is the biggest it can be on the interval. Here,*

$$\begin{aligned} \mathbf{AE(t)} &= | - (1/6)(1.2 \times 10^{-5})^3 e^{-1.2\times 10^{-5} c}\, t^3| \le | - (1/6)(1.2 \times 10^{-5})^3 e^{-1.2\times 10^{-5} c}\, t^3| \\ &\le (1/6)(1.2 \times 10^{-5})^3\, (1)\, (5)^3 = (1/6)\, 1.728 \times 10^{-15}\, (125) = \mathbf{ME} \end{aligned}$$

Example 3.1.7 Do this same problem on the interval $[0, T]$.

Solution *The approximations are the same,* $0 < c < T$ *and*

$$\begin{aligned} \mathbf{AE} &= | - (1/6)(1.2 \times 10^{-5})^3\, e^{-1.2\times 10^{-5} c}\, t^3| \le | - (1/6)(1.2 \times 10^{-5})^3\, e^{-1.2\times 10^{-5} c}\, t^3| \\ &\le (1/6)(1.2 \times 10^{-5})^3\, T^3 = (1/6)\, 1.728 \times 10^{-15}\, (T^3) = \mathbf{ME}. \end{aligned}$$

We can find higher order Taylor polynomials and remainders using these arguments as long as f has higher order derivatives. But, for our purposes, we can stop here.

3.2 Euler's Method with Time Independence

Let's try to approximate the solution to the model $x' = f(x)$ with $x(0) = x_0$. Note the dynamics does not depend on time t. The solution $x(t)$ can be written

$$x(t) = x(0) + x'(0)(t - 0) + x''(c_t)(t - 0)^2/2$$

where c_t is some number between 0 and t. To approximate the solution, we will divide the interval $[0, T]$ into pieces of length h. We call h the **stepsize**. If h does not evenly divide T, we just use the last subinterval even though it may be a bit short. Let N be the number of subintervals we get by doing this.

- Example: Divide $[0, 5]$ using $h = 0.4$. Then $5/0.4 = 12.5$ so we create 13 subintervals with the last one of length 0.2 instead of 0.4. So $N = 13$.
- Example: Divide $[0, 10]$ using $h = 0.2$. Then $10/0.2 = 50$ and we get $N = 50$.

To approximate the **true** solution $x(h)$ we then have

$$\begin{aligned} x(h) &= x(0) + x'(0)(h-0) + x''(c_h)(h-0)^2/2 \\ &= x_0 + x'(0)h + x''(c_h)h^2/2, \end{aligned}$$

where c_h is between 0 and h. We can rewrite this more. Note $x' = f(x)$ tells us we can replace $x'(0)$ by $f(x(0)) = f(x_0)$. Also, since $x' = f(x)$, the chain rule tells us $x'' = f'(x)\, x' = (df/dx)\, f$ where we let $f'(x) = (df/dx)(x)$. So $x''(c_h) = f'(x(c_h))\, f(x(c_h))$ and we have

$$x(h) = x_0 + f(x_0)h + f'(x(c_h))\, f(x(c_h))\, h^2/2$$

Let x_1 be the true solution $x(h)$ and let $\hat{x}_0$ be the starting or zeroth Euler approximate which is defined by $\hat{x}_0 = x_0$. Hence, we make no error at first. Further, let the first Euler approximate $\hat{x}_1$ be defined by $\hat{x}_1 = x_0 + f(x_0)\, h = \hat{x}_0 + f(\hat{x}_0)\, h$. Which is the tangent line approximation to x at the point $t = 0$! Then we have

$$x_1 = \hat{x}_1 + f'(x(c_h))\, f(x(c_h))\, h^2/2.$$

Define the error at the first step by $E_1 = |x_1 - \hat{x}_1|$. Thus,

$$E_1 = |x_1 - \hat{x}_1| = |f'(x(c_h))|\, |f(x(c_h))|\, h^2/2.$$

Now let's do bounds. Now x is continuous on $[0, T]$ so x is bounded. Hence, $||x||_\infty = \max_{0 \le t \le T} |x(t)|$ is some finite number. Call it D. We see $x(t)$ lives in the interval $[-D, D]$ which is on the x axis. Then

- f is continuous on $[-D, D]$, so $||f||_\infty = \max_{-D \le x \le D} |f(x)|$ is some finite number.
- f' is continuous on $[-D, D]$, so $||f'||_\infty = \max_{-D \le x \le D} |f'(x)|$ is some finite number.

The specific numbers we used for the example $y' = 3y$ are an example of these bounds. Using these bounds, we have

$$\begin{aligned} E_1 = |x_1 - \hat{x}_1| &= |f'(x(c_h))|\, |f(x(c_h))|\, h^2/2 \\ &\le ||f||_\infty\, ||f'||_\infty\, h^2/2 = B\, h^2/2 \le C\, h^2/2. \end{aligned}$$

where we let $A = ||f||_\infty$, $B = ||f||_\infty\, ||f'||_\infty$ and C be the maximum of A and B.

Now let's do the approximation for $x(2h)$. We will let $x_2 = x(2h)$ and we will define the second Euler approximate by $\hat{x}_2 = \hat{x}_1 + f(\hat{x}_1)\, h$. The tangent line approximation to x at h gives

$$\begin{aligned} x(2h) &= x(h) + x'(h)\, h + x''(c_{2h})\, h^2/2 \\ &= x(h) + f(x(h))h + f'(x(c_{2h}))\, f(x(c_{2h}))\, h^2/2 \\ &= x_1 + f(x_1)h + f'(x(c_{2h}))\, f(x(c_{2h}))\, h^2/2. \end{aligned}$$

Now add and subtract $\hat{x}_2 = \hat{x}_1 + f(\hat{x}_1)\,h$ to this equation.

$$
\begin{aligned}
x(2h) &= x_1 + f(x_1)h + f'(x(c_{2h}))\,f(x(c_{2h}))\,h^2/2 \\
&= x_1 + f(x_1)h + \hat{x}_2 - \hat{x}_2 + f'(x(c_{2h}))\,f(x(c_{2h}))\,h^2/2 \\
&= (x_1 + f(x_1)h - \hat{x}_1 - f(\hat{x}_1)\,h) + \hat{x}_2 \\
&\quad + f'(x(c_{2h}))\,f(x(c_{2h}))\,h^2/2 \\
&= \hat{x}_2 + (x_1 - \hat{x}_1) + (f(x_1) - f(\hat{x}_1))\,h \\
&\quad + f'(x(c_{2h}))\,f(x(c_{2h}))\,h^2/2.
\end{aligned}
$$

We are almost there! Next, we can apply the Mean Value Theorem to the difference $f(x_1) - f(\hat{x}_1)$ and find $f(x_1) - f(\hat{x}_1) = f'(x_d)(x_1 - \hat{x}_1)$ with x_d between x_1 and $\hat{x}_1$. Plugging this in, we have

$$
\begin{aligned}
x_2 &= \hat{x}_2 + (x_1 - \hat{x}_1) + (f(x_1) - f(\hat{x}_1))\,h \\
&\quad + f'(x(c_{2h}))\,f(x(c_{2h}))\,h^2/2 \\
&= \hat{x}_2 + (x_1 - \hat{x}_1) + f'(x_d)\,(x_1 - \hat{x}_1)\,h \\
&\quad + f'(x(c_{2h}))\,f(x(c_{2h}))\,h^2/2 \\
&= \hat{x}_2 + (x_1 - \hat{x}_1)\,(1 + f'(x_d)h) + f'(x(c_{2h}))\,f(x(c_{2h}))\,h^2/2
\end{aligned}
$$

Thus

$$
x_2 - \hat{x}_2 = (x_1 - \hat{x}_1)\,(1 + f'(x_d)h) + f'(x(c_{2h}))\,f(x(c_{2h}))\,h^2/2.
$$

Now $E_2 = |x_2 - \hat{x}_2|$, so we can overestimate

$$
\begin{aligned}
E_2 = |x_2 - \hat{x}_2| &\le |x_1 - \hat{x}_1|\,(1 + |f'(x_d)|h) \\
&\quad + |f'(x(c_{2h}))|\,|f(x(c_{2h}))|\,h^2/2 \\
&\le E_1\,(1 + ||f'||_\infty h) + ||f'||_\infty\,||f||_\infty\,h^2/2 \\
&= E_1\,(1 + Ah) + B\,h^2/2 \le E_1(1 + Ch) + Ch^2/2.
\end{aligned}
$$

Since $E_1 \le Ch^2/2$, we find

$$
E_2 \le C\,h^2/2\,(1 + Ch) + C\,h^2/2 = (C\,h^2/2)\,(1 + (1 + Ch))
$$

Let's look at e^{1+Ch}. We know using our approximations

$$
e^{ut} = 1 + ut + u^2 e^{uc} t^2/2
$$

for some c between 0 and u. But the error term is positive so we know $e^{ut} \ge 1 + ut$. Letting $u = Ch$ and using $t = 1$, we have

$$1 + (1 + Ch) \leq 1 + e^{Ch}.$$

and so

$$E_2 \leq (C\, h^2/2)\, (1 + e^{Ch})$$

Now let's do the approximation for $x(3h)$. We will let $x_3 = x(3h)$ and we will define the third Euler approximate by $\hat{x}_3 = \hat{x}_2 + f(\hat{x}_2)\, h$. The tangent line approximation to x at $2h$ gives

$$\begin{aligned} x(3h) &= x(2h) + x'(2h)\, h + x''(c_{3h})\, h^2/2 \\ &= x(2h) + f(x(2h))h + f'(x(c_{3h}))\, f(x(c_{3h}))\, h^2/2 \\ &= x_2 + f(x_2)h + f'(x(c_{3h}))\, f(x(c_{3h}))\, h^2/2. \end{aligned}$$

Now add and subtract $\hat{x}_3 = \hat{x}_2 + f(\hat{x}_2)\, h$ to this equation.

$$\begin{aligned} x(3h) &= x_2 + f(x_2)h + f'(x(c_{3h}))\, x(c_{3h})\, h^2/2 \\ &= x_2 + f(x_2)h + \hat{x}_3 - \hat{x}_3 + f'(x(c_{3h}))\, f(x(c_{3h}))\, h^2/2 \\ &= (x_2 + f(x_2)h - \hat{x}_2 - f(\hat{x}_2)\, h) + \hat{x}_3 \\ &\quad + f'(x(c_{3h}))\, f(x(c_{3h}))\, h^2/2 \\ &= \hat{x}_3 + (x_2 - \hat{x}_2) + (f(x_2) - f(\hat{x}_2))\, h \\ &\quad + f'(x(c_{3h}))\, f(x(c_{3h}))\, h^2/2. \end{aligned}$$

But we can apply the Mean Value Theorem to the difference $f(x_2) - f(\hat{x}_2)$. We find $f(x_2) - f(\hat{x}_2) = f'(x_u)(x_2 - \hat{x}_2)$ with x_u between x_2 and $\hat{x}_2$. Plugging this in, we find

$$\begin{aligned} x_3 &= \hat{x}_3 + (x_2 - \hat{x}_2) + (f(x_2) - f(\hat{x}_2))\, h \\ &\quad + f'(x(c_{3h}))\, f(x(c_{3h}))\, h^2/2 \\ &= \hat{x}_3 + (x_2 - \hat{x}_2) + f'(x_u)\, (x_2 - \hat{x}_2)\, h \\ &\quad + f'(x(c_{3h}))\, f(x(c_{3h}))\, h^2/2 \\ &= \hat{x}_3 + (x_2 - \hat{x}_2)\, (1 + f'(x_u)h) + f'(x(c_{3h}))\, f(x(c_{3h}))\, h^2/2 \end{aligned}$$

Thus

$$x_3 - \hat{x}_3 = (x_2 - \hat{x}_2)\, (1 + f'(x_u)h) + f'(x(c_{3h}))\, f(x(c_{3h}))\, h^2/2.$$

Now $E_3 = |x_3 - \hat{x}_3|$, so we can overestimate

$$\begin{aligned}
E_3 = |x_3 - \hat{x}_3| &\le |x_2 - \hat{x}_2|\,(1 + |f'(x_u)|h) \\
&\quad + |f'(x(c_{3h}))|\,|f(x(c_{3h}))|\,h^2/2 \\
&\le E_2\,(1 + ||f'||_\infty h) + ||f'||_\infty\,||f||_\infty\,h^2/2 \\
&= E_2\,(1 + Ah) + B\,h^2/2.
\end{aligned}$$

Since $E_2 \le (C\,h^2/2)\,(2 + Ch)$, we find

$$\begin{aligned}
E_3 &\le \left((C\,h^2/2)\,(2 + Ch)\right)\,(1 + Ah) + B\,h^2/2 \\
&\le \left((C\,h^2/2)\,(2 + Ch)\right)\,(1 + Ch) + C\,h^2/2 \\
&= C\,\left(1 + (1 + Ch) + (1 + Ch)^2\right)\,h^2/2
\end{aligned}$$

We have already shown that $1 + u \le e^u$ for any u. It follows $(1 + u)^2 \le (e^u)^2 = e^{2u}$. So we have

$$\begin{aligned}
E_3 &\le C\,\left(1 + (1 + Ch) + (1 + Ch)^2\right)\,h^2/2 \\
&\le C\,\left(1 + e^{Ch} + e^{2Ch}\right)\,h^2/2.
\end{aligned}$$

We also know $1 + u + u^2 = (u^3 - 1)/(u - 1)$. So we have

$$E_3 \le C\,\frac{e^{3Ch} - 1}{e^{Ch} - 1}\,h^2/2.$$

Continuing we find after N steps

$$E_N \le C\,\frac{e^{NCh} - 1}{e^{Ch} - 1}\,h^2/2.$$

Now after N steps, we reach the end of the interval T. So $Nh = T$. Rewriting, we have the absolute error we make after N Euler approximation steps is

$$E_N \le C\,\frac{e^{CT} - 1}{e^{Ch} - 1}\,h^2/2.$$

The total error is what we get when we add up $E_1 + \cdots + E_N \le N E_N$. This gives the estimate

$$\begin{aligned}
E_1 + \cdots + E_N &\le N\,C\,\frac{e^{CT} - 1}{e^{Ch} - 1}\,h^2/2 \\
&\le \frac{N}{2}\,(Ch)\,\frac{e^{CT} - 1}{Ch}\,h \\
&= \frac{N}{2}\,(e^{CT} - 1)\,h
\end{aligned}$$

The total or global error thus has the form of a constant times h and hence, the solution on the entire interval is of order h. Of course, the **local error** at each step is on the order of h^2.

3.3 Euler's Method with Time Dependence

If the dynamics function depends on time, we will still be able to look carefully at the Euler approximates. Let's look carefully at the errors we make when we use Euler's method in this case. We start with a preliminary result called a *Lemma*.

Lemma 3.3.1 (Estimating The Exponential Function)
For all x, $1 + x \leq e^x$ *which implies* $(1 + x)^n \leq e^{nx}$.

Proof We know if the base point is 0, then since e^x is twice differentiable, there is a number ξ between 0 and x so that $e^x = 1 + x + e^{\xi}\,\frac{x^2}{2}$. However, the last term is always nonnegative and so dropping that term we obtain $e^x \geq 1 + x$ from which the final result follows. ∎

3.3.1 Lipschitz Functions and Taylor Expansions

We need another basic idea about functions which is called *Lipschitzity*. This is a definition.

Definition 3.3.1 (*Functions Satisfying Lipschitz Conditions*)
If f is a function defined on the finite interval $[a, b]$, we say f satisfies a Lipschitz condition on $[a, b]$ if there is a positive constant K so that

$$|f(x) - f(y)| \leq K|x - y|$$

for all x and y in $[a, b]$.

Many functions satisfy a Lipschitz condition. For example, if f has a derivative that is bounded by the constant K on $[a, b]$, this means $|f'(x)| \leq K$ for all x in $[a, b]$. We can then apply the Mean Value Theorem to f on any interval $[x, y]$ in $[a, b]$ to see there is a number ξ between x and y so that

$$|f(x) - f(y)| = |f'(\xi)|\;|x - y| \leq K\;|x - y|.$$

We see f is Lipschitz on $[a, b]$.

Next, if we want to solve the equation $x'(t) = f(t, x), x(t_0) = x_0$, one way we can attack it is to assume the solution x has enough smoothness to allow us to expand it in a Taylor series expansion about the base point (t_0, x_0). We assume f is continuous on a rectangle D which is the set of (t, x) with $t \in [a, b]$ and $x \in [c, d]$ with the point (t_0, x_0) in D for some finite intervals $[a, b]$ and $[c, d]$. The theory of the solutions to our models allows us to make sure the value of d is large enough to hold the image of the solution x; i.e., $x(t) \in [c, d]$ for all $t \in [t_0, b]$. Also, for convenience, we are assuming x is a scalar variable although the arguments we present can easily be modified to handle x being a vector. Using the Taylor Remainder theorem, this gives for a given time point t:

$$x(t) = x(t_0) + \frac{dx}{dt}(t_0)\,(t - t_0) + \frac{1}{2}\,\frac{d^2x}{dt^2}(t_0)\,(t - t_0)^2 + \frac{1}{6}\,\frac{d^3x}{dt^3}(\xi)\,(t - t_0)^3$$

where ξ is some point in the interval $[t_0, t]$. We also know by the chain rule that since x' is f that

$$\frac{d^2x}{dt^2} = \frac{\partial f}{\partial t} + \frac{\partial f}{\partial t}\,\frac{dx}{dt}$$
$$\frac{d^2x}{dt^2} = \frac{\partial f}{\partial t} + \frac{\partial f}{\partial x}\,f$$

which implies, switching to a standard subscript notation for partial derivatives (yes, these calculations are indeed yucky[2]!)

$$\frac{d^3x}{dt^3} = (f_{tt} + f_{tx}\,f) + (f_{xt} + f_{xx}\,f)\,f + f_x\,(f_t + f_x\,f)$$

The important thing to note is that this third order derivative is made up of algebraic combinations on f and its various partial derivatives. We typically assume that on the interval $[t_0, b]$, that all of these functions are continuous and bounded. Thus, letting $||g||$ represent the maximum value of the continuous function $g(s, u)$ on the interval $[t_0, b] \times [c, d]$, we know there is a constant B so that

$$|x''(t)| \le ||f_x||_\infty\,||f||_\infty + ||f_t||_\infty = B.$$

implying $||x''||_\infty \le B$ on $[t_0, b]$. Further, there is a constant C so that

$$\begin{aligned}|x'''(\xi)| &\le (||f_{tt}||_\infty + ||f_{tx}||_\infty\,||f||_\infty) + (||f_{xt}||_\infty + ||f_{xx}||_\infty\,||f||_\infty)\,||f||_\infty \\ &\quad + ||f_x||_\infty\,(||f_t||_\infty + ||f_x||_\infty\,||f||_\infty) \\ &= C.\end{aligned}$$

That is, $||x'''||_\infty \le C$ on $[t_0, b]$. Of course, if f has sufficiently smooth higher order partial derivatives, we can find bounds on higher derivatives of x as well. Now, using the standard abbreviations $f^0 = f(t_0, x_0)$, $f_t^0 = \frac{\partial f}{\partial t}(t_0, x_0)$ and $f_x^0 = \frac{\partial f}{\partial x}(t_0, x_0)$ with similar notations for the second order partials, we see our solution can be written as

$$\begin{aligned}
x(t) &= x_0 + f^0\,(t-t_0) + \frac{1}{2}(f_t^0 + f_x^0 f^0)\,(t-t_0)^2 + \frac{1}{6}\frac{d^3x}{dt^3}(\xi)\,(t-t_0)^3 \\
&= x_0 + f^0\,(t-t_0) + \frac{1}{2}\,(f_t^0 + f_x^0 f^0)\,(t-t_0)^2 \\
&\quad + \frac{1}{6}\left(f_{tt} + f_{tx} f) + (f_{xt} + f_{xx} f)\, f + f_x\,(f_t + f_x f \right)\Bigg|_{\xi}\,(t-t_0)^3.
\end{aligned}$$

We can now state a result which tells us how much error we make with Euler's method. From the remarks above, the assumption $||x''||_\infty$ is bounded is not an unreasonable for many models we wish to solve. Our discussions above allow us to be fairly quantitative about how much *local* error and *global* error we make using Euler's approximations. We state this as a theorem.

Theorem 3.3.2 (Error Estimates For Euler's Method)
Assume x is a solution to $x'(t) = f(t, x(t)),\ \ x(t_0) = x_0$ on the interval $[t_0, b]$ for some positive b. Assume f and its first order partials are continuous on a rectangle D which is the set of (t, x) with $t \in [a, b]$ and $x \in [c, d]$ with the point (t_0, x_0) in D for some finite intervals $[a, b]$ and $[c, d]$ with $[c, d]$ containing the range of the solution x. Let $\{t_0, t_1, \ldots, t_N\}$ denote the steps of discrete times obtained using the step size h in the Euler method. Then the Euler approximates $\hat{x}_n$ satisfy

$$|x_n - \hat{x}_n| \leq e^{b-a} K\, |e_0| + \left[\frac{e^{(b-a)K} - 1}{K}\right] \frac{h}{2}\, ||x''||_\infty$$

where x_n is the value of the true *solution $x(t_n)$ and $e_0 = x_0 - \hat{x}_0$. If we also know $e_0 = 0$ (the usual state of affairs), then, letting the constant B be defined by*

$$B = \left[\frac{e^{(b-a)K} - 1}{2K}\right] ||x''||_\infty,$$

we can say $\left| x(b) - \hat{x}_{N(h)} \right| \leq B\, h$ where $N(h)$ is the index at which $t_{N_h} = b$.

Proof From our remarks earlier, our assumptions on f and its first order partials tell us f satisfies a Lipschitz condition with Lipschitz constant K in D and that the solution x has a bounded second derivative on the interval $[t_0, b]$. Let $e_n = x_n - \hat{x}_n$ and $\tau_n = \frac{h}{2}\, x''(\xi_n)$. Then, the usual Taylor series expansion gives

$$\begin{aligned}
x(t_{n+1}) &= x(t_n) + h\, f(t_n, x(t_n)) + h\, x''(\xi_n)\, \frac{h}{2} \\
&= x(t_n) + h\, f(t_n, x(t_n)) + h\, \tau_n
\end{aligned}$$

where ξ_n is between t_{n+1} and t_n and the Euler approximates satisfy

$$\hat{x}_{n+1} = \hat{x}_n + h\, f(t_n, \hat{x}_n).$$

Subtracting, we find

$$x_{n+1} - \hat{x}_{n+1} = x_n - \hat{x}_n + h \left(f(t_n, x(t_n)) - f(t_n, \hat{x}_n) \right) + h\, \tau_n.$$

Thus,

$$e_{n+1} = e_n + h \left(f(t_n, x(t_n)) - f(t_n, \hat{x}_n) \right) + h\, \tau_n.$$

This leads to the estimate

$$|e_{n+1}| \leq |e_n| + h\, |f(t_n, x(t_n)) - f(t_n, \hat{x}_n)| + h\, |\tau_n|.$$

Now apply the Lipschitz condition on f to rewrite the above as

$$\begin{aligned} |e_{n+1}| &\leq |e_n| + h\ K\ |x_n - \hat{x}_n| + h\ |\tau_n| \\ &\leq (1 + K)|e_n| + h\ |\tau_n| \end{aligned}$$

It is easy to see that

$$|\tau_n| = \frac{h}{2}\, x''(\xi_n) \leq \frac{h}{2}\, ||x''||_\infty.$$

For convenience, let $\tau(h) = \frac{h}{2}\, ||x''||$. Then, we have the estimate

$$|e_{n+1}| \leq (1 + K)|e_n| + h\ |\tau(h)|.$$

This is a recursion relation. We can easily see what is happening by working out a few terms.

$$\begin{aligned} |e_1| &\leq |e_0|(1 + hK) + h|\tau(h)| \\ |e_2| &\leq |e_1|(1 + hK) + h|\tau(h)| \\ &\leq \left(|e_0|(1 + hK) + h|\tau(h)| \right)(1 + hK) + h|\tau(h)| \\ &\leq |e_0|(1 + hK)^2 + h\tau(h)\left(1 + (1 + hK) \right) \\ |e_3| &\leq |e_2|(1 + hK) + h|\tau(h)| \\ &\leq |e_0|(1 + hK)^3 + h|\tau(h)|\left(1 + (1 + hK) + (1 + hK)^2 \right). \end{aligned}$$

It is easy to see that after n steps, we find

$$|e_n| \le |e_0|(1+hK)^n + h|\tau(h)| \sum_{i=0}^{n-1} (1+hK)^i.$$

It is well-known that for any value of $r \neq 1$, we have the identity

$$1 + r + r^2 + r^3 + \cdots + r^{n-1} = \frac{1-r^n}{1-r}.$$

Letting $r = 1 + hK$, we can rewrite our error estimate as

$$|e_n| \le |e_0|(1+hK)^n + |\tau(h)| \frac{(1+hK)^n - 1}{K}.$$

Then, using Lemma 3.3.1, we have

$$|e_n| \le |e_0|\, e^{nhK} + h|\tau(h)| \frac{(1+hK)^n - 1}{hK}.$$

But since $t_n = t_0 + nh$, we know $nh \le b - t_0$ leading to

$$|e_n| \le |e_0|\, e^{(b-t_0)K} + |\tau(h)| \frac{(1+hK)^n - 1}{K}.$$

But $(1+hK)^n \le e^{nhK}$ and since $nh = b - a$, we have

$$|e_n| \le |e_0|\, e^{(b-t_0)K} + |\tau(h)| \frac{e^{(b-a)K} - 1}{K}.$$

Now if $e_0 = 0$ (as it normally would be), we have

$$|e_n| \le \frac{e^{(b-a)K} - 1}{2K} \,||x''||_\infty\, h$$

and letting

$$B = \frac{e^{(b-a)K} - 1}{2K} \,||x''||_\infty,$$

we have $|e_{N(h)}| \le Bh$ as required. ■

Comment 3.3.1 *Note the local error we make at each step is proportional to h^2 but the global error after we reach $t = b$ is proportional to h. Hence, Euler's method is an order 1 method.*

3.4 Euler's Algorithm

Here then is **Euler's Algorithm** to approximate the solution to $x' = f(t, x)$ with $x(0) = x_0$ using step size h for as many steps as we want.

- $\hat{x}_0 = x_0$ so $E_0 = 0$.
- $\hat{x}_1 = \hat{x}_0 + f(t, \hat{x}_0)\, h;\quad E_1 = |x_1 - \hat{x}_1|$.
- $\hat{x}_2 = \hat{x}_1 + f(t, \hat{x}_1)\, h;\quad E_2 = |x_2 - \hat{x}_2|$.
- $\hat{x}_3 = \hat{x}_2 + f(t, \hat{x}_2)\, h;\quad E_3 = |x_3 - \hat{x}_3|$.
- $\hat{x}_4 = \hat{x}_3 + f(t, \hat{x}_3)\, h;\quad E_4 = |x_4 - \hat{x}_4|$.
- Continue as many steps as you want.

Recursively:

- $\hat{x}_0 = x_0$ so $E_0 = 0$.
- $\hat{x}_{n+1} = \hat{x}_n + f(t, \hat{x}_n)\, h;\quad E_{n+1} = |x_{n+1} - \hat{x}_{n+1}|$ for $n = 0, 1, 2, 3, \ldots$.

The approximation scheme above is called **Euler's Method**.

3.4.1 Examples

We need to do some examples by hand so here goes.

Example 3.4.1 Find the first three Euler approximates for $x' = -2x, x(0) = 3$ using $h = 0.3$. Find the true solution values and errors also.

Solution *Here* $f(x) = -2x$ *and the true solution is* $x(t) = 3e^{-2t}$.

Step 0: $\hat{x}_0 = x_0 = 3$ *so* $E_0 = 0$.
Step 1:

$$\begin{aligned}
\hat{x}_1 &= \hat{x}_0 + f(\hat{x}_0)\, h = 3 + f(3)\,(0.3) \\
&= 3 + (-2(3))\,(0.3) = 3 - 6(0.3) = 3 - 1.8 = 1.2. \\
x_1 &= x(h) = 3e^{-2h} = 3e^{-0.6} = 1.646. \\
E_1 &= |x_1 - \hat{x}_1| = |1.646 - 1.2| = 0.446.
\end{aligned}$$

Step 2:

$$\begin{aligned}
\hat{x}_2 &= \hat{x}_1 + f(\hat{x}_1)\, h = 1.2 + f(1.2)\,(0.3) \\
&= 1.2 + (-2(1.2))\,(0.3) = 1.2 - 2.4(0.3) = 0.48. \\
x_2 &= x(2h) = 3e^{-2(2h)} = 3e^{-4h} = 3e^{-1.2} = 0.9036. \\
E_2 &= |x_2 - \hat{x}_2| = |0.9036 - 0.48| = 0.4236.
\end{aligned}$$

Step 3:

$$\begin{aligned}
\hat{x}_3 &= \hat{x}_2 + f(\hat{x}_2)\,h = 0.48 + f(0.48)\,(0.3) \\
&= 0.48 + (-2(0.48))\,(0.3) = 0.48 - 0.96(0.3) = 0.192. \\
x_3 &= x(3h) = 3e^{-2(3h)} = 3e^{-6h} = 3e^{-1.8} = 0.4959. \\
E_3 &= |x_3 - \hat{x}_3| = |0.4959 - 0.192| = 0.3039.
\end{aligned}$$

Example 3.4.2 Find the first three Euler approximates for $x' = 2x$, $x(0) = 4$ using $h = 0.2$. Find the true solution values and errors also.

Solution *Here* $f(x) = 2x$ *and the true solution is* $x(t) = 4e^{2t}$.

Step 0: $\hat{x}_0 = x_0 = 4$ *so* $E_0 = 0$.
Step 1:

$$\begin{aligned}
\hat{x}_1 &= \hat{x}_0 + f(\hat{x}_0)\,h = 4 + f(4)\,(0.3) \\
&= 4 + (2(4))\,(0.2) = 4 + 8(0.2) = 5.6. \\
x_1 &= x(h) = 4e^{2h} = 4e^{0.4} = 5.9673. \\
E_1 &= |x_1 - \hat{x}_1| = |5.9673 - 5.6| = 0.3673.
\end{aligned}$$

Step 2:

$$\begin{aligned}
\hat{x}_2 &= \hat{x}_1 + f(\hat{x}_1)\,h = 5.6 + f(5.6)\,(0.2) \\
&= 5.6 + (2(5.6))\,(0.2) = 5.6 + 11.2(0.2) = 7.84. \\
x_2 &= x(2h) = 4e^{2(2h)} = 4e^{4h} = 4e^{0.8} = 8.9032. \\
E_2 &= |x_2 - \hat{x}_2| = |8.9032 - 7.84| = 1.0622.
\end{aligned}$$

Step 3:

$$\begin{aligned}
\hat{x}_3 &= \hat{x}_2 + f(\hat{x}_2)\,h = 7.84 + f(7.84)\,(0.2) \\
&= 7.84 + (2(7.84))\,(0.2) = 7.84 + 15.68(0.2) = 10.976. \\
x_3 &= x(3h) = 4e^{2(3h)} = 4e^{6h} = 4e^{1.2} = 13.2805. \\
E_3 &= |x_3 - \hat{x}_3| = |13.2805 - 10.976| = 2.3045.
\end{aligned}$$

3.5 Runge–Kutta Methods

These methods are based on more sophisticated ways of approximating the solution y'. These methods use multiple function evaluations at different time points around a given t^* to approximate $y(t^*)$. In more advanced classes, we can show this technique generates a sequence $\{y_n\}$ starting at y_0 using the following recursion equation:

$$y_{n+1} = y_n + h \times F^o(t_n, y_n, h, f)$$
$$y_0 = y0$$

where h is the step size we use for our underlying partition of the time space giving

$$t_i = t_0 + i \times h$$

for appropriate indices and F^o is a fairly complicated function of the previous approximate solution, the step size and the right hand side function f. The Runge–Kutta methods are available for various choices of the superscript o which is called the order of the method. We will not discuss much about F^o in this course, as it is best served up in a more advanced class. What we can say is this: For order o, the **local** error is like h^{o+1}. So

Order One: **Local** error is h^2 and this method is the same as the Euler Method. The **global** error then goes down linearly with h.

Order Two: **Local** error is h^3 and this method is better than the Euler Method. If the **global** error for a given stepsize h is then E, halving the stepsize to $h/2$ gives a new global error of $E/4$. Thus, the global error goes down quadratically. This means halving the stepsize has a dramatic effect of the global error.

Order Three: **Local** error is h^4 and this method is better than the Euler Method. If the **global** error for a given stepsize h is E, then halving the stepsize to $h/2$ gives a new global error of $E/8$. Thus, the global error goes down as a cubic power. This means halving the stepsize has an even more dramatic effect of the global error.

Order Four: **Local** error is h^5 and this method is better than the order three Method. If the **global** error for a given stepsize h is E, then halving the stepsize to $h/2$ gives a new global error of $E/16$. Thus, the global error goes down as a fourth power! This means halving the stepsize has huge effect of the global error.

We will now look at MatLab code that allows us to solve our differential equation problems using the Runge–Kutta method instead of the Euler method of Sect. 3.3.

3.5.1 The MatLab Implementation

The basic code to implement the Runge–Kutta methods is broken into two pieces. The first one, **`RKstep.m`** implements the evaluation of the next approximation solution at point (t_n, y_n) given the old approximation at (t_{n-1}, y_{n-1}). We then loop through all the steps to get to the chosen final time using the code in **`FixedRK.m`**. The details of these algorithms are beyond the scope of this text and so we will not go into them here. In this code, we are allowing for the dynamic functions to depend on time also. Previously, we have used dynamics like **f = @(x) 3*x** and we expect the dynamics functions to have that form in **DoEuler**. However, we want to have more complicated dynamics now—at least the possibility of it!—so we will adapt what we have done before. We will now define our dynamics as if they depend on time. So from now on, we would write **f=@(t,x) 3*x** even though there is no time dependence. We then rewrite our **DoEuler** to **DoEulerTwo** so that we can use these more general dynamics. This code is in **DoEulerTwo.m** and we have discussed it in the first text on starting your calculus journey. You can review this function in that text. The Runge–Kutta code uses the new dynamics functions. We have gone over this code in the previous text, but we will show it to you again for completeness. In this code, you see the lines like **feval(fname,t,x)** which means take the function **fname** passed in as an argument and **evaluate** it as the pair **(t,x)**. Hence, **fname(t,x)** is the same as **f(t,x)**.

Listing 3.1: RKstep.m: Runge–Kutta Codes

```
function [tnew,ynew,fnew] = RKstep(fname,tc,yc,fc,h,k)
%
% fname    the name of the right hand side function f(t,y)
%          t is a scalar usually called time and
%          y is a vector of size d
% yc       approximate solution to y'(t) = f(t,y(t)) at t=tc
% fc       f(tc,yc)
% h        The time step
% k        The order of the Runge-Kutta Method 1<= k <= 4
%
% tnew     tc+h
% ynew     approximate solution at tnew
% fnew     f(tnew,ynew)
%
if k==1
  k1 = h*fc;
  ynew = yc+k1;
elseif k==2
  k1 = h*fc;
  k2 = h*feval(fname,tc+(h/2),yc+(k1/2));
  ynew = yc + k2;
elseif k==3
  k1 = h*fc;
  k2 = h*feval(fname,tc+(h/2),yc+(k1/2));
  k3 = h*feval(fname,tc+h,yc-k1+2*k2);
  ynew = yc+(k1+4*k2+k3)/6;
elseif k==4
  k1 = h*fc;
  k2 = h*feval(fname,tc+(h/2),yc+(k1/2));
  k3 = h*feval(fname,tc+(h/2),yc+(k2/2));
  k4 = h*feval(fname,tc+h,yc+k3);
  ynew = yc+(k1+2*k2+2*k3+k4)/6;
else
  disp(sprintf('The RK method %2d order is not allowed!',k));
end
tnew = tc+h;
fnew = feval(fname,tnew,ynew);
end
```

The code above does all the work. It manages all of the multiple tangent line calculations that Runge–Kutta needs at each step. We loop through all the steps to get to the chosen final time using the code in **FixedRK.m** which is shown below.

Listing 3.2: FixedRK.m: The Runge–Kutta Solution

```
function [tvals,yvals,fcvals] = FixedRK(fname,t0,y0,h,k,n)
%
%            Gives approximate solution to
%              y'(t) = f(t,y(t))
%              y(t0) = y0
%            using a kth order RK method
%
% t0         initial time
% y0         initial state
% h          stepsize
% k          RK order   1<= k <= 4
% n          Number of steps to take
%
% tvals      time values of form
%            tvals(j) = t0 + (j-1)*h,  1 <= j <= n
% yvals      approximate solution
%            yvals(:j) = approximate solution at
%            tvals(j),   1 <= j <= n
%
tc = t0;
yc = y0;
tvals = tc;
yvals = yc;
fc = feval(fname,tc,yc);
for j=1:n-1
  [tc,yc,fc] = RKstep(fname,tc,yc,fc,h,k);
  yvals = [yvals yc];
  tvals = [tvals tc];
  fcvals = [fcvals fc];
end
end
```

Here is an example where we solve a specific model using all four Runge–Kutta choices and plot them all together. Note when we use **RKstep**, we only return the first two outputs; that is, for our returned variables, we write **[htime1,xhat1] = FixedRK(f,0,20,0.06,1,N1);** instead of returning the full list of outputs which includes function evaluations **[htime1,xhat1,fhat1] = FixedRK (f,0,20,0.06,1,N1);**. We can do this as it is all right to not return the third output. However, you still have to return the arguments in the order stated when the function is defined. For example, if we used the command **[htime1,fhat1] = FixedRK(f,0,20,0.06,1,N1);**, this would return the approximate values and place them in the variable **fhat1** which is not what we would want to do.

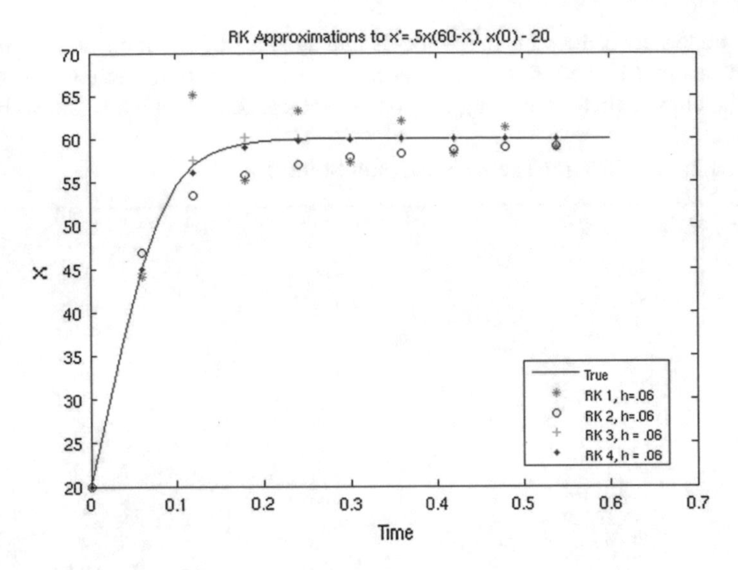

Fig. 3.1 True versus Euler for $x' = 0.5x(60 - x), x(0) = 20$

Listing 3.3: True versus All Four Runge–Kutta Approximations

```
f = @(t,x) .5*x.*(60-x);
true = @(t) 60./( 1 + (60/20 - 1)*exp(-.5*60*t) ) ;
T = .6;
time = linspace(0,T,31);
h1 = .06;
N1 = ceil(T/h1);
[htime1,xhat1] = FixedRK(f,0,20,.06,1,N1);
[htime2,xhat2] = FixedRK(f,0,20,.06,2,N1);
[htime3,xhat3] = FixedRK(f,0,20,.06,3,N1);
[htime4,xhat4] = FixedRK(f,0,20,.06,4,N1);
% the ... at the end of the line allows us to continue
% a long line to the start of the next line
plot(time,true(time),htime1,xhat1,'*',htime2,xhat2,'o',...
     htime3,xhat3,'+',htime4,xhat4,'.');
xlabel('Time');
ylabel('x');
% We want to use the derivative symbol x' so
% since Matlab treat ' as the start and stop of the label
% we write ''. That way Matlab will treat '' as a single quote
% that is our differentiation symbol
title('RK Approximations to x''=.5x(60-x), x(0) - 20');
% Notice we continue this line too
legend('True','RK 1, h=.06','RK 2, h=.06','RK 3, h = .06',...
'RK 4, h = .06','Location','Best');
```

This generates Fig. 3.1
Note Runge–Kutta Order 4 does a great job even with a large step size.

Chapter 4
Multivariable Calculus

Now let's review functions of more than one variable since it may have been some time since you looked at these ideas. This is gone over carefully in the first text on starting calculus ideas, but it is a good idea to talk about them again.

4.1 Functions of Two Variables

Let's start by looking at the $x-y$ plane as a collection of two dimensional vectors. Each vector is rooted at the origin and the head of the vector corresponds to our usual coordinate pair (x, y). The set of all such x and y determines the $x-y$ plane which we will also call $\Re^2$. The superscript two is used because we are now explicitly acknowledging that we can think of these ordered pairs as vectors also with just a slight identification on our part. Since we know about vectors, note if we have a vector we can rewrite it, using our standard rules for vector arithmetic and scaling of vectors as

$$\begin{bmatrix} 6 \\ 7 \end{bmatrix} = 6 \begin{bmatrix} 1 \\ 0 \end{bmatrix} + 7 \begin{bmatrix} 0 \\ 1 \end{bmatrix}$$

A little thought will let you see we can do this for any vector and so we define special vectors $\boldsymbol{i} = \boldsymbol{e_1}$ and $\boldsymbol{j} = \boldsymbol{e_2}$ as follows:

$$\boldsymbol{i} = \boldsymbol{e_1} = \begin{bmatrix} 1 \\ 0 \end{bmatrix} \text{ and } \boldsymbol{j} = \boldsymbol{e_2} = \begin{bmatrix} 0 \\ 1 \end{bmatrix}$$

J.K. Peterson, *Calculus for Cognitive Scientists*,
Cognitive Science and Technology, DOI 10.1007/978-981-287-877-9_4

Thus, any vector can be written as

$$\begin{bmatrix} x \\ y \end{bmatrix} = x\, \boldsymbol{e_1} + y\, \boldsymbol{e_1}$$
$$= x\, \boldsymbol{i} + y\, \boldsymbol{j}$$

Now let's start looking at functions that map each ordered pair (x, y) into a number. Let's begin with an example. Consider the function $f(x, y) = x^2 + y^2$ defined for all x and y. Hence, for each x and y we pick, we calculate a number we can denote by z whose value is $f(x, y) = x^2 + y^2$. Using the same ideas we just used for the $x-y$ plane, we see the set of all such triples $(x, y, z) = (x, y, x^2 + y^2)$ defines a **surface** in $\Re^3$ which is the collection of all ordered triples (x, y, z). Each of these triples can be identified with a three dimensional vector whose tail is the origin and whose head is the triple (x, y, z). We note any three dimensional vector can be written as

$$\begin{bmatrix} x \\ y \\ z \end{bmatrix} = x\, \boldsymbol{e_1} + y\, \boldsymbol{e_2} + yz\, \boldsymbol{e_3}$$
$$= x\, \boldsymbol{i} + y\, \boldsymbol{j} + z\, \boldsymbol{k}$$

where we define the special vectors used in this representation by

$$\boldsymbol{i} = \boldsymbol{e_1} = \begin{bmatrix} 1 \\ 0 \\ 0 \end{bmatrix}, \ \boldsymbol{j} = \boldsymbol{e_2} = \begin{bmatrix} 0 \\ 1 \\ 0 \end{bmatrix} \text{ and } \boldsymbol{k} = \boldsymbol{e_3} = \begin{bmatrix} 0 \\ 0 \\ 1 \end{bmatrix}$$

We can plot this surface in MatLab with fairly simple code. As discussed in the first volume, the utility function **`DrawSimpleSurface`** which manages how to draw such a surface using boolean variables like **`DoGrid`** to turn a graph on or off. The surface is drawn using a **grid**, a **mesh**, **traces**, **patches** and **columns** and a **base** all of which contribute to a somewhat cluttered figure. Hence, the boolean variables allows us to select how much *clutter* we want to see! So if the boolean variable **`DoGrid`** is set to one, the grid is drawn. The code is self-explanatory so we just lay it out here. We haven't shown all the code for the individual drawing functions, but we think you'll find it interesting to see how we manage the pieces in this one piece of code. So check this out.

Listing 4.1: DrawSimpleSurface

```
function DrawSimpleSurface(f,delx,nx,dely,ny,x0,y0,domesh,dotraces,dogrid,dopatch,
    docolumn,dobase)
% f is the function defining the surface
% delx is the size of the x step
% nx is the number of steps left and right from x0
% dely is the size of the y step
% ny is the number of steps left and right from y0
% (x0,y0) is the location of the column rectangle base
% domesh = 1 if we want to do the mesh
% dogrid = 1 if we want to do the grid
% dopatch = 1 if we want the patch above the column
% dobase = 1 if we want the base of the column
% docolumn = 1 if we want the column
% dotraces = 1 if we want the traces
%
hold on
if dotraces==1
  % set up x trace for x0, y trace for y0
  DrawTraces(f,delx,nx,dely,ny,x0,y0);
end
if domesh==1 % plot the surface
  DrawMesh(f,delx,nx,dely,ny,x0,y0);
end
if dogrid==1 %plot x, y grid
  DrawGrid(f,delx,nx,dely,ny,x0,y0);
end
if dopatch==1
  % draw patch for top of column
  DrawPatch(f,delx,nx,dely,ny,x0,y0);
end
if dobase==1
  % draw patch for top of column
  DrawBase(f,delx,nx,dely,ny,x0,y0);
end
if docolumn==1
  %draw column
  DrawColumn(f,delx,nx,dely,ny,x0,y0);
end
hold off
end
```

Hence, to draw everything for this surface, we would use the session:

Listing 4.2: Drawing a simple surface

```
f = @(x,y) x.^2+y.^2;
DrawSimpleSurface(f,0.5,2,0.5,2,0.5,0.5,1,1,1,1,1,1);
```

This surface has circular cross sections for different positive values of z and it is called a *circular paraboloid*. If you used $f(x, y) = 4x^2 + 3y^2$, the cross sections for positive z would be ellipses and we would call the surface an *elliptical paraboloid*. Now this code is not perfect. However, as an exploratory tool it is not bad! Now it is time for you to play with it a bit in the exercises below.

4.1.1 Homework

Exercise 4.1.1 *Explore the surface graph of the* circular paraboloid $f(x, y) = x^2 + y^2$ *for different values of* (x_0, y_0) *and* Δx *and* Δy. *Experiment with the 3D rotated view to make sure you see everything of interest.*

Exercise 4.1.2 *Explore the surface graph of the* elliptical paraboloid $f(x, y) = 2x^2 + y^2$ *for different values of* (x_0, y_0) *and* Δx *and* Δy. *Experiment with the 3D rotated view to make sure you see everything of interest.*

Exercise 4.1.3 *Explore the surface graph of the* elliptical paraboloid $f(x, y) = 2x^2 + 3y^2$ *for different values of* (x_0, y_0) *and* Δx *and* Δy. *Experiment with the 3D rotated view to make sure you see everything of interest.*

4.2 Continuity

Let's recall the ideas of continuity for a function of one variable. Consider these three versions of a function f defined on $[0, 2]$.

$$f(x) = \begin{cases} x^2, & \text{if } 0 \leq x < 1 \\ 10, & \text{if } x = 1 \\ 1 + (x-1)^2 & \text{if } 1 < x \leq 2. \end{cases}$$

The first version is not continuous at $x = 1$ because although the $\lim_{x \to 1} f(x)$ exists and equals 1 ($\lim_{x \to 1^-} f(x) = 1$ and $\lim_{x \to 1^+} f(x) = 1$), the value of $f(1)$ is 10 which does not match the limit. Hence, we know f here has a removable discontinuity at $x = 1$. Note continuity failed because the limit existed but the value of the function did not match it. The second version of f is given below.

$$f(x) = \begin{cases} x^2, & \text{if } 0 \leq x \leq 1 \\ (x-1)^2 & \text{if } 1 < x \leq 2. \end{cases}$$

In this case, the $\lim_{x \to 1^-} = 1$ and $f(1) = 1$, so f is continuous from the left. However, $\lim_{x \to 1^+} = 0$ which does not match $f(1)$ and so f is not continuous from the right. Also, since the right and left hand limits do not match at $x = 1$, we know $\lim_{x \to 1}$ does not exist. Here, the function fails to be continuous because the limit does not exist. The final example is below:

$$f(x) = \begin{cases} x^2, & \text{if } 0 \leq x < 1 \\ x + (x-1)^2 & \text{if } 1 < x \leq 2. \end{cases}$$

Here, the limit and the function value at 1 both match and so f is continuous at $x = 1$. To extend these ideas to two dimensions, the first thing we need to do is to look at the meaning of the limiting process. What does $\lim_{(x,y)\to(x_0,y_0)}$ mean? Clearly, in one dimension we can approach a point x_0 from x in two ways: from the left or from the right or jump around between left and right. Now, it is apparent that we can approach a given point (x_0, y_0) in an infinite number of ways. Draw a point on a piece of paper and convince yourself that there are many ways you can draw a curve from another point (x, y) so that the curve ends up at (x_0, y_0)! We still want to define continuity in the same way; i.e. f is continuous at the point (x_0, y_0) if $\lim_{(x,y)\to(x_0,y_0)} f(x, y) = f(x_0, y_0)$. If you look at the graphs of the surface $z = x^2 + y^2$ we have done previously, we clearly see that we have this kind of behavior. There are no jumps, tears or gaps in the surface we have drawn. Let's make this formal.

Definition 4.2.1 (*Continuity*)
Let $z = f(x, y)$ be a function of the two independent variables x and y defined on some domain. At each pair (x, y) where f is defined in a circle of some finite radius r,

$$B_r(x_0, y_0) = \{(x, y) \mid \sqrt{(x - x_0)^2 + (y - y_0)^2} < r\},$$

if $\lim_{(x,y)\to(x_0,y_0)} f(x, y)$ exists and matches $f(x_0, y_0)$, we say f is continuous at (x_0, y_0).

Here is an example of a function which is not continuous at the point $(0, 0)$. Let

$$f(x, y) = \begin{cases} \frac{2x}{\sqrt{x^2+y^2}}, & \text{if } (x, y) \neq (0, 0) \\ 0, & \text{if } (x, y) = (0, 0). \end{cases}$$

If we show the limit as we approach $(0, 0)$ does not exist, then we will know f is not continuous at $(0, 0)$. If this limit exists, we should get the same value for the limit no matter what path we take to reach $(0, 0)$. Let the first path be given by $x(t) = t$ and $y(t) = 2t$. Then, as $t \to 0$, $(x(t), y(t)) \to (0, 0)$ as desired. Plugging in to f, we find for $t \neq 0$, $f(t, 2t) = 2t/\sqrt{t^2 + 4t^2} = 2/\sqrt{5}$ and hence the limit along this path is this constant value $2/\sqrt{5}$. On the other hand, along the path $x(t) = t$ and $y(t) = -3t$, for $t \neq 0$, we have $f(t, -3t) = 2/3$ which is not the same. Since the limiting value differs on two paths, the limit can't exist. Hence, f is not continuous at $(0, 0)$.

4.3 Partial Derivatives

Let's go back to our simple surface example and look at the traces again. In Fig. 4.1, we show the traces for the base point $x_0 = 0.5$ and $y_0 = 0.5$. We have also drawn vertical lines down from the traces to the $x-y$ plane to further emphasize the placement of

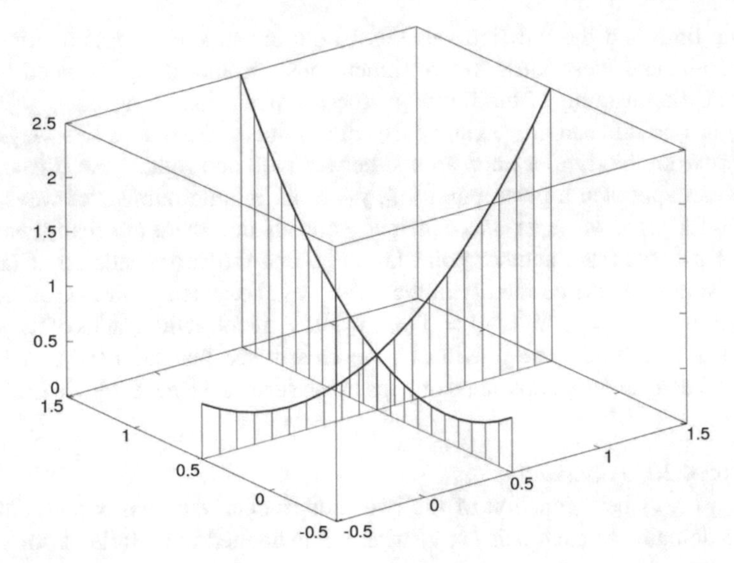

Fig. 4.1 The traces $f(x_0, y)$ and $f(x, y_0)$ for the surface $z = x^2 + y^2$ for $x_0 = 0.5$ and $y_0 = 0.5$

the traces on the surface. The surface itself is not shown as it is somewhat distracting and makes the illustration too busy.

You can generate this type of graph yourself with the function **DrawFullTraces** as follows:

Listing 4.3: Drawing a full trace

```
DrawFullTraces(f,0.5,2,0.5,2,0.5,0.5);
```

Note, that each trace has a well-defined tangent line and derivative at the points x_0 and y_0. We have

$$\begin{aligned} \frac{d}{dx} f(x, y_0) &= \frac{d}{dx}(x^2 + y_0^2) \\ &= 2x \end{aligned}$$

as the value y_0 in this expression is a constant and hence its derivative with respect to x is zero. We denote this new derivative as $\frac{\partial f}{\partial x}$ which we read as *the partial derivative of f with respect to x*. It's value as the point (x_0, y_0) is $2x_0$ here. For any value of (x, y), we would have $\frac{\partial f}{\partial x} = 2x$. We also have

$$\begin{aligned}\frac{d}{dy} f(x_0, y) &= \frac{d}{dy}(x_0^2 + y^2) \\ &= 2y\end{aligned}$$

We then denote this new derivative as $\frac{\partial f}{\partial y}$ which we read as *the partial derivative of f with respect to y*. It's value as the point (x_0, y_0) is then $2y_0$ here. For any value of (x, y), we would have $\frac{\partial f}{\partial y} = 2y$.

The tangent lines for these two traces are then

$$\begin{aligned}T(x, y_0) &= f(x_0, y_0) + \frac{d}{dx} f(x, y_0)\Big|_{x_0} (x - x_0) \\ &= (x_0^2 + y_0^2) + 2x_0(x - x_0) \\ T(x_0, y) &= f(x_0, y_0) + \frac{d}{dy} f(x_0, y)\Big|_{y_0} (y - y_0) \\ &= (x_0^2 + y_0^2) + 2y_0(y - y_0).\end{aligned}$$

We can also write these tangent line equations like this using our new notation for partial derivatives.

$$\begin{aligned}T(x, y_0) &= f(x_0, y_0) + \frac{\partial f}{\partial x}(x_0, y_0)\,(x - x_0) \\ &= (x_0^2 + y_0^2) + 2x_0(x - x_0) \\ T(x_0, y) &= f(x_0, y_0) + \frac{\partial f}{\partial y}(x_0, y_0)\,(y - y_0) \\ &= (x_0^2 + y_0^2) + 2y_0(y - y_0).\end{aligned}$$

We can draw these tangent lines in 3D. To draw $T(x, y_0)$, we fix the y value to be y_0 and then we draw the usual tangent line in the $x-z$ plane. This is a copy of the $x-z$ plane translated over to the value y_0; i.e. it is parallel to the $x-z$ plane we see at the value $y = 0$. We can do the same thing for the tangent line $T(x, y_0)$; we fix the x value to be x_0 and then draw the tangent line in the copy of the $y - z$ plane translated to the value x_0. We show this in Fig. 4.3. Note the $T(x, y_0)$ and the $T(x_0, y)$ lines are determined by vectors as shown below.

$$\boldsymbol{A} = \begin{bmatrix} 1 \\ 0 \\ \frac{d}{dx} f(x, y_0)\Big|_{x_0} \end{bmatrix} = \begin{bmatrix} 1 \\ 0 \\ 2x_0 \end{bmatrix} \text{ and } \boldsymbol{B} = \begin{bmatrix} 0 \\ 1 \\ \frac{d}{dy} f(x_0, y)\Big|_{y_0} \end{bmatrix} = \begin{bmatrix} 0 \\ 1 \\ 2y_0 \end{bmatrix}$$

Note that if we connect the lines determined by the vectors $\boldsymbol{A}$ and $\boldsymbol{B}$, we determine a *flat* sheet which you can interpret as a piece of paper laid on top of these two lines.

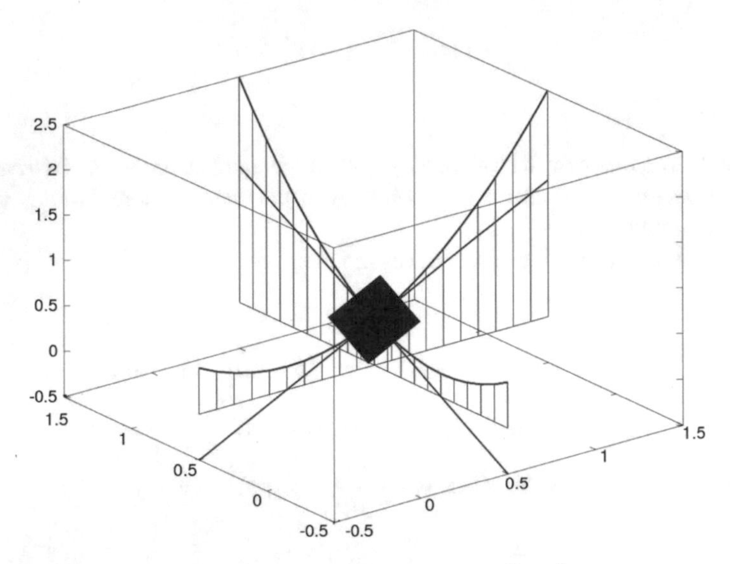

Fig. 4.2 The traces $f(x_0, y)$ and $f(x, y_0)$ for the surface $z = x^2 + y^2$ for $x_0 = 0.5$ and $y_0 = 0.5$ with added *tangent lines*. We have added the tangent plane determined by the *tangent lines*

Of course, we can only envision a small finite subset of this sheet of paper as you can see in Fig. 4.2. Imagine that the sheet extends infinitely in all directions! The sheet of paper we are plotting is called the **tangent plane** to our surface at the point (x_0, y_0). We will talk about this more formally later.

To draw this picture with the tangent lines, the traces and the tangent plane, we use **DrawTangentLines** which has arguments **(f,fx,fy,delx,nx,dely,ny, r,x0,y0)**. There are three new arguments: **fx** which is $\partial f/\partial x$, **fy** which is $\partial f/\partial y$ and **r** which is the size of the tangent plane that is plotted. For the picture shown in Fig. 4.3, we've removed the tangent plane because the plot was getting pretty busy. We did this by commenting out the line that plots the tangent plane. It is easy for you to go into the code and add it back in if you want to play around. The MatLab command line is

Listing 4.4: Drawing Tangent Lines

```
fx = @(x,y) 2*x;
fy = @(x,y) 2*y;
%
DrawTangentLines(f,fx,fy,0.5,2,0.5,2,.3,0.5,0.5);
```

If you want to see the tangent plane as well as the tangent lines, all you have to do is look at the following lines in **DrawTangentLines.m**.

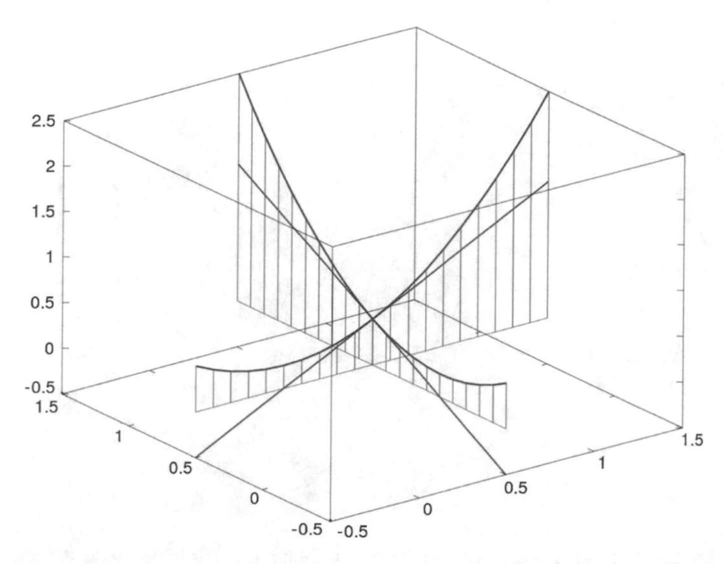

Fig. 4.3 The traces $f(x_0, y)$ and $f(x, y_0)$ for the surface $z = x^2 + y^2$ for $x_0 = 0.5$ and $y_0 = 0.5$ with added *tangent lines*

Listing 4.5: Drawing Tangent Lines

```
% set up a new local mesh grid near (x0,y0)
[U,V] = meshgrid(u,v)
% set up the tangent plane at (x0,y0)
W = f(x0,y0) + fx(x0,y0)*(U-x0) + fy(x0,y0)*(V-y0)
% plot the tangent plane
surf(U,V,W,'EdgeColor','blue');
```

These lines setup the tangent plane and the tangent plane is turned off if there is a % in front of **surf(U,V,W,'EdgeColor','blue');**. We edited the file to take the % out so we can see the tangent plane. We then see the plane in Fig. 4.4 as we saw before.

The ideas we have been discussing can be made more general. When we take the derivative with respect to one variable while holding the other variable constant (as we do when we find the normal derivative along a trace), we say we are taking a **partial derivative of f**. Here there are two flavors: the partial derivative with respect to x and the partial derivative with respect to y. We can now state some formal definitions and introduce the notations and symbols we use for these things. We define the process of partial differentiation carefully below.

Definition 4.3.1 (*Partial Derivatives*)
Let $z = f(x, y)$ be a function of the two independent variables x and y defined on some domain. At each pair (x, y) where f is defined in a circle of some finite radius r, $B_r(x_0, y_0) = \{(x, y) \mid \sqrt{(x - x_0)^2 + (y - y_0)^2} < r\}$, it makes sense to try to find the limits

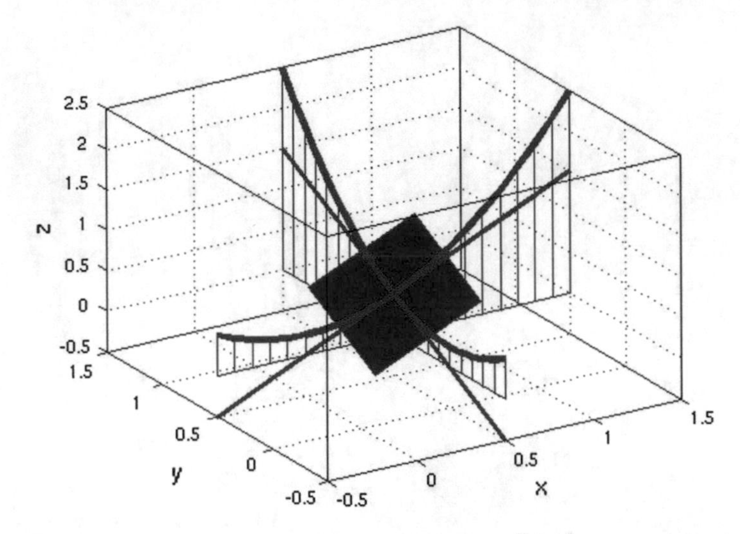

Fig. 4.4 The traces $f(x_0, y)$ and $f(x, y_0)$ for the surface $z = x^2 + y^2$ for $x_0 = 0.5$ and $y_0 = 0.5$ with added *tangent lines*

$$\lim_{x \to x_0, y = y_0} \frac{f(x, y) - f(x, y_0)}{x - x_0}$$
$$\lim_{x = x_0, y \to y_0} \frac{f(x, y) - f(x_0, y)}{y - y_0}$$

If these limits exists, they are called the partial derivatives of f with respect to x and y at (x_0, y_0), respectively.

Comment 4.3.1 *For these partial derivatives, we use the symbols*

$$f_x(x_0, y_0),\ \frac{\partial f}{\partial x}(x_0, y_0),\ z_x(x_0, y_0),\ \frac{\partial z}{\partial x}(x_0, y_0)$$

and

$$f_y(x_0, y_0),\ \frac{\partial f}{\partial y}(x_0, y_0),\ z_y(x_0, y_0),\ \frac{\partial z}{\partial y}(x_0, y_0)$$

Comment 4.3.2 *We often use another notation for partial derivatives. The function f of two variables x and y can be thought of as having two arguments or slots into which we place values. So another useful notation is to let the symbol $D_1 f$ be f_x and $D_2 f$ be f_y. We will be using this notation later when we talk about the* **chain rule**.

Comment 4.3.3 *It is easy to take partial derivatives. Just imagine the one variable held constant and take the derivative of the resulting function just like you did in your earlier calculus courses.*

Example 4.3.1 Let $z = f(x, y) = x^2 + 4y^2$ be a function of two variables. Find $\frac{\partial z}{\partial x}$ and $\frac{\partial z}{\partial y}$.

Solution *Thinking of y as a constant, we take the derivative in the usual way with respect to x. This gives*

$$\frac{\partial z}{\partial x} = 2x$$

as the derivative of $4y^2$ *with respect to x is 0. So, we know* $f_x = 2x$.
In a similar way, we find $\frac{\partial z}{\partial y}$. *We see*

$$\frac{\partial z}{\partial y} = 8y$$

as the derivative of x^2 *with respect to y is 0. So* $f_y = 8y$.

Example 4.3.2 Let $z = f(x, y) = 4x^2y^3$. Find $\frac{\partial z}{\partial x}$ and $\frac{\partial z}{\partial y}$.

Solution *Thinking of y as a constant, take the derivative in the usual way with respect to x: This gives*

$$\frac{\partial z}{\partial x} = 8xy^3$$

as the term $4y^3$ *is considered a "constant" here. So* $f_x = 8xy^3$.
Similarly,

$$\frac{\partial z}{\partial y} = 12x^2y^2$$

as the term $4x^2$ *is considered a "constant" here. So* $f_y = 12x^2y^2$.

Now let's do some without spelling out each step. Make sure you see what we are doing!

Example 4.3.3 $f(x, y) = \frac{x^4+1}{y^3+2}$.

Solution

$$\frac{\partial f}{\partial x} = \frac{(4x^3)}{y^3 + 2}$$
$$\frac{\partial f}{\partial y} = -\frac{(x^4 + 1)(3y^2)}{(y^3 + 2)^2}$$

Example 4.3.4 $f(x, y) = \frac{x^4y^2+2}{y^3x^5+20}$.

Solution

$$\frac{\partial f}{\partial x} = \frac{(4x^3y^2)(y^3x^5 + 20) - (x^4y^2 + 2)(5y^3x^4)}{(y^3x^5 + 20)^2}$$
$$\frac{\partial f}{\partial y} = \frac{(2x^4y)(y^3x^5 + 20) - (x^4y^2 + 2)(3y^2x^5)}{(y^3x^5 + 20)^2}$$

Example 4.3.5 $f(x, y) = \sin(x^3y + 2)$.

Solution

$$\frac{\partial f}{\partial x} = \cos(x^3y + 2)\ (3x^2y)$$
$$\frac{\partial f}{\partial y} = \cos(x^3y + 2)\ (x^3)$$

Example 4.3.6 $f(x, y) = e^{-(x^2+y^4)}$

Solution

$$\frac{\partial f}{\partial x} = e^{-(x^2+y^4)}\ (-2x)$$
$$\frac{\partial f}{\partial y} = e^{-(x^2+y^4)}\ (-4y^3)$$

Example 4.3.7 $f(x, y) = \ln(\sqrt{x^2 + 2y^2})$.

Solution

$$\frac{\partial f}{\partial x} = \frac{1}{2}\,\frac{2x}{x^2 + 2y^2}$$
$$\frac{\partial f}{\partial y} = \frac{1}{2}\,\frac{4y}{x^2 + 2y^2}$$

4.3.1 Homework

These are for you: for each of these functions, find f_x and f_y.

First, functions with no cross terms.

Exercise 4.3.1 $f(x, y) = x^2 + 3y^2$.

Exercise 4.3.2 $f(x, y) = 4x^2 + 5y^4$.

Exercise 4.3.3 $f(x, y) = -3x + 2y^8$.

Next, functions with cross terms.

Exercise 4.3.4 $f(x, y) = x^2 y^2$.

Exercise 4.3.5 $f(x, y) = 2x^3 y^2 + 5x$.

Exercise 4.3.6 $f(x, y) = 3x y^2$.

Exercise 4.3.7 $f(x, y) = x^2 y^5$.

Next, functions with fractions.

Exercise 4.3.8 $f(x, y) = \frac{x^2}{y^5}$.

Exercise 4.3.9 $f(x, y) = \frac{x^2+2y}{5x+y^3}$.

Exercise 4.3.10 $f(x, y) = \frac{x+2y}{5x+y}$.

Exercise 4.3.11 $f(x, y) = \frac{x^2+2}{5+y}$.

Now, sin and cos things.

Exercise 4.3.12 $f(x, y) = \sin(xy)$.

Exercise 4.3.13 $f(x, y) = \sin(x^2 y)$.

Exercise 4.3.14 $f(x, y) = \cos(x + 3y)$.

Exercise 4.3.15 $f(x, y) = \sin(\sqrt{x + y})$.

Exercise 4.3.16 $f(x, y) = \cos^2(x + 4y)$.

Exercise 4.3.17 $f(x, y) = \sqrt{\sin(xy)}$.

Now, let's add ln and exp.

Exercise 4.3.18 $f(x, y) = e^{xy}$

Exercise 4.3.19 $f(x, y) = e^{x+4y}$

Exercise 4.3.20 $f(x, y) = e^{-3xy}$.

Exercise 4.3.21 $f(x, y) = \ln(x^2 + 4y^2)$.

Exercise 4.3.22 $f(x, y) = \ln(\sqrt{1 + xy})$.

Exercise 4.3.23 $f(x, y) = e^{\sin(3x+5y)}$.

Exercise 4.3.24 $f(x, y) = e^{\cos(3x^2+5y)}$.

Exercise 4.3.25 $f(x, y) = \ln(1 + 3x + 8y)$.

4.4 Tangent Planes

Before we discuss tangent planes to a function f again, let's digress to the ideas of planes in general in 3D. We define a plane as follows.

Definition 4.4.1 (*Planes*)
A plane in 3D through the point (x_0, y_0, z_0) is defined as the set of all points (x, y, z) so that the angle between the vectors $\boldsymbol{D}$ and $\boldsymbol{N}$ is zero where $\boldsymbol{D}$ is the vector we get by connecting the point (x_0, y_0, z_0) to the point (x, y, z). Hence, for

$$\boldsymbol{D} = \begin{bmatrix} x - x_0 \\ y - y_0 \\ z - z_0 \end{bmatrix} \text{ and } \boldsymbol{N} = \begin{bmatrix} N_1 \\ N_2 \\ N_3 \end{bmatrix}$$

the plane is the set of points (x, y, z) so that $< \boldsymbol{D}, \boldsymbol{N} > = 0$. The vector $\boldsymbol{N}$ is called the **normal vector** to the plane.

Comment 4.4.1 *A little thought shows that any plane crossing through the origin is a two dimensional subspace of* $\Re^3$.

Example 4.4.1 The equation $2x + 3y - 5z = 0$ defines the plane whose normal vector is $\boldsymbol{N} = [2, 3, -5]^T$ which passes through the origin $(0, 0, 0)$.

Example 4.4.2 The equation $2(x - 2) + 3(y - 1) - 5(z + 3) = 0$ defines the plane whose normal vector is $\boldsymbol{N} = [2, 3, -5]^T$ which passes through the point $(2, 1, -3)$. Note this can be rewritten as $2x + 3y - 5z = 4 + 3 + 15 = 22$ after a simple manipulation.

Example 4.4.3 The equation $2x + 3y - 5z = 11$ corresponds to a plane with normal vector $\boldsymbol{N} = [2, 3, -5]^T$ which passes through some point (x_0, y_0, z_0). There are an infinite number of choices for this base point: any triple which solves $2x_0 + 3y_0 - 5z_0 = 11$ will do the job. An easy way to pick one is to pick two and solve for the third. So for example, if $z_0 = 0$ and $y_0 = 4$, we find $2x_0 + 12 = 11$ which gives $x_0 = -1/2$. Thus, this plane could be rewritten as $2(x + 1/2) + 3(y - 4) - 5z = 0$.

4.4.1 The Vector Cross Product

There is another very useful way to define a plane which we did not discuss in the first volume. As long as the vectors $\boldsymbol{A}$ and $\boldsymbol{B}$ point in different directions, they determine a new vector $\boldsymbol{A} \times \boldsymbol{B}$ which is perpendicular to both of them and can serve as the normal to a plane. Note, saying the vectors $\boldsymbol{A}$ and $\boldsymbol{B}$ point in different directions is the same as saying they are linear dependent.

Definition 4.4.2 (*Planes Again*)
The plane in 3D determined by the vectors $\boldsymbol{A}$ and $\boldsymbol{B}$ containing the point (x_0, y_0, z_0) is defined as the plane whose normal vector is $\boldsymbol{N} = \boldsymbol{A} \times \boldsymbol{B}$.

To find the vector $\boldsymbol{C} = \boldsymbol{A} \times \boldsymbol{B}$ requires a bit of calculation. Assume $\boldsymbol{C} = [C_1, C_2, C_3]^T$, $\boldsymbol{A} = [A_1, A_2, A_3]^T$ and $\boldsymbol{B} = [B_1, B_2, B_3]^T$. Then, we want $< \boldsymbol{C}, \boldsymbol{A} >= 0$ and $< \boldsymbol{C}, \boldsymbol{B} >= 0$. For convenience, we will assume all of the components of $\boldsymbol{A}$ and $\boldsymbol{B}$ are nonzero. This gives the equations

$$\begin{aligned} A_1C_1 + A_2C_2 + A_3C_3 &= 0 \\ B_1C_1 + B_2C_2 + B_3C_3 &= 0 \end{aligned}$$

Solving both equations for C_1, we have

$$C_1 = -\frac{A_2}{A_1}C_2 - \frac{A_3}{A_1}C_3 = -\frac{B_2}{B_1}C_2 - \frac{B_3}{B_1}C_3.$$

Solving for C_2, we have

$$\begin{aligned} C_2 &= \left(\frac{-\frac{B_3}{B_1} - \frac{A_3}{A_1}}{-\frac{B_2}{B_1} - \frac{A_2}{A_1}} \right) C_3 \\ &= \left(\frac{-\frac{B_3}{B_1} - \frac{A_3}{A_1}}{-\frac{B_2}{B_1} - \frac{A_2}{A_1}} \right) \frac{A_1B_1}{A_1B_1} C_3 \\ &= \frac{B_1A_3 - A_1B_3}{A_1B_2 - B_1A_2} C_3. \end{aligned}$$

We know $A_1C_1 = -A_2C_2 - A_3C_3$ and so substituting for C_2 we obtain

$$\begin{aligned} C_1 &= -\left(\frac{A_2}{A_1} \frac{B_1A_3 - A_1B_3}{A_1B_2 - B_1A_2} + \frac{A_3}{A_1} \right) C_3 \\ &= -\left(\frac{A_2(B_1A_3 - A_1B_3) + A_3(A_1B_2 - B_1A_2)}{A_1(A_1B_2 - B_1A_2)} \right) C_3 \\ &= -\frac{A_2B_1A_3 - A_1A_2B_3 + A_3A_1B_2 - A_3B_1A_2}{A_1(A_1B_2 - B_1A_2)} C_3 \\ &= -\frac{-A_1A_2B_3 + A_3A_1B_2}{A_1(A_1B_2 - B_1A_2)} C_3 \\ &= \frac{A_2B_3 - A_3B_2}{A_1B_2 - B_1A_2} C_3. \end{aligned}$$

We have found the components of $\boldsymbol{C}$ can be written in terms of the parameter C_3:

$$C_1 = \frac{A_2B_3 - A_3B_2}{A_1B_2 - B_1A_2}\, C_3$$
$$C_2 = \frac{B_1A_3 - A_1B_3}{A_1B_2 - B_1A_2}\, C_3.$$

Choosing $C_3 = A_1B_2 - B_1A_2$, we find that the vector we are looking for has the components

$$C_1 = A_2B_3 - A_3B_2$$
$$C_2 = B_1A_3 - A_1B_3$$
$$C_3 = A_1B_2 - B_1A_2$$

We can reorganize these components by recognizing they are the determinants of certain 2×2 matrices; i.e.

$$C_1 = \det \begin{bmatrix} A_2 & A_3 \\ B_2 & B_3 \end{bmatrix}$$
$$C_2 = -\det \begin{bmatrix} A_1 & A_3 \\ B_1 & B_3 \end{bmatrix}$$
$$C_3 = \det \begin{bmatrix} A_1 & A_2 \\ B_1 & B_2 \end{bmatrix}$$

Then using the standard basis vectors for $\Re^3$, $\boldsymbol{i}$, $\boldsymbol{j}$ and $\boldsymbol{k}$, we see the vector $\boldsymbol{C} = \boldsymbol{A} \times \boldsymbol{B}$ can be written as

$$\boldsymbol{A} \times \boldsymbol{B} = \boldsymbol{i} \det \begin{bmatrix} A_2 & A_3 \\ B_2 & B_3 \end{bmatrix} - \boldsymbol{j} \det \begin{bmatrix} A_1 & A_3 \\ B_1 & B_3 \end{bmatrix} + \boldsymbol{k} \det \begin{bmatrix} A_1 & A_2 \\ B_1 & B_2 \end{bmatrix}$$

This leads us to the following definition.

Definition 4.4.3 (*The Vector Cross Product*)
The cross product of the two nonzero vectors $\boldsymbol{A}$ and $\boldsymbol{B}$ is defined to be the vector $\boldsymbol{A} \times \boldsymbol{B}$ where

$$\boldsymbol{A} \times \boldsymbol{B} = \boldsymbol{i} \det \begin{bmatrix} A_2 & A_3 \\ B_2 & B_3 \end{bmatrix} - \boldsymbol{j} \det \begin{bmatrix} A_1 & A_3 \\ B_1 & B_3 \end{bmatrix} + \boldsymbol{k} \det \begin{bmatrix} A_1 & A_2 \\ B_1 & B_2 \end{bmatrix}$$

This calculation is performed so often we define a special use of the determinant to help us remember. We define the matrix

$$\boldsymbol{C} = \begin{bmatrix} \boldsymbol{i} & \boldsymbol{j} & \boldsymbol{k} \\ A_1 & A_2 & A_3 \\ B_1 & B_2 & B_3 \end{bmatrix}$$

and we define the determinant of this matrix to coincide with the definition of $\boldsymbol{A} \times \boldsymbol{B}$.

Comment 4.4.2 *This is easy to remember. Start with the* $\boldsymbol{i}$ *in row one. Cross out the first row and first column of* $\boldsymbol{C}$*. The first term in the cross product is then* $\boldsymbol{i}$ *times the determinant of the* 2×2 *submatrix that is left over. This is the matrix*

$$\begin{bmatrix} A_1 & A_2 \\ B_1 & B_2 \end{bmatrix}$$

The first term is then

$$\boldsymbol{i} \ \det \begin{bmatrix} A_1 & A_2 \\ B_1 & B_2 \end{bmatrix}$$

The second term in row one is $\boldsymbol{j}$*. Associate this term with a* **minus** *sign and since it is in row one, column two cross that row and column out of* $\boldsymbol{C}$ *to obtain the submatrix*

$$\begin{bmatrix} A_1 & A_3 \\ B_1 & B_3 \end{bmatrix}$$

We now have the second term

$$-\boldsymbol{j} \ \det \begin{bmatrix} A_1 & A_3 \\ B_1 & B_3 \end{bmatrix}$$

The last term is associated with the row one, column three entry $\boldsymbol{k}$*. Cross out that row and column in* $\boldsymbol{C}$ *to obtain the submatrix*

$$\begin{bmatrix} A_2 & A_3 \\ B_2 & B_3 \end{bmatrix}$$

This gives the last term

$$\boldsymbol{k} \ \det \begin{bmatrix} A_2 & A_3 \\ B_2 & B_3 \end{bmatrix}$$

The cross product is then the sum of these three terms. This is also called expanding a 3×3 *determinant by the first row, but that is another story.*

4.4.1.1 Homework

For each of these problems, graph the two vectors as well as the cross product.

Exercise 4.4.1 *Find* $\boldsymbol{i} \times \boldsymbol{j}$, $\boldsymbol{i} \times \boldsymbol{k}$ *and* $\boldsymbol{j} \times \boldsymbol{k}$,

Exercise 4.4.2 *Find* $\boldsymbol{A} \times \boldsymbol{B}$ *for* $\boldsymbol{A} = \begin{bmatrix}1 & -2 & 3\end{bmatrix}^T$ *and* $\boldsymbol{B} = \begin{bmatrix}-1 & 5 & 3\end{bmatrix}^T$.

Exercise 4.4.3 *Find* $\boldsymbol{A} \times \boldsymbol{B}$ *for* $\boldsymbol{A} = \begin{bmatrix}-2 & 4 & -3\end{bmatrix}^T$ *and* $\boldsymbol{B} = \begin{bmatrix}6 & -1 & 2\end{bmatrix}^T$.

Exercise 4.4.4 *Find* $\boldsymbol{A} \times \boldsymbol{B}$ *for* $\boldsymbol{A} = \begin{bmatrix}0 & -3 & 1\end{bmatrix}^T$ *and* $\boldsymbol{B} = \begin{bmatrix}2 & 0 & 8\end{bmatrix}^T$.

4.4.2 Back to Tangent Planes

Recall the tangent plane to a surface $z = f(x, y)$ at the point (x_0, y_0) was the plane determined by the tangent lines $T(x, y_0)$ and $T(x_0, y)$. The $T(x, y_0)$ line was determined by the vector

$$\boldsymbol{A} = \begin{bmatrix} 1 \\ 0 \\ \frac{d}{dx} f(x, y_0)\Big|_{x_0} \end{bmatrix} = \begin{bmatrix} 1 \\ 0 \\ 2x_0 \end{bmatrix}$$

and the $T(x_0, y)$ line was determined by the vector

$$\boldsymbol{B} = \begin{bmatrix} 0 \\ 1 \\ \frac{d}{dy} f(x_0, y)\Big|_{y_0} \end{bmatrix} = \begin{bmatrix} 0 \\ 1 \\ 2y_0 \end{bmatrix}$$

We know now that we can write these vectors more generally as

$$\boldsymbol{A} = \begin{bmatrix} 1 \\ 0 \\ \frac{\partial f}{\partial x}(x_0, y_0) \end{bmatrix}$$

$$\boldsymbol{B} = \begin{bmatrix} 0 \\ 1 \\ \frac{\partial f}{\partial y}(x_0, y_0) \end{bmatrix}$$

The plane determined by these vectors has normal $\boldsymbol{A} \times \boldsymbol{B}$ which is

$$\boldsymbol{A} \times \boldsymbol{B} = \begin{bmatrix} \boldsymbol{i} & \boldsymbol{j} & \boldsymbol{k} \\ 1 & 0 & f_x(x_0, y_0) \\ 0 & 1 & f_y(x_0, y_0) \end{bmatrix}$$

or

$$\begin{aligned} \boldsymbol{A} \times \boldsymbol{B} &= \boldsymbol{i} \det \begin{bmatrix} 0 & f_x(x_0, y_0) \\ 1 & f_y(x_0, y_0) \end{bmatrix} - \boldsymbol{j} \det \begin{bmatrix} 1 & f_x(x_0, y_0) \\ 0 & f_y(x_0, y_0) \end{bmatrix} + \boldsymbol{k} \det \begin{bmatrix} 0 & 1 \\ 0 & 1 \end{bmatrix} \\ &= -f_x(x_0, y_0)\boldsymbol{i} - f_y(x_0, y_0)\boldsymbol{j} + \boldsymbol{k} \\ &= \begin{bmatrix} -f_x(x_0, y_0) \\ -f_y(x_0, y_0) \\ 1 \end{bmatrix}. \end{aligned}$$

The tangent plane to the surface $z = f(x, y)$ at the point (x_0, y_0) is then given by

$$-f_x(x_0, y_0)(x - x_0) - f_y(x_0, y_0)(y - y_0) + (z - f(x_0, y_0) = 0.$$

This then gives the traditional equation of the tangent plane:

$$z = f(x_0, y_0) + f_x(x_0, y_0)(x - x_0) + f_y(x_0, y_0)(y - y_0). \tag{4.1}$$

We can use another compact definition at this point. We can define the **gradient** of the function f to be the vector $\boldsymbol{\nabla} f$. The gradient is defined as follows.

Definition 4.4.4 (*The Gradient*)
The gradient of the scalar function $z = f(x, y)$ is defined to be the vector $\boldsymbol{\nabla} f$ where

$$\boldsymbol{\nabla} f(x_0, y_0) = \begin{bmatrix} f_x(x_0, y_0) \\ f_y(x_0, y_0) \end{bmatrix}.$$

Note the gradient takes a scalar function argument and returns a vector answer.

Using the gradient, Eq. 4.1 can be rewritten as

$$\begin{aligned} f(x, y) &= f(x_0, y_0) + < \boldsymbol{\nabla} f, \boldsymbol{X} - \boldsymbol{X_0} > \\ &= f(x_0, y_0) + \boldsymbol{\nabla} f^T (\boldsymbol{X} - \boldsymbol{X_0}) \end{aligned}$$

where $\boldsymbol{X} - \boldsymbol{X_0} = \left[x - x_0 \; y - y_0\right]^T$. The obvious question to ask now is how much of a discrepancy is there between the value $f(x, y)$ and the value of the tangent plane?

Example 4.4.4 Find the gradient of $f(x, y) = x^2 + 4xy + 9y^2$ and the equation of the tangent plane to this surface at the point $(1, 2)$.

Solution

$$\nabla f(x, y) = \begin{bmatrix} 2x + 4y \\ 4x + 18y \end{bmatrix}.$$

The equation of the tangent plane at $(1, 2)$ *is then*

$$\begin{aligned} z &= f(1, 2) + f_x(1, 2)(x - 2) + f_y(1, 2)(y - 2) \\ &= 45 + 10(x - 2) + 40(y - 2) \\ &= -55 + 10x + 40y. \end{aligned}$$

Note this can also be written as $10x + 40y + z = 55$ *which is also a standard form. However, in this form, the attachment point* $(1, 2, 45)$ *is hidden from view.*

4.4.3 Homework

Exercise 4.4.5 *Find the gradient of* $f(x, y) = x^2 - xy^2 + y^2$ *and the equation of the tangent plane to this surface at the point* $(1, 1)$.

Exercise 4.4.6 *Find the gradient of* $f(x, y) = -x^2 + y^2 + 4y$ *and the equation of the tangent plane to this surface at the point* $(-1, 1)$.

Exercise 4.4.7 *Find the gradient of* $f(x, y) = \sin(xy)$ *and the equation of the tangent plane to this surface at the point* $(\pi/4, -\pi/4)$.

4.4.4 Computational Results

We can use MatLab/Octave to draw tangent planes and tangent lines to a surface. Consider the function `DrawTangentPlanePackage`. The source code is similar to what we have done in previous functions. This time, we send in the function `f` and the two partial derivatives `fx` and `fy`. First, we plot the traces and draw vertical lines from the traces to the $x-y$ plane. Note this code will not do very well on surfaces where the z values become negative! But then, this code is just for exploration and it is easy enough to alter it for other jobs. And it is a good exercise! After the traces and their *shadow lines* are drawn, we draw the tangent lines. Finally, we draw the tangent plane. The tangent plane calculation uses the partial derivatives we sent into this function as arguments.

Listing 4.6: DrawTangentPlanePackage

```
function DrawTangentPlanePackage(f,fx,fy,delx,nx,dely,ny,r,x0,y0)
% f is the function defining the surface
% delx is the size of the x step
% nx is the number of steps left and right from x0
% dely is the size of the y step
% ny is the number of steps left and right from y0
% r is the size of the drawn tangent plane
% (x0,y0) is the location of the column rectangle base
%
% set up x and y stuff
x = x0-nx*delx:delx/5:x0+nx*delx;
y = y0-ny*dely:dely/5:y0+ny*dely;
[rows,sx] = size(x);
[rows,sy] = size(y);
hold on
  % set up x trace for x = x0
  % set up y trace for y = y0
  xtrace = f(x0,y);
  ytrace = f(x,y0);
  fixedx = x0*ones(1,sx);
  fixedy = y0*ones(1,sy);
  plot3(fixedx,y,xtrace,'LineWidth',4,'Color','red');
  plot3(x,fixedy,ytrace,'LineWidth',4,'Color','red');
  % now draw x0, y0 line in xy plane
  U = [x0;x0];
  V = [y0-ny*dely;y0+ny*delx];
  W = [0;0];
  plot3(U,V,W);
  U = [y0-ny*dely;y0+ny*delx];
  V = [y0;y0];
  W = [0;0];
  plot3(U,V,W);
  % now fill in planes formed by x0, y0 lines
  for i=1:sy
    U = [x0;x0];
    V = [y(i);y(i)];
    W = [0;f(x0,y(i))];
    plot3(U,V,W,'LineWidth',1,'Color','red');
  end
  for i=1:sx
    U = [x(i);x(i)];
    V = [y0;y0];
    W = [0;f(x(i),y0)];
    plot3(U,V,W,'LineWidth',1,'Color','red');
  end
  % now draw tangent lines
  % set up new local variables centered at (x0,y0)
  TX = @(x) (f(x0,y0) + fx(x0,y0)*(x-x0));
  TY = @(y) (f(x0,y0) + fy(x0,y0)*(y-y0));
  U = [x0-nx*delx;x0+nx*delx];
  V = [y0;y0];
  W = [TX(x0-nx*delx);TX(x0+nx*delx)];
  plot3(U,V,W,'LineWidth',3,'Color','blue');
  U = [x0;x0];
  V = [y0-ny*dely;y0+ny*dely];
  W = [TY(y0-ny*dely);TY(y0+ny*dely)];
  plot3(U,V,W,'LineWidth',3,'Color','blue');
  % plot tangent plane
  u = [x0-r*delx;x0+r*delx];
  v = [y0-r*dely;y0+r*dely];
  % set up a new local mesh grid near (x0,y0)
  [U,V] = meshgrid(u,v);
  % set up the tangent plane at (x0,y0)
  w = @(u,v) f(x0,y0) + fx(x0,y0)*(u-x0) + fy(x0,y0)*(v-y0);
  W = w(U,V);
  % plot the tangent plane
  surf(U,V,W,'EdgeColor','blue');
hold off
end
```

The illustrations this code produces have already been used in Fig. 4.2. Practice with this code and draw other pictures! A typical session to generate this figure would look like

Listing 4.7: Drawing Tangent Planes

```
f = @(x,y) x.^2+y.^2;
fx = @(x,y) 2*x;
fy = @(x,y) 2*y;
DrawTangentPlanePackage(f,fx,fy,0.5,2,0.5,2,.3,0.5,0.5);
```

4.4.5 Homework

Exercise 4.4.8 *Draw tangent lines and planes for the surface* $f(x, y) = x^2 + 3y^2$ *for various points* (x_0, y_0).

Exercise 4.4.9 *Draw tangent lines and planes for the surface* $f(x, y) = -x^2 - 3y^2$ *for various points* (x_0, y_0). *You will need to modify the code to make this work!*

Exercise 4.4.10 *Draw tangent lines and planes for the surface* $f(x, y) = x^2 - 3y^2$ *for various points* (x_0, y_0). *You will need to modify the code to make this work! Make sure you try the point* $(0, 0)$.

4.5 Derivatives in Two Dimensions!

Let's look at the partial derivatives of $f(x, y)$. As long as $f(x, y)$ is defined locally at (x_0, y_0), we can say $f_x(x_0, y_0)$ and $f_y(x_0, y_0)$ exist if and only if there are error functions $E_1(x, y, x_0, y_0)$ and $E_2(x, y, x_0, y_0)$ so that

$$
\begin{aligned}
f(x, y_0) &= f(x_0, y_0) + f_x(x_0, y_0)(x - x_0) + E_1(x, x_0, y_0)\\
f(x_0, y) &= f(x_0, y_0) + f_y(x_0, y_0)(y - y_0) + E_2(y, x_0, y_0)
\end{aligned}
$$

with $E_1 \to 0$ and $E_1/(x - x_0) \to 0$ as $x \to x_0$ and $E_2 \to 0$ and $E_2/(y - x_0) \to 0$ as $y \to y_0$. Using the ideas we have presented here, we can come up with a way to define the differentiability of a function of two variables.

Definition 4.5.1 (*Error Form of Differentiability For Two Variables*)
If $f(x, y)$ is defined locally at (x_0, y_0), then f is differentiable at (x_0, y_0) if there are two numbers L_1 and L_2 so that the error function $E(x, y, x_0, y_0) = f(x, y) - f(x_0, y_0) - L_1(x - x_0) - L_2(y - y_0)$ satisfies $\lim_{(x,y)\to(x_0,y_0)} E(x, y, x_0, y_0) = 0$

and $\lim_{(x,y)\rightarrow(x_0,y_0)} E(x, y, x_0, y_0)/||(x - x_0, y - y_0)|| = 0$. Here, the term $||(x - x_0, y - y_0)|| = \sqrt{(x - x_0)^2 + (y - y_0)^2}$.

Note if f is differentiable at (x_0, y_0), f must be continuous at (x_0, y_0). The argument is simple:

$$f(x, y) = f(x_0, y_0) + L_1\,(x_0, y_0)(x - x_0) + L_2\,(y - y_0) + E(x, y, x_0, y_0)$$

and as $(x, y) \rightarrow (x_0, y_0)$, we have $f(x, y) \rightarrow f(x_0, y_0)$ which is the definition of f being continuous at (x_0, y_0). Hence, we can say

Theorem 4.5.1 (Differentiable Implies Continuous: Two Variables)
If f is differentiable at (x_0, y_0) then f is continuous at (x_0, y_0).

Proof We have sketched the argument already. ■

From Definition 4.5.1, we can show if f is differentiable at the point (x_0, y_0), then $L_1 = f_x(x_0, y_0)$ and $L_2 = f_y(x_0, y_0)$. The argument goes like this: since f is differentiable at (x_0, y_0), we can say

$$\lim_{(x,y)\rightarrow(x_0,y_0)} \frac{f(x, y) - f(x_0, y_0) - L_1(x - x_0) - L_2(y - y_0)}{\sqrt{(x - x_0)^2 + (y - y_0)^2}} = 0.$$

We can rewrite this using $\Delta x = x - x_0$ and $\Delta y = y - y_0$ as

$$\lim_{(\Delta x,\Delta y)\rightarrow(0,0)} \frac{f(x_0 + \Delta x, y_0 + \Delta y) - f(x_0, y_0) - L_1\Delta x - L_2\Delta y}{\sqrt{(\Delta x)^2 + (\Delta y)^2}} = 0.$$

In particular, for $\Delta y = 0$, we find

$$\lim_{(\Delta x)\rightarrow 0} \frac{f(x_0 + \Delta x, y_0) - f(x_0, y_0) - L_1\Delta x}{\sqrt{(\Delta x)^2}} = 0.$$

For $\Delta x > 0$, we find $\sqrt{(\Delta x)^2} = \Delta x$ and so

$$\lim_{\Delta x\rightarrow 0^+} \frac{f(x_0 + \Delta x, y_0) - f(x_0, y_0)}{\Delta x} = L_1.$$

Thus, the right hand partial derivative $f_x(x_0, y_0)^+$ exists and equals L_1. On the other hand, if $\Delta x < 0$, then $\sqrt{(\Delta x)^2} = -\Delta x$ and we find, with a little manipulation, that we still have

$$\lim_{(\Delta x)\to 0^-} \frac{f(x_0+\Delta x, y_0) - f(x_0, y_0)}{\Delta x} = L_1.$$

So the left hand partial derivative $f_x(x_0, y_0)^-$ exists and equals L_1 also. Combining, we see $f_x(x_0, y_0) = L_1$. A similar argument shows that $f_y(x_0, y_0) = L_2$. Hence, we can say if f is differentiable at (x_0, y_0) then f_x and f_y exist at this point and we have

$$f(x, y) = f(x_0, y_0) + f_x(x_0, y_0)(x - x_0) + f_y(x_0, y_0)(y - y_0) + E_f(x, y, x_0, y_0)$$

where $E_f(x, y, x_0, y_0) \to 0$ and $E_f(x, y, x_0, y_0)/||(x - x_0, y - y_0)|| \to 0$ as $(x, y) \to (x_0, y_0)$. Note this argument is a pointwise argument. It only tells us that differentiability at a point implies the existence of the partial derivatives at that point.

4.6 The Chain Rule

Now that we know a bit about two dimensional derivatives, let's go for gold and figure out the new version of the chain rule. The argument we make here is very similar in spirit to the one dimensional one. You should go back and check it out! We will do this argument carefully but without tedious rigor. At least that is our hope. You'll have to let us know how we did!

We assume there are two functions $u(x, y)$ and $v(x, y)$ defined locally about (x_0, y_0) and that there is a third function $f(u, v)$ which is defined locally around $(u_0 = u(x_0, y_0), v_0 = v(x_0, y_0))$. Now assume $f(u, v)$ is differentiable at (u_0, v_0) and $u(x, y)$ and $v(x, y)$ are differentiable at (x_0, y_0). Then we can say

$$u(x, y) = u(x_0, y_0) + u_x(x_0, y_0)(x - x_0) + u_y(x_0, y_0)(y - y_0) + E_u(x, y, x_0, y_0)$$
$$v(x, y) = v(x_0, y_0) + v_x(x_0, y_0)(x - x_0) + v_y(x_0, y_0)(y - y_0) + E_v(x, y, x_0, y_0)$$
$$f(u, v) = f(u_0, v_0) + f_u(u_0, v_0)(u - u_0) + f_v(u_0, v_0)(v - v_0) + E_f(u, v, u_0, v_0)$$

where all the error terms behave as usual as $(x, y) \to (x_0, y_0)$ and $(u, v) \to (u_0, v_0)$. Note that as $(x, y) \to (x_0, y_0)$, $u(x, y) \to u_0 = u(x_0, y_0)$ and $v(x, y) \to v_0 = v(x_0, y_0)$ as u and v are continuous at the (u_0, v_0) since they are differentiable there. Let's consider the partial of f with respect to x. Let $\Delta u = u(x_0 + \Delta x, y_0) - u(x_0, y_0)$ and $\Delta v = v(x_0 + \Delta x, y_0) - v(x_0, y_0)$. Thus, $u_0 + \Delta u = u(x_0 + \Delta x, y_0)$ and $v_0 + \Delta v = v(x_0 + \Delta x, y_0)$. Hence

$$\begin{aligned}
&\frac{f(u_0 + \Delta u, v_0 + \Delta v) - f(u_0, v_0)}{\Delta x} \\
&= \frac{f_u(u_0, v_0)(u - u_0) + f_v(u_0, v_0)(v - v_0) + E_f(u, v, u_0, v_0)}{\Delta x} \\
&= f_u(u_0, v_0)\,\frac{u - u_0}{\Delta x} + f_v(u_0, v_0)\,\frac{v - v_0}{\Delta x} + \frac{E_f(u, v, u_0, v_0)}{\Delta x} \\
&= f_u(u_0, v_0)\,\frac{u_x(x_0, y_0)(x - x_0) + E_u(x, x_0, y_0)}{\Delta x} \\
&\quad + f_v(u_0, v_0)\,\frac{v_x(x_0, y_0)(x - x_0) + E_v(x, x_0, y_0)}{\Delta x} + \frac{E_f(u, v, u_0, v_0)}{\Delta x} \\
&= f_u(u_0, v_0)\,u_x(x_0, y_0) + f_v(u_0, v_0)\,v_x(x_0, y_0) \\
&\quad + \frac{E_u(x, x_0, y_0)}{\Delta x} + \frac{E_v(x, x_0, y_0)}{\Delta x} + \frac{E_f(u, v, u_0, v_0)}{\Delta x}.
\end{aligned}$$

As $(x, y) \rightarrow (x_0, y_0)$, $(u, v) \rightarrow (u_0, v_0)$ and so $E_f(u, v, u_0, v_0)/\Delta x \rightarrow 0$. The other two error terms go to zero also as $(x, y) \rightarrow (x_0, y_0)$. Hence, we conclude

$$\frac{\partial f}{\partial x} = \frac{\partial f}{\partial u}\,\frac{\partial u}{\partial x} + \frac{\partial f}{\partial v}\,\frac{\partial v}{\partial x}.$$

A similar argument shows

$$\frac{\partial f}{\partial y} = \frac{\partial f}{\partial u}\,\frac{\partial u}{\partial y} + \frac{\partial f}{\partial v}\,\frac{\partial v}{\partial y}.$$

This result is known as the **Chain Rule**.

Theorem 4.6.1 (The Chain Rule)
Assume there are two functions $u(x, y)$ and $v(x, y)$ defined locally about (x_0, y_0) and that there is a third function $f(u, v)$ which is defined locally around $(u_0 = u(x_0, y_0), v_0 = v(x_0, y_0))$. Further assume $f(u, v)$ is differentiable at (u_0, v_0) and $u(x, y)$ and $v(x, y)$ are differentiable at (x_0, y_0). Then f_x and f_y exist at (x_0, y_0) and are given by

$$\begin{aligned}
\frac{\partial f}{\partial x} &= \frac{\partial f}{\partial u}\,\frac{\partial u}{\partial x} + \frac{\partial f}{\partial v}\,\frac{\partial v}{\partial x} \\
\frac{\partial f}{\partial y} &= \frac{\partial f}{\partial u}\,\frac{\partial u}{\partial y} + \frac{\partial f}{\partial v}\,\frac{\partial v}{\partial y}.
\end{aligned}$$

Proof We have sketched the argument already. □

4.6.1 Examples

Example 4.6.1 Let $f(x, y) = x^2 + 2x + 5y^4$. Then if $x = r\cos(\theta)$ and $y = r\sin(\theta)$, using the chain rule, we find

$$\frac{\partial f}{\partial r} = \frac{\partial f}{\partial x}\frac{\partial x}{\partial r} + \frac{\partial f}{\partial y}\frac{\partial y}{\partial r}$$
$$\frac{\partial f}{\partial \theta} = \frac{\partial f}{\partial x}\frac{\partial x}{\partial \theta} + \frac{\partial f}{\partial y}\frac{\partial y}{\partial \theta}$$

This becomes

$$\frac{\partial f}{\partial r} = \left(2x + 2\right)\cos(\theta) + \left(20y^3\right)\sin(\theta)$$
$$\frac{\partial f}{\partial \theta} = \left(2x + 2\right)\left(-r\sin(\theta)\right) + \left(20y^3\right)\left(r\cos(\theta)\right)$$

You can then substitute in for x and y to get the final answer in terms of r and θ (kind of ugly though!).

Example 4.6.2 Let $f(x, y) = 10x^2y^4$. Then if $u = x^2 + 2y^2$ and $v = 4x^2 - 5y^2$, using the chain rule, we find $f(u, v) = 10u^2v^4$ and so

$$\frac{\partial f}{\partial x} = \frac{\partial f}{\partial u}\frac{\partial u}{\partial x} + \frac{\partial f}{\partial v}\frac{\partial v}{\partial x}$$
$$\frac{\partial f}{\partial y} = \frac{\partial f}{\partial u}\frac{\partial u}{\partial y} + \frac{\partial f}{\partial v}\frac{\partial v}{\partial y}$$

This becomes

$$\frac{\partial f}{\partial x} = \left(20uv^4\right)2x + \left(40u^2v^3\right)8x$$
$$\frac{\partial f}{\partial \theta} = \left(20uv^4\right)4y + \left(40u^2v^3\right)(-10y)$$

You can then substitute in for u and v to get the final answer in terms of x and y (even more ugly though!).

Example 4.6.3 In the discussion of Hamilton's Rule from the first course on calculus for biologists, we discuss a fitness function w for a model of altruism which depends on P which is the probability of giving aid and Q which is the probability of receiving aid. The model is

$$w = w_0 + b\,Q - c\,P$$

where w_0 is a baseline fitness amount. Note, the chain rule gives us

$$\begin{aligned}\frac{\partial w}{\partial P} &= \frac{\partial w}{\partial P}\frac{\partial P}{\partial P} + \frac{\partial w}{\partial Q}\frac{\partial Q}{\partial P}\\ &= -c + b\,\frac{\partial Q}{\partial P}\end{aligned}$$

Let $\partial_P\, Q$ be denoted by r, the **coefficient of relatedness**. The parameter r is very hard to understand even though it was introduced in 1964 to study altruism. Altruism occurs if fitness increases or $\frac{\partial w}{\partial P} > 0$. So altruism occurs if $-c + br > 0$ or $rb > c$. This inequality is **Hamilton's Rule**, but what counts is we understand what these terms mean biologically.

4.6.1.1 Homework

Exercise 4.6.1 *Let* $f(x, y) = x^2 + 5x^2y^3$ *and let* $u(x, y) = 2xy$ *and* $v(x, y) = x^2 - y^2$. *Find* $\partial_x f(u, v)$ *and* $\partial_y f(u, v)$.

Exercise 4.6.2 *Let* $f(x, y) = 2x^2 - 5x^2y^5$ *and let* $u(s, t) = st^2$ *and* $v(s, t) = s^2 + t^4$. *Find* f_s *and* f_t.

Exercise 4.6.3 *Let* $f(x, y) = 5x^2y^3 + 10$ *and let* $u(x, y) = sin(st)$ *and* $v(s, t) = cos(st)$. *Find* f_s *and* f_t.

4.7 Tangent Plane Approximation Error

We are now ready to give you a whirlwind tour of what you can call second order ideas in calculus for two variables. Or as some would say, let's drink from the fountain of knowledge with a fire hose! Well, maybe not that intense....

We will use these ideas for some practical things. Recall from the first class on calculus for biologist's, we used these ideas to to find the minimum and maximum of functions of two variables and we applied those ideas to the problem of finding the best straight line that fits a collection of data points. This **regression line** is of great importance to you in your career as biologists! We also introduced the ideas of **average or mean**, **covariance** and **variance** when we worked out how to find the regression line. The slope of the regression line has many important applications and we showed you some of them in our Hamilton's rule model.

Once we have the chain rule, we can quickly develop other results such as how much error we make when we approximate our surface $f(x, y)$ using a tangent plane at a point (x_0, y_0). To finish our arguments, we need an analog of the Mean Value Theorem from Calculus. The first thing we need is to know when a function of two variables is differentiable. Just because it's partials exist at a point is not enough to guarantee that! But we can prove that if the partials are continuous around that point, then the derivative does exist. And that means we can write the function in terms of its tangent plane plus an error. The arguments to do this are not terribly hard, so let's go through them. We will need a version of the **Mean Value Theorem** for functions of two variables. Here it is:

Theorem 4.7.1 (Mean Value Theorem)
Assume the partials of $f(x, y)$ exist at (x_0, y_0) and that f is defined locally around (x_0, y_0). Then given any (x, y) where f is locally defined, there is a point on the line between (x_0, y_0) and (x, y), (x_c, y_c) with $x_c = x_0 + c(x - x_0)$ and $y_c c = y_0 + c(y - y_0)$ so that

$$f(x, y) - f(x_0, y_0) = f_x(x_c, y_c)(x - x_0) + f_y(x_c, y_c)(y - y_0).$$

Proof The argument that shows this is pretty straightforward. We apply the chain rule using the simpler functions $u(t) = x_0 + t(x - x_0)$ and $v(t) = y_0 + t(y - y_0)$. Then u and v are differentiable with $u'(t) = x - x_0$ and $v'(t) = y - y_0$. Hence, if $h(t) = f(u(t), v(t))$, we have

$$h'(t) = f_x(u(t))\,(x - x_0) + f_y(v(t))\,(y - y_0)$$

and from the usual calculus mean value theorem,

$$h(1) - h(0) = h'(c)$$

where c is between 0 and 1. Using the definition of h and h', we have

$$f(x, y) - f(x_0, y_0) = f_x(x_c, y_c)(x - x_0) + f_y(x_c, y_c)(y - y_0).$$

□

Now it is not true that just because a function f has partial derivatives at a point (x_0, y_0) that f is differentiable. There are many examples where partials can exist at a point and the function itself does not satisfy the definition of differentiability. However, if we know the partials are themselves continuous locally at the point (x_0, y_0) then it is true that f is differentiable there. Once we know f is differentiable there we can apply chain rule type ideas. Let's assume f is defined locally around (x_0, y_0) and consider the difference

$$\begin{aligned} f(x_0 + t\Delta x, y_0 + t\Delta y) - f(x_0, y_0) &= f(x_0 + t\Delta x, y_0 + t\Delta y) - f(x_0 + t\Delta x, y_0) \\ &\quad + f(x_0 + t\Delta x, y_0) - f(x_0, y_0). \end{aligned}$$

From the Mean Value Theorem, we know we can write

$$\begin{aligned} f(x_0 + t\Delta x, y_0 + t\Delta y) - f(x_0 + t\Delta x, y_0) &= f_y(x_0 + t\Delta x, y_0 + t_1\Delta y)(t\Delta y) \\ f(x_0 + t\Delta x, y_0) - f(x_0, y_0) &= f_x(x_0 + t_2\Delta x, y_0)(t\Delta x) \end{aligned}$$

for some numbers t_1 and t_2 between 0 and t. Hence, at $t = 1$, we have

$$\begin{aligned} &f(x_0 + \Delta x, y_0 + \Delta y) - f(x_0, y_0) - f_x(x_0, y_0)\Delta x - f_y(x_0, y_0)\Delta y \\ &\quad = \left(f_x(x_0 + t_2\Delta x, y_0) - f_x(x_0, y_0) \right) \Delta x \\ &\quad + \left(f_y(x_0 + t\Delta x, y_0 + t_1\Delta y) - f_y(x_0, y_0) \right) \Delta y \end{aligned}$$

where t_1 and t_2 are between 0 and 1. Let

$$E(x, y, x_0, y_0) = f(x_0 + \Delta x, y_0 + \Delta y) - f(x_0, y_0) - f_x(x_0, y_0)\Delta x - f_y(x_0, y_0)\Delta y$$

like usual. Then we have

$$\begin{aligned} E(x, y, x_0, y_0) = &\left(f_x(x_0 + t_2\Delta x, y_0) - f_x(x_0, y_0) \right) \Delta x \\ &+ \left(f_y(x_0 + t\Delta x, y_0 + t_1\Delta y) - f_y(x_0, y_0) \right) \Delta y \end{aligned}$$

We know as $(\Delta x, \Delta y) \rightarrow (0, 0)$, the numbers (t_1, t_2) we found using the Mean Value Theorem will also go to $(0, 0)$ and so $(x_0 + t_2\Delta x, y_0) \rightarrow (x_0, y_0)$ and $(x_0 + \Delta x, y_0 + t_1\Delta y) \rightarrow (x_0, y_0)$. Then the continuity of f_x and f_y at (x_0, y_0) tells us

$$\begin{aligned} &\left(f_x(x_0 + t_2\Delta x, y_0) - f_x(x_0, y_0) \right) \Delta x \\ &\quad + \left(f_y(x_0 + t\Delta x, y_0 + t_1\Delta y) - f_y(x_0, y_0) \right) \Delta y \rightarrow 0. \end{aligned}$$

We conclude $E(x, y, x_0, y_0) \rightarrow 0$ as $(\Delta x, \Delta y) \rightarrow 0$. Further,

$$\frac{E(x, y, x_0, y_0)}{\sqrt{(\Delta x)^2 + (\Delta y)^2}} = \left(f_x(x_0 + t_2 \Delta x, y_0) - f_x(x_0, y_0) \right) \frac{\Delta x}{\sqrt{(\Delta x)^2 + (\Delta y)^2}}$$
$$+ \left(f_y(x_0 + t \Delta x, y_0 + t_1 \Delta y) - f_y(x_0, y_0) \right) \frac{\Delta y}{\sqrt{(\Delta x)^2 + (\Delta y)^2}}$$

But the terms $|\Delta x|/\sqrt{(\Delta x)^2 + (\Delta y)^2} \leq 1$ and $|\Delta y|/\sqrt{(\Delta x)^2 + (\Delta y)^2} \leq 1$ and so as $(\Delta x, \Delta y) \to (0, 0)$, we must have $E(x, y, x_0, y_0)/\sqrt{(\Delta x)^2 + (\Delta y)^2} \to 0$ as well. These two limits show that f is differentiable at (x_0, y_0). We can state this as a theorem. We use this idea a lot in two dimensional calculus.

Theorem 4.7.2 (Continuous Partials Imply Differentiability)
Assume the partials of $f(x, y)$ exist at (x_0, y_0) and that f is defined locally around (x_0, y_0). Further, assume the partials are continuous locally at (x_0, y_0). Then f is differentiable at (x_0, y_0).

Now let's go back to the old idea of a tangent plane to a surface. For the surface $z = f(x, y)$ if its partials are continuous functions (they usually are for our work!) then f is differentiable and hence we know that

$$f(x, y) = f(x_0, y_0) + f_x(x_0, y_0)(x - x_0) + f_y(x_0, y_0)(y - y_0) + E(x, y, x_0, y_0)$$

and $E(x, y, x_0, y_0) \to 0$ and $E(x, y, x_0, y_0)/\sqrt{(x - x_0)^2 + (y - y_0)^2} \to 0$ as $(x, y) \to (x_0, y_0)$.

4.8 Second Order Error Estimates

We can characterize the error must better if we have access to what are called the second order partial derivatives of f. Roughly speaking, we take the partials of f_x and f_y to obtain the second order terms. We can make this discussion brief. Assuming f is defined locally as usual near (x_0, y_0), we can ask about the partial derivatives of the functions f_x and f_y with respect to x and y also. We define the second order partials of f as follows.

Definition 4.8.1 (*Second Order Partials*)
If $f(x, y)$, f_x and f_y are defined locally at (x_0, y_0), we can attempt to find following limits:

$$\lim_{x \to x_0, y = y_0} \frac{f_x(x, y) - f_x(x, y_0)}{x - x_0} = \partial_x(f_x)$$
$$\lim_{x = x_0, y \to y_0} \frac{f_x(x, y) - f_x(x_0, y)}{y - y_0} = \partial_y(f_x)$$

$$\lim_{x \to x_0, y = y_0} \frac{f_y(x, y) - f_y(x, y_0)}{x - x_0} = \partial_x(f_y)$$
$$\lim_{x = x_0, y \to y_0} \frac{f_y(x, y) - f_y(x_0, y)}{y - y_0} = \partial_y(f_y)$$

Comment 4.8.1 *When these second order partials exist at (x_0, y_0), we use the following notations interchangeably: $f_{xx} = \partial_x(f_x)$, $f_{xy} = \partial_y(f_x)$, $f_{yx} = \partial_y(f_x)$ and $f_{yy} = \partial_y(f_y)$.*

The second order partials are often organized into a matrix called the **Hessian**.

Definition 4.8.2 (*The Hessian*)
If $f(x, y)$, f_x and f_y are defined locally at (x_0, y_0), if the second order partials exist at (x_0, y_0), we define the Hessian, $\boldsymbol{H}(x_0, y_0)$ at (x_0, y_0) to be the matrix

$$\boldsymbol{H}(x_0, y_0) = \begin{bmatrix} f_{xx}(x_0, y_0) & f_{xy}(x_0, y_0) \\ f_{yx}(x_0, y_0) & f_{yy}(x_0, y_0) \end{bmatrix}$$

Comment 4.8.2 *It is possible to prove that if the first order partials are continuous locally near (x_0, y_0) then the mixed order partials f_{xy} and f_{yx} must match at the point (x_0, y_0). Most of our surfaces have this property. Hence, for these* **smooth** *surfaces, the Hessian is a symmetric matrix!*

Example 4.8.1 Let $f(x, y) = 2x - 8xy$. Find the first and second order partials of f and its Hessian.

Solution *The partials are*

$$\begin{aligned} f_x(x, y) &= 2 - 8y \\ f_y(x, y) &= -8x \\ f_{xx}(x, y) &= 0 \\ f_{xy}(x, y) &= -8 \\ f_{yx}(x, y) &= -8 \\ f_{yy}(x, y) &= 0. \end{aligned}$$

and so the Hessian is

$$\boldsymbol{H}(x, y) = \begin{bmatrix} f_{xx}(x, y) & f_{xy}(x, y) \\ f_{yx}(x, y) & f_{yy}(x, y) \end{bmatrix} = \begin{bmatrix} 0 & -8 \\ -8 & 0 \end{bmatrix}$$

4.8.1 Homework

Exercise 4.8.1 *Let* $f(x, y) = 5x - 2xy$. *Find the first and second order partials of* f *and its Hessian.*

Exercise 4.8.2 *Let* $f(x, y) = -8y + 9xy - 2y^2$. *Find the first and second order partials of* f *and its Hessian.*

Exercise 4.8.3 *Let* $f(x, y) = 4x - 6xy - x^2$. *Find the first and second order partials of* f *and its Hessian.*

Exercise 4.8.4 *Let* $f(x, y) = 4x^2 - 6xy - x^2$. *Find the first and second order partials of* f *and its Hessian.*

4.8.2 Hessian Approximations

We can now explain the most common approximation result for tangent planes. Let

$$h(t) = f(x_0 + t\Delta x, y_0 + t\Delta y)$$

as usual. Then we know we can write

$$h(t) = h(0) + h'(0)t + h''(c)\frac{t^2}{2}.$$

Using the chain rule, we find

$$h'(t) = f_x(x_0 + t\Delta x, y_0 + t\Delta y)\Delta x + f_y(x_0 + t\Delta x, y_0 + t\Delta y)\Delta y$$

and

$$\begin{aligned} h''(t) &= \partial_x \Big(f_x(x_0 + t\Delta x, y_0 + t\Delta y)\Delta x + f_y(x_0 + t\Delta x, y_0 + t\Delta y)\Delta y \Big) \Delta x \\ &\quad + \partial_y \Big(f_x(x_0 + t\Delta x, y_0 + t\Delta y)\Delta x + f_y(x_0 + t\Delta x, y_0 + t\Delta y)\Delta y \Big) \Delta y \\ &= f_{xx}(x_0 + t\Delta x, y_0 + t\Delta y)(\Delta x)^2 + f_{yx}(x_0 + t\Delta x, y_0 + t\Delta y)(\Delta y)(\Delta x) \\ &\quad + f_{xy}(x_0 + t\Delta x, y_0 + t\Delta y)(\Delta x)(\Delta y) + f_{yy}(x_0 + t\Delta x, y_0 + t\Delta y)(\Delta y)^2 \end{aligned}$$

We can rewrite this in matrix–vector form as

$$h''(t) = \begin{bmatrix} \Delta x & \Delta y \end{bmatrix} \begin{bmatrix} f_{xx}(x_0 + t\Delta x, y_0 + t\Delta y) & f_{yx}(x_0 + t\Delta x, y_0 + t\Delta y) \\ f_{xy}(x_0 + t\Delta x, y_0 + t\Delta y) & f_{yy}(x_0 + t\Delta x, y_0 + t\Delta y) \end{bmatrix} \begin{bmatrix} \Delta x \\ \Delta y \end{bmatrix}$$

Of course, using the definition of $\boldsymbol{H}$, this can be rewritten as

$$h''(t) = \begin{bmatrix} \Delta x \\ \Delta y \end{bmatrix}^T \boldsymbol{H}(x_0 + t\Delta x, y_0 + t\Delta y) \begin{bmatrix} \Delta x \\ \Delta y \end{bmatrix}$$

Thus, our tangent plane approximation can be written as

$$h(1) = h(0) + h'(0)(1 - 0) + h''(c)\frac{1}{2}$$

for some c between 0 and 1. Substituting for the h terms, we find

$$\begin{aligned} f(x_0 + \Delta x, y_0 + \Delta y) &= f(x_0, y_0) + f_x(x_0, y_0)\Delta x + f_y(x_0, y_0)\Delta y \\ &\quad + \frac{1}{2}\begin{bmatrix} \Delta x \\ \Delta y \end{bmatrix}^T \boldsymbol{H}(x_0 + c\Delta x, y_0 + c\Delta y) \begin{bmatrix} \Delta x \\ \Delta y \end{bmatrix} \end{aligned}$$

Clearly, we have shown how to express the error in terms of second order partials. There is a point c between 0 and 1 so that

$$E(x_0, y_0, \Delta x, \Delta y) = \frac{1}{2}\begin{bmatrix} \Delta x \\ \Delta y \end{bmatrix}^T \boldsymbol{H}(x_0 + c\Delta x, y_0 + c\Delta y) \begin{bmatrix} \Delta x \\ \Delta y \end{bmatrix}$$

Note the error is a quadratic expression in terms of the Δx and Δy.

4.9 Extrema Ideas

To understand how to think about finding places where the minimum and maximum of a function to two variables might occur, all you have to do is realize it is a common sense thing. We already know that the tangent plane attached to the surface which represents our function of two variables is a way to approximate the function near the point of attachment. We have seen in our pictures what happens when the tangent plane is **flat**. This flatness occurs at the minimum and maximum of the function. It also occurs in other situations, but we will leave that more complicated event for other courses. The functions we want to deal with are quite nice and have great minima and maxima. However, we do want you to know there are more things in the world and we will touch on them only briefly.

To see what to do, just recall the equation of the tangent plane error to our function of two variables $f(x, y)$.

$$\begin{aligned} f(x, y) &= f(x_0, y_0) + \nabla(f)(x_0, y_0)[x - x_0, y - y_0]^T \\ &+ (1/2)[x - x_0, y - y_0]H(x_0 + c(x - x_0), y_0 + c(y - y_0))[x - x_0, y - y_0]^T \end{aligned}$$

where c is some number between 0 and 1 that is different for each x. We also know that the equation of the tangent plane to $f(x, y)$ at the point (x_0, y_0) is

$$f(x, y) = f(x_0, y_0) + < \nabla f, \boldsymbol{X} - \boldsymbol{X_0} > .$$

Now let's assume the tangent plane is flat at (x_0, y_0). Then the gradient ∇f is the zero vector and we have $\frac{\partial f}{\partial x}(x_0, y_0) = 0$ and $\frac{\partial f}{\partial y}(x_0, y_0) = 0$. So the tangent plane error equation simplifies to

$$\begin{aligned} f(x, y) = f(x_0, y_0) &+ (1/2)[x - x_0, y - y_0] \\ &\times H(x_0 + c(x - x_0), y_0 + c(y - y_0))[x - x_0, y - y_0]^T \end{aligned}$$

Now let's simplify this. The Hessian is just a 2×2 matrix whose components are the second order partials of f. Let

$$\begin{aligned} A(c) &= \frac{\partial^2 f}{\partial x^2}(x_0 + c(x - x_0), y_0 + c(y - y_0)) \\ B(c) &= \frac{\partial^2 f}{\partial x \, \partial y}(x_0 + c(x - x_0), y_0 + c(y - y_0)) \\ &= \frac{\partial^2 f}{\partial y \, \partial x}(x_0 + c(x - x_0), y_0 + c(y - y_0)) \\ D(c) &= \frac{\partial^2 f}{\partial y^2}(x_0 + c(x - x_0), y_0 + c(y - y_0)) \end{aligned}$$

Then, we have

$$f(x, y) = f(x_0, y_0) + (1/2) \begin{bmatrix} x - x_0 & y - y_0 \end{bmatrix} \begin{bmatrix} A(c) & B(c) \\ B(c) & D(c) \end{bmatrix} \begin{bmatrix} x - x_0 \\ y - y_0 \end{bmatrix}$$

We can multiply this out (a nice simple pencil and paper exercise!) to find

$$f(x, y) = f(x_0, y_0) + 1/2\ \big(A(c)(x - x_0)^2 \\ + 2B(c)(x - x_0)\ (y - y_0) + D(c)(y - y_0)^2\big)$$

Now it is time to remember an old technique from high school—completing the square. Remember if we had a quadratic like $u^2 + 3uv + 6v^2$, to complete the square we take half of the number in front of the mixed term uv and square it and add and subtract it times v^2 as follows.

$$u^2 + 3uv + 6v^2 = u^2 + 3uv + (3/2)^2 v^2 - (3/2)^2 v^2 + 6v^2.$$

Now group the first three terms together and combine the last two terms into one term.

$$u^2 + 3uv + 6v^2 = \left(u^2 + 3uv + (3/2)^2 v^2\right) + \left(6 - (3/2)^2\right) v^2.$$

The first three terms are a *perfect square*, $(u + (3/2)v)^2$. Simplifying, we find

$$u^2 + 3uv + 6v^2 = \left(u + (3/2)v\right)^2 + (135/4)\ v^2.$$

This is called *completely the square*! Now let's do this with the Hessian quadratic we have. First, factor our the $A(c)$. We will assume it is not zero so the divisions are fine to do. Also, for convenience, we will replace $x - x_0$ by Δx and $y - y_0$ by Δy. This gives

$$f(x, y) = f(x_0, y_0) + \frac{A(c)}{2}\left((\Delta x)^2 + 2\frac{B(c)}{A(c)}\Delta x\ \Delta y + \frac{D(c)}{A(c)}(\Delta y)^2\right).$$

One half of the $\Delta x \Delta y$ coefficient is $\frac{B(c)}{A(c)}$ so add and subtract $(B(c)/A(c))^2(\Delta y)^2$. We find

$$f(x, y) = f(x_0, y_0) + \frac{A(c)}{2}\left((\Delta x)^2 + 2\frac{B(c)}{A(c)}\Delta x\ \Delta y \right. \\ \left. + \left(\frac{B(c)}{A(c)}\right)^2 (\Delta y)^2 - \left(\frac{B(c)}{A(c)}\right)^2 (\Delta y)^2 + \frac{D(c)}{A(c)}(\Delta y)^2\right).$$

Now group the first three terms together—the perfect square and combine the last two terms into one. We have

$$f(x,y) = f(x_0,y_0) \\ + \frac{A(c)}{2}\left(\left(\Delta x + \frac{B(c)}{A(c)}\,\Delta y\right)^2 + \left(\frac{A(c)\,D(c) - (B(c))^2}{(A(c))^2}\right)(\Delta y)^2\right).$$

Now we need this common sense result which says that if a function g is continuous at a point (x_0, y_0) and positive or negative , then it is positive or negative in a circle of radius r centered at (x_0, y_0). Here is the formal statement.

Theorem 4.9.1 (Nonzero Values and Continuity)
If $f(x_0, y_0)$ is a place where the function is positive or negative in value, then there is a radius r so that $f(x, y)$ is positive or negative in a circle of radius r around the center (x_0, y_0).

Proof This can be argued carefully using limits.

- If $f(x_0, y_0) > 0$ and f is continuous at (x_0, y_0), then if no matter now close we were to (x_0, y_0), we could find a point (x_r, y_r) where $f(x_r, y_r) = 0$, then that set of points would define a path to (x_0, y_0) and the limiting value of f on that path would be 0.
- But we know the value at (x_0, y_0) is positive and we know f is continuous there. Hence, the limiting values for all paths should match.
- So we can't find such points for all values of r. We see there will be a first r where we can't do this and so inside the circle determined by that r, f will be nonzero.
- You might think we haven't ruled out the possibility that f could be negative at some points. But the only way a continuous f could switch between positive and negative is to pass through zero. And we have already ruled that out. So f is positive inside this circle of radius r. □

Now getting back to our problem. We have at this point where the partials are zero, the following expansion

$$f(x,y) = f(x_0,y_0) + \frac{A(c)}{2}\left(\left((\Delta x)^2 + 2\frac{B(c)}{A(c)}\Delta x\,\Delta y + \left(\frac{B(c)}{A(c)}\right)^2(\Delta y)^2\right) \\ + \left(\frac{A(c)\,D(c) - (B(c))^2}{(A(c))^2}\right)(\Delta y)^2\right).$$

The algebraic sign of the terms after the function value $f(x_0, y_0)$ are completely determined by the terms which are not squared. We have two simple cases:

- $A(c) > 0$ and $A(c)\,D(c) - (B(c))^2 > 0$ which implies the term after $f(x_0, y_0)$ is positive.
- $A(c) < 0$ and $A(c)\,D(c) - (B(c))^2 > 0$ which implies the term after $f(x_0, y_0)$ is negative.

Now let's assume all the second order partials are continuous at (x_0, y_0). We know $A(c) = \frac{\partial^2 f}{\partial x^2}(x_0 + c(x - x_0), y_0 + c(y - y_0))$ and from Theorem 4.9.1, if $\frac{\partial^2 f}{\partial x^2}(x_0, y_0) > 0$, then so is $A(c)$ in a circle around (x_0, y_0). The other term $A(c)\,D(c) - (B(c))^2 > 0$ will also be positive is a circle around (x_0, y_0) as long as $\frac{\partial^2 f}{\partial x^2}(x_0, y_0)\,\frac{\partial^2 f}{\partial y^2}(x_0, y_0) - \frac{\partial^2 f}{\partial x \partial y}(x_0, y_0) > 0$. We can say similar things about the negative case. Now to save typing let $\frac{\partial^2 f}{\partial x^2}(x_0, y_0) = f^0_{xx}$, $\frac{\partial^2 f}{\partial y^2}(x_0, y_0) = f^0_{yy}$ and $\frac{\partial^2 f}{\partial x \partial y}(x_0, y_0) = f^0_{xy}$. So we can restate our two cases as

- $f^0_{xx} > 0$ and $f^0_{xx}\, f^0_{yy} - (f^0_{xy})^2 > 0$ which implies the term after $f(x_0, y_0)$ is positive. This implies that $f(x, y) > f(x_0, y_0)$ in a circle of some radius r which says $f(x_0, y_0)$ is a minimum value of the function locally at that point.
- $f^0_{xx} < 0$ and $f^0_{xx}\, f^0_{yy} - (f^0_{xy})^2 > 0$ which implies the term after $f(x_0, y_0)$ is negative. This implies that $f(x, y) < f(x_0, y_0)$ in a circle of some radius r which says $f(x_0, y_0)$ is a maximum value of the function locally at that point.

where, for convenience, we use a superscript 0 to denote we are evaluating the partials at (x_0, y_0). So we have come up with a great condition to verify if a place where the partials are zero is a minimum or a maximum. If you think about it a bit, you'll notice we left out the case where $f^0_{xx}\, f^0_{yy} - (f^0_{xy})^2 < 0$ which is important but we will not do that in this class. That is for later courses to pick up, however it is the test for the analog of the behavior we see in the cubic $y = x^3$. The derivative is 0 but there is neither a minimum or maximum at $x = 0$. In two dimensions, the situation is more interesting of course. This kind of behavior is called a **saddle**. We have another Theorem!

Theorem 4.9.2 (Extrema Test)
If the partials of f are zero at the point (x_0, y_0), we can determine if that point is a local minimum or local maximum of f using a second order test. We must assume the second order partials are continuous at the point (x_0, y_0).

- *If $f^0_{xx} > 0$ and $f^0_{xx}\, f^0_{yy} - (f^0_{xy})^2 > 0$ then $f(x_0, y_0)$ is a local minimum.*
- *$f^0_{xx} < 0$ and $f^0_{xx}\, f^0_{yy} - (f^0_{xy})^2 > 0$ then $f(x_0, y_0)$ is a local maximum.*

We just don't know anything if the test $f^0_{xx}\, f^0_{yy} - (f^0_{xy})^2 = 0$. If the test gives $f^0_{xx}\, f^0_{yy} - (f^0_{xy})^2 < 0$, we have a saddle.

Proof We have sketched out the reasons for this above. □

Now the second order test fails if $\det(\boldsymbol{H}(x_0, y_0)) = 0$ at the critical point as a few examples show. First, the function $f(x, y) = x^4 + y^4$ has a global minimum at $(0, 0)$ but at that point

$$H(x, y) = \begin{bmatrix} 12x^2 & 0 \\ 0 & 12y^2 \end{bmatrix} \Longrightarrow \det(H(x_0, y_0)) = 144x^2y^2.$$

and hence, $\det(H(x_0, y_0)) = 0$. Secondly, the function $f(x, y) = -x^4 - y^4$ has a global maximum at (0, 0) but at that point

$$H(x, y) = \begin{bmatrix} -12x^2 & 0 \\ 0 & -12y^2 \end{bmatrix} \Longrightarrow \det(H(x_0, y_0)) = 144x^2y^2.$$

and hence, $\det(H(x_0, y_0)) = 0$ as well. Finally, $f(x, y) = x^4 - y^4$ has a saddle at (0, 0) but at that point

$$H(x, y) = \begin{bmatrix} 12x^2 & 0 \\ 0 & -12y^2 \end{bmatrix} \Longrightarrow \det(H(x_0, y_0)) = -144x^2y^2.$$

and hence, $\det(H(x_0, y_0)) = 0$ again. So if the $\det(H(x_0, y_0)) = 0$, we just don't know what the behavior is.

Let's finish this section with a more careful discussion of the idea of a saddle. Recall at a critical point (x_0, y_0), we found that

$$f(x, y) = f(x_0, y_0) + \frac{A(c)}{2}\left(\left(\Delta x + \frac{B(c)}{A(c)}\Delta y\right)^2 + \left(\frac{A(c)\,D(c) - (B(c))^2}{(A(c))^2}\right)(\Delta y)^2\right).$$

Now suppose we knew $A(c)\,D(c) - (B(c))^2 < 0$. Then, using the usual continuity argument, we know that there is a circle around the critical point (x_0, y_0) so that $A(c)\,D(c) - (B(c))^2 < 0$ when $c = 0$. This is the same as saying $\det(H(x_0, y_0)) < 0$. But notice that on the line going through the critical point having $\Delta y = 0$, this gives

$$f(x, y) = f(x_0, y_0) + \frac{A(c)}{2}\left(\Delta x\right)^2.$$

and on the line through the critical point with $\Delta x + \frac{B(c)}{A(c)}\Delta y = 0$. we have

$$f(x, y) = f(x_0, y_0) + \frac{A(c)}{2}\left(\frac{A(c)\,D(c) - (B(c))^2}{(A(c))^2}\right)(\Delta y)^2$$

Now, if $A(c) > 0$, the first case gives $f(x, y) = f(x_0, y_0) +$ **a positive number** showing f has a minimum on that trace. However, the second case gives $f(x, y) = f(x_0, y_0) -$ **a positive number** which shows f has a maximum on that trace. The fact that f is minimized in one direction and maximized in another direction gives rise to the expression that we consider f to behave like a saddle at this critical point. The analysis is virtually the same if $A(c) < 0$, except the first trace has the maximum and the second trace has the minimum. Hence, the test for a saddle point is to see if $\det(\boldsymbol{H}(x_0, y_0)) < 0$ as we stated in Theorem 4.9.2.

4.9.1 Examples

Example 4.9.1 Use our tests to show $f(x, y) = x^2 + 3y^2$ has a minimum at $(0, 0)$.

Solution *The partials here are $f_x = 2x$ and $f_y = 6y$. These are zero at $x = 0$ and $y = 0$. The Hessian at this critical point is*

$$H(x, y) = \begin{bmatrix} 2 & 0 \\ 0 & 6 \end{bmatrix} = H(0, 0).$$

as H is constant here. Our second order test says the point $(0, 0)$ corresponds to a minimum because $f_{xx}(0, 0) = 2 > 0$ and $f_{xx}(0, 0)\, f_{yy}(0, 0) - (f_{xy}(0, 0))^2 = 12 > 0$.

Example 4.9.2 Use our tests to show $f(x, y) = x^2 + 6xy + 3y^2$ has a saddle at $(0, 0)$.

Solution *The partials here are $f_x = 2x + 6y$ and $f_y = 6x + 6y$. These are zero at when*

$$\begin{aligned} 2x + 6y &= 0 \\ 6x + 6y &= 0 \end{aligned}$$

which has solution $x = 0$ and $y = 0$. The Hessian at this critical point is

$$H(x, y) = \begin{bmatrix} 2 & 6 \\ 6 & 6 \end{bmatrix} = H(0, 0).$$

as H is again constant here. Our second order test says the point $(0, 0)$ corresponds to a saddle because $f_{xx}(0, 0) = 2 > 0$ and $f_{xx}(0, 0)\, f_{yy}(0, 0) - (f_{xy}(0, 0))^2 = 12 - 36 < 0$.

Example 4.9.3 Show our tests fail on $f(x, y) = 2x^4 + 4y^6$ even though we know there is a minimum value at $(0, 0)$.

Solution *For $f(x, y) = 2x^4 + 4y^6$, you find that the critical point is $(0, 0)$ and all the second order partials are 0 there. So all the tests fail. Of course, a little common sense tells you $(0, 0)$ is indeed the place where this function has a minimum value. Just think about how it's surface looks. But the tests just fail. This is much like the curve $f(x) = x^4$ which has a minimum at $x = 0$ but all the tests fail on it also.*

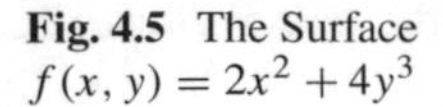

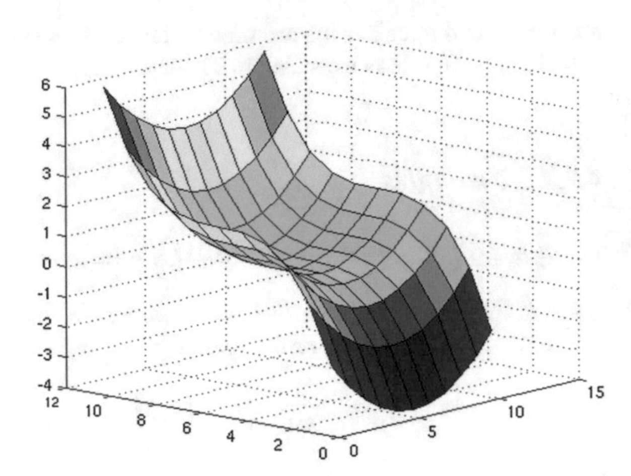

Fig. 4.5 The Surface $f(x, y) = 2x^2 + 4y^3$

Example 4.9.4 Show our tests fail on $f(x, y) = 2x^2 + 4y^3$ and the surface does not have a minimum or maximum at the critical point $(0, 0)$.

Solution *For* $f(x, y) = 2x^2 + 4y^3$, *the critical point is again* $(0, 0)$ *and* $f_{xx}(0, 0) = 4$, $f_{yy}(0, 0) = 0$ *and* $f_{xy}(0, 0) = f_{yx}(0, 0) = 0$. *So* $f_{xx}(0, 0)\ f_{yy}(0, 0) - (f_{xy}(0, 0))^2 = 0$ *so the test fails. Note the* $x = 0$ *trace is* $4y^3$ *which is a cubic and so is negative below* $y = 0$ *and positive above* $y = 0$. *Not much like a minimum or maximum behavior on this trace! But the trace for* $y = 0$ *is* $2x^2$ *which is a nice parabola which does reach its minimum at* $x = 0$. *So the behavior of the surface around* $(0, 0)$ *is not a maximum or a minimum. The surface acts a lot like a cubic. Do this in MatLab.*

Listing 4.8: The surface $f(x, y) = 2x^2 + 4y^3$

```
[X,Y] = meshgrid(-1:.2:1);
Z = 2*X.^2 + 4*Y.^3;
surf(Z);
```

This will give you a surface. In the plot that is shown go to the tool menu and click of the rotate 3D option and you can spin it around. Clearly like a cubic! You can see the plot in Fig. 4.5.

4.9.2 Homework

Exercise 4.9.1 *Use our tests to show* $f(x, y) = 4x^2 + 2y^2$ *has a minimum at* $(0, 0)$. *Feel free to draw a surface plot to help you see what is going on.*

Exercise 4.9.2 *Use our tests to find where* $f(x, y) = 2x^2 + 3x + 3y^2 + 8y$ *has a minimum. Feel free to draw a surface plot to help you see what is going on.*

Exercise 4.9.3 *Use our tests to find where* $f(x, y) = 100 - 2x^2 + 3x - 3y^2 + 8y$ *has a maximum. Feel free to draw a surface plot to help you see what is going on.*

Exercise 4.9.4 *Use our tests to find where* $f(x, y) = 2x^2 + x + 8 + 4y^2 + 8y + 20$ *has a minimum. Feel free to draw a surface plot to help you see what is going on.*

Exercise 4.9.5 *Show our tests fail on* $f(x, y) = 6x^4 + 8y^8$ *even though we know there is a minimum value at* $(0, 0)$. *Feel free to draw a surface plot to help you see what is going on.*

Exercise 4.9.6 *Show our tests fail on* $f(x, y) = 10x^2 + 5y^5$ *and the surface does not have a minimum or maximum at the critical point* $(0, 0)$. *Feel free to draw a surface plot to help you see what is going on.*

Part III
The Main Event

Chapter 5
Integration

To help us with our modeling tasks, we need to explore two new ways to compute antiderivatives and definite integrals. These methods are called *Integration By Parts* and *Partial Fraction Decompositions*. You should also recall our discussions about antiderivatives or primitives and Riemann integration from Peterson (2015) where we go over topics such as how we define the Riemann Integral, the Fundamental Theorem Calculus and the use of the Cauchy Fundamental Theorem Calculus. you should also review the basic ideas of continuity and differentiability.

5.1 Integration by Parts

This technique is based on the product rule for differentiation. Let's assume that the functions f and g are both differentiable on the finite interval $[a, b]$. Then the product fg is also differentiable on $[a, b]$ and

$$(f(t)\,g(t))' = f'(t)\,g(t) + f(t)\,g'(t)$$

Now, we know the antiderivative of $(fg)'$ is $fg + C$, where for convenience of notation, we don't write the usual (t) in each term. So, if we compute the definite integral of both sides of this equation on $[a, b]$, we find

$$\int_a^b (f(t)\,g(t))'\,dt = \int_a^b f'(t)\,g(t)\,dt + \int_a^b f(t)\,g'(t)\,dt$$

The left hand side is simply $(fg)\mid_a^b = f(b)g(b) - f(a)g(a)$. Hence,

$$\int_a^b f'(t)\,g(t)\,dt + \int_a^b f(t)\,g'(t)\,dt = (fg)\mid_a^b$$

J.K. Peterson, *Calculus for Cognitive Scientists*,
Cognitive Science and Technology, DOI 10.1007/978-981-287-877-9_5

This is traditionally written as

$$\int_a^b f(t)\, g'(t)\, dt = (fg)\,|_a^b - \int_a^b g(t)\, f'(t)\, dt \tag{5.1}$$

We usually write this in an even more abbreviated form. If we let $u(t) = f(t)$, then $du = f'(t)\, dt$. Also, if $v(t) = g(t)$, then $dv = g'(t)\, dt$. Then, we can rephrase Eq. 5.1 as Eq. 5.2.

$$\int_a^b u\, dv = uv\,|_a^b - \int_a^b v\, du \tag{5.2}$$

We can also develop the integration by parts formula as an indefinite integral. When we do that we obtain the version in Eq. 5.3

$$\int u\, dv = uv - \int_a^b v\, du + C \tag{5.3}$$

Equation 5.2 gives what is commonly called the *Integration By Parts* formula.

5.1.1 How Do We Use Integration by Parts?

Let's work through some problems carefully. As usual, we will give many details at first and gradually do the problems faster with less written down. You need to work hard at understanding this technique.

Example 5.1.1 Evaluate $\int \ln(t)\, dt$.

Solution *Let $u(t) = \ln(t)$ and $dv = dt$. Then $du = \frac{1}{t}\, dt$ and $v = \int dt = t$. When we find the antiderivative v, at this stage we don't need to carry around an arbitrary constant C as we will add one at the end. Applying Integration by Parts, we have*

$$\begin{aligned}
\int \ln(t)\, dt &= \int u dv \\
&= uv - \int v du \\
&= \ln(t)\, t - \int t\, \frac{1}{t}\, dt \\
&= \ln(t)\, t - \int dt \\
&= t\, \ln(t) - t + C
\end{aligned}$$

Example 5.1.2 Evaluate $\int t\, \ln(t)\, dt$.

Solution *Let* $u(t) = \ln(t)$ *and* $dv = tdt$. *Then* $du = \frac{1}{t}\, dt$ *and* $v = \int tdt = t^2/2$. *Applying Integration by Parts, we have*

$$\begin{aligned}\int t\ \ln(t)\, dt &= \int udv \\ &= uv - \int vdu \\ &= \ln(t)\, t^2/2 - \int t^2/2\, \frac{1}{t}\, dt \\ &= \frac{t^2}{2}\ \ln(t) - \int t/2\, dt \\ &= \frac{t^2}{2}\ \ln(t) - \frac{t^2}{4} + C\end{aligned}$$

Example 5.1.3 Evaluate $\int t^3\ \ln(t)\, dt$.

Solution *Let* $u(t) = \ln(t)$ *and* $dv = t^3 dt$. *Then* $du = \frac{1}{t}\, dt$ *and* $v = \int t^3 dt = t^4/4$. *Applying Integration by Parts, we have*

$$\begin{aligned}\int t^3\ \ln(t)\, dt &= \int udv \\ &= uv - \int vdu \\ &= \ln(t)\, t^4/4 - \int t^4/4\, \frac{1}{t}\, dt \\ &= \frac{t^4}{4}\ \ln(t) - \int t^3/4\, dt \\ &= \frac{t^2}{2}\ \ln(t) - \frac{t^4}{16} + C\end{aligned}$$

Example 5.1.4 Evaluate $\int t\ e^t\ dt$.

Solution *Let* $u(t) = t$ *and* $dv = e^t dt$. *Then* $du = dt$ *and* $v = \int e^t dt = e^t$. *Applying Integration by Parts, we have*

$$\begin{aligned}\int t\ e^t\ dt &= \int udv \\ &= uv - \int vdu \\ &= e^t\ t - \int e^t\ dt \\ &= t\ e^t - \int e^t\ dt \\ &= t\ e^t - e^t + C\end{aligned}$$

Example 5.1.5 Evaluate $\int t^2 e^t \, dt$.

Solution *Let* $u(t) = t^2$ *and* $dv = e^t dt$. *Then* $du = 2t dt$ *and* $v = \int e^t dt = e^t$. *Applying Integration by Parts, we have*

$$\begin{aligned}\int t^2 e^t \, dt &= \int u dv \\ &= uv - \int v du \\ &= e^t t^2 - \int e^t \, 2t \, dt \\ &= t^2 e^t - \int 2t \, e^t \, dt\end{aligned}$$

Now the integral $\int 2t \, e^t \, dt$ *also requires the use of integration by parts. So we integrate again using this technique Let* $u(t) = 2t$ *and* $dv = e^t dt$. *Then* $du = 2dt$ *and* $v = \int e^t dt = e^t$. *Applying Integration by Parts again, we have*

$$\begin{aligned}\int 2t \, e^t \, dt &= \int u dv \\ &= uv - \int v du \\ &= e^t \, 2t - \int e^t \, 2 \, dt \\ &= 2t \, e^t - \int 2 \, e^t \, dt \\ &= 2t \, e^t - 2 \, e^t + C\end{aligned}$$

It is very awkward to do these multiple integration by parts in two separate steps like we just did. It is much more convenient to repackage the computation like this:

$$\begin{aligned}&\int t^2 e^t \, dt && = uv - \int v du \\ &\boxed{\begin{array}{ll} u = t^2 & dv = e^t dt \\ du = 2t dt & v = e^t \end{array}} && = e^t t^2 - \int e^t \, 2t \, dt \\ & && = e^t t^2 - \int e^t \, 2t \, dt \\ & && = t^2 e^t - \int 2t \, e^t \, dt \\ & && = \quad \boxed{\begin{array}{ll} u = 2t & dv = e^t dt \\ du = 2dt & v = e^t \end{array}} \\ & && = t^2 e^t - \left\{ e^t \, 2t - \int e^t \, 2 \, dt \right\} \\ & && = t^2 e^t - \left\{ 2t \, e^t - 2 \, e^t \right\} + C\end{aligned}$$

The framed boxes are convenient for our explanation, but this is still a bit awkward (and long!) to write out for our problem solution. So let's try this:

$$\int t^2\, e^t\, dt = e^t\, t^2 - \int e^t\, 2t\, dt$$

$$\boxed{u = t^2;\ dv = e^t dt;\ du = 2t dt;\ v = e^t}$$

$$\begin{aligned}
&= e^t\, t^2 - \int e^t\, 2t\, dt \\
&= t^2\, e^t - \int 2t\, e^t\, dt
\end{aligned}$$

$$\boxed{u = 2t;\ dv = e^t dt;\ du = 2dt\ v = e^t}$$

$$\begin{aligned}
&= t^2\, e^t - \left\{ e^t\, 2t - \int e^t\, 2\, dt \right\} \\
&= t^2\, e^t - \{2t\, e^t - 2\, e^t\} + C \\
&= t^2\, e^t - 2t\, e^t + 2\, e^t + C
\end{aligned}$$

Example 5.1.6 Evaluate $\int t^2 \sin(t)\, dt$.

Solution *We will do this one the short way:*

$$\int t^2\ \sin(t)\, dt = -t^2\ \cos(t) - \int -\cos(t)\, 2t\, dt$$

$$\boxed{u = t^2;\ dv = \sin(t) dt;\ du = 2t dt;\ v = -\cos(t)}$$

$$= -t^2\ \cos(t) + \int 2t\ \cos(t)\, dt$$

$$\boxed{u = 2t;\ dv = \cos(t) dt;\ du = 2dt\ v = \sin(t)}$$

$$\begin{aligned}
&= -t^2\ \cos(t) + \left\{ 2t\ \sin(t) - \int 2\ \sin(t)\, dt \right\} \\
&= -t^2\ \cos(t) + \{2t\ \sin(t) + 2\ \cos(t)\} + C \\
&= -t^2\ \cos(t) + 2t\ \sin(t) + 2\ \cos(t) + C
\end{aligned}$$

Example 5.1.7 Evaluate $\int_1^3 t^2 \sin(t)\, dt$.

Solution *We will do this one the short way: first do the indefinite integral just like the last problem.*

$$\int t^2 \sin(t)\,dt = -t^2 \cos(t) - \int -\cos(t)\, 2t\, dt$$

$$\boxed{u = t^2;\ dv = \sin(t)dt;\ du = 2tdt;\ v = -\cos(t)}$$

$$= -t^2 \cos(t) + \int 2t \cos(t)\, dt$$

$$\boxed{u = 2t;\ dv = \cos(t)dt;\ du = 2dt;\ v = \sin(t)}$$

$$= -t^2 \cos(t) + \left\{2t \sin(t) - \int 2 \sin(t)\, dt\right\}$$

$$= -t^2 \cos(t) + \{2t \sin(t) + 2 \cos(t)\} + C$$

$$= -t^2 \cos(t) + 2t \sin(t) + 2 \cos(t) + C$$

Then, we see

$$\begin{aligned}\int_1^3 t^2 \sin(t)\, dt &= \{-t^2 \cos(t) + 2t \sin(t) + 2 \cos(t)\}(3) \\ &\quad - \{-t^2 \cos(t) + 2t \sin(t) + 2 \cos(t)\}(1) \\ &= \{-9 \cos(3) + 6 \sin(3) + 2 \cos(3)\} \\ &\quad - \{-\cos(1) + 2 \sin(1) + 2 \cos(1)\}\end{aligned}$$

And it is not clear we can do much to simplify this expression except possibly just use our calculator to actually compute a value!

5.1.2 Homework

Exercise 5.1.1 *Evaluate* $\int \ln(5t)\, dt$.

Exercise 5.1.2 *Evaluate* $\int 2t \ln(t^2)\, dt$.

Exercise 5.1.3 *Evaluate* $\int (t + 1)^2 \ln(t + 1)\, dt$.

Exercise 5.1.4 *Evaluate* $\int t^2 e^{2t}\, dt$.

Exercise 5.1.5 *Evaluate* $\int_0^2 t^2 e^{-3t}\, dt$.

Exercise 5.1.6 *Evaluate* $\int 10t \sin(4t)\, dt$.

Exercise 5.1.7 *Evaluate* $\int 6t \cos(8t)\, dt$.

Exercise 5.1.8 *Evaluate* $\int (6t + 4) \cos(8t)\, dt$.

Exercise 5.1.9 *Evaluate* $\int_2^5 (t^2 + 5t + 3) \ln(t)\, dt$.

Exercise 5.1.10 *Evaluate* $\int (t^2 + 5t + 3)\, e^{2t}\, dt$.

5.2 Partial Fraction Decompositions

Suppose we wanted to integrate a function like $\frac{1}{(x+2)\,(x-3)}$. This does not fit into a simple substitution method at all. The way we do this kind of problem is to split the fraction $\frac{1}{(x+2)\,(x-3)}$ into the sum of the two simpler fractions $\frac{1}{x+2}$ and $\frac{1}{x-3}$. This is called the *Partial Fractions Decomposition* approach Hence, we want to find numbers A and B so that

$$\frac{1}{(x+2)\,(x-3)} = \frac{A}{x+2} + \frac{B}{x-3}$$

If we multiply both sides of this equation by the term $(x+2)\,(x-3)$, we get the new equation

$$1 = A\,(x-3) + B\,(x+2)$$

Since this equation holds for $x = 3$ and $x = -2$, we can evaluate the equation twice to get

$$\begin{aligned} 1 &= \{A\,(x-3) + B\,(x+2)\}\;|_{x=3} \\ &= 5\,B \\ 1 &= \{A\,(x-3) + B\,(x+2)\}\;|_{x=-2} \\ &= -5\,A \end{aligned}$$

Thus, we see A is $-1/5$ and B is $1/5$. Hence,

$$\frac{1}{(x+2)\,(x-3)} = \frac{-1/5}{x+2} + \frac{1/5}{x-3}$$

We could then integrate as follows

$$\begin{aligned} \int \frac{1}{(x+2)\,(x-3)}\,dx &= \int \left(\frac{-1/5}{x+2} + \frac{1/5}{x-3}\right)\,dx \\ &= \int \frac{-1/5}{x+2}\,dx + \int \frac{1/5}{x-3}\,dx \\ &= -1/5 \int \frac{1}{x+2}\,dx + 1/5 \int \frac{1}{x-3}\,dx \\ &= -1/5\;\ln\left(|\,x+2\,|\right) + 1/5\;\ln\left(|\,x-3\,|\right) + C \\ &= 1/5\;\ln\left(\frac{|\,x-3\,|}{|\,x+2\,|}\right) + C \\ &= \ln\left(\frac{|\,x-3\,|}{|\,x+2\,|}\right)^{1/5} + C \end{aligned}$$

where it is hard to say which of these equivalent forms is the most useful. In general, in later chapters, as we work out various modeling problems, we will choose whichever of the forms above is best for our purposes.

5.2.1 How Do We Use Partial Fraction Decompositions to Integrate?

Let's do some examples:

Example 5.2.1 Evaluate

$$\int \frac{5}{(t+3)\,(t-4)}\,dt.$$

Solution *We start with the decomposition:*

$$\frac{5}{(t+3)\,(t-4)} = \frac{A}{t+3} + \frac{B}{t-4}$$
$$5 = A\,(t-4) + B\,(t+3)$$

Now evaluate at $t = 4$ *and* $t = -3$*, to get*

$$\begin{aligned} 5 &= (A\,(t-4) + B\,(t+3))\,|_{t=4} \\ &= 7\,B \\ 5 &= (A\,(t-4) + B\,(t+3))\,|_{t=-3} \\ &= -7\,A \end{aligned}$$

Thus, we know $A = -5/7$ *and* $B = 5/7$*. We can now evaluate the integral:*

$$\begin{aligned} \int \frac{5}{(t+3)\,(t-4)}\,dt &= \int \left(\frac{-5/7}{t+3} + \frac{5/7}{t-4}\right) dt \\ &= \int \frac{-5/7}{t+3}\,dt + \int \frac{5/7}{t-4}\,dt \\ &= -5/7 \int \frac{1}{t+3}\,dt + 5/7 \int \frac{1}{t-4}\,dt \\ &= -5/7\,\ln(|\,t+3\,|) + 5/7\,\ln(|\,t-4\,|) + C \\ &= \frac{5}{7}\,\ln\left(\frac{|\,t-4\,|}{|\,t+3\,|}\right) + C \\ &= \ln\left(\frac{|\,t-4\,|}{|\,t+3\,|}\right)^{5/7} + C \end{aligned}$$

Example 5.2.2 Evaluate

$$\int \frac{10}{(2t-3)\,(8t+5)}\,dt.$$

Solution *We start with the decomposition:*

$$\frac{10}{(2t-3)\,(8t+5)} = \frac{A}{2t-3} + \frac{B}{8t+5}$$
$$10 = A\,(8t+5) + B\,(2t-3)$$

Now evaluate at $t = -5/8$ *and* $t = 3/2$*, to get*

$$\begin{aligned} 10 &= (A\,(8t+5) + B\,(2t-3))\,|_{t=-5/8} \\ &= (-10/8 - 3)\,B = -34/8B \\ 10 &= (A\,(8t+5) + B\,(2t-3))\,|_{t=3/2} \\ &= (24/2 + 5)\,A = 17\,A \end{aligned}$$

Thus, we know $B = -80/34 = -40/17$ *and* $A = 10/17$*. We can now evaluate the integral:*

$$\begin{aligned} \int \frac{10}{(2t-3)\,(8t+5)}\,dt &= \int \left(\frac{10/17}{2t-3} + \frac{-40/17}{8t+5}\right) dt \\ &= \int \frac{10/17}{2t-3}\,dt + \int \frac{-40/17}{8t+5}\,dt \\ &= 10/17 \int \frac{1}{2t-3}\,dt - 40/17 \int \frac{1}{8t+5}\,dt \\ &= 10/17\,(1/2)\,\ln(\mid 2t-3 \mid) - 40/17\,(1/8)\,\ln(\mid 8t+5 \mid) + C \\ &= \frac{5}{17}\,\ln\left(\frac{\mid 2t-3 \mid}{\mid 8t+5 \mid}\right) + C \\ &= \ln\left(\frac{\mid 2t-3 \mid}{\mid 8t+5 \mid}\right)^{5/17} + C \end{aligned}$$

Example 5.2.3 Evaluate

$$\int \frac{6}{(4-t)\,(9+t)}\,dt.$$

Solution *We start with the decomposition:*

$$\frac{6}{(4-t)\,(9+t)} = \frac{A}{4-t} + \frac{B}{9+t}$$
$$6 = A\,(9+t) + B\,(4-t)$$

Now evaluate at $t = 4$ *and* $t = -9$*, to get*

$$\begin{aligned} 6 &= (A\,(4-t) + B\,(9+t))\,|_{t=4} \\ &= 13\,B \\ 6 &= (A\,(4-t) + B\,(9+t))\,|_{t=-9} \\ &= 13\,A \end{aligned}$$

Thus, we know $A = 6/13$ *and* $B = 6/13$*. We can now evaluate the integral:*

$$\begin{aligned} \int \frac{6}{(4-t)\,(9+t)}\,dt &= \int \left(\frac{6/13}{4-t} + \frac{6/13}{9+t}\right)\,dt \\ &= \int \left(\frac{-6/13}{t-4} + \frac{6/13}{t+9}\right)\,dt \\ &= \int \frac{-6/13}{t-4}\,dt + \int \frac{6/13}{t+9}\,dt \\ &= 6/13 \int \frac{1}{t-4}\,dt + 6/13 \int \frac{1}{t+9}\,dt \\ &= 6/13\,\ln(|\,t-4\,|) + 6/13\,\ln(|\,t+9\,|) + C \\ &= \frac{6}{13}\,\ln(|\,(2t-3)\,(8t+5)\,|) + C \\ &= \ln(|\,(2t-3)\,(8t+5)\,|)^{6/13} + C \end{aligned}$$

Example 5.2.4 Evaluate

$$\int_4^7 \frac{-6}{(t-2)\,(2t+8)}\,dt.$$

Solution *Again, we start with the decomposition:*

$$\frac{-6}{(t-2)\,(2t+8)} = \frac{A}{t-2} + \frac{B}{2t+8}$$
$$-6 = A\,(2t+8) + B\,(t-2)$$

Now evaluate at $t = 2$ and $t = -4$, to get

$$\begin{aligned} -6 &= (A\,(2t+8) + B\,(t-2))\,|_{t=2} \\ &= 12\,A \\ -6 &= (A\,(2t+8) + B\,(t-2))\,|_{t=-4} \\ &= -6\,B \end{aligned}$$

Thus, we know $A = -1/2$ and $B = 1$. We can now evaluate the indefinite integral:

$$\begin{aligned} \int \frac{-6}{(t-2)\,(2t+8)}\,dt &= \int \left(\frac{-1/2}{t-2} + \frac{1}{2t+8}\right) dt \\ &= \int \left(\frac{-1/2}{t-2} + \frac{1}{2t+8}\right) dt \\ &= \int \frac{-1/2}{t-2}\,dt + \int \frac{1}{2t+8}\,dt \\ &= -1/2 \int \frac{1}{t-2}\,dt + \int \frac{1}{2t+8}\,dt \\ &= -1/2\,\ln(|\,t-2\,|) + 1/2\,\ln(|\,2t+8\,|) + C \\ &= \frac{1}{2}\ln\left(\frac{|\,2t+8\,|}{|\,t-2\,|}\right) + C \\ &= \frac{1}{2}\ln\left(|\,\frac{2t+8}{t-2}\,|\right) + C \end{aligned}$$

Then, the definite integral becomes

$$\begin{aligned} \int_4^7 \frac{-6}{(t-2)\,(2t+8)}\,dt &= \frac{1}{2}\ln\left(|\,\frac{2t+8}{t-2}\,|\right)\Bigg|_{t=7} - \frac{1}{2}\ln\left(|\,\frac{2t+8}{t-2}\,|\right)\Bigg|_{t=4} \\ &= \frac{1}{2}\left(\ln\left(\frac{22}{5}\right) - \ln\left(\frac{16}{2}\right)\right) \\ &= \frac{1}{2}\ln\left(\frac{22}{5}\frac{2}{16}\right) \\ &= \frac{1}{2}\ln\left(\frac{44}{80}\right) = \frac{1}{2}\ln\left(\frac{11}{20}\right) \end{aligned}$$

Note that this evaluation would not make sense if the on an interval $[a, b]$ is any of these two natural logarithm functions were undefined at some point in $[a, b]$. Here, both natural logarithm functions are nicely defined on $[4, 7]$.

5.2.2 Homework

Exercise 5.2.1 *Evaluate*

$$\int \frac{16}{(u+2)\,(u-4)}\,du.$$

Exercise 5.2.2 *Evaluate*

$$\int \frac{35}{(2z-6)\,(z+6)}\,dz.$$

Exercise 5.2.3 *Evaluate*

$$\int \frac{-2}{(s+4)\,(s-5)}\,ds.$$

Exercise 5.2.4 *Evaluate*

$$\int \frac{6}{(x^2-16)}\,dx.$$

Exercise 5.2.5 *Evaluate*

$$\int_4^6 \frac{9}{(2y-4)\,(3y-9)}\,dy.$$

Exercise 5.2.6 *Evaluate*

$$\int \frac{-8}{(w+7)\,(w-10)}\,dw.$$

Exercise 5.2.7 *Evaluate*

$$\int \frac{3}{(t+1)\,(t-3)}\,dt.$$

Reference

J. Peterson, *Calculus for Cognitive Scientists: Derivatives, Integration and Modeling*, Springer Series on Cognitive Science and Technology (Springer Science+Business Media Singapore Pte Ltd, Singapore, 2015 in press)

Chapter 6
Complex Numbers

In the chapters to come, we will need to use the idea of a *complex number*. When we use the quadratic equation to find the roots of a polynomial like $f(t) = t^2 + t + 1$, we find

$$t = \frac{-1 \pm \sqrt{1-4}}{2}$$
$$= -\frac{1}{2} \pm \sqrt{-3}$$

Since, it is well known that there are no numbers in our world whose squares can be negative, it was *easy* to think that the term $\sqrt{-3}$ represented some sort of *imaginary* quantity. But, it seemed reasonable that the usual properties of the $\sqrt{}$ function should hold. Thus, we can write

$$\sqrt{-3} = \sqrt{-1}\,\sqrt{3}$$

and the term $\sqrt{-1}$ had the amazing property that when you squared it, you got back -1! Thus, the square root of any negative number $\sqrt{-c}$ for a positive c could be rewritten as $\sqrt{-1} \times \sqrt{c}$. It became clear to the people studying the roots of polynomials such as our simple one above, that if the set of real numbers was augmented to include numbers of the form $\sqrt{-1} \times \sqrt{c}$, there would be a nice way to represent any root of a polynomial. Since a number like $\sqrt{-1} \times \sqrt{4}$ or $2 \times \sqrt{-1}$ is also possible, it seemed like two copies of the real numbers were needed: one that was the usual real numbers and another copy which was any real number times this strange quantity $\sqrt{-1}$. It became very convenient to label the set of all traditional real numbers as the x axis and the set of numbers prefixed by $\sqrt{-1}$ as the y axis.

Since the prefix $\sqrt{-1}$ was the only real difference between the usual real numbers and the new numbers with the prefix $\sqrt{-1}$, this prefix $\sqrt{-1}$ seemed the the quintessential representative of this difference. Historically, since these new prefixed numbers were already thought of as *imaginary*, it was decided to start labeling $\sqrt{-1}$ as the simple letter $\boldsymbol{i}$ where $\boldsymbol{i}$ is *short* for *imaginary*! Thus, a number of this

J.K. Peterson, *Calculus for Cognitive Scientists*,
Cognitive Science and Technology, DOI 10.1007/978-981-287-877-9_6

sort could be represented as $a + b\,\boldsymbol{i}$ where a and b are any ordinary real numbers. In particular, the roots of our polynomial could be written as

$$t = -\frac{1}{2} \pm \boldsymbol{i}\sqrt{3}.$$

6.1 The Complex Plane

A generic complex number $z = a + b\,\boldsymbol{i}$ can thus be graphed as a point in the plane which has as ordinate axis the usual x axis and as abscissa, the new $\boldsymbol{i}y$ axis. We call this coordinate system the Complex Plane. The magnitude of the complex number z is defined to be the length of the vector which connects the ordered pair (a, b) in this plane with the origin $(0, 0)$. This is labeled as r in Fig. 6.1. This magnitude is called the *modulus* or *magnitude* of z and is represented by the same symbol we use for the absolute value of a number, $\mid z \mid$. However, here this magnitude is the length of the hypotenuse of the triangle consisting of the points $(0, 0)$, $(a, 0)$ and $a, b)$ in the complex plane. Hence,

$$\mid z \mid = \sqrt{(a)^2 + (b)^2}$$

The angle measured from the positive x axis to this vector is called the *angle* associated with the complex number z and is commonly denoted by the symbol θ or $Arg(z)$. Hence, there are two equivalent ways we can represent a complex number z. We can use coordinate information and write $z = a + b\,\boldsymbol{i}$ or we can use magnitude and angle information. In this case, if you look at Fig. 6.1, you can clearly see that $a = r\ \cos(\theta)$ and $b = r\ \sin(\theta)$. Thus,

$$\begin{aligned} z &= \mid z \mid\ (\cos(\theta) + \boldsymbol{i}\ \sin(\theta)) \\ &= r\ (\cos(\theta) + i\ \sin(\theta)) \end{aligned}$$

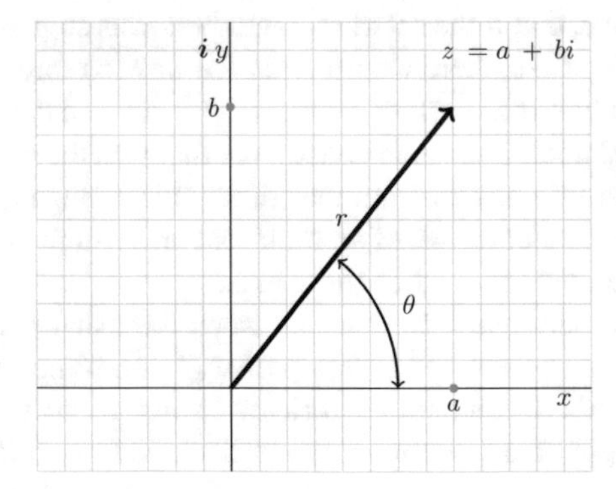

A complex number $a+bi$ has real part a and imaginary part b. The coordinate (a, b) is graphed in the usual Cartesian manner as an ordered pair in the $x-iy$ complex plane. The magnitude of z is $\sqrt{(a)^2 + (b)^2}$ which is shown on the graph as r. The angle associated with z is drawn as an arc of angle θ

Fig. 6.1 Graphing complex numbers

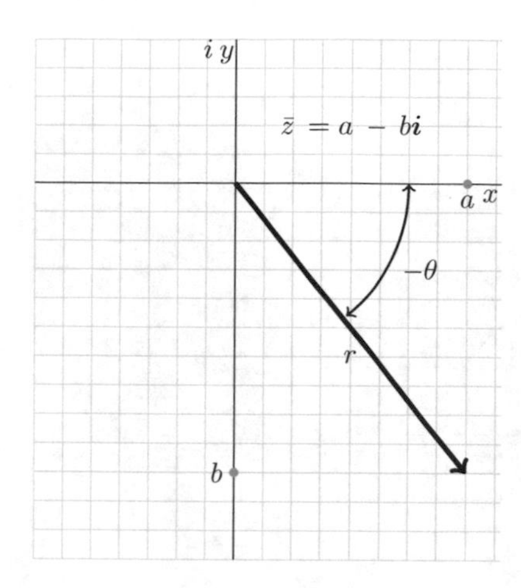

A complex number $a + \ b\,\boldsymbol{i}$ has real part a and imaginary part b. Its complex conjugate is $a - \ b\,\boldsymbol{i}$ The coordinate $(a, -b)$ is graphed in the usual Cartesian manner as an ordered pair in the complex plane. The magnitude of $\bar{z}$ is $\sqrt{(a)^2 + (b)^2}$ which is shown on the graph as r. The angle associated with $\bar{z}$ is drawn as an arc of angle $-\theta$

Fig. 6.2 Graphing the conjugate of a complex numbers

We can interpret the number $\cos(\theta) + \boldsymbol{i}\ \sin(\theta)$ in a different way. Given a complex number $z = a + \ b\,\boldsymbol{i}$, we define the *complex conjugate* of z to be $\bar{z} = a - \ b\,\boldsymbol{i}$. It is easy to see that

$$z\,\bar{z} = |\,z\,|^2$$
$$z^{-1} = \frac{\bar{z}}{|\,z\,|^2}$$

Now look at Fig. 6.1 again. In this figure, z is graphed in Quadrant 1 of the complex plane. Now imagine that we replace z by $\bar{z}$. Then the imaginary component changes to -5 which is a *reflection* across the positive x axis. The magnitude of $\bar{z}$ and z will then be the same but $Arg(\bar{z})$ is $-\theta$. We see this illustrated in Fig. 6.2.

6.1.1 Complex Number Calculations

Example 6.1.1 For the complex number $z = 2\ + 4\,\boldsymbol{i}$

1. Find its magnitude
2. Write it in the form $r\ (\cos(\theta) + \boldsymbol{i}\ \sin(\theta))$ using radians
3. Graph it in the complex plane showing angle in degrees

Solution *The complex number* $2\ + 4\,\boldsymbol{i}$ *has magnitude* $\sqrt{(2)^2 + (4)^2}$ *which is* $\sqrt{20}$. *Since both the real and imaginary component of the complex number are positive, this complex number is graphed in the first quadrant of the complex plane. Hence, the angle associated with* z *should be between* 0 *and* $\pi/2$. *We can easily calculate the angle* θ *associated with* z *to be* $\tan^{-1}(4/2) = \tan^{-1}(4/2)$.

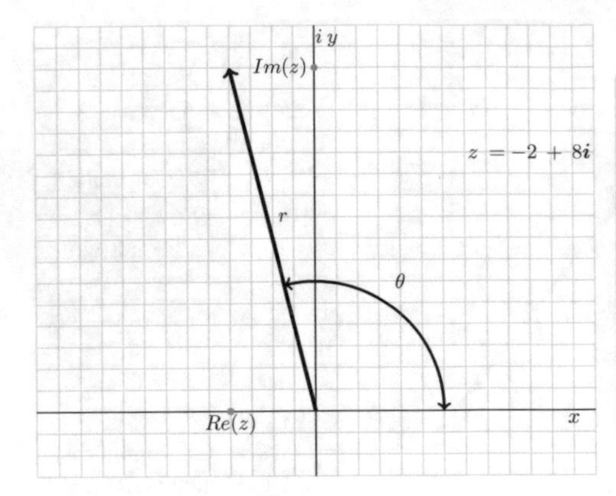

A complex number $-2 + 8\,i$ has real part -2 and imaginary part 8. The coordinate $(-2, 8)$ is graphed in the usual Cartesian manner as an ordered pair in the $x - iy$ complex plane. The magnitude of z is $\sqrt{(-2)^2 + (8)^2}$ which is shown on the graph as r. The angle associated with z is drawn as an arc of angle θ

Fig. 6.3 The complex number $-2 + 8i$

Example 6.1.2 For the complex number $-2 + 8\,\boldsymbol{i}$

1. Find its magnitude
2. Write it in the form $r\ (\cos(\theta) + i\ \sin(\theta))$ using radians
3. Graph it in the complex plane showing angle in degrees

Solution *The complex number $-2 + 8\,\boldsymbol{i}$ has magnitude $\sqrt{(-2)^2 + (8)^2}$ which is $\sqrt{68}$. This complex number is graphed in the second quadrant of the complex plane because the real part is negative and the imaginary part is positive. Hence, the angle associated with z should be between $\pi/2$ and π. The angle θ associated with z should be $\pi - \tan^{-1}(|8/(-2)| = 1.82$ rad or $104.04°$. The graph in Fig. 6.3 is instructive.*

We can summarize what we know about complex numbers in Proposition 6.1.1

Proposition 6.1.1 (Complex Numbers)
If z is $a + b\,\boldsymbol{i}$, then

1. *The magnitude of z is denoted $|\ z\ |$ and is defined to be $\sqrt{(a)^2 + (b)^2}$.*
2. *The complex conjugate of z is $a - b\,\boldsymbol{i}$.*
3. *The argument or angle associated with the complex number z is the angle θ defined as the angle measured from the positive x axis to the radius vector which points from the origin $0 + 0\,\boldsymbol{i}$ to $a + b\,\boldsymbol{i}$. This angle is also called $Arg(z)$. Sometimes we measure this angle clockwise and sometimes counterclockwise.*
4. *$Arg(\bar{z}) = -Arg(z)$.*
5. *The magnitude of z is often denoted by r since it is the length of the radius vector described above.*
6. *The polar coordinate version of z is given by $z = r\ (\cos(\theta) + \boldsymbol{i}\ \sin(\theta))$.*

6.1.2 Homework

Exercise 6.1.1 *For the complex number* $-3 + 6\,\boldsymbol{i}$

1. *Find its magnitude*
2. *Write it in the form* $r\,(\cos(\theta) + \boldsymbol{i}\,\sin(\theta))$ *using radians*
3. *Graph it in the complex plane showing angle in degrees*

Exercise 6.1.2 *For the complex number* $-3 - 6\,\boldsymbol{i}$

1. *Find its magnitude*
2. *Write it in the form* $r\,(\cos(\theta) + \boldsymbol{i}\,\sin(\theta))$ *using radians*
3. *Graph it in the complex plane showing angle in degrees*

Exercise 6.1.3 *For the complex number* $3 - 2\,\boldsymbol{i}$

1. *Find its magnitude*
2. *Write it in the form* $r\,(\cos(\theta) + \boldsymbol{i}\,\sin(\theta))$ *using radians*
3. *Graph it in the complex plane showing angle in degrees*

Exercise 6.1.4 *For the complex number* $5 + 3\,\boldsymbol{i}$

1. *Find its magnitude*
2. *Write it in the form* $r\,(\cos(\theta) + \boldsymbol{i}\,\sin(\theta))$ *using radians*
3. *Graph it in the complex plane showing angle in degrees*

Exercise 6.1.5 *For the complex number* $-4 - 3\,\boldsymbol{i}$

1. *Find its magnitude*
2. *Write it in the form* $r\,(\cos(\theta) + \boldsymbol{i}\,\sin(\theta))$ *using radians*
3. *Graph it in the complex plane showing angle in degrees*

Exercise 6.1.6 *For the complex number* $5 - 7\,\boldsymbol{i}$

1. *Find its magnitude*
2. *Write it in the form* $r\,(\cos(\theta) + \boldsymbol{i}\,\sin(\theta))$ *using radians*
3. *Graph it in the complex plane showing angle in degrees*

6.2 Complex Functions

Let $z = r\,(\cos(\theta) + \boldsymbol{i}\,\sin(\theta))$ and let's think of θ are a variable now. We have seen the function $f(\theta) = r\,(\cos(\theta) + \boldsymbol{i}\,\sin(\theta))$ arises when we interpret a complex number in terms of the triangle formed by its angle and its magnitude. Now let's think of θ as a variable. Let's find $f'(\theta)$. We have

$$\begin{aligned} f'(\theta) &= -r\sin(\theta) + \boldsymbol{i}\, r\cos(\theta) \\ &= \boldsymbol{i}^2\, r\sin(\theta) + \boldsymbol{i}\, r\cos(\theta) \\ &= \boldsymbol{i}\left(r\boldsymbol{i}\sin(\theta) + r\cos(\theta)\right) \\ &= \boldsymbol{i}\, f(\theta). \end{aligned}$$

So $f'(\theta) = \boldsymbol{i}\, f(\theta)$ or $\frac{f'(\theta)}{f(\theta)} = \boldsymbol{i}$. Taking the antiderivative of both sides, this suggests

$$\ln(f(\theta)) = \boldsymbol{i} \Rightarrow f(\theta) = e^{\boldsymbol{i}\theta}.$$

Our antiderivative argument suggests we define $e^{\boldsymbol{i}\,\theta} = r\,(\cos(\theta) + \boldsymbol{i}\sin(\theta))$ and by direct calculation, we had $(e^{\boldsymbol{i}\,\theta})' = \boldsymbol{i}\, e^{\boldsymbol{i}\,\theta}$. This motivational argument is not quite right, of course, and there is deeper mathematics at work here, but it helps to explain why we define $e^{\boldsymbol{i}\theta} = \cos(\theta) + \boldsymbol{i}\sin(\theta)$. Hence, we *extend* the exponential function to *complex* numbers as follows:

Definition 6.2.1 (*The Extension Of e^b to e^{ib}*)
We extend the exponential function exp to complex number arguments as follows: for any real numbers a and b, define

$$\begin{aligned} e^{ib} &\equiv \cos(b) + i\ \sin(b) \\ e^{a+i\,b} &\equiv e^a\ e^{ib} \\ &= e^a\ (\cos(b) + i\ \sin(b)) \end{aligned}$$

Definition 6.2.2 (*The Extension Of e^t to $e^{(a+b\,i)\,t}$*)
We extend the exponential function exp to the complex argument $a + b\,\boldsymbol{i}\,t$ as follows

$$\begin{aligned} e^{ibt} &\equiv \cos(bt) + \boldsymbol{i}\ \sin(bt) \\ e^{(a+i\,b)\,t} &\equiv e^{at}\ e^{ibt} \\ &= e^{at}\ (\cos(bt) + \boldsymbol{i}\ \sin(bt)) \end{aligned}$$

Thus, we can rephrase the polar coordinate form of the complex number $z = a + b\,\boldsymbol{i}$ as $z = r\, e^{\boldsymbol{i}\,\theta}$.

6.2.1 Calculations with Complex Functions

Example 6.2.1 For the complex function

$$e^{(-2+8\,i)t}$$

1. Find its magnitude
2. Write it in its fully expanded form

Solution *We know that*

$$\exp\left((a + b\,\boldsymbol{i})\,t\right) = e^{at}\ \left(\cos(bt) + \boldsymbol{i}\ \sin(bt)\right).$$

Hence,

$$\begin{aligned}\exp\left((-2 + 8\,\boldsymbol{i})\,t\right) &= e^{-2t}\ e^{8it}\\ &= e^{-2t}\ \left(\cos(8t) + \boldsymbol{i}\ \sin(8t)\right).\end{aligned}$$

Since the complex magnitude of e^{8it} *is always one, we see*

$$|\ e^{(-2+8\,\boldsymbol{i})t}\ | = e^{-2t}$$

Example 6.2.2 For the complex function

$$e^{(-1+2\,\boldsymbol{i})t}$$

1. Find its magnitude
2. Write it in its fully expanded form

Solution *We have*

$$\begin{aligned}\exp\left((-1 + 2\,\boldsymbol{i})\,t\right) &= e^{-t}\ e^{2it}\\ &= e^{-t}\ \left(\cos(2t) + \boldsymbol{i}\ \sin(2t)\right).\end{aligned}$$

Since the complex magnitude of e^{2it} *is always one, we see*

$$|\ e^{(-1+2\,\boldsymbol{i})t}\ | = e^{-t}$$

Example 6.2.3 For the complex function

$$e^{2i\,t}$$

1. Find its magnitude
2. Write it in its fully expanded form

Solution *We have*

$$\begin{aligned}\exp\left((0 + 2\,\boldsymbol{i})\,t\right) &= e^{0t}\ e^{2it}\\ &= \cos(2t) + \boldsymbol{i}\ \sin(2t).\end{aligned}$$

Since the complex magnitude of e^{2it} is always one, we see

$$| \, e^{(-1+2i)t} \, | = 1$$

6.2.2 Homework

For each of the following complex numbers, find its magnitude, write it in the form $r\, e^{i\theta}$ using radians, and graph it in the complex plane showing angle in degrees.

Exercise 6.2.1 $z = -3 + 6\,\boldsymbol{i}$.

Exercise 6.2.2 $z = -3 - 6\,\boldsymbol{i}$.

Exercise 6.2.3 $z = 3 - 6\,\boldsymbol{i}$.

Exercise 6.2.4 $z = 2 + 8\,\boldsymbol{i}$.

Exercise 6.2.5 $z = 5 + 1\,\boldsymbol{i}$.

For each of the following complex functions, find its magnitude and write it in its fully expanded form.

Exercise 6.2.6 $e^{(-2+3\,\boldsymbol{i})\,t}$.

Exercise 6.2.7 $e^{(-1+4\,\boldsymbol{i})\,t}$.

Exercise 6.2.8 $e^{(0+14\,\boldsymbol{i})\,t}$.

Exercise 6.2.9 $e^{(-8-9\,\boldsymbol{i})\,t}$.

Exercise 6.2.10 $e^{(-2+\pi\,\boldsymbol{i})\,t}$.

Chapter 7
Linear Second Order ODEs

We now turn our attention to *Linear Second Order Differential Equations*. The way we solve these is built from our understanding of exponential growth and the models built out of that idea. The idea of a half life is also important though it is not used as much in these second order models. A great example also comes from Protein Modeling and its version of half life called Response Time.

These have the general form

$$a\,u''(t) + b\,u'(t) + c\,u(t) = 0 \tag{7.1}$$
$$u(0) = u_0, \quad u'(0) = u_1 \tag{7.2}$$

where we assume a is not zero. Here are some examples just so you can get the feel of these models.

$$\begin{aligned}
x''(t) \; - 3\,x'(t) \; - 4\,x(t) &= 0\\
x(0) &= 3\\
x'(0) &= 8\\
4\,y''(t) + 8\,y'(t) + \; y(t) &= 0\\
y(0) &= -1\\
y'(0) &= 2\\
z''(t) \; + 5\,z'(t) \; + 3\,z(t) &= 0\\
z(0) &= 2\\
z'(0) &= -3
\end{aligned}$$

We have already seen that for a first order problem like $u'(t) = r\,u(t)$ with $u(0) = u_0$ has the solution $u(t) = u_0\,e^{rt}$. It turns out that the solutions to Eqs. 7.1 and 7.2 will also have this kind of form. To see how this happens, let's look at a new concept called an *operator*. For us, an operator is something that takes as input a function and then transforms the function into another one. A great example is the indefinite integral

J.K. Peterson, *Calculus for Cognitive Scientists*,
Cognitive Science and Technology, DOI 10.1007/978-981-287-877-9_7

operator we might call $\mathcal{I}$ which takes a nice continuous function f and outputs the new function $\int f$. Another good one is the differentiation operator $\boldsymbol{D}$ which takes a differentiable function f and creates the new function f'. Hence, if we let $\boldsymbol{D}$ denote differentiation with respect to the independent variable and $\boldsymbol{c}$ be *the multiply by a constant operator* defined by $\boldsymbol{c}(f) = cf$, we could rewrite $u'(t) = ru(t)$ as

$$\boldsymbol{D}u = \boldsymbol{r}\, u \tag{7.3}$$

where we suppress the (t) notation for simplicity of exposition. In fact, we could rewrite the model again as

$$\left(\boldsymbol{D} - \boldsymbol{r}\right)u = 0 \tag{7.4}$$

where we let $\boldsymbol{D} - \boldsymbol{r}$ act on u to create $u' - ru$. We can apply this idea to Eq. 7.1. Next, let $\boldsymbol{D}^2$ be the second derivative operator: this means $\boldsymbol{D}^2 u = u''$. Then we can rewrite Eq. 7.1 as

$$a\; \boldsymbol{D}^2 u + b\; \boldsymbol{D}u + \boldsymbol{c}\, u = 0 \tag{7.5}$$

where we again suppress the time variable t. For example, if we had the problem

$$u''(t) + 5\, u'(t) + 6\, u(t) = 0$$

it could be rewritten as

$$\boldsymbol{D}^2 u + 5\boldsymbol{D}u + \boldsymbol{6}\, u = 0.$$

A little more practice is good at this point. These models convert to the operator form indicated We ignore the initial conditions for the moment.

$$\begin{aligned} x''(t)\; - 3\, x'(t)\; - 4\, x(t) = 0 &\Longleftrightarrow (\boldsymbol{D}^2 - \boldsymbol{3D} - \boldsymbol{4})(x) = 0 \\ 4\, y''(t)\; + 8\, y'(t)\; +\; y(t) = 0 &\Longleftrightarrow (\boldsymbol{4D}^2 + \boldsymbol{8D} + \boldsymbol{1})(y) = 0 \\ z''(t)\; + 5\, z'(t)\; + 3\, z(t) = 0 &\Longleftrightarrow (\boldsymbol{D}^2 + \boldsymbol{5D} + \boldsymbol{3})(z) = 0. \end{aligned}$$

In fact, we should begin to think of the models and their operator forms as interchangeable. The model most useful to us now is the first order linear model. We now can write

$$x' = 3\, x;\;\; x(0) = A \;\;\Longleftrightarrow\;\; (\boldsymbol{D} - \boldsymbol{3})(x) = 0;\;\; x(0) = A.$$

and

$$x' = -2\, x;\;\; x(0) = A \;\;\Longleftrightarrow\;\; (\boldsymbol{D} + \boldsymbol{2})(x) = 0;\;\; x(0) = A.$$

Now consider the model

$$y'' + 5y' + 6y = 0 \Longleftrightarrow (\boldsymbol{D}^2 + \boldsymbol{5D} + \boldsymbol{6})(y) = 0.$$

Let's try factoring: consider for an function f and do the computations with the factors in both orders.

$$\begin{aligned}(\boldsymbol{D}+\boldsymbol{2})\,(\boldsymbol{D}+\boldsymbol{3})(f) &= (\boldsymbol{D}+\boldsymbol{2})(f'+3f)\\ &= \boldsymbol{D}(f'+3f)+2\,(f'+3f)\\ &= f''+3f'+2f'+6f = f''+5f'+6f.\end{aligned}$$

and

$$\begin{aligned}(\boldsymbol{D}+\boldsymbol{3})\,(\boldsymbol{D}+\boldsymbol{2})(f) &= (\boldsymbol{D}+\boldsymbol{3})(f'+2f)\\ &= \boldsymbol{D}(f'+2f)+3\,(f'+2f)\\ &= f''+2f'+3f'+6f = f''+5f'+6f.\end{aligned}$$

We see that

$$(\boldsymbol{D}^2+\boldsymbol{5D}+\boldsymbol{6})(y) = 0 \Longleftrightarrow (\boldsymbol{D}+\boldsymbol{2})\,(\boldsymbol{D}+\boldsymbol{3})(y) = 0 \Longleftrightarrow (\boldsymbol{D}+\boldsymbol{3})\,(\boldsymbol{D}+\boldsymbol{2})(y) = 0.$$

Now we can figure out how to find the most general solution to this model.

- The general solution to $(\boldsymbol{D}+\boldsymbol{3})(y) = 0$ is $y(t) = Ae^{-3t}$.
- The general solution to $(\boldsymbol{D}+\boldsymbol{2})(y) = 0$ is $y(t) = Be^{-2t}$.
- Let our most general solution be $y(t) = Ae^{-3t} + Be^{-2t}$.

We know from our study of first order equations that a problem of the form $(\boldsymbol{D}+\boldsymbol{r})u = 0$ has a solution of the form e^{rt}. This suggests that there are two possible solutions to the problem above. One satisfies $(\boldsymbol{D}+\boldsymbol{3})u = 0$ and the other $(\boldsymbol{D}+\boldsymbol{2})u = 0$. Hence, it seems that any combination of the functions e^{-3t} and e^{-2t} should work. Thus, a general solution to our problem would have the form $A\,e^{-3t} + B\,e^{-2t}$ for arbitrary constants A and B. With this intuition established, let's try to solve this more formally.

For the problem Eq. 7.1, let's assume that e^{rt} is a solution and try to find what values of r might work. We see for $u(t) = e^{rt}$, we find

$$\begin{aligned}0 &= a\,u''(t) + b\,u'(t) + c\,u(t)\\ &= a\,r^2\,e^{rt} + b\,r\,e^{rt} + c\,e^{rt}\\ &= \left(a\,r^2 + b\,r + c\right)e^{rt}.\end{aligned}$$

Since e^{rt} can never be 0, we must have

$$0 = a\,r^2 + b\,r + c.$$

The roots of the quadratic equation above are the only values of r that will work as the solution e^{rt}. We call this quadratic equation the *Characteristic Equation of a linear second order differential equation* Eq. 7.1. To find these values of r, we can either factor the quadratic or use the quadratic formula. If you remember from your earlier algebra course, there are three types of roots:

(i): the roots are both real and distinct. We let r_1 be the smallest root and r_2 the bigger one.
(ii): the roots are both the same. We let $r_1 = r_2 = r$ in this case.
(iii): the roots are a complex pair of the form $a\ \pm b\,\boldsymbol{i}$.

Example 7.0.1 Consider this model.

$$u''(t) + 7\,u'(t) + 10\,u(t) = 0$$

Note the operator form of this model is

$$(D^2 + 7D + 10)(x) = 0.$$

Let's derive the characteristic equation. We assume the model has a solution of the form e^{rt} for some value of r. Then, plugging $u(t) = e^{rt}$ into the model, we find

$$(r^2 + 7r + 10)(e^{rt}) = 0.$$

Since e^{rt} is never 0 no matter what r's value is, we see this implies we need values of r that satisfy

$$r^2 + 7r + 10 = 0.$$

This factors as

$$(r + 2)(r + 5) = 0.$$

Thus, the roots of the model's characteristic equation are $r_1 = -5$ and $r_2 = -2$.

7.1 Homework

For these models,

(i): Write the model in operator form.
(ii): Derive the characteristic equation.
(iii): Find the roots of the characteristic equation.

Exercise 7.1.1

$$u''(t) + 50\,u'(t) - 6\,u(t) = 0.$$

Exercise 7.1.2

$$u''(t) + 7\,u'(t) + 2\,u(t) = 0.$$

Exercise 7.1.3

$$u''(t) - 8\,u'(t) + 3\,u(t) = 0.$$

Exercise 7.1.4

$$u''(t) + 6\,u'(t) + 10\,u(t) = 0.$$

Exercise 7.1.5

$$u''(t) + 5\,u'(t) - 24\,u(t) = 0.$$

Now, let's figure out what to do in each of these three cases.

7.2 Distinct Roots

Our general problem

$$\begin{aligned} a\,u''(t) + b\,u'(t) + c\,u(t) &= 0 \\ u(0) = u_0, \quad u'(0) &= u_1 \end{aligned}$$

which has characteristic equation

$$0 = a\,r^2 + b\,r + c,$$

factors as

$$0 = (r - r_1)\,(r - r_2).$$

We thus know that the general solution u has the form

$$u(t) = A\, e^{r_1\, t} + B\, e^{r_2\, t}.$$

We are told that initially, $u(0) = u_0$ and $u'(0) = u_1$. Taking the derivative of u, we find

$$u'(t) = r_1\, A\, e^{r_1\, t} + r_2\, B\, e^{r_2\, t}.$$

Hence, to satisfy the initial conditions, we must find A and B to satisfy the two equations in two unknowns below:

$$\begin{aligned} u(0) &= A\, e^{r_1\, 0} + B\, e^{r_2\, 0} \;=\; A \,+\, B \;=\; u_0 \\ u'(0) &= r_1\, A\, e^{r_1\, 0} + r_2\, B\, e^{r_2\, 0} \;=\; r_1\, A + r_2\, B \;=\; u_1. \end{aligned}$$

Thus to find the appropriate A and B, we must solve the system of two equations in two unknowns:

$$\begin{aligned} A + B &= u_0, \\ r_1\, A + r_2\, B &= u_1. \end{aligned}$$

Example 7.2.1 For this model,

- Derive the characteristic equation.
- Find the roots of the characteristic equation.
- Find the general solution.
- Solve the IVP.
- Plot and print the solution using MatLab.

$$\begin{aligned} x''(t) \;-\; 3\, x'(t) \;-\; 10\, x(t) &= 0 \\ x(0) &= -10 \\ x'(0) &= 10 \end{aligned}$$

Solution *(i): To derive the characteristic equation, we assume the solution has the form* e^{rt} *and plug that into the problem. We find*

$$\left(r^2 - 3\, r - 10\right) e^{rt} = 0.$$

Since e^{rt} is never zero, we see this implies that

$$r^2 - 3\,r - 10 = 0.$$

This is the characteristic equation for this problem.
(ii): We find the roots of the characteristic equation by either factoring it or using the quadratic formula. This one factors nicely giving

$$(r+2)\,(r-5) = 0.$$

Hence, the roots of this characteristic equation are $r_1 = -2$ and $r_2 = 5$.
(iii): The general solution is thus

$$u(t) = A\,e^{-2t} + B\,e^{5t}.$$

(iv): Next, we find the values of A and B which will let the solution satisfy the initial conditions. We have

$$\begin{aligned} u(0) &= A\,e^0 + B\,e^0 \;=\; A + B \;=\; -10 \\ u'(0) &= -2\,A\,e^0 + 5\,B\,e^0 \;=\; -2\,A\,+5\,B \;=\; 20. \end{aligned}$$

This gives the system of two equations in the two unknowns A and B

$$\begin{aligned} A + B &= -10 \\ -2\,A + 5\,B &= 10. \end{aligned}$$

Multiplying the first equation by 2 and adding, we get $B = -10/7$. It then follows that $A = -60/7$. Thus, the solution to this initial value problem is

$$u(t) = -60/7\,e^{-2t} \;-\, 10/7\,e^{5t}.$$

(v): Finally, we can graph this solution in MatLab as follows:

Listing 7.1: Solution to $x'' - 3x' - 10x = 0;\, x(0) = -10, x'(0) = 10$

```
time = linspace(0,1,101);
A = -60.0/7.0;
B = -10.0/7.0;
u = @(t) A*exp(-2.0*t) + B*exp(5*t);
plot(time,u(time));
```

which generates the plot we see in Fig. 7.1.

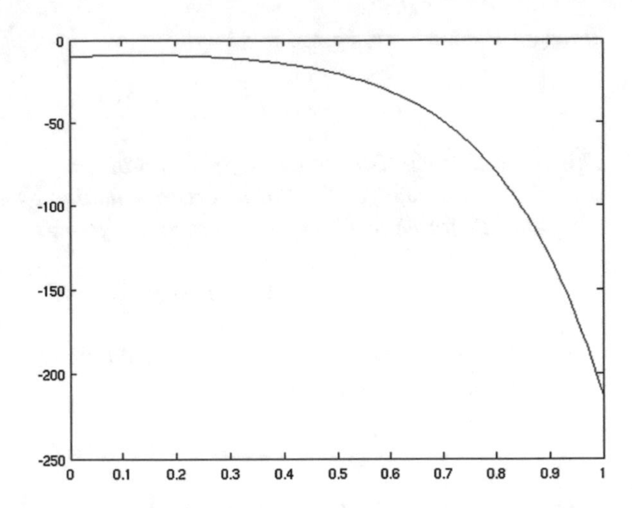

Fig. 7.1 Linear second order problem: two distinct roots

7.2.1 Homework

For the models below,

- Find the characteristic equation.
- Find the general solution.
- Solve the IVP.
- Plot and print the solution using MatLab.

Exercise 7.2.1 *For the ODE below*

$$\begin{aligned} y''(t) \ + 5\, y'(t) \ + 6\, y(t) &= 0 \\ y(0) &= 1 \\ y'(0) &= -2. \end{aligned}$$

Exercise 7.2.2 *For the ODE below*

$$\begin{aligned} z''(t) \ + 9\, z'(t) \ + 14\, z(t) &= 0 \\ z(0) &= -1 \\ z'(0) &= 1. \end{aligned}$$

Exercise 7.2.3 *For the ODE below*

$$\begin{aligned} P''(t) \ - 2\, P'(t) \ - 8\, P(t) &= 0 \\ P(0) &= 1 \\ P'(0) &= 2. \end{aligned}$$

Exercise 7.2.4 *For the ODE below*

$$\begin{aligned} u''(t) \; + 3\, u'(t) \; - 10\, u(t) &= 0 \\ u(0) &= -1 \\ u'(0) &= -2. \end{aligned}$$

Exercise 7.2.5 *For the ODE below*

$$\begin{aligned} x''(t) \; + 3\, x'(t) \; + 2\, x(t) &= 0 \\ x(0) &= -10 \\ x'(0) &= 200. \end{aligned}$$

7.3 Repeated Roots

Now our general problem

$$\begin{aligned} a\, u''(t) + b\, u'(t) + c\, u(t) &= 0 \\ u(0) \;=\; u_0, \quad u'(0) &= u_1 \end{aligned}$$

which has characteristic equation

$$0 = a\, r^2 + b\, r + c,$$

factors as

$$0 = (r - r_1)^2.$$

We have one solution $u_1(t) \;=\; e^{r_1 t}$, but we don't know if there are others. Let's assume that another solution is a product $f(t)e^{r_1 t}$. We know we want

$$\begin{aligned} 0 &= \left(\boldsymbol{D} - \boldsymbol{r_1}\right)\left(\boldsymbol{D} - \boldsymbol{r_1}\right)\left(f(t)\, e^{r_1 t}\right) \\ &= \left(\boldsymbol{D} - \boldsymbol{r_1}\right)\left(\left\{f'(t) \; + r_1 f(t) - r_1 f(t)\right\}\, e^{r_1 t}\right) \\ &= \left(\boldsymbol{D} - \boldsymbol{r_1}\right)\left(f'(t) e^{r_1 t}\right) \\ &= \left(f''(t) + r_1 f'(t) - r_1 f'(t)\right)\, e^{r_1 t} \\ &= f''(t)\, e^{r_1 t}. \end{aligned}$$

Since $e^{r_1 t}$ is never zero, we must have $f'' = 0$. This tells us that $f(t) = \alpha\, t + \beta$ for any α and β. Thus, a second solution has the form

$$\begin{aligned} v(t) &= (\alpha\, t + \beta)\, e^{r_1 t} \\ &= \alpha\, t\, e^{r_1 t} + \beta\, e^{r_1 t} \end{aligned}$$

The only new function in this second solution is $u_2(t) = te^{r_1 t}$. Hence, our general solution in the case of repeated roots will be

$$\begin{aligned} u(t) &= A\, e^{r_1\, t} + B\, t\, e^{r_1\, t} \\ &= \left(A + B\, t \right) e^{r_1\, t}. \end{aligned}$$

We are told that initially, $u(0) = u_0$ and $u'(0) = u_1$. Taking the derivative of u, we find

$$u'(t) = B\; e^{r_1\, t} + r_1 \left(A + B\, t \right) e^{r_1\, t}.$$

Hence, to satisfy the initial conditions, we must find A and B to satisfy the two equations in two unknowns below:

$$\begin{aligned} u(0) &= \left(A + B\; (0) \right) e^{r_1\, 0} \;=\; A \;=\; u_0 \\ u'(0) &= B\; e^{r_1\, (0)} + r_1 \left(A + B\; (0) \right) e^{r_1\, (0)} \;=\; B + r_1\, A \;=\; u_1. \end{aligned}$$

Thus to find the appropriate A and B, we must solve the system of two equations in two unknowns:

$$\begin{aligned} A &= u_0, \\ B + r_1\, A &= u_1. \end{aligned}$$

Example 7.3.1 Now let's look at a problem with repeated roots. We want to

- Find the characteristic equation.
- Find the roots of the characteristic equation.
- Find the general solution.
- Solve the IVP.
- Plot and print the solution using MatLab.

$$\begin{aligned} u''(t)\; + 16\, u'(t)\; + 64\, u(t) &= 0 \\ u(0) &= 1 \\ u'(0) &= 8 \end{aligned}$$

Solution *(i): To find the characteristic equation, we assume the solution has the form e^{rt} and plug that into the problem. We find*

$$\left(r^2 + 16\,r + 64\right) e^{rt} = 0.$$

Since e^{rt} is never zero, we see this implies that

$$r^2 + 16\,r + 64 = 0.$$

This is the characteristic equation for this problem.
(ii): We find the roots of the characteristic equation by either factoring it or using the quadratic formula. This one factors nicely giving

$$(r + 8)\,(r + 8) = 0.$$

Hence, the roots of this characteristic equation are repeated: $r_1 = -8$ and $r_2 = -8$.
(iii): The general solution is thus

$$u(t) = A\,e^{-8t} + B\,t\,e^{-8t}$$

(iv): Next, we find the values of A and B which will let the solution satisfy the initial conditions. We have

$$\begin{aligned} u(0) &= A\,e^0 + B\,0\,e^0 \;=\; A \;=\; 1 \\ u'(0) &= \left(-8Ae^{-8t} + Be^{-8t} - 8Bte^{-8t}\right)\Big|_{t=0} \;=\; -8\,A + B \;=\; 8. \end{aligned}$$

This gives the system of two equations in the two unknowns A and B

$$\begin{aligned} A &= 1 \\ -8\,A\,+\,B &= 8. \end{aligned}$$

This tells us that $B = 16$. Thus, the solution to this initial value problem is

$$u(t) = e^{-8t} + 16\,t\,e^{-8t}.$$

(v): Finally, we can graph this solution in MatLab as follows:

Listing 7.2: Solution to $u'' + 16u' + 64u = 0$; $u(0) = 1$, $u'(0) = 8$

```
time = linspace(0,3,101);
A = 1.0;
B = 16.0;
u = @(t)  A*exp(-8*t)+ 16.0*t.*exp(-8*t);
plot(time,u(time))
```

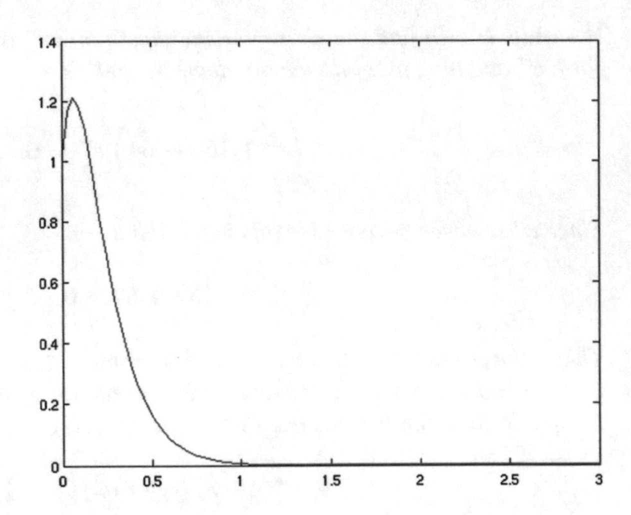

Fig. 7.2 Linear second order problem: repeated roots

which generates the plot we see in Fig. 7.2.

7.3.1 Homework

For the models below,

- Find the characteristic equation.
- Find the general solution.
- Solve the IVP.
- Plot and print the solution using MatLab.

Exercise 7.3.1

$$\begin{aligned} x''(t) - 12\,x'(t) + 36\,x(t) &= 0 \\ x(0) &= 1 \\ x'(0) &= -2. \end{aligned}$$

Exercise 7.3.2

$$\begin{aligned} y''(t) + 14\,y'(t) + 49\,y(t) &= 0 \\ y(0) &= -1 \\ y'(0) &= 1. \end{aligned}$$

Exercise 7.3.3

$$\begin{aligned} w''(t) + 6\,w'(t) + 9\,w(t) &= 0 \\ w(0) &= 1 \\ w'(0) &= 2. \end{aligned}$$

Exercise 7.3.4

$$\begin{aligned} Q''(t) + 4\,Q'(t) + 4\,Q(t) &= 0 \\ Q(0) &= -1 \\ Q'(0) &= -2. \end{aligned}$$

Exercise 7.3.5

$$\begin{aligned} \Phi''(t) + 2\,\Phi'(t) + \Phi(t) &= 0 \\ \Phi(0) &= -10 \\ \Phi'(0) &= 200. \end{aligned}$$

7.4 Complex Roots

In this last case, our general problem

$$\begin{aligned} a\,u''(t) + b\,u'(t) + c\,u(t) &= 0 \\ u(0) = u_0, \quad u'(0) &= u_1 \end{aligned}$$

which has characteristic equation

$$0 = a\,r^2 + b\,r + c,$$

factors as

$$0 = a\left(r - (c + di)\right)\left(r - (c - di)\right).$$

because the roots are complex. Now we suspect the solutions are

$$u_1(t) = e^{(c+di)t} \quad \text{and} \quad u_2(t) = e^{(c-di)t}.$$

We have already seen how to interpret the complex functions $e^{(c+di)t}$ and $e^{(c-di)t}$ (see Chap. 6, Definition 6.2.2). Let's try to find out what the derivative of such a function might be. First, it seems reasonable that if $f(t)$ has a derivative $f'(t)$ at t, then multiplying by i to get $if(t)$ only changes the derivative to $if'(t)$. In fact, the derivative of $(c+id)f(t)$ should be $(c+id)f'(t)$. Thus,

$$\begin{aligned}\left(e^{(c+id)t}\right)' &= \left(e^{ct}\{\cos(dt) + i\sin(dt)\}\right)' \\ &= ae^{ct}\ \cos(dt) - d\,e^{ct}\ \sin(dt) + i\,ce^{ct}\ \sin(dt) + i\,d\,e^{ct}\ \cos(dt) \\ &= e^{ct}\cos(dt)\Big(c + i\,d\Big) + e^{ct}\sin(dt)\Big(i\,c - d\Big).\end{aligned}$$

We also know that $i^2 = -1$, so replacing $-d$ by $i^2 d$ in the last equation, we find

$$\begin{aligned}\left(e^{(c+id)t}\right)' &= e^{ct}\cos(dt)\Big(c + i\,d\Big) + e^{ct}\sin(dt)\Big(i\,c + i^2\,d\Big) \\ &= e^{ct}\cos(dt)\Big(c + i\,d\Big) + i\,e^{ct}\sin(dt)\Big(c + i\,d\Big) \\ &= (c + i\,d)e^{ct}\Big(\cos(dt) + i\sin(dt)\Big) \\ &= (c + i\,d)\,e^{(c+id)t}.\end{aligned}$$

We conclude that we can now take the derivative of $e^{(c\pm id)t}$ to get $(c \pm id)\,e^{(c\pm id)t}$. We can now test to see if $Ae^{(c+id)t} + Be^{(c-id)t}$ solves our problem. We see

$$\left(e^{(c\,\pm\,id)t}\right)'' = (c\,\pm\,id)^2\,e^{(c\,\pm\,id)t}.$$

Thus,

$$\begin{aligned}&a\left(Ae^{(c+id)t} + Be^{(c-id)t}\right)'' + b\left(Ae^{(c+id)t} + Be^{(c-id)t}\right)' + c\left(Ae^{(c+id)t} + Be^{(c-id)t}\right) \\ &= Ae^{(c+id)t}\Big(a\,(c+id)^2 + b\,(c+id) + c\Big) + Be^{(c-id)t}\Big(a\,(c-id)^2 + b\,(c-id) + c\Big).\end{aligned}$$

Now since $c+id$ and $c-id$ are roots of the characteristic equation, we know

$$\begin{aligned}a\,(c+id)^2 + b\,(c+id) + c &= 0 \\ a\,(c-id)^2 + b\,(c-id) + c &= 0.\end{aligned}$$

Thus,

$$\begin{aligned}&a\left(Ae^{(c+id)t} + Be^{(c-id)t}\right)'' + b\left(Ae^{(c+id)t} + Be^{(c-id)t}\right)' + c\left(Ae^{(c+id)t} + Be^{(c-id)t}\right) \\ &= Ae^{(c+id)t}\Big(0\Big) + Be^{(c-id)t}\Big(0\Big) = 0.\end{aligned}$$

In fact, you can see that all of the calculations above would work even if the constants A and B were complex numbers. So we have shown that any combination of the two complex solutions $e^{(c+id)t}$ and $e^{(c-id)t}$ is a solution of our problem. Of course, a solution that actually has complex numbers in it doesn't seem that useful for our world. After all, we can't even graph it! So we have to find a way to construct solutions which are always real valued and use them as our solution. Note

$$\begin{aligned}
u_1(t) &= \frac{1}{2}\left(e^{(c+id)t} + e^{(c-id)t}\right)\\
&= \frac{1}{2}e^{ct}\left(\cos(dt) + i\ \sin(dt) + \cos(dt) - i\ \sin(dt)\right)\\
&= \frac{1}{2}\,2\ \cos(dt)\ e^{ct}\\
&= e^{ct}\ \cos(dt)
\end{aligned}$$

and

$$\begin{aligned}
u_2(t) &= \frac{1}{2i}\left(e^{(c+id)t} - e^{(c-id)t}\right)\\
&= \frac{1}{2i}e^{ct}\left(\cos(dt) + i\ \sin(dt) - \cos(dt) + i\ \sin(dt)\right)\\
&= \frac{1}{2i}\,2\,i\ \sin(dt)\ e^{ct}\\
&= e^{ct}\ \sin(dt)
\end{aligned}$$

are both real valued solutions! So we will use as general solution to this problem

$$u(t) = A\ e^{ct}\ \cos(dt) + B\ e^{ct}\ \sin(dt)$$

where the constants A and B are now restricted to be real numbers. To solve the initial value problem, then we have

$$\begin{aligned}
u(0) &= A\ e^{c0}\ \cos(d0) + B\ e^{c0}\ \sin(d0)\ =\ A\ =\ u_0\\
u'(0) &= \left(c\ A\ e^{ct}\ \cos(dt) - d\ A\ e^{ct}\ \sin(dt) + c\ B\ e^{ct}\ \sin(dt) + d\ B\ e^{ct}\ \cos(dt)\right)\Big|_{t=0}\\
&= c\ A + d\ B\ =\ u_1.
\end{aligned}$$

Hence, to solve the initial value problem, we find A and B by solving the two equations in two unknowns below:

$$\begin{aligned}
A &= u_0\\
c\ A + d\ B &= u_1.
\end{aligned}$$

Example 7.4.1 For this model,

- Find the characteristic equation.
- Find the roots of the characteristic equation.
- Find the general complex solution.
- Find the general real solution.
- Solve the IVP.
- Plot and print the solution using MatLab.

$$\begin{aligned} u''(t) \ + 8\,u'(t) \ + 25\,u(t) &= 0 \\ u(0) &= 3 \\ u'(0) &= 4 \end{aligned}$$

Solution *(i): To find the characteristic equation, we assume the solution has the form e^{rt} and plug that into the problem. We find*

$$\left(r^2 + 8\,r + 25\right) e^{rt} = 0.$$

Since e^{rt} is never zero, we see this implies that

$$r^2 + 8\,r + 25 = 0.$$

This is the characteristic equation for this problem.
(ii): We find the roots of this characteristic equation using the quadratic formula. We find

$$r = \frac{-8 \pm \sqrt{64 - 100}}{2} = -4 \pm 3\,i.$$

Hence, the roots of this characteristic equation occur as the complex pair: $r_1 = -4 + 3\,i$ and $r_2 = -4 - 3\,i$.
(iii): The general complex solution is thus

$$u(t) = A\,e^{(-4+3\,i)t} + B\,e^{(-4-3\,i)t}.$$

(iv): The real solutions we want are then $e^{-4t}\cos(3t)$ and $e^{-4t}\sin(3t)$. The general real solution is thus

$$u(t) = A\,e^{-4t}\cos(3t) + B\,e^{-4t}\sin(3t).$$

for arbitrary real numbers A and B. (v): Next, we find the values of A and B which will let the solution satisfy the initial conditions. We have

$$
\begin{aligned}
u(0) &= \left(A\,e^{-4t}\,\cos(3t) + B\,e^{-4t}\,\sin(3t)\right)\Big|_{t=0} \\
&= A \,=\, 3, \\
u'(0) &= \left(-4Ae^{-4t}\cos(3t) - 3Ae^{-4t}\sin(3t) - 4Be^{-4t}\sin(3t) + 3Be^{-4t}\cos(3t)\right)\Big|_{t=0} \\
&= -4A + 3B \,=\, 4.
\end{aligned}
$$

This gives the system of two equations in the two unknowns A and B

$$
\begin{aligned}
A &= 3 \\
-4\,A \,+\, 3\,B &= 4.
\end{aligned}
$$

It then follows that $B = 16/3$. *Thus, the solution to this initial value problem is*

$$
u(t) = 3\,e^{-4t}\,\cos(3t) + 16/3\,e^{-4t}\,\sin(3t).
$$

(v): Finally, we can graph this solution in MatLab as follows:

Listing 7.3: Solution to $u'' + 8u' + 25u = 0$; $u(0) = 3$, $u'(0) = 4$

```
t = linspace(0,2,101);
A = 3;
B = 16.0/3.0;
u = A*exp(-4*t).*cos(3*t) + B*exp(-4*t).*sin(3*t);
plot(t,u);
```

which generates the plot we see in Fig. 7.3.

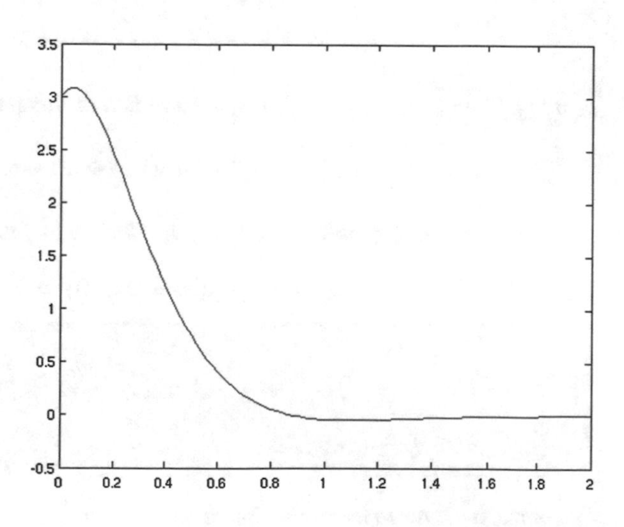

Fig. 7.3 Linear second order problem: complex roots

Example 7.4.2 For

$$u''(t) \;-8\,u'(t)\;+20\,u(t) = 0$$
$$u(0) = 3$$
$$u'(0) = 4$$

- Derive the characteristic equation and find its roots.
- Find the general complex and general real solution.
- Solve the IVP and plot and print the solution using MatLab.

Solution *(i): To find the characteristic equation, we assume the solution is* e^{rt}. *We find* $(r^2 - 8\,r + 20)\,e^{rt} = 0$. *Since* e^{rt} *is never zero, we must have* $r^2 - 8\,r\, + 20 = 0$. *Using the quadratic formula, we find* $r = \frac{8 \pm \sqrt{64-80}}{2} = 4 \pm 2\,i$.
(ii): The general complex solution is $\phi(t) = c_1\,e^{(4+2\,i)t} + c_2\,e^{(4-2\,i)t}$.
(iii): The general real solution is thus $u(t) = A\,e^{4t}\;\cos(2t) + B\,e^{4t}\;\sin(2t)$ *for arbitrary real numbers A and B. Thus,*

$$u'(t) = 4A\,e^{4t}\;\cos(2t) - 2Ae4t\sin(2t)$$
$$+4B\,e^{4t}\;\sin(2t) + 2Be^{4t}\cos(2t).$$

(iv): Next, we find the values of A and B which will let the solution satisfy the initial conditions. We have $u(0) = 3$ *and* $u'(0) = 4A + 2B \;=\; 4$. *This gives the system of two equations in the two unknowns A and B*

$$A = 3$$
$$4\,A\;+\;2\,B = 4 \Rightarrow 2B = 4 - 4A = -8.$$

It then follows that $B = -4$. *Thus, the solution to this initial value problem is*

$$u(t) = 3\,e^{4t}\;\cos(2t)\;\;-4\;\,e^{4t}\;\sin(2t).$$

(v): Finally, we can graph this solution in MatLab as follows:

Listing 7.4: Solution to $u'' + 8u' + 20u = 0$; $u(0) = 3$, $u'(0) = 4$

```
t = linspace(0,2,101);
A = 3;
B = -4;
f = @(t) exp(4.0*t).*( A*cos(2*t)  + B*sin(2*t) );
plot(t,f(t));
```

Of course, we have to judge the time interval to choose for the **`linspace`** *command. We see the plot in Fig. 7.4. Note as t gets large,* $u(t)$ *oscillates out of control.*

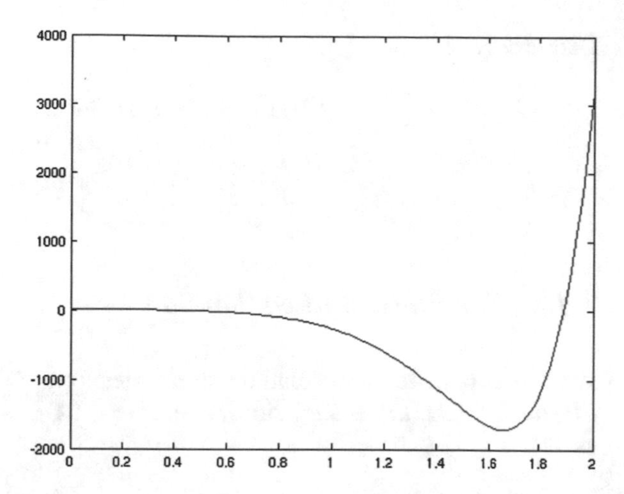

Fig. 7.4 Linear second order problem: complex roots two

7.4.1 Homework

For these models,

- Find the characteristic equation and roots.
- Find the general complex solution.
- Find the general real solution.
- Solve the IVP.
- Plot and print the solution using MatLab.

Exercise 7.4.1

$$\begin{aligned} y''(t) \; - 8\, y'(t) \; + 41\, y(t) &= 0 \\ y(0) &= +3 \\ y'(0) &= -2. \end{aligned}$$

Exercise 7.4.2

$$\begin{aligned} x''(t) \; - 2\, x'(t) \; + 2\, x(t) &= 0 \\ x(0) &= -5 \\ x'(0) &= 1. \end{aligned}$$

Exercise 7.4.3

$$\begin{aligned} u''(t) \; - 2\, u'(t) \; + 10\, u(t) &= 0 \\ u(0) &= +1 \\ u'(0) &= -2. \end{aligned}$$

Exercise 7.4.4

$$\begin{aligned} P''(t) \ -6\,P'(t) \ +13\,P(t) &= 0 \\ P(0) &= 1 \\ P'(0) &= 2. \end{aligned}$$

7.4.2 The Phase Shifted Solution

We can also write these solutions in another form. Our solutions here look like $u(t) = A\,e^{ct}\ \cos(dt) + B\,e^{ct}\ \sin(dt)$ Let $R = \sqrt{(A)^2\ +\ (B)^2}$. Rewrite the solution as

$$u(t) = R\,e^{ct}\left(\frac{A}{R}\cos(dt) + \frac{B}{R}\ \sin(dt)\right).$$

Define the angle δ by $\tan(\delta) = \frac{B}{A}$. Then the angle's value will depend on where A and B just like when we find angles for complex numbers and vectors. So $\cos(\delta) = \frac{A}{R}$ and $\sin(\delta) = \frac{B}{R}$. Now, there is a trigonometric identity $\cos(E - F) = \cos(E)\cos(F) + \sin(E)\sin(F)$. Here we have

$$\begin{aligned} u(t) &= R\,e^{ct}\left(\cos(\delta)\cos(dt) + \sin(\delta)\ \sin(dt)\right) \\ &= R\,e^{ct}\ \cos(dt - \delta). \end{aligned}$$

The angle δ is called the **phase shift**. When written in this form, the solution is said to be in **phase shifted cosine form**.

Example 7.4.3 Consider the solution $u(t) = 3\,e^{-4t}\ \cos(3t) + (16/3)\,e^{-4t}\ \sin(3t)$. which gives $u(0) = 3$ and $u'(0) = 4$. Find the phase shifted cosine solution.

Solution *Let* $R = \sqrt{(3)^2\ +\ (16/3)^2} = \sqrt{265}/3$. *A and B are positive so they are in Quadrant 1. So* $\delta = tan^{-1}(16/9)$. *We have* $u(t) = \sqrt{(3)^2\ +\ (16/3)^2}e^{-4t}\cos(3t-\delta)$. *To draw this solution by hand, do the following:*

- *On your graph, draw the curve* $(\sqrt{265}/3)e^{-4t}$. *This is the* **top** *curve that bounds our solution called the top envelope.*
- *On your graph, draw the curve* $-(\sqrt{265}/3)e^{-4t}$. *This is the* **bottom** *curve that bounds our solution called the bottom envelope.*
- *Draw the point* (0, 3) *and from it draw an arrow pointing up as the initial slope is positive.*

- *The solution starts at $t = 0$ and points up. It hits the* **top** *curve when the* cos *term hits its maximum of* 1. *It then flips and moves towards its minimum value of* -1 *where it hits the* **bottom** *curve.*
- *Keep drawing the curve as it hits top and bottom in a cycle. This graph is exponential decay that is oscillating towards zero.*

Example 7.4.4 Consider the solution $u(t) = 3\,e^{2t}\,\cos(4t) - 5\,e^{2t}\,\sin(4t)$. This gives $u(0) = 3$ and $u'(0) = 6 - 20 = -14$. Convert to phase shifted form.

Solution *Let* $R = \sqrt{(3)^2 + (-5)^2} = \sqrt{34}$. *Since* $A = 3$ *and* $B = -5$, *this is quadrant* 4. *So we use* $\delta = 2\pi - tan^{-1}(5/3)$. *We have* $u(t) = \sqrt{34}\,e^{2t}\,\cos(4t - \delta)$. *To draw the solution by hand, do this:*

- *On your graph, draw the curve* $\sqrt{34}e^{2t}$. *This is the* **top** *curve that bounds our solution.*
- *On your graph, draw the curve* $-\sqrt{34}e^{2t}$. *This is the* **bottom** *curve that bounds our solution.*
- *Draw the point* $(0, 3)$ *and from it draw an arrow pointing down as the initial slope is negative.*
- *The solution starts at $t = 0$ and points down. It hits the* **bottom** *curve when the* cos *term hits its minimum of* -1. *It then flips and moves towards its maximum value of* 1 *where it hits the* **top** *curve.*
- *Keep drawing the curve as it hits bottom and top in a cycle. This graph is exponential growth that is oscillating out of control.*

Example 7.4.5 Consider the solution $u(t) = -8\,e^{-2t}\,\cos(2t) - 6\,e^{-2t}\,\sin(2t)$. Here $u(0) = -8$ and $u'(0) = 4$. Find the phase shifted form.

Solution *Let* $R = \sqrt{(-8)^2 + (-6)^2} = 10$. *A is negative and B is negative so they are in Quadrant 3. So the angle is* $\delta = \pi + tan^{-1}(6/8)$. *We have* $u(t) = 10\,e^{-2t}\,\cos(2t - \delta)$. *Then draw the solution like so:*

- *On your graph, draw the curve* $10e^{-2t}$. *This is the* **top** *curve that bounds our solution.*
- *On your graph, draw the curve* $-10e^{2t}$. *This is the* **bottom** *curve that bounds our solution.*
- *Draw the point* $(0, -8)$ *and from it draw an arrow pointing up as the initial slope is positive.*
- *The solution starts at $t = 0$ and points up. It hits the* **top** *curve when the* cos *term hits its maximum of* 1. *It then flips and moves towards its minimum value of* -1 *where it hits the* **bottom** *curve.*
- *Keep drawing the curve as it hits top and bottom in a cycle. This graph is exponential decay that is oscillating towards zero.*

7.4.2.1 Homework

For the models below,

- Find the characteristic equation and roots.
- Find the general solution and solve the IVP.
- Convert the solution to phase shifted cosine form.

Exercise 7.4.5

$$\begin{aligned} x''(t) \ + 4\,x'(t) \ + 29\,x(t) &= 0 \\ x(0) &= 8 \\ x'(0) &= -2. \end{aligned}$$

Exercise 7.4.6

$$\begin{aligned} y''(t) \ + 6\,y'(t) \ + 45\,y(t) &= 0 \\ y(0) &= 6 \\ y'(0) &= 1. \end{aligned}$$

Exercise 7.4.7

$$\begin{aligned} u''(t) \ - 4\,u'(t) \ + 8\,u(t) &= 0 \\ u(0) &= -3 \\ u'(0) &= 2. \end{aligned}$$

Chapter 8
Systems

We are now ready to solve what are called *Linear Systems of differential equations*. These have the form

$$x'(t) = a\,x(t) + b\,y(t) \tag{8.1}$$
$$y'(t) = c\,x(t) + d\,y(t) \tag{8.2}$$
$$x(0) = x_0 \tag{8.3}$$
$$y(0) = y_0 \tag{8.4}$$

for any numbers a, b, c and d and *initial conditions* x_0 and y_0. The full problem is called, as usual, an *Initial Value Problem* or **IVP** for short. The two initial conditions are just called the **IC**'s for the problem to save writing. For example, we might be interested in the system

$$\begin{aligned} x'(t) &= -2\,x(t) + 3\,y(t) \\ y'(t) &= 4\,x(t) + 5\,y(t) \\ x(0) &= 5 \\ y(0) &= -3 \end{aligned}$$

Here the **IC**'s are $x(0) = 5$ and $y(0) = -3$. Another sample problem might be the one below.

$$\begin{aligned} x'(t) &= 14\,x(t) + 5\,y(t) \\ y'(t) &= -4\,x(t) + 8\,y(t) \\ x(0) &= 2 \\ y(0) &= 7 \end{aligned}$$

We are interested in learning how to solve these problems.

J.K. Peterson, *Calculus for Cognitive Scientists*,
Cognitive Science and Technology, DOI 10.1007/978-981-287-877-9_8

8.1 Finding a Solution

For linear first order problems like $u' = 3u$ and so forth, we have found the solution has the form $u(t) = \alpha\, e^{3t}$ for some number α. We would then determine the value of α to use by looking at the initial condition. To see what to do with Eqs. 8.1 and 8.2, first let's rewrite the problem in terms of matrices and vectors. In this form, Eqs. 8.1 and 8.2 can be written as

$$\begin{bmatrix} x'(t) \\ y'(t) \end{bmatrix} = \begin{bmatrix} a & b \\ c & d \end{bmatrix} \begin{bmatrix} x(t) \\ y(t) \end{bmatrix}.$$

The initial conditions Eqs. 8.3 and 8.4 can then be redone in vector form as

$$\begin{bmatrix} x(0) \\ y(0) \end{bmatrix} = \begin{bmatrix} x_0 \\ y_0 \end{bmatrix}.$$

8.1.1 Worked Out Examples

Here are some examples of the conversion of a system of two linear differential equations into matrix–vector form.

Example 8.1.1 Convert

$$\begin{aligned} x'(t) &= 6\, x(t) + 9\, y(t) \\ y'(t) &= -10\, x(t) + 15\, y(t) \\ x(0) &= 8 \\ y(0) &= 9 \end{aligned}$$

into a matrix–vector system.

Solution *The new form is seen to be*

$$\begin{bmatrix} x'(t) \\ y'(t) \end{bmatrix} = \begin{bmatrix} 6 & 9 \\ -10 & 15 \end{bmatrix} \begin{bmatrix} x(t) \\ y(t) \end{bmatrix}$$
$$\begin{bmatrix} x(0) \\ y(0) \end{bmatrix} = \begin{bmatrix} 8 \\ 9 \end{bmatrix}.$$

Example 8.1.2 Convert

$$\begin{aligned} x'(t) &= 2\, x(t) + 4\, y(t) \\ y'(t) &= -\, x(t) + 7\, y(t) \\ x(0) &= 2 \\ y(0) &= -3 \end{aligned}$$

into a matrix–vector system.

Solution *The new form is seen to be*

$$\begin{bmatrix} x'(t) \\ y'(t) \end{bmatrix} = \begin{bmatrix} 2 & 4 \\ -1 & 7 \end{bmatrix} \begin{bmatrix} x(t) \\ y(t) \end{bmatrix}$$
$$\begin{bmatrix} x(0) \\ y(0) \end{bmatrix} = \begin{bmatrix} 2 \\ -3 \end{bmatrix}.$$

8.1.2 A Judicious Guess

Now that we know how to do this conversion, it seems reasonable to believe that is a constant times e^{rt} solve a first order linear problem like $u' = ru$, perhaps a *vector* times e^{rt} will work here. Let's make this formal. We'll work with a specific system first because numbers are always easier to make sense of in the initial exposure to a technique. So let's look at the problem below

$$\begin{aligned} x'(t) &= 3\,x(t) + 2\,y(t) \\ y'(t) &= -4\,x(t) + 5\,y(t) \\ x(0) &= 2 \\ y(0) &= -3 \end{aligned}$$

Let's assume the solution has the form $\boldsymbol{V}\,e^{rt}$ because by our remarks above since this is a vector system it seems reasonable to move to using a vector rather than a constant. Let's denote the components of $\boldsymbol{V}$ as follows:

$$\boldsymbol{V} = \begin{bmatrix} V_1 \\ V_2 \end{bmatrix}.$$

Then, it is easy to see that the derivative of $\boldsymbol{V}\,e^{rt}$ is

$$\begin{aligned} \left(\boldsymbol{V}\,e^{rt}\right)' &= \left(\begin{bmatrix} V_1 \\ V_2 \end{bmatrix} e^{rt}\right)' \\ &= \left(\begin{bmatrix} V_1\,e^{rt} \\ V_2\,e^{rt} \end{bmatrix}\right)' \\ &= \begin{bmatrix} V_1 \left(e^{rt}\right)' \\ V_2 \left(e^{rt}\right)' \end{bmatrix} \\ &= \begin{bmatrix} V_1\,r\,e^{rt} \\ V_2\,r\,e^{rt} \end{bmatrix} \end{aligned}$$

$$= \begin{bmatrix} V_1 \\ V_2 \end{bmatrix} r\, e^{rt}$$
$$= r\, \boldsymbol{V}\, e^{rt}$$

Let's plug in our possible solution into the original problem. That is, we assume the solution is

$$\begin{bmatrix} x(t) \\ y(t) \end{bmatrix} = \boldsymbol{V}\, e^{rt}.$$

Hence,

$$\begin{bmatrix} x'(t) \\ y'(t) \end{bmatrix} = r\, \boldsymbol{V}\, e^{rt}.$$

When, we plug these terms into the matrix–vector form of the problem, we find

$$r\, \boldsymbol{V}\, e^{rt} = \begin{bmatrix} 3 & 2 \\ -4 & 5 \end{bmatrix} \boldsymbol{V}\, e^{rt}.$$

We can rewrite this as

$$r\, \boldsymbol{V}\, e^{rt} - \begin{bmatrix} 3 & 2 \\ -4 & 5 \end{bmatrix} \boldsymbol{V}\, e^{rt} = \begin{bmatrix} 0 \\ 0 \end{bmatrix}.$$

Since one of these terms is a matrix and one is a vector, we need to write all the terms in terms of matrices if possible. Recall, the two by two identity matrix is

$$I = \begin{bmatrix} 1 & 0 \\ 0 & 1 \end{bmatrix}$$

and $I\, \boldsymbol{V} = \boldsymbol{V}$ always. Thus, we can rewrite our system as

$$r \begin{bmatrix} 1 & 0 \\ 0 & 1 \end{bmatrix} \boldsymbol{V}\, e^{rt} - \begin{bmatrix} 3 & 2 \\ -4 & 5 \end{bmatrix} \boldsymbol{V}\, e^{rt} = \begin{bmatrix} 0 \\ 0 \end{bmatrix}.$$

Now, factor out the common e^{rt} term to give

$$\left(r \begin{bmatrix} 1 & 0 \\ 0 & 1 \end{bmatrix} \boldsymbol{V} - \begin{bmatrix} 3 & 2 \\ -4 & 5 \end{bmatrix} \boldsymbol{V} \right) e^{rt} = \begin{bmatrix} 0 \\ 0 \end{bmatrix}.$$

Even though we don't know yet what values of r will work for this problem, we do know that the term e^{rt} is never zero no matter what value r has. Hence, we can say that we are looking for a value of r and a vector $\boldsymbol{V}$ so that

$$r \begin{bmatrix} 1 & 0 \\ 0 & 1 \end{bmatrix} \boldsymbol{V} - \begin{bmatrix} 3 & 2 \\ -4 & 5 \end{bmatrix} \boldsymbol{V} = \begin{bmatrix} 0 \\ 0 \end{bmatrix}.$$

For convenience, let the matrix of coefficients determined by our system of differential equations be denoted by A, i.e.

$$A = \begin{bmatrix} 3 & 2 \\ -4 & 5 \end{bmatrix}.$$

Then, the equation that r and $\boldsymbol{V}$ must satisfy becomes

$$r\, I\, \boldsymbol{V} - A\, \boldsymbol{V} = \begin{bmatrix} 0 \\ 0 \end{bmatrix}.$$

Finally, noting the vector $\boldsymbol{V}$ is common, we factor again to get our last equation

$$\left(r\, I - A \right) \boldsymbol{V} = \begin{bmatrix} 0 \\ 0 \end{bmatrix}.$$

We can then plug in the value of I and A to get the system of equations that r and $\boldsymbol{V}$ must satisfy in order for $\boldsymbol{V}\, e^{rt}$ to be a solution.

$$\begin{bmatrix} r-3 & -2 \\ -(-4) & r-5 \end{bmatrix} \begin{bmatrix} V_1 \\ V_2 \end{bmatrix} = \begin{bmatrix} 0 \\ 0 \end{bmatrix}.$$

To finish this discussion, note that for any value of r, this is a system of two linear equations in the two unknowns V_1 and V_2. **If** we choose a value of r for which $det\,(r I - A)$ was non zero, the theory we have so carefully gone over in Sect. 2.4 tells us the two lines determined by row 1 and row 2 of this system have different slopes. This means this system of equations has only one solution. Since both equations cross through the origin, this unique solution must be $V_1 = 0$ and $V_2 = 0$. But, of course, this tells us the solution is $x(t) = 0$ and $y(t) = 0$! We will not be able to solve for the initial conditions $x(0) = 2$ and $y(0) = -3$ with this solution. So we must reject any choice of r which gives us $det\,(r I - A) \neq 0$.

This leaves only one choice: the values of r where $det\,(r I - A) = 0$. Now, go back to Sect. 2.10 where we discussed the **eigenvalues** and **eigenvectors** of a matrix A. The values of r where $det\,(r I - A) = 0$ are what we called the **eigenvalues** of our matrix A and for these values of r, we must find **non zero** vectors $\boldsymbol{V}$ (non zero because otherwise, we can't solve the IC's!) so that

$$\begin{bmatrix} r-3 & -2 \\ 4 & r-5 \end{bmatrix} \begin{bmatrix} V_1 \\ V_2 \end{bmatrix} = \begin{bmatrix} 0 \\ 0 \end{bmatrix}.$$

Note, as we did in Sect. 2.10, the system above is the same as

$$\begin{bmatrix} 3 & 2 \\ -4 & 5 \end{bmatrix} \begin{bmatrix} V_1 \\ V_2 \end{bmatrix} = r \begin{bmatrix} V_1 \\ V_2 \end{bmatrix}.$$

Then, for each eigenvalue we find, we should have a solution of the form

$$\begin{bmatrix} x(t) \\ y(t) \end{bmatrix} = \begin{bmatrix} V_1 \\ V_2 \end{bmatrix} e^{rt}.$$

In general, for a system of two linear models like this, there are three choices for the eigenvalues.

- Two real and distinct eigenvalues r_1 and r_2 which the eigenvectors E_1 and E_2. This has been discussed thoroughly. We can now say more about this type of solution. The two eigenvectors E_1 and E_2 are linearly independent vectors in $\Re^2$ and the two solutions $e^{r_1 tt}$ and $e^{r_2 t}$ are **linear independent functions**. Hence, the set of all possible solutions which is denoted by the general solution

$$\begin{bmatrix} x(t) \\ y(t) \end{bmatrix} = a \ E_1 \, e^{r_1 t} + b \ E_2 \, e^{r_2 t}$$

  represents the *span* of these two linearly independent functions. In fact, these two linearly independent solutions to the model are the basis *vectors* of the two dimensional vector space that consists of the solutions to this model. Note, we are not saying anything new here, but we are saying it with new terminology and a higher level of abstraction. Thus, the general solution will be

$$\begin{bmatrix} x(t) \\ y(t) \end{bmatrix} = a \, \boldsymbol{E}_1 \, e^{r_1 t} + b \, \boldsymbol{E}_2 \, e^{r_2 t},$$

  where $\boldsymbol{E}_1$ is the eigenvector for eigenvalue r_1 and $\boldsymbol{E}_2$ is the eigenvector for eigenvalue r_2 and a and b are arbitrary real numbers chosen to satisfy the IC's.
- The eigenvalues are repeated so that $r_1 = r_2 = \alpha$ for some real number. We are not yet sure what to do in this case. There are two possibilities:

  1. The eigenvalue of value α when plugged into the eigenvalue–eigenvector equation

$$\begin{bmatrix} \alpha - a & -b \\ c & \alpha - d \end{bmatrix} \begin{bmatrix} V_1 \\ V_2 \end{bmatrix} = \begin{bmatrix} 0 \\ 0 \end{bmatrix}.$$

     turns out to be as usual and the two rows of this matrix are multiples. Hence, we use either the top or bottom row to find our choice of nonzero eigenvector E_1.

$$\begin{bmatrix} \alpha - a & -b \\ c & \alpha - d \end{bmatrix} \begin{bmatrix} V_1 \\ V_2 \end{bmatrix} = \begin{bmatrix} 0 \\ 0 \end{bmatrix}.$$

For example, if

$$A = \begin{bmatrix} 3 & 1 \\ -1 & 1 \end{bmatrix}$$

the characteristic equation if $(r-2)^2 = 0$ which gives the repeated eigenvalue $\alpha = 2$. We then find the eigenvalue–eigenvector equation is

$$\begin{bmatrix} 2-3 & -1 \\ 1 & 2-1 \end{bmatrix} \begin{bmatrix} V_1 \\ V_2 \end{bmatrix} = \begin{bmatrix} 0 \\ 0 \end{bmatrix}.$$

The top row and the bottom row are multiples and we find $E_1 = \begin{bmatrix} 1 & -1 \end{bmatrix}^T$. Note in this case, the set of all V_1 and V_2 we can use are all multiples of E_1. Hence, this set of numbers forms a line through the origin in $\Re^2$. Another way of saying this is that the set of all possible V_1 and V_2 here is a one dimensional subspace of $\Re^2$. We know one solution to our model is $E_1 e^{2t}$ but what is the other one?

2. The other possibility is that A is a multiple of the identity, say $A = 2I$. Then, the characteristic equation is the same as the first case: $(r-2)^2$. However, the eigenvalue–eigenvector equation is very different. We find

$$\begin{bmatrix} 2-2 & 0 \\ 0 & 2-2 \end{bmatrix} \begin{bmatrix} V_1 \\ V_2 \end{bmatrix} = \begin{bmatrix} 0 \\ 0 \end{bmatrix}.$$

which is a very strange system as both the top and bottom equation give $0V_1 + 0V_2 = 0$. This says there are no restrictions on the values of V_1 and V_2 but they can be picked independently. So pick $V_1 = 1$ *and* $V_2 = 0$ to give one choice of eigenvector: $E_1 = \begin{bmatrix} 1 & 0 \end{bmatrix}^T$. Then pick $V_2 = 0$ *and* $V_2 = 1$ to give a second choice of eigenvector: $E_2 = \begin{bmatrix} 0 & 1 \end{bmatrix}^T$. Another way of looking at this is the set of all possible V_1 and V_2 is just $\Re^2$ and so we are free to pick any basis of $\Re^2$ for our eigenvectors we want. Hence, we might as well pick the simplest one: $E_1 = \boldsymbol{i}$ and $E_2 = \boldsymbol{j}$. We actually have two linearly independent solutions to our model in this case. They are $E_1 e^{2t}$ and $E_2 e^{2t}$ a

- In the last case, the eigenvalues are complex number. If we let the eigenvalue be $r = \alpha + \beta i$, note the corresponding eigenvector could be a complex vector. So let's write it as $\boldsymbol{V} = \boldsymbol{E} + i\boldsymbol{F}$ where $\boldsymbol{E}$ and $\boldsymbol{F}$ have only real valued components. Then we know

$$\begin{bmatrix} a & b \\ c & d \end{bmatrix} (\boldsymbol{E} + i\boldsymbol{F}) = r\,(\boldsymbol{E} + i\boldsymbol{F}).$$

Now take the complex conjugate of each side to get

$$\overline{\begin{bmatrix} a & b \\ c & d \end{bmatrix}}\, \overline{\boldsymbol{E} + i\boldsymbol{F}} = \overline{r}\, \overline{\boldsymbol{E} + i\boldsymbol{F}}.$$

But all the entries of the matrix are real, so complex conjugation does not change them. The other conjugations then give

$$\begin{bmatrix} a & b \\ c & d \end{bmatrix} (\boldsymbol{E} - i\boldsymbol{F}) = \overline{r}\,(\boldsymbol{E} - i\boldsymbol{F}).$$

This says that $\overline{r} = \alpha - \beta i$ is also an eigenvalue with eigenvector the complex conjugate of the eigenvector for r, i.e. $\boldsymbol{E} - i\boldsymbol{F}$. So eigenvalues and eigenvectors here occur in complex conjugate pairs and the general complex solution is

$$\begin{aligned} \begin{bmatrix} x(t) \\ y(t) \end{bmatrix} &= c_1\ (\boldsymbol{E} + i\boldsymbol{F})\, e^{(\alpha+\beta i)t} + c_2\ (\boldsymbol{E} - i\boldsymbol{F})\, e^{(\alpha-\beta i)t} \\ &= e^{\alpha t}\left((\boldsymbol{E} + i\boldsymbol{F})e^{i\beta t} + (\boldsymbol{E} - i\boldsymbol{F})e^{i\beta t}\right) \end{aligned}$$

where c_1 and c_2 are complex numbers. Since we are interested in real solutions, from this general complex solution, we will extract two linearly independent real solutions which will form the basis for our two dimensional subspace of solutions. We will return to this case later.

We are now ready for some definitions for *Characteristic Equation of the linear system*.

Definition 8.1.1 (*The Characteristic Equation of a Linear System of ODEs*)
For the system

$$\begin{aligned} x'(t) &= a\,x(t) + b\,y(t) \\ y'(t) &= c\,x(t) + d\,y(t) \\ x(0) &= x_0 \\ y(0) &= y_0, \end{aligned}$$

the characteristic equation is defined by

$$det\left(r\,I - A\right) = 0$$

where A is the coefficient matrix

$$A = \begin{bmatrix} a & b \\ c & d \end{bmatrix}.$$

We can then define the eigenvalue of a linear system of differential equations.

Definition 8.1.2 (*The Eigenvalues of a Linear System of ODEs*)
The roots of the characteristic equation of the linear system are called its eigenvalues and any nonzero vector $\boldsymbol{V}$ satisfying

$$\left(r\, I\, V - A \right) V = \begin{bmatrix} 0 \\ 0 \end{bmatrix}.$$

for an eigenvalue r is called an eigenvector associated with the eigenvalue r.

Finally, the general solution of this system can be built from its eigenvalues and associated eigenvectors. It is pretty straightforward when the two eigenvalues are real and distinct numbers and a bit more complicated in the other two cases. But don't worry, we'll cover all of it soon enough. Before we go on, let's do some more characteristic equation derivations.

8.1.3 Sample Characteristic Equation Derivations

Let's do some more examples. Here is the first one.

Example 8.1.3 Derive the characteristic equation for the system below

$$\begin{aligned} x'(t) &= 8\, x(t) + 9\, y(t) \\ y'(t) &= 3\, x(t) - 2\, y(t) \\ x(0) &= 12 \\ y(0) &= 4 \end{aligned}$$

Solution *First, note the matrix–vector form is*

$$\begin{bmatrix} x'(t) \\ y'(t) \end{bmatrix} = \begin{bmatrix} 8 & 9 \\ 3 & -2 \end{bmatrix} \begin{bmatrix} x(t) \\ y(t) \end{bmatrix}.$$
$$\begin{bmatrix} x(0) \\ y(0) \end{bmatrix} = \begin{bmatrix} 12 \\ 4 \end{bmatrix}.$$

The coefficient matrix A is thus

$$A = \begin{bmatrix} 8 & 9 \\ 3 & -2 \end{bmatrix}.$$

We assume the solution has the form $\boldsymbol{V}\, e^{rt}$ *and plug this into the system. This gives*

$$r\, \boldsymbol{V}\, e^{rt} - \begin{bmatrix} 8 & 9 \\ 3 & -2 \end{bmatrix} \boldsymbol{V}\, e^{rt} = \begin{bmatrix} 0 \\ 0 \end{bmatrix}.$$

Now rewrite using the identity matrix I and factor to obtain

$$\left(r\, I - \begin{bmatrix} 8 & 9 \\ 3 & -2 \end{bmatrix} \right) \boldsymbol{V}\, e^{rt} = \begin{bmatrix} 0 \\ 0 \end{bmatrix}.$$

Then, since e^{rt} can never be zero no matter what value r is, we find the values of r and the vectors $\mathbf{V}$ *we seek satisfy*

$$\left(r\,I - \begin{bmatrix} 8 & 9 \\ 3 & -2 \end{bmatrix}\right) \mathbf{V} = \begin{bmatrix} 0 \\ 0 \end{bmatrix}.$$

Now, if r is chosen so that $det\ (rI - A) \neq 0$, the only solution to this system of two linear equations in the two unknowns V_1 and V_2 is $V_1 = 0$ and $V_2 = 0$. This leads to the solution $x(t) = 0$ and $y(t) = 0$ always and this solution does not satisfy the initial conditions. Hence, we must find values r which give $det\ (rI - A) = 0$. The resulting polynomial is

$$\begin{aligned} det\left(r\,I - \begin{bmatrix} 8 & 9 \\ 3 & -2 \end{bmatrix}\right) &= det \begin{bmatrix} r-8 & -9 \\ -3 & r+2 \end{bmatrix} \\ &= (r-8)(r+2) - 27 = r^2 - 6r - 43. \end{aligned}$$

This is the characteristic equation of this system.

The next one is very similar. We will expect you to be able to do this kind of derivation also.

Example 8.1.4 Derive the characteristic equation for the system below

$$\begin{aligned} x'(t) &= -10\,x(t) - 7\,y(t) \\ y'(t) &= 8\,x(t) + 5\,y(t) \\ x(0) &= -1 \\ y(0) &= -4 \end{aligned}$$

Solution *We see the matrix–vector form is*

$$\begin{bmatrix} x'(t) \\ y'(t) \end{bmatrix} = \begin{bmatrix} -10 & -7 \\ 8 & 5 \end{bmatrix} \begin{bmatrix} x(t) \\ y(t) \end{bmatrix}.$$
$$\begin{bmatrix} x(0) \\ y(0) \end{bmatrix} = \begin{bmatrix} -1 \\ -4 \end{bmatrix}.$$

with coefficient matrix A given by

$$A = \begin{bmatrix} -10 & -7 \\ 8 & 5 \end{bmatrix}.$$

Assume the solution has the form $\mathbf{V}\ e^{rt}$ *and plug this into the system giving*

$$r\,\mathbf{V}\,e^{rt} - \begin{bmatrix} -10 & -7 \\ 8 & 5 \end{bmatrix} \mathbf{V}\,e^{rt} = \begin{bmatrix} 0 \\ 0 \end{bmatrix}.$$

Rewriting using the identity matrix I and factoring, we obtain

$$\left(r\, I - \begin{bmatrix} -10 & -7 \\ 8 & 5 \end{bmatrix} \right) \boldsymbol{V}\, e^{rt} = \begin{bmatrix} 0 \\ 0 \end{bmatrix}.$$

*Then, since e^{rt} can never be zero no matter what value r is, we find the values of r and the vectors **V** we seek satisfy*

$$\left(r\, I - \begin{bmatrix} -10 & -7 \\ 8 & 5 \end{bmatrix} \right) \boldsymbol{V} = \begin{bmatrix} 0 \\ 0 \end{bmatrix}.$$

Again, if r is chosen so that det $(rI - A) \neq 0$, the only solution to this system of two linear equations in the two unknowns V_1 and V_2 is $V_1 = 0$ and $V_2 = 0$. This gives us the solution $x(t) = 0$ and $y(t) = 0$ always and this solution does not satisfy the initial conditions. Hence, we must find values r which give det $(rI - A) = 0$. The resulting polynomial is

$$\begin{aligned} det \left(r\, I - \begin{bmatrix} -10 & -7 \\ 8 & 5 \end{bmatrix} \right) &= det \begin{bmatrix} r + 10 & 7 \\ -8 & r - 5 \end{bmatrix} \\ &= (r + 10)(r - 5) + 56 = r^2 + 5r + 6. \end{aligned}$$

This is the characteristic equation of this system.

8.1.4 Homework

For each of these problems,

- Write the matrix–vector form.
- Derive the characteristic equation but you don't have to find the roots.

Exercise 8.1.1

$$\begin{aligned} x' &= 2\, x + 3\, y \\ y' &= 8\, x - 2\, y \\ x(0) &= 3 \\ y(0) &= 5. \end{aligned}$$

Exercise 8.1.2

$$\begin{aligned} x' &= -4\, x + 6\, y \\ y' &= 9\, x + 2\, y \\ x(0) &= 4 \\ y(0) &= -6. \end{aligned}$$

8.2 Two Distinct Eigenvalues

Next, let's do the simple case of two distinct real eigenvalues. Before you look at these calculations, you should review how we found eigenvectors in Sect. 2.10. We worked out several examples there, however, now let's put them in this system context. Each eigenvalue r has a corresponding eigenvector $\boldsymbol{E}$. Since in this course, we want to concentrate on the situation where the two roots of the characteristic equation are distinct real numbers, we will want to find the eigenvector, $\boldsymbol{E}_1$, corresponding to eigenvalue r_1 and the eigenvector, $\boldsymbol{E}_2$, corresponding to eigenvalue r_2. The general solution will then be of the form

$$\begin{bmatrix} x(t) \\ y(t) \end{bmatrix} = a\,\boldsymbol{E}_1\,e^{r_1 t} + b\,\boldsymbol{E}_2\,e^{r_2 t}$$

where we will use the IC's to choose the correct values of a and b. Let's do a complete example now. We start with the system

$$\begin{aligned} x'(t) &= -3\,x(t) + 4\,y(t) \\ y'(t) &= -1\,x(t) + 2\,y(t) \\ x(0) &= 2 \\ y(0) &= -4 \end{aligned}$$

First, note the matrix–vector form is

$$\begin{aligned} \begin{bmatrix} x'(t) \\ y'(t) \end{bmatrix} &= \begin{bmatrix} -3 & 4 \\ -1 & 2 \end{bmatrix} \begin{bmatrix} x(t) \\ y(t) \end{bmatrix}. \\ \begin{bmatrix} x(0) \\ y(0) \end{bmatrix} &= \begin{bmatrix} 2 \\ -4 \end{bmatrix}. \end{aligned}$$

The coefficient matrix A is thus

$$A = \begin{bmatrix} -3 & 4 \\ -1 & 2 \end{bmatrix}.$$

The characteristic equation is thus

$$\begin{aligned} det\left(r\,I - \begin{bmatrix} -3 & 4 \\ -1 & 2 \end{bmatrix}\right) &= det\begin{bmatrix} r+3 & -4 \\ 1 & r-2 \end{bmatrix} \\ &= (r+3)(r-2)+4 \;=\; r^2 + r - 2 \\ &= (r+2)(r-1). \end{aligned}$$

Thus, the eigenvalues of the coefficient matrix A are $r_1 = -2$ and $r = 1$. The general solution will then be of the form

$$\begin{bmatrix} x(t) \\ y(t) \end{bmatrix} = a\, \boldsymbol{E}_1\, e^{-2t} + b\, \boldsymbol{E}_2\, e^{t}$$

Next, we find the **eigenvectors** associated with these eigenvalues.

1. For eigenvalue $r_1 = -2$, substitute the value of this eigenvalue into

$$\begin{bmatrix} r+3 & -4 \\ 1 & r-2 \end{bmatrix}$$

This gives

$$\begin{bmatrix} 1 & -4 \\ 1 & -4 \end{bmatrix}$$

The two rows of this matrix should be multiples of one another. If not, we made a mistake and we have to go back and find it. Our rows are indeed multiples, so pick one row to solve for the eigenvector. We need to solve

$$\begin{bmatrix} 1 & -4 \\ 1 & -4 \end{bmatrix} \begin{bmatrix} v_1 \\ v_2 \end{bmatrix} = \begin{bmatrix} 0 \\ 0 \end{bmatrix}$$

Picking the top row, we get

$$\begin{aligned} v_1 - 4\, v_2 &= 0 \\ v_2 &= \frac{1}{4} v_1 \end{aligned}$$

Letting $v_1 = a$, we find the solutions have the form

$$\begin{bmatrix} v_1 \\ v_2 \end{bmatrix} = a \begin{bmatrix} 1 \\ \frac{1}{4} \end{bmatrix}$$

The vector

$$\boldsymbol{E}_1 = \begin{bmatrix} 1 \\ 1/4 \end{bmatrix}$$

is our choice for an eigenvector corresponding to eigenvalue $r_1 = -2$. Thus, the first solution to our system is

$$\begin{bmatrix} x_1(t) \\ y_1(t) \end{bmatrix} = \boldsymbol{E}_1\, e^{-2t} = \begin{bmatrix} 1 \\ 1/4 \end{bmatrix} e^{-2t}.$$

2. For eigenvalue $r_2 = 1$, substitute the value of this eigenvalue into

$$\begin{bmatrix} r+3 & -4 \\ 1 & r-2 \end{bmatrix}$$

This gives

$$\begin{bmatrix} 4 & -4 \\ 1 & -1 \end{bmatrix}$$

Again, the two rows of this matrix should be multiples of one another. If not, we made a mistake and we have to go back and find it. Our rows are indeed multiples, so pick one row to solve for the eigenvector. We need to solve

$$\begin{bmatrix} 4 & -4 \\ 1 & -1 \end{bmatrix} \begin{bmatrix} v_1 \\ v_2 \end{bmatrix} = \begin{bmatrix} 0 \\ 0 \end{bmatrix}$$

Picking the bottom row, we get

$$\begin{aligned} v_1 - v_2 &= 0 \\ v_2 &= v_1 \end{aligned}$$

Letting $v_1 = b$, we find the solutions have the form

$$\begin{bmatrix} v_1 \\ v_2 \end{bmatrix} = b \begin{bmatrix} 1 \\ 1 \end{bmatrix}$$

The vector

$$\boldsymbol{E}_2 = \begin{bmatrix} 1 \\ 1 \end{bmatrix}$$

is our choice for an eigenvector corresponding to eigenvalue $r_2 = 1$. Thus, the second solution to our system is

$$\begin{bmatrix} x_2(t) \\ y_2(t) \end{bmatrix} = \boldsymbol{E}_2 \, e^t = \begin{bmatrix} 1 \\ 1 \end{bmatrix} e^t.$$

The general solution is therefore

$$\begin{aligned} \begin{bmatrix} x(t) \\ y(t) \end{bmatrix} &= a \begin{bmatrix} x_1(t) \\ y_1(t) \end{bmatrix} + b \begin{bmatrix} x_2(t) \\ y_2(t) \end{bmatrix} \\ &= a \begin{bmatrix} 1 \\ 1/4 \end{bmatrix} e^{-2t} + b \begin{bmatrix} 1 \\ 1 \end{bmatrix} e^t. \end{aligned}$$

Finally, we solve the IVP. Given the IC's, we find two equations in two unknowns for a and b:

$$\begin{bmatrix} x(0) \\ y(0) \end{bmatrix} = \begin{bmatrix} 2 \\ -4 \end{bmatrix} = a \begin{bmatrix} 1 \\ 1/4 \end{bmatrix} e^0 + b \begin{bmatrix} 1 \\ 1 \end{bmatrix} e^0$$
$$= \begin{bmatrix} a+b \\ (1/4)a+b \end{bmatrix}.$$

This is the usual system

$$a + b = 2$$
$$(1/4)a + b = -4.$$

This easily solves to give $a = 24/5$ and $b = -14/5$. Hence, the solution to this IVP is

$$\begin{bmatrix} x(t) \\ y(t) \end{bmatrix} = a \begin{bmatrix} x_1(t) \\ y_1(t) \end{bmatrix} + b \begin{bmatrix} x_2(t) \\ y_2(t) \end{bmatrix}$$
$$= (24/5) \begin{bmatrix} 1 \\ 1/4 \end{bmatrix} e^{-2t} + (-14/5) \begin{bmatrix} 1 \\ 1 \end{bmatrix} e^t.$$

This can be rewritten as

$$x(t) = (24/5)\, e^{-2t} - (14/5)\, e^t$$
$$y(t) = (6/5)\, e^{-2t} - (14/5)\, e^t.$$

Note when t is very large, the only terms that matter are the ones which grow fastest. Hence, we could say

$$x(t) \approx -\, (14/5)\, e^t$$
$$y(t) \approx -\, (14/5)\, e^t.$$

or in vector form

$$\begin{bmatrix} x(t) \\ y(t) \end{bmatrix} \approx -(14/5) \begin{bmatrix} 1 \\ 1 \end{bmatrix} e^t.$$

This is just a multiple of the eigenvector $\boldsymbol{E}_2$! Note the graph of $x(t)$ and $y(t)$ on an x–y plane will get closer and closer to the straight line determined by this eigenvector. So we will call $\boldsymbol{E}_2$ the dominant eigenvector direction for this system.

8.2.1 Worked Out Solutions

You need some additional practice. Let's work out a few more. Here is the first one.

Example 8.2.1 For the system below

$$\begin{bmatrix} x'(t) \\ y'(t) \end{bmatrix} = \begin{bmatrix} -20 & 12 \\ -13 & 5 \end{bmatrix} \begin{bmatrix} x(t) \\ y(t) \end{bmatrix}$$
$$\begin{bmatrix} x(0) \\ y(0) \end{bmatrix} = \begin{bmatrix} -1 \\ 2 \end{bmatrix}$$

- Find the characteristic equation
- Find the general solution
- Solve the IVP

Solution *The characteristic equation is*

$$det\left(r \begin{bmatrix} 1 & 0 \\ 0 & 1 \end{bmatrix} - \begin{bmatrix} -20 & 12 \\ -13 & 5 \end{bmatrix}\right) = 0$$

or

$$\begin{aligned} 0 &= det\left(\begin{bmatrix} r+20 & -12 \\ 13 & r-5 \end{bmatrix}\right) \\ &= (r+20)(r-5)+156 \\ &= r^2+15r+56 \\ &= (r+8)(r+7) \end{aligned}$$

Hence, **eigenvalues**, *of the characteristic equation are $r_1 = -8$ and $r_2 = -7$. We need to find the associated* **eigenvectors** *for these eigenvalues.*

1. *For eigenvalue $r_1 = -8$, substitute the value of this eigenvalue into*

$$\begin{bmatrix} r+20 & -12 \\ 13 & r-5 \end{bmatrix}$$

This gives

$$\begin{bmatrix} 12 & -12 \\ 13 & -13 \end{bmatrix}$$

Again, the two rows of this matrix should be multiples of one another. If not, we made a mistake and we have to go back and find it. Our rows are indeed multiples, so pick one row to solve for the eigenvector. We need to solve

$$\begin{bmatrix} 12 & -12 \\ 3 & -13 \end{bmatrix} \begin{bmatrix} v_1 \\ v_2 \end{bmatrix} = \begin{bmatrix} 0 \\ 0 \end{bmatrix}$$

Picking the top row, we get

$$12v_1 - 12\,v_2 = 0$$
$$v_2 = v_1$$

Letting $v_1 = a$, we find the solutions have the form

$$\begin{bmatrix} v_1 \\ v_2 \end{bmatrix} = a \begin{bmatrix} 1 \\ 1 \end{bmatrix}$$

The vector

$$\boldsymbol{E}_1 = \begin{bmatrix} 1 \\ 1 \end{bmatrix}$$

is our choice for an eigenvector corresponding to eigenvalue $r_1 = -8$.

2. *For eigenvalue $r_2 = -7$, substitute the value of this eigenvalue into*

$$\begin{bmatrix} r + 20 & -12 \\ 13 & r - 5 \end{bmatrix}$$

This gives

$$\begin{bmatrix} 13 & -12 \\ 13 & -12 \end{bmatrix}$$

Again,the two rows of this matrix should be multiples of one another. Picking the bottom row, we get

$$13v_1 - 12v_2 = 0$$
$$v_2 = (13/12)v_1$$

Letting $v_1 = b$, we find the solutions have the form

$$\begin{bmatrix} v_1 \\ v_2 \end{bmatrix} = b \begin{bmatrix} 1 \\ 13/12 \end{bmatrix}$$

The vector

$$\boldsymbol{E}_2 = \begin{bmatrix} 1 \\ 13/12 \end{bmatrix}$$

is our choice for an eigenvector corresponding to eigenvalue $r_2 = -7$.

The general solution to our system is thus

$$\begin{bmatrix} x(t) \\ y(t) \end{bmatrix} = a \begin{bmatrix} 1 \\ 1 \end{bmatrix} e^{-2t} + b \begin{bmatrix} 1 \\ 13/12 \end{bmatrix} e^{t}$$

We solve the IVP by finding the a and b that will give the desired initial conditions. This gives

$$\begin{bmatrix} -1 \\ 2 \end{bmatrix} = a \begin{bmatrix} 1 \\ 1 \end{bmatrix} + b \begin{bmatrix} 1 \\ 13/12 \end{bmatrix}$$

or

$$\begin{aligned} -1 &= a + b \\ 2 &= a + (13/12)b \end{aligned}$$

This is easily solved using elimination to give $a = -37$ *and* $b = 36$. *The solution to the IVP is therefore*

$$\begin{aligned} \begin{bmatrix} x(t) \\ y(t) \end{bmatrix} &= -37 \begin{bmatrix} 1 \\ 1 \end{bmatrix} e^{-8t} + 36 \begin{bmatrix} 1 \\ 13/12 \end{bmatrix} e^{-7t} \\ &= \begin{bmatrix} -37\, e^{-8t} + 36\, e^{-7t} \\ -37\, e^{-8t} - 36(13/12)\, e^{-7t} \end{bmatrix} \end{aligned}$$

Note when t is very large, the only terms that matter are the ones which grow fastest or, in this case, the ones which decay the slowest. Hence, we could say

$$\begin{bmatrix} x(t) \\ y(t) \end{bmatrix} \approx 36 \begin{bmatrix} 1 \\ (13/12) \end{bmatrix} e^{-7t}.$$

This is just a multiple of the eigenvector $\mathbf{E}_2$*! Note the graph of* $x(t)$ *and* $y(t)$ *on an x–y plane will get closer and closer to the straight line determined by this eigenvector. So we will call* $\mathbf{E}_2$ *the dominant eigenvector direction for this system.*

We now have all the information needed to analyze the solutions to this system graphically.

Here is another example is great detail. Again, remember you will have to know how to do these steps yourselves.

Example 8.2.2 For the system below

$$\begin{aligned} \begin{bmatrix} x'(t) \\ y'(t) \end{bmatrix} &= \begin{bmatrix} 4 & 9 \\ -1 & -6 \end{bmatrix} \begin{bmatrix} x(t) \\ y(t) \end{bmatrix} \\ \begin{bmatrix} x(0) \\ y(0) \end{bmatrix} &= \begin{bmatrix} 4 \\ -2 \end{bmatrix} \end{aligned}$$

- Find the characteristic equation
- Find the general solution
- Solve the IVP

Solution *The characteristic equation is*

$$det\left(r\begin{bmatrix}1 & 0\\ 0 & 1\end{bmatrix} - \begin{bmatrix}4 & 9\\ -1 & -6\end{bmatrix}\right) = 0$$

or

$$\begin{aligned} 0 &= det\left(\begin{bmatrix}r-4 & -9\\ 1 & r+6\end{bmatrix}\right)\\ &= (r-4)(r+6)+9\\ &= r^2+2r-15\\ &= (r+5)(r-3)\end{aligned}$$

Hence, **eigenvalues** *of the characteristic equation are $r_1 = -5$ and $r_2 = 3$. Next, we find the* **eigenvectors**.

1. *For eigenvalue $r_1 = -5$, substitute the value of this eigenvalue into*

$$\begin{bmatrix}r-4 & -9\\ 1 & r+6\end{bmatrix}$$

This gives

$$\begin{bmatrix}-9 & -9\\ 1 & 1\end{bmatrix}$$

We need to solve

$$\begin{bmatrix}-9 & -9\\ 1 & 1\end{bmatrix}\begin{bmatrix}v_1\\ v_2\end{bmatrix} = \begin{bmatrix}0\\ 0\end{bmatrix}$$

Picking the bottom row, we get

$$\begin{aligned} v_1 + v_2 &= 0\\ v_2 &= -v_1\end{aligned}$$

Letting $v_1 = a$, we find the solutions have the form

$$\begin{bmatrix}v_1\\ v_2\end{bmatrix} = a\begin{bmatrix}1\\ -1\end{bmatrix}$$

The vector

$$\boldsymbol{E}_1 = \begin{bmatrix} 1 \\ -1 \end{bmatrix}$$

is our choice for an eigenvector corresponding to eigenvalue $r_1 = -5$.

2. *For eigenvalue* $r_2 = 3$, *substitute the value of this eigenvalue into*

$$\begin{bmatrix} r-4 & -9 \\ 1 & r+6 \end{bmatrix}$$

This gives

$$\begin{bmatrix} -1 & -9 \\ 1 & 9 \end{bmatrix}$$

This time, we need to solve

$$\begin{bmatrix} -1 & -9 \\ 1 & 9 \end{bmatrix} \begin{bmatrix} v_1 \\ v_2 \end{bmatrix} = \begin{bmatrix} 0 \\ 0 \end{bmatrix}$$

Picking the bottom row, we get

$$v_1 + 9\, v_2 = 0$$
$$v_2 = \frac{-1}{9}\, v_1$$

Letting $v_1 = b$, *we find the solutions have the form*

$$\begin{bmatrix} v_1 \\ v_2 \end{bmatrix} = b \begin{bmatrix} 1 \\ -1/9 \end{bmatrix}$$

The vector

$$\begin{bmatrix} 1 \\ -1/9 \end{bmatrix}$$

is our choice for an eigenvector corresponding to eigenvalue $r_2 = 3$.
The general solution to our system is thus

$$\begin{bmatrix} x(t) \\ y(t) \end{bmatrix} = a \begin{bmatrix} 1 \\ -1 \end{bmatrix} e^{-5t} + b \begin{bmatrix} 1 \\ -1/9 \end{bmatrix} e^{3t}$$

We solve the IVP by finding the a and b that will give the desired initial conditions. This gives

$$\begin{bmatrix} 4 \\ -2 \end{bmatrix} = a \begin{bmatrix} 1 \\ -1 \end{bmatrix} + b \begin{bmatrix} 1 \\ -1/9 \end{bmatrix}$$

or

$$\begin{aligned} 4 &= a + b \\ -2 &= -a + \frac{-1}{9}\, b \end{aligned}$$

This is easily solved using elimination to give $a = 7/4$ *and* $B = 9/4$. *The solution to the IVP is therefore*

$$\begin{aligned} \begin{bmatrix} x(t) \\ y(t) \end{bmatrix} &= \frac{7}{4} \begin{bmatrix} 1 \\ -1 \end{bmatrix} e^{-5t} + \frac{9}{4} \begin{bmatrix} 1 \\ \frac{-1}{9} \end{bmatrix} e^{3t} \\ &= \begin{bmatrix} \frac{7}{4}\, e^{-5t} + \frac{9}{4}\, e^{3t} \\ \frac{-7}{4}\, e^{-5t} + \frac{-1}{4}\, e^{3t} \end{bmatrix} \end{aligned}$$

Again, the dominant eigenvector is $\boldsymbol{E}_2$.

8.2.2 Homework

For these problems:

1. Write matrix, vector form.
2. Find characteristic equation. No derivation needed this time.
3. Find the two eigenvalues.
4. Find the two associated eigenvectors in glorious detail.
5. Write general solution.
6. Solve the IVP.

Exercise 8.2.1

$$\begin{aligned} x' &= 3\,x + y \\ y' &= 5\,x - y \\ x(0) &= 4 \\ y(0) &= -6. \end{aligned}$$

The eigenvalues should be -2 *and* 4.

Exercise 8.2.2

$$\begin{aligned} x' &= x + 4\,y \\ y' &= 5\,x + 2\,y \\ x(0) &= 4 \\ y(0) &= -5. \end{aligned}$$

The eigenvalues should be -3 *and* 6.

Exercise 8.2.3

$$\begin{aligned} x' &= -3\,x + y \\ y' &= -4\,x + 2\,y \\ x(0) &= 1 \\ y(0) &= 6. \end{aligned}$$

The eigenvalues should be -2 *and* 1.

8.2.3 Graphical Analysis

Let's try to analyze these systems graphically. We are interested in what these solutions look like for many different initial conditions. So let's look at the problem

$$\begin{aligned} x'(t) &= -3\,x(t) + 4\,y(t) \\ y'(t) &= -x(t) + 2\,y(t) \\ x(0) &= x_0 \\ y(0) &= y_0 \end{aligned}$$

8.2.3.1 Graphing the Nullclines

The set of (x, y) pairs where $x' = 0$ is called the **nullcline** for x; similarly, the points where $y' = 0$ is the **nullcline** for y. The x' equation can be set equal to zero to get $-3x + 4y = 0$. This is the same as the straight line $y = 3/4\,x$. This straight line divides the x–y plane into three pieces: the part where $x' > 0$; the part where $x' = 0$; and, the part where $x' < 0$. In Fig. 8.1, we show the part of the x–y plane where $x' > 0$ with one shading and the part where it is negative with another. Similarly, the y' equation can be set to 0 to give the equation of the line $-x + 2y = 0$. This gives the straight line $y = 1/2\,x$. In Fig. 8.2, we show how this line also divided the x–y plane into three pieces .

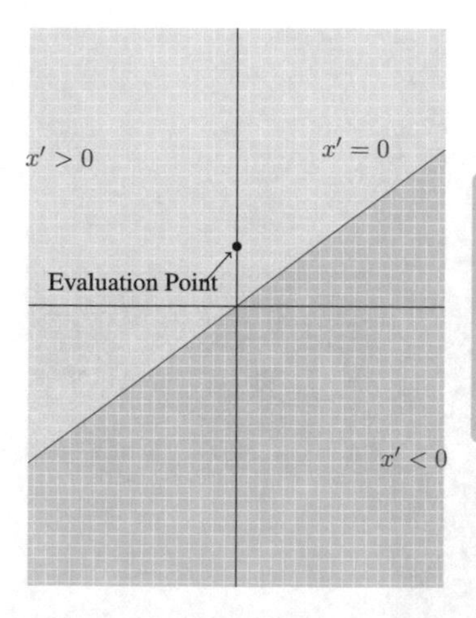

The x' equation for our system is $x' = -3x + 4y$. Setting this to 0, we get $y = 3/4\ x$ whose graph is shown. At the point $(0, 1)$, $x' = -3 \cdot 0 + 4 \cdot 1 = +4$. Hence, every point in the $x - y$ plane on this side of the $x' = 0$ line will give $x' > 0$. Hence, we have shaded the part of the plane where $x' > 0$ as shown.

Fig. 8.1 Finding where $x' < 0$ and $x' > 0$

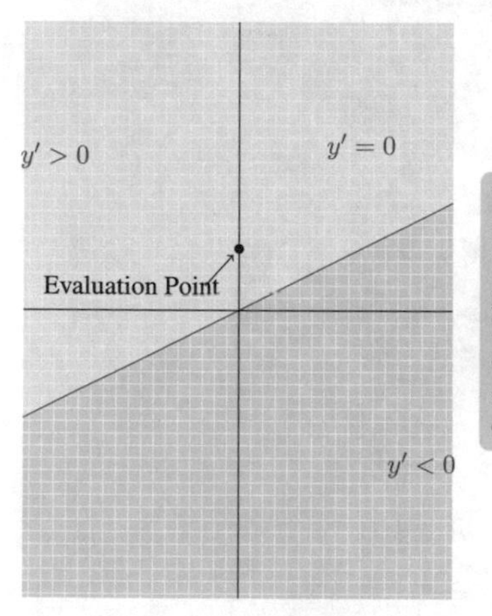

The y' equation for our system is $y' = -x + 2y$. Setting this to 0, we get $y = 1/2\ x$ whose graph is shown. At the point $(0, 1)$, $y' = -1 \cdot 0 + 2 \cdot 1 = +2$. Hence, every point in the $x - y$ plane on this side of the $y' = 0$ line will give $y' > 0$. Hence, we have shaded the part of the plane where $y' > 0$ as shown.

Fig. 8.2 Finding where $y' < 0$ and $y' > 0$

The shaded areas shown in Figs. 8.1 and 8.2 can be combined into Fig. 8.3. In this figure, we divide the x–y plane into four regions marked with a I, II, III or IV. In each region, x' and y' are either positive or negative. Hence, each region can be marked with an ordered pair, $(x'\pm,\ y'\pm)$.

8.2.3.2 Homework

For each of these problems,

- Find the x' nullcline and determine the plus and minus regions in the plane.
- Find the y' nullcline and determine the plus and minus regions in the plane.
- Assemble the two nullclines into one picture showing the four regions that result.

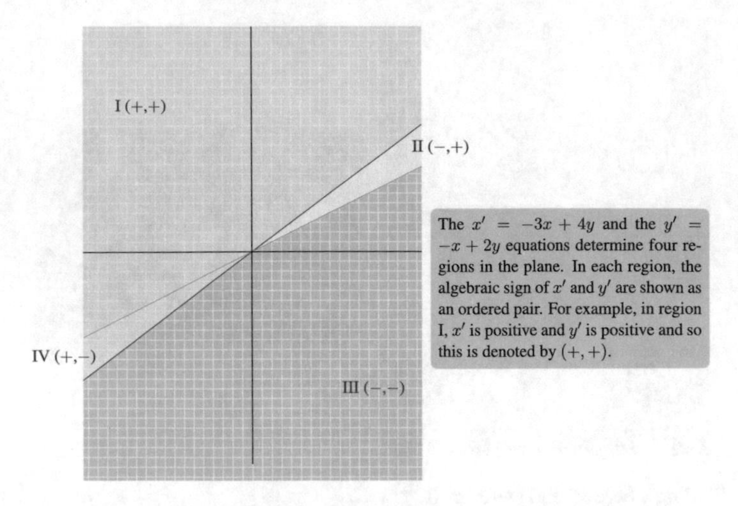

The $x' = -3x + 4y$ and the $y' = -x + 2y$ equations determine four regions in the plane. In each region, the algebraic sign of x' and y' are shown as an ordered pair. For example, in region I, x' is positive and y' is positive and so this is denoted by $(+, +)$.

Fig. 8.3 Combining the x' and y' algebraic sign regions

Exercise 8.2.4

$$\begin{aligned} x' &= 2\,x + 3\,y \\ y' &= 8\,x - 2\,y \\ x(0) &= 3 \\ y(0) &= 5. \end{aligned}$$

Exercise 8.2.5

$$\begin{aligned} x' &= -4\,x + 6\,y \\ y' &= 9\,x + 2\,y \\ x(0) &= 4 \\ y(0) &= -6. \end{aligned}$$

8.2.3.3 Graphing the Eigenvector Lines

Now we add the eigenvector lines. In Sect. 8.2, we found that this system has eigenvalues $r_1 = -2$ and $r_2 = 1$ with associated eigenvectors

$$\boldsymbol{E}_1 = \begin{bmatrix} 1 \\ 1/4 \end{bmatrix}, \ \boldsymbol{E}_2 = \begin{bmatrix} 1 \\ 1 \end{bmatrix}.$$

Recall a vector $\boldsymbol{V}$ with components a and b,

$$\boldsymbol{V} = \begin{bmatrix} a \\ b \end{bmatrix}$$

determines a straight line with slope b/a. Hence, these eigenvectors each determine a straight line. The $\boldsymbol{E}_1$ line has slope $1/4$ and the $\boldsymbol{E}_2$ line has slope 1. We can graph these two lines overlaid on the graph shown in Fig. 8.3.

8.2.3.4 Homework

For each of these problems,

- Find the x' nullcline and determine the plus and minus regions in the plane.
- Find the y' nullcline and determine the plus and minus regions in the plane.
- Assemble the two nullclines into one picture showing the four regions that result.
- Draw the Eigenvector lines on the same picture (Fig. 8.4).

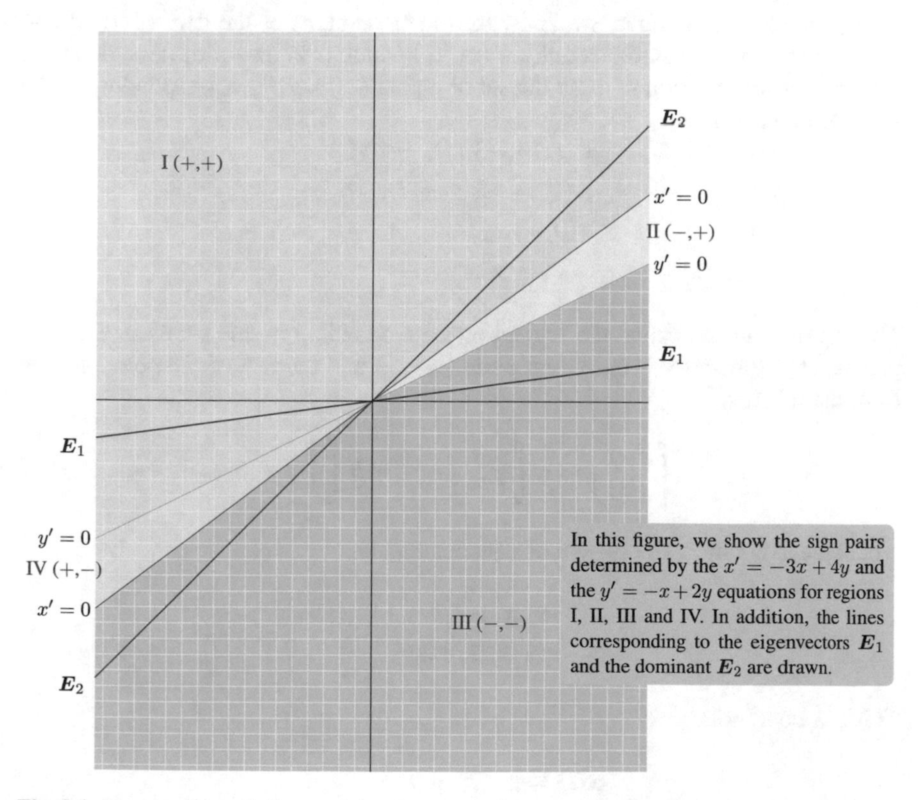

Fig. 8.4 Drawing the nullclines and the eigenvector lines on the same graph

Exercise 8.2.6

$$\begin{aligned} x' &= 2\,x + 3\,y \\ y' &= 8\,x - 2\,y \\ x(0) &= 3 \\ y(0) &= 5. \end{aligned}$$

Exercise 8.2.7

$$\begin{aligned} x' &= -4\,x + 6\,y \\ y' &= 9\,x + 2\,y \\ x(0) &= 4 \\ y(0) &= -6. \end{aligned}$$

8.2.3.5 Graphing Region I Trajectories

In each of the four regions, we know the algebraic signs of the derivatives x' and y'. If we are given an initial condition (x_0, y_0) which is in one of these regions, we can use this information to draw the set of points $(x(t), y(t))$ corresponding to the solution to our system

$$\begin{aligned} x'(t) &= -3\,x(t) + 4\,y(t) \\ y'(t) &= -x(t) + 2\,y(t) \\ x(0) &= x_0 \\ y(0) &= y_0. \end{aligned}$$

This set of points is called the **trajectory** corresponding to this solution. The first point on the trajectory is the initial point (x_0, y_0) and the rest of the points follow from the solution

$$\begin{bmatrix} x(t) \\ y(t) \end{bmatrix} = a \begin{bmatrix} 1 \\ 1/4 \end{bmatrix} e^{-2t} + b \begin{bmatrix} 1 \\ 1 \end{bmatrix} e^{t}.$$

where a and b satisfy the system of equations

$$\begin{aligned} x_0 &= a + b \\ y_0 &= (1/4)a + b \end{aligned}$$

This can be rewritten as

$$\begin{aligned} x(t) &= a\,e^{-2t} + b\,e^{t} \\ y(t) &= (1/4)a\,e^{-2t} + b\,e^{t}. \end{aligned}$$

Hence,

$$\begin{aligned}\frac{dy}{dx} &= \frac{y'(t)}{x'(t)}\\ &= \frac{(-2/4)a\,e^{-2t} + b\,e^{t}}{-2a\,e^{-2t} + b\,e^{t}}\end{aligned}$$

when t is large, as long as b is not zero, the terms involving e^{-2t} are negligible and so we have

$$\begin{aligned}\frac{dy}{dx} &= \frac{y'(t)}{x'(t)}\\ &\approx \frac{b\,e^{t}}{b\,e^{t}}\\ &\approx \boldsymbol{E}_2.\end{aligned}$$

Hence, when t is large, the slopes of the trajectory approach 1, the slope of $\boldsymbol{E}_2$. So, we can conclude that for large t, as long as b is not zero, the trajectory either parallels the line determined by $\boldsymbol{E}_2$ or approaches it asymptotically.

Of course, if an initial condition is chosen that lies on the line determined by $\boldsymbol{E}_1$, then a little thought will tell you that b is zero in this case and we have

$$\begin{aligned}\frac{dy}{dx} &= \frac{y'(t)}{x'(t)}\\ &= \frac{(-2/4)a\,e^{-2t}}{-2a\,e^{-2t}}\\ &= \boldsymbol{E}_1.\end{aligned}$$

In this case,

$$\begin{aligned}x(t) &= a\,e^{-2t}\\ y(t) &= (1/4)a\,e^{-2t}.\end{aligned}$$

and so the coordinates $(x(t), y(t))$ go to $(0, 0)$ along the line determined by $\boldsymbol{E}_1$.

We conclude that unless an initial condition is chosen exactly on the line determined by $\boldsymbol{E}_1$, all trajectories eventually begin to either parallel the line determined by $\boldsymbol{E}_2$ or approach it asymptotically. If the initial condition is chosen on the line determined by $\boldsymbol{E}_1$, then the trajectories stay on this line and approach the origin where they stop as that is a place where both x' and y' become 0. In Fig. 8.5, we show three trajectories which begin in Region I. They all have a $(+, +)$ sign pattern for x' and y', so the x and y components should both increase. We draw the trajectories with the concavity as shown because that is the only way they can smoothly approach the eigenvector line $\boldsymbol{E}_2$. We show this in Fig. 8.5.

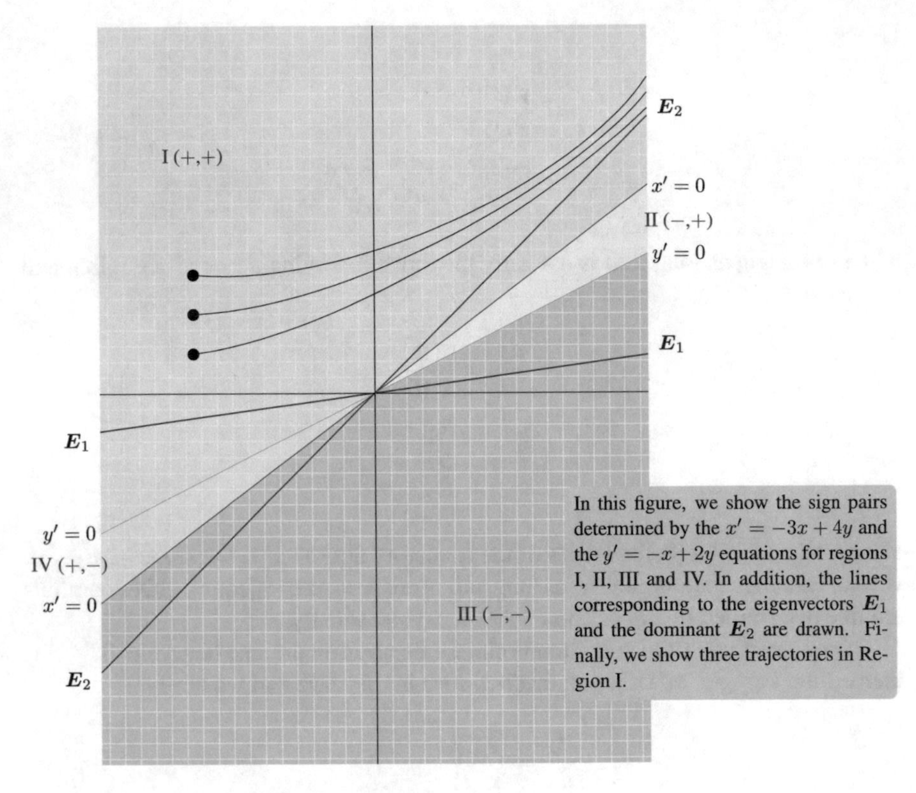

Fig. 8.5 Trajectories in Region I

8.2.3.6 Can Trajectories Cross?

Is it possible for two trajectories to cross? Consider the trajectories shown in Fig. 8.6. These two trajectories cross at some point. The two trajectories correspond to different initial conditions which means that the a and b associated with them will be different. Further, these initial conditions don't start on eigenvector $\boldsymbol{E}_1$ or eigenvector $\boldsymbol{E}_2$, so the a and b values for both trajectories will be non zero. If we label these trajectories by (x_1, y_1) and (x_2, y_2), we see

$$x_1(t) = a_1\, e^{-2t} + b_1\, e^t$$
$$y_1(t) = (1/4)a_1\, e^{-2t} + b_1\, e^t.$$

and

$$x_2(t) = a_2\, e^{-2t} + b_2\, e^t$$
$$y_2(t) = (1/4)a_2\, e^{-2t} + b_2\, e^t.$$

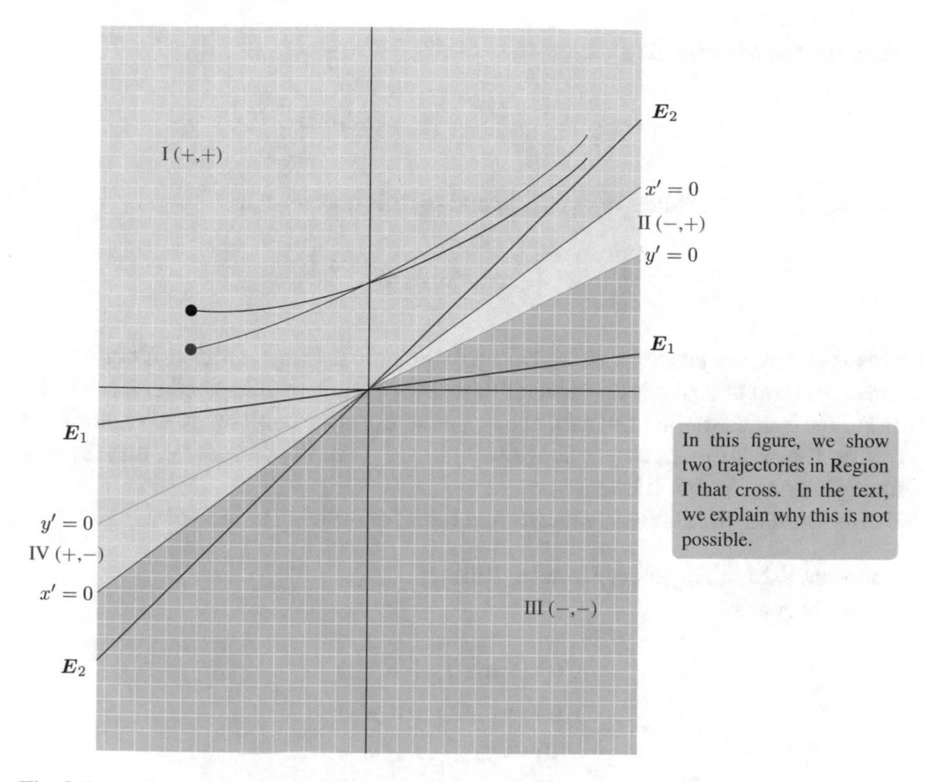

Fig. 8.6 Combining the x' and y' algebraic sign regions

Since we assume they cross, there has to be a time point, t^*, so that $(x_1(t^*), y_1(t^*))$ and $(x_2(t^*), y_2(t^*))$ match. This means, using vector notation,

$$\begin{bmatrix} x_1(t) \\ y_1(t) \end{bmatrix} = a_1 \begin{bmatrix} 1 \\ 1/4 \end{bmatrix} e^{-2t} + b_1 \begin{bmatrix} 1 \\ 1 \end{bmatrix} e^{t},$$
$$\begin{bmatrix} x_2(t) \\ y_2(t) \end{bmatrix} = a_2 \begin{bmatrix} 1 \\ 1/4 \end{bmatrix} e^{-2t} + b_2 \begin{bmatrix} 1 \\ 1 \end{bmatrix} e^{t}.$$

Setting these two equal at t^*, then gives

$$a_1 \begin{bmatrix} 1 \\ 1/4 \end{bmatrix} e^{-2t^*} + b_1 \begin{bmatrix} 1 \\ 1 \end{bmatrix} e^{t^*} = a_2 \begin{bmatrix} 1 \\ 1/4 \end{bmatrix} e^{-2t^*} + b_2 \begin{bmatrix} 1 \\ 1 \end{bmatrix} e^{t^*}.$$

This is a bit messy, so for convenience, let the number e^{t^*} be denoted by U and the number e^{-2t^*} be V. Then, we can rewrite as

$$a_1 \begin{bmatrix} 1 \\ 1/4 \end{bmatrix} V + b_1 \begin{bmatrix} 1 \\ 1 \end{bmatrix} U = a_2 \begin{bmatrix} 1 \\ 1/4 \end{bmatrix} V + b_2 \begin{bmatrix} 1 \\ 1 \end{bmatrix} U.$$

Next, we can combine like vectors to find

$$(a_1 - a_2)V \begin{bmatrix} 1 \\ 1/4 \end{bmatrix} = (b_2 - b_1)U \begin{bmatrix} 1 \\ 1 \end{bmatrix}.$$

No matter what the values of a_1, a_2, b_1 and b_2, this tells us that

$$\begin{bmatrix} 1 \\ 1/4 \end{bmatrix} = \text{a multiple of } \begin{bmatrix} 1 \\ 1 \end{bmatrix}.$$

This is clearly not possible, so we have to conclude that trajectories can't cross. We can do this sort of analysis for trajectories that start in any region, whether it is I, II, III or IV. Further, a similar argument shows that a trajectory can't cross an eigenvector line as if it did, the argument above would lead us to the conclusion that $\boldsymbol{E}_1$ is a multiple of $\boldsymbol{E}_2$, which it is not.

We can state the results here as formal rules for drawing trajectories.

Theorem 8.2.1 (Trajectory Drawing Rules)
Given the system

$$\begin{aligned} x'(t) &= a\,x(t) + b\,y(t) \\ y'(t) &= c\,x(t) + d\,y(t) \\ x(0) &= x_0 \\ y(0) &= y_0, \end{aligned}$$

assume the eigenvalues r_1 and r_2 are different with either both negative or one negative and one positive. Let $\boldsymbol{E}_1$ and $\boldsymbol{E}_2$ be the associated eigenvectors. Then, the trajectories of this system corresponding to different initial conditions can not cross each other. In particular, trajectories can not cross eigenvector lines.

8.2.3.7 Graphing Region II Trajectories

In region II, trajectories start where $x' < 0$ and $y' > 0$. Hence, the x values must decrease and the y values, increase in this region. We draw the trajectory in this way, making sure it curves in such a way that it has no corners or kinks, until it hits the nullcline $x' = 0$. At that point, the trajectory moves into region I. Now $x' > 0$ and $y' > 0$, so the trajectory moves upward along the eigenvector $\boldsymbol{E}_2$ line like we showed in the Region I trajectories. We show this in Fig. 8.7. Note although the trajectories seem to overlap near the $\boldsymbol{E}_2$ line, they actually do not because trajectories can not cross as was be explained in Sect. 8.2.3.6.

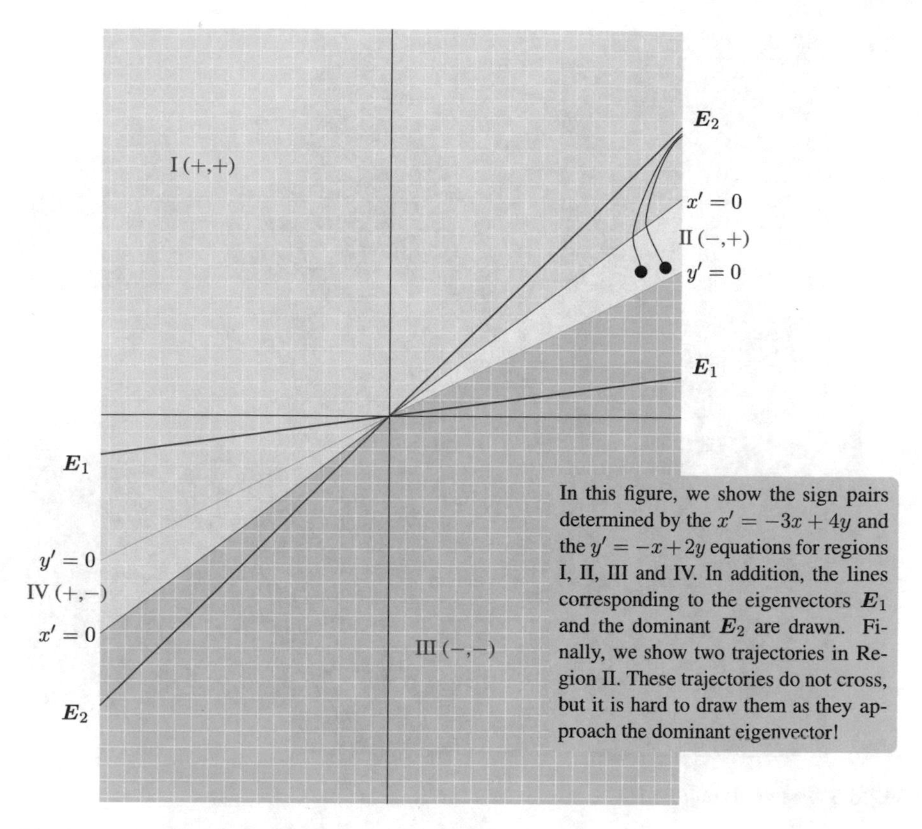

Fig. 8.7 Trajectories in Region II

8.2.3.8 Graphing Region III Trajectories

Next, we examine trajectories that begin in Region III. Here x' and y' are negative, so the x and y values will decrease and the trajectories will approach the dominant eigenvector $\boldsymbol{E}_2$ line from the right side as is shown in Fig. 8.8. The initial condition that starts in Region III above the eigenvector $\boldsymbol{E}_2$ line will move towards the $y' = 0$ line following $x' < 0$ and $y' < 0$ until it hits the line $x' = 0$ using $x' < 0$ and $y' > 0$. Then it moves upward towards the eigenvector $\boldsymbol{E}_2$ line as shown. It is easier to see this in a magnified view as shown in Fig. 8.9.

8.2.3.9 Graphing Region IV Trajectories

Finally, we examine trajectories that begin in region IV. Here x' is positive and y' is negative, so the x values will grow and y values will decrease. The trajectories will behave in this manner until they intersect the $x' = 0$ nullcline. Then, they will cross

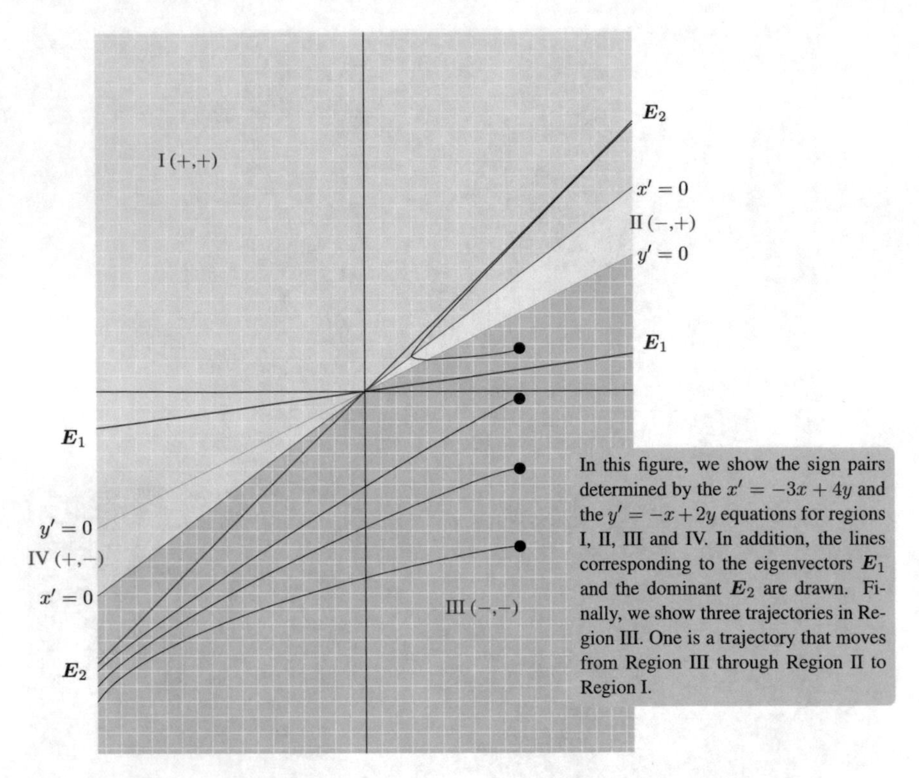

Fig. 8.8 Region III trajectories

into Region III and approach the dominant eigenvector $\boldsymbol{E}_2$ line from the left side as is shown in Fig. 8.10.

8.2.3.10 The Combined Trajectories

In Fig. 8.11, we show all the region trajectories on one plot. We can draw more, but these should be enough to give you an idea of how to draw them. In addition, there is a type of trajectory we haven't drawn yet. Recall, the general solution is

$$\begin{bmatrix} x_1(t) \\ y_1(t) \end{bmatrix} = a \begin{bmatrix} 1 \\ 1/4 \end{bmatrix} e^{-2t} + b \begin{bmatrix} 1 \\ 1 \end{bmatrix} e^{t}.$$

If an initial condition was chosen to lie on eigenvector $\boldsymbol{E}_1$ line, then $b = 0$. Hence, for these initial conditions, we have

$$\begin{bmatrix} x_1(t) \\ y_1(t) \end{bmatrix} = a \begin{bmatrix} 1 \\ 1/4 \end{bmatrix} e^{-2t}.$$

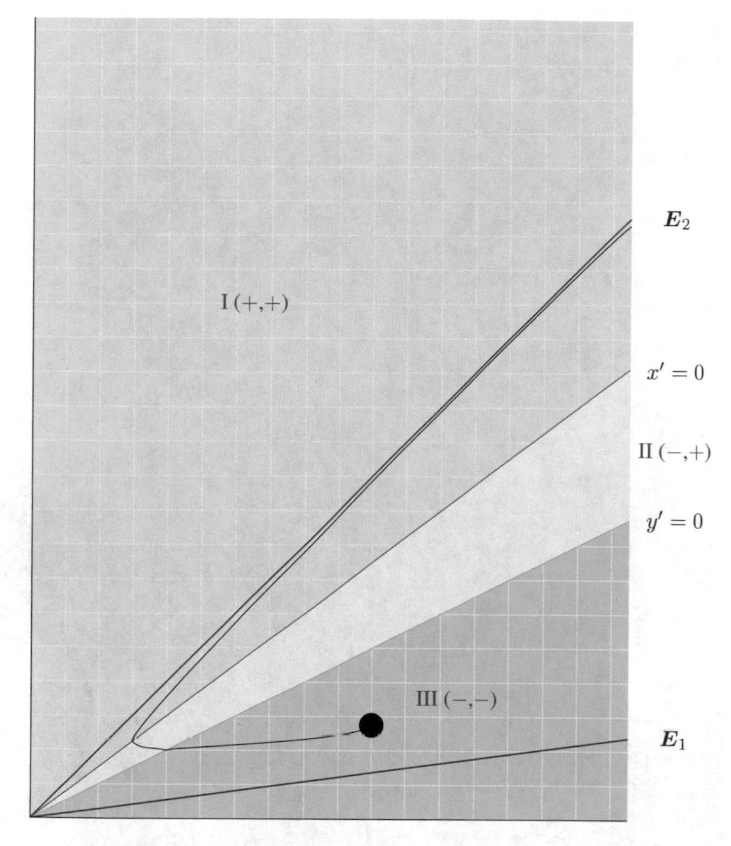

Fig. 8.9 A magnified Region III trajectory

Thus, these trajectories start somewhere on the eigenvector $\boldsymbol{E}_1$ line and then as t increases, $x(t)$ and $y(t)$ go to $(0, 0)$ along this eigenvector. You can easily imagine these trajectories by placing a dot on the $\boldsymbol{E}_1$ line with an arrow pointing towards the origin.

We can do this sort of qualitative analysis for the three cases:

- One eigenvalue negative and one eigenvalue positive: example $r_1 = -2$ and $r_2 = 1$ which we have just completed.
- Both eigenvalues negative: example $r_1 = -2$ and $r_2 = -1$ which we have not done.
- Both eigenvalues positive: example $r_1 = 1$ and $r_2 = 2$ which we have not done.
- In each case, we have two eigenvectors $\boldsymbol{E_1}$ and $\boldsymbol{E_2}$. The way we label our eigenvalues will **always** make most trajectories approach the $\boldsymbol{E_2}$ line as t increases because r_2 is always the largest eigenvalue.

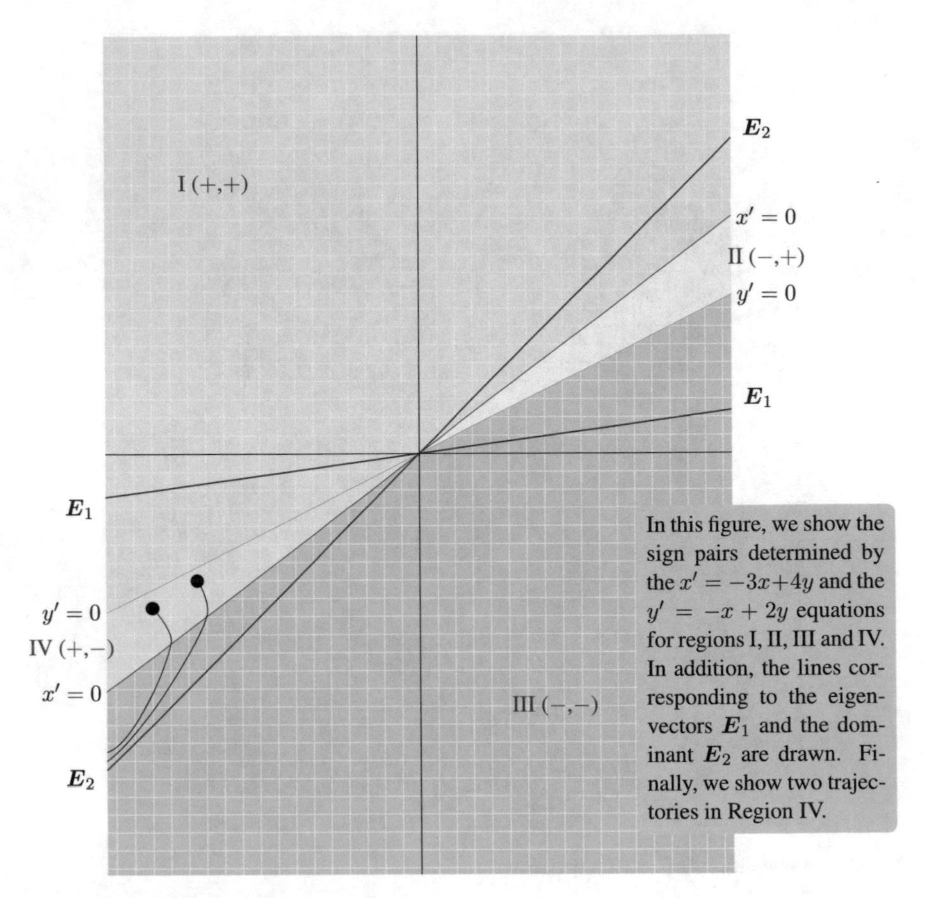

Fig. 8.10 Region IV trajectories

8.2.3.11 Mixed Sign Eigenvalues

Here we have one negative eigenvalue and one positive eigenvalue. The positive one is the dominant one: example, $r_1 = -2$ and $r_2 = 1$ so the dominant eigenvalue is $r_2 = 1$.

- Trajectories that start on the $\boldsymbol{E_1}$ line go towards $(0, 0)$ along that line.
- Trajectories that start on the $\boldsymbol{E_2}$ line move outward along that line.
- All other ICs give trajectories that move outward from $(0, 0)$ and approach the dominant eigenvector line, the $\boldsymbol{E_2}$ line as t increases.
- The $(+, +)$, $(+, -)$, $(-, +)$ and $(-, -)$ regions tells us the details of how this is done. We find these regions using the nullcline analysis.

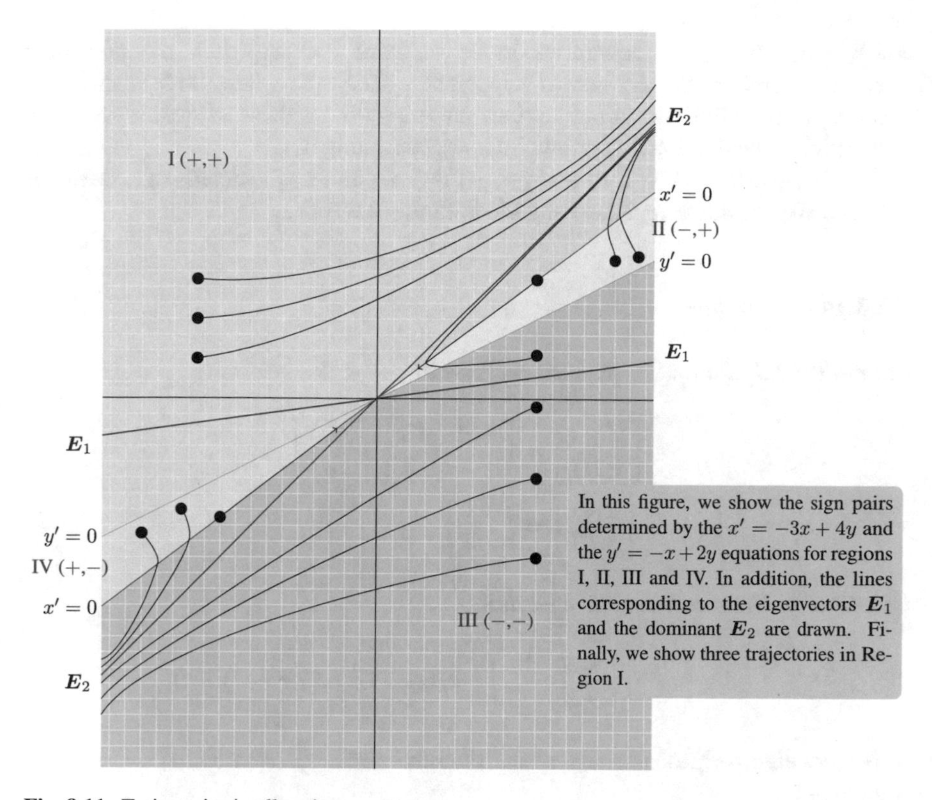

Fig. 8.11 Trajectories in all regions

8.2.3.12 Two Negative Eigenvalues

Here we have two negative eigenvalues. The least negative one is the dominant one: example, $r_1 = -2$ and $r_2 = -1$ so the dominant eigenvalue is $r_2 = -1$.

- Trajectories that start on the $\boldsymbol{E_1}$ line go towards $(0, 0)$ along that line.
- Trajectories that start on the $\boldsymbol{E_2}$ line go towards $(0, 0)$ along that line.
- All other ICs give trajectories move towards $(0, 0)$ and approach the dominant eigenvector line, the $\boldsymbol{E_2}$ line as t increases.
- The $(+, +)$, $(+, -)$, $(-, +)$ and $(-, -)$ regions tells us the details of how this is done. We find these regions using the nullcline analysis.

8.2.3.13 Two Positive Eigenvalues

Now we have two positive eigenvalues. The positive one is the dominant one: example, $r_1 = 2$ and $r_2 = 3$ so the dominant eigenvalue is $r_2 = 3$.

- Trajectories that start on the $\boldsymbol{E_1}$ line move outward along that line.
- Trajectories that start on the $\boldsymbol{E_2}$ line move outward along that line.

- All other ICs give trajectories move outward and approach the dominant eigenvector line, the $\boldsymbol{E_2}$ line as t increases. This case where both eigenvalues are positive is the hardest one to draw and many times these trajectories become parallel to the dominant line rather than approaching it.
- The $(+,+)$, $(+,-)$, $(-,+)$ and $(-,-)$ regions tells us the details of how this is done. We find these regions using the nullcline analysis.

8.2.3.14 Examples

Example 8.2.3 Do the phase plane analysis for

$$\begin{bmatrix} x'(t) \\ y'(t) \end{bmatrix} = \begin{bmatrix} -20 & 12 \\ -13 & 5 \end{bmatrix} \begin{bmatrix} x(t) \\ y(t) \end{bmatrix}$$
$$\begin{bmatrix} x(0) \\ y(0) \end{bmatrix} = \begin{bmatrix} -1 \\ 2 \end{bmatrix}$$

Solution • *The characteristic equation is*

$$det\left(r\boldsymbol{I} - \begin{bmatrix} -20 & 12 \\ -13 & 5 \end{bmatrix}\right) = 0 \Rightarrow (r+8)(r+7) = 0.$$

- *Hence,* **eigenvalues** *or* **roots** *of the characteristic equation are* $r_1 = -8$ *and* $r_2 = -7$.
- *The vector*

$$\boldsymbol{E_1} = \begin{bmatrix} 1 \\ 1 \end{bmatrix}$$

is our choice for an **eigenvector** *corresponding to eigenvalue* $r_1 = -8$.
- *The vector*

$$\boldsymbol{E_2} = \begin{bmatrix} 1 \\ \frac{13}{12} \end{bmatrix}$$

is our choice for an **eigenvector** *corresponding to eigenvalue* $r_2 = -7$.
- *Since both eigenvalues are negative we have* $r_2 = -7$ *is the dominant one.*
- *Trajectories that start on the* $\boldsymbol{E_1}$ *line go towards* $(0, 0)$ *along that line.*
- *Trajectories that start on the* $\boldsymbol{E_2}$ *line go towards* $(0, 0)$ *along that line.*
- *All other ICs give trajectories that move towards* $(0, 0)$ *and approach the dominant eigenvector line, the* $\boldsymbol{E_2}$ *line as t increases.*
- *The* $(+,+)$, $(+,-)$, $(-,+)$ *and* $(-,-)$ *regions tells us the details of how this is done. We find these regions using the nullcline analysis. Here* $x' = 0$ *gives* $-20x + 12y = 0$ *or* $y = 20/12x$ *while* $y' = 0$ *gives* $-13x + 5y = 0$ *or* $y = 13/5x$.

Example 8.2.4 Do the phase plane analysis for

$$\begin{bmatrix} x'(t) \\ y'(t) \end{bmatrix} = \begin{bmatrix} 4 & 9 \\ -1 & -6 \end{bmatrix} \begin{bmatrix} x(t) \\ y(t) \end{bmatrix}$$
$$\begin{bmatrix} x(0) \\ y(0) \end{bmatrix} = \begin{bmatrix} 4 \\ -2 \end{bmatrix}$$

Solution • *The characteristic equation is*

$$det\left(r\boldsymbol{I} - \begin{bmatrix} 4 & 9 \\ -1 & -6 \end{bmatrix}\right) = 0 \Rightarrow (r+5)(r-3) = 0.$$

- *Hence,* **eigenvalues** *or* **roots** *of the characteristic equation are* $r_1 = -5$ *and* $r_2 = 3$.
- *The vector*

$$\boldsymbol{E_1} = \begin{bmatrix} 1 \\ -1 \end{bmatrix}$$

is our choice for an **eigenvector** *corresponding to eigenvalue* $r_1 = -5$.
- *The vector*

$$\boldsymbol{E_2} = \begin{bmatrix} 1 \\ -\frac{1}{9} \end{bmatrix}$$

is our choice for an **eigenvector** *corresponding to eigenvalue* $r_2 = 3$.
- *Since one eigenvalue is positive and one is negative, we have* $r_2 = 3$ *is the dominant one.*
- *Trajectories that start on the* $\boldsymbol{E_1}$ *line go towards* $(0, 0)$ *along that line.*
- *Trajectories that start on the* $\boldsymbol{E_2}$ *line move outward from* $(0, 0)$ *along that line.*
- *All other ICs give trajectories that move outward from* $(0, 0)$ *and approach the dominant eigenvector line, the* $\boldsymbol{E_2}$ *line as t increases.*
- *The* $(+,+)$, $(+,-)$, $(-,+)$ *and* $(-,-)$ *regions tells us the details of how this is done. We find these regions using the nullcline analysis. Here* $x' = 0$ *gives* $4x + 9y = 0$ *or* $y = -4/9x$ *while* $y' = 0$ *gives* $-x - 6y = 0$ *or* $y = -1/6x$.

Example 8.2.5 Do the phase plane analysis for

$$\begin{bmatrix} x'(t) \\ y'(t) \end{bmatrix} = \begin{bmatrix} 4 & -2 \\ 3 & -1 \end{bmatrix} \begin{bmatrix} x(t) \\ y(t) \end{bmatrix}$$
$$\begin{bmatrix} x(0) \\ y(0) \end{bmatrix} = \begin{bmatrix} 14 \\ -22 \end{bmatrix}$$

Solution • *The characteristic equation is*

$$det\left(r\boldsymbol{I} - \begin{bmatrix} 4 & -2 \\ 3 & -1 \end{bmatrix}\right) = 0 \Rightarrow (r-1)(r-2) = 0.$$

- *Hence,* **eigenvalues** *or* **roots** *of the characteristic equation are* $r_1 = 1$ *and* $r_2 = 2$.
- *The vector*

$$\boldsymbol{E_1} = \begin{bmatrix} 1 \\ \frac{3}{2} \end{bmatrix}$$

is our choice for an **eigenvector** *corresponding to eigenvalue* $r_1 = 1$.
- *The vector*

$$\boldsymbol{E_2} = \begin{bmatrix} 1 \\ 1 \end{bmatrix}$$

is our choice for an **eigenvector** *corresponding to eigenvalue* $r_2 = 2$.
- *Since both eigenvalues are positive, the larger one,* $r_2 = 2$*, is the dominant one.*
- *Trajectories that start on the* $\boldsymbol{E_1}$ *line move outward from* $(0, 0)$ *along that line.*
- *Trajectories that start on the* $\boldsymbol{E_2}$ *line move outward from* $(0, 0)$ *along that line.*
- *All other ICs give trajectories that move outward from* $(0, 0)$ *and approach the dominant eigenvector line, the* $\boldsymbol{E_2}$ *line as t increases.*
- *The* $(+,+)$, $(+,-)$, $(-,+)$ *and* $(-,-)$ *regions tells us the details of how this is done. We find these regions using the nullcline analysis. Here* $x' = 0$ *gives* $4x - 2y = 0$ *or* $y = 2x$ *while* $y' = 0$ *gives* $3x - y = 0$ *or* $y = 3x$.

Finally, here is how we would work out a problem by hand in Figs. 8.12, 8.13 and 8.14; note the wonderful handwriting displayed on these pages.

8.2.3.15 Homework

You are now ready to do some problems on your own. For the problems below

- Find the characteristic equation
- Find the general solution
- Solve the IVP
- On the same x–y graph,

1. draw the $x' = 0$ line
2. draw the $y' = 0$ line
3. draw the eigenvector one line

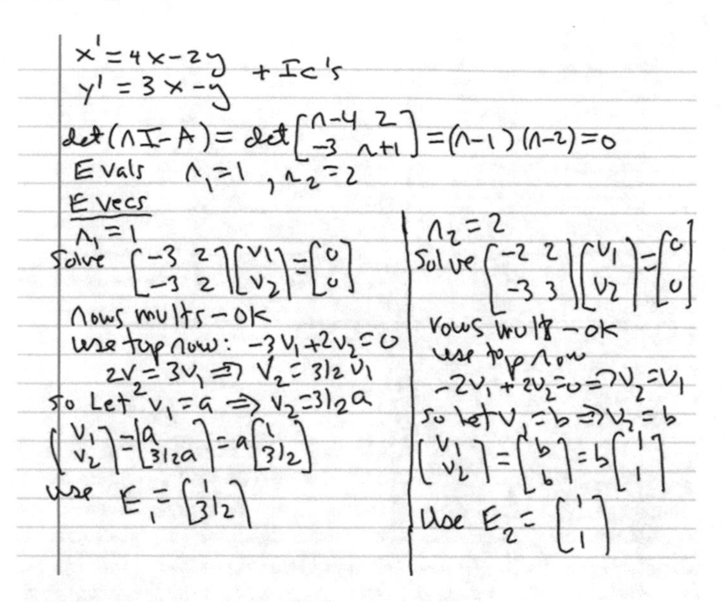

Fig. 8.12 Example $x' = 4x - 2y,\ y' = 3x - y$, Page 1

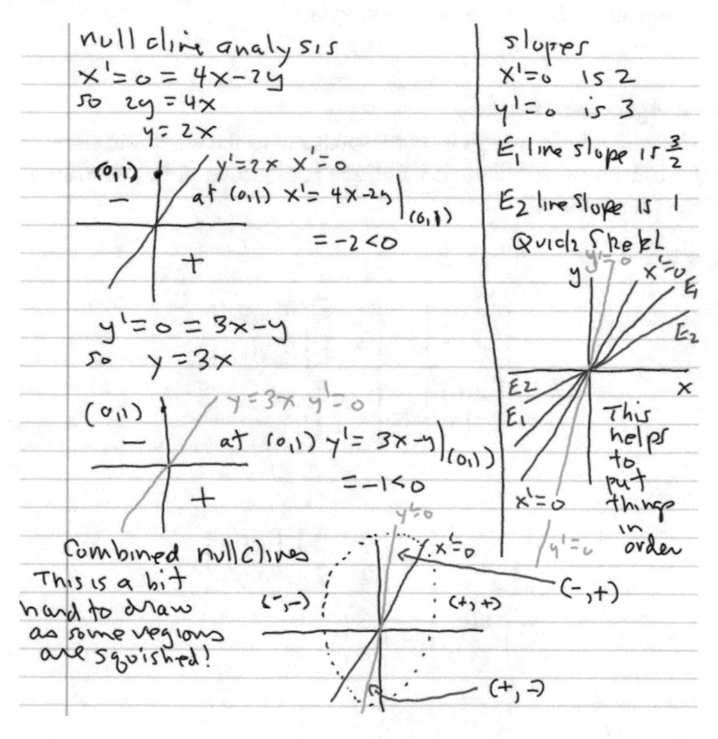

Fig. 8.13 Example $x' = 4x - 2y,\ y' = 3x - y$, Page 2

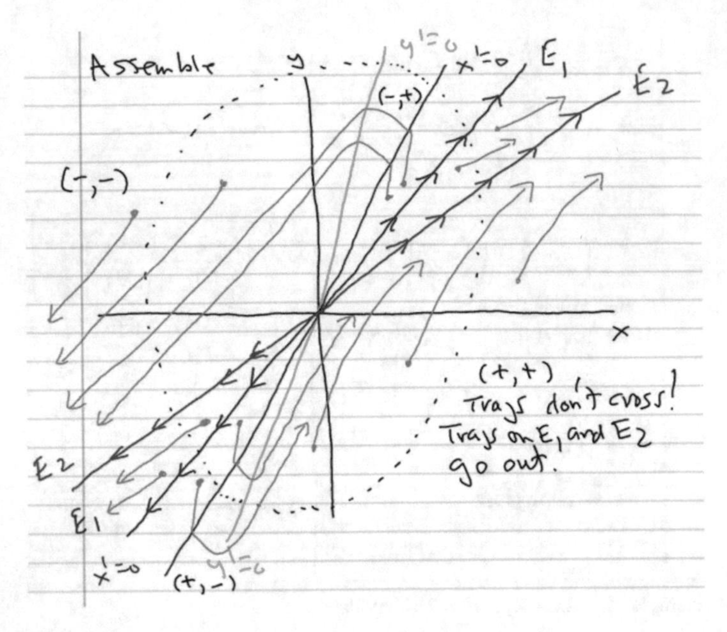

Fig. 8.14 Example $x' = 4x - 2y,\ y' = 3x - y$, Page 3

4. draw the eigenvector two line
5. divide the x–y into four regions corresponding to the algebraic signs of x' and y'
6. draw the trajectories of enough solutions for various initial conditions to create the phase plane portrait

Exercise 8.2.8

$$\begin{bmatrix} x'(t) \\ y'(t) \end{bmatrix} = \begin{bmatrix} 1 & 3 \\ 3 & 1 \end{bmatrix} \begin{bmatrix} x(t) \\ y(t) \end{bmatrix}$$
$$\begin{bmatrix} x(0) \\ y(0) \end{bmatrix} = \begin{bmatrix} -3 \\ 1 \end{bmatrix}.$$

Exercise 8.2.9

$$\begin{bmatrix} x'(t) \\ y'(t) \end{bmatrix} = \begin{bmatrix} 3 & 12 \\ 2 & 1 \end{bmatrix} \begin{bmatrix} x(t) \\ y(t) \end{bmatrix}$$
$$\begin{bmatrix} x(0) \\ y(0) \end{bmatrix} = \begin{bmatrix} 6 \\ 1 \end{bmatrix}.$$

Exercise 8.2.10

$$\begin{bmatrix} x'(t) \\ y'(t) \end{bmatrix} = \begin{bmatrix} -1 & 1 \\ -2 & -4 \end{bmatrix} \begin{bmatrix} x(t) \\ y(t) \end{bmatrix}$$
$$\begin{bmatrix} x(0) \\ y(0) \end{bmatrix} = \begin{bmatrix} 3 \\ 8 \end{bmatrix}.$$

Exercise 8.2.11

$$\begin{bmatrix} x'(t) \\ y'(t) \end{bmatrix} = \begin{bmatrix} 3 & 4 \\ -7 & -8 \end{bmatrix} \begin{bmatrix} x(t) \\ y(t) \end{bmatrix}$$
$$\begin{bmatrix} x(0) \\ y(0) \end{bmatrix} = \begin{bmatrix} -2 \\ 4 \end{bmatrix}.$$

Exercise 8.2.12

$$\begin{bmatrix} x'(t) \\ y'(t) \end{bmatrix} = \begin{bmatrix} -1 & 1 \\ -3 & -5 \end{bmatrix} \begin{bmatrix} x(t) \\ y(t) \end{bmatrix}$$
$$\begin{bmatrix} x(0) \\ y(0) \end{bmatrix} = \begin{bmatrix} 2 \\ -4 \end{bmatrix}.$$

Exercise 8.2.13

$$\begin{bmatrix} x'(t) \\ y'(t) \end{bmatrix} = \begin{bmatrix} -5 & 2 \\ -4 & 1 \end{bmatrix} \begin{bmatrix} x(t) \\ y(t) \end{bmatrix}$$
$$\begin{bmatrix} x(0) \\ y(0) \end{bmatrix} = \begin{bmatrix} 21 \\ 5 \end{bmatrix}.$$

8.2.3.16 Adding Inputs

Now consider the sample problem

$$\begin{aligned} x'(t) &= -2\,x(t) + 3\,y(t) + u \\ y'(t) &= 4\,x(t) + 5\,y(t) + v \\ x(0) &= 1 \\ y(0) &= 2 \end{aligned}$$

for some constants u and v. In matrix–vector form, we have

$$\left(\begin{bmatrix} x(t) \\ y(t) \end{bmatrix}\right)' = \begin{bmatrix} -2 & 3 \\ 4 & 5 \end{bmatrix} \begin{bmatrix} x(t) \\ y(t) \end{bmatrix} + \begin{bmatrix} u \\ v \end{bmatrix}$$

We call the vector whose components are u and v, the inputs to this model. In general, if the external inputs are both zero, we call the model a *homogeneous model* and otherwise, it is a *nonhomogeneous model*. Now we don't know how to solve this, but let's try a guess. Let's assume there is a solution $x_{inputs}(t) = x^*$ and $y_{inputs}(t) = y^*$, where x^* and y^* are constants. Then,

$$\begin{bmatrix} x_{inputs}(t) \\ y_{inputs}(t) \end{bmatrix}' = \begin{bmatrix} x^* \\ y^* \end{bmatrix}' = \begin{bmatrix} 0 \\ 0 \end{bmatrix}$$

Now plugging this assumed solution into the original model, we find

$$\begin{bmatrix} 0 \\ 0 \end{bmatrix} = \begin{bmatrix} -2 & 3 \\ 4 & 5 \end{bmatrix} \begin{bmatrix} x^* \\ y^* \end{bmatrix} + \begin{bmatrix} u \\ v \end{bmatrix}$$

Since $det(\boldsymbol{A}) = -22$ which is not zero, we know $\boldsymbol{A}^{-1}$ exists. Manipulating a bit, our original equation becomes

$$\begin{bmatrix} -2 & 3 \\ 4 & 5 \end{bmatrix} \begin{bmatrix} x^* \\ y^* \end{bmatrix} = -\begin{bmatrix} u \\ v \end{bmatrix}$$

and so multiplying both sides by $\boldsymbol{A}^{-1}$, we find

$$\begin{bmatrix} x^* \\ y^* \end{bmatrix} = -\begin{bmatrix} -2 & 3 \\ 4 & 5 \end{bmatrix}^{-1} \begin{bmatrix} u \\ v \end{bmatrix}$$

This is easy to solve as we can calculate

$$\begin{bmatrix} -2 & 3 \\ 4 & 5 \end{bmatrix}^{-1} = \frac{-1}{22} \begin{bmatrix} 5 & -3 \\ -4 & -2 \end{bmatrix}$$

We conclude a solution to the model is

$$\begin{bmatrix} x_{inputs} \\ y_{inputs} \end{bmatrix} = \begin{bmatrix} x^* \\ y^* \end{bmatrix} = -\frac{-1}{22} \begin{bmatrix} 5 & -3 \\ -4 & -2 \end{bmatrix} \begin{bmatrix} u \\ v \end{bmatrix}$$

for any inputs u and v. For this sample problem, the characteristic equation is

$$det(r\boldsymbol{I} - \boldsymbol{A}) = r^2 - 3r - 22$$

which has roots

$$r = \left(\frac{3 \pm \sqrt{9 - 4(-22)}}{2} \right) = \frac{3}{2} \pm \frac{\sqrt{97}}{2}$$

The eigenvalues are approximately $r_1 = -3.42$ and $r_2 = 6.42$. The associated eigenvectors can easily be found to be

$$\boldsymbol{E}_1 = \begin{bmatrix} 1.0 \\ -0.47 \end{bmatrix}, \quad \boldsymbol{E}_2 = \begin{bmatrix} 1.0 \\ 2.81 \end{bmatrix}$$

and the general solution to this model with **no** inputs would be

$$\begin{bmatrix} x_{no\ input}(t) \\ y_{no\ input}(t) \end{bmatrix} = a\boldsymbol{E}_1 e^{-3.42t} + b\boldsymbol{E}_2 e^{6.42t}$$

for arbitrary a and b. A little thought then shows that adding $[x_{no\ input}(t), y_{no\ input}(t)]^T$ to the solution $[x_{inputs}(t), y_{inputs}(t)]^T$ will always solve the model with inputs. So the most general solution to the model with constant inputs must be of the form

$$\begin{aligned} \begin{bmatrix} x(t) \\ y(t) \end{bmatrix} &= \begin{bmatrix} x_{no\ input}(t) \\ y_{no\ input}(t) \end{bmatrix} + \begin{bmatrix} x_{inputs}(t) \\ y_{inputs}(t) \end{bmatrix} \\ &= a\boldsymbol{E}_1 e^{-3.42t} + b\boldsymbol{E}_2 e^{6.42t} - \boldsymbol{A}^{-1} \begin{bmatrix} u \\ v \end{bmatrix} \end{aligned}$$

where $\boldsymbol{A}$ is the coefficient matrix of our model. The solution $[x_{no\ input}(t), y_{no\ input}(t)]^T$ occurs so often it is called the **homogeneous solution** to the model (who knows why!) and the solution $[x_{inputs}(t), y_{inputs}(t)]^T$ because it actually works for the model with these particular inputs, is called the **particular solution**. To save *subscripting*, we label the homogeneous solution $[x_h(t), y_h(t)]^T$ and the particular solution $[x_p(t), y_p(t)]^T$. Finally, any model with inputs that are not zero is called an **non-homogeneous model**. We are thus ready for a definition.

Definition 8.2.1 (*The Nonhomogeneous Model*)
Any solution to the general non-homogeneous model

$$\begin{bmatrix} x(t) \\ y(t) \end{bmatrix}' = \begin{bmatrix} -2 & 3 \\ 4 & 5 \end{bmatrix} \begin{bmatrix} x(t) \\ y(t) \end{bmatrix} + \begin{bmatrix} f(t) \\ g(t) \end{bmatrix}$$

for input functions $f(t)$ and $g(t)$ is called the **particular solution** and is labeled $[x_p(t), y_p(t)]^T$. The model with no inputs is called the homogeneous model and its solutions are called **homogeneous solutions** and labeled $[x_h(t), y_h(t)]^T$. The general solution to the model with nonzero inputs is then

$$\begin{bmatrix} x(t) \\ y(t) \end{bmatrix} = \begin{bmatrix} x_h(t) \\ y_h(t) \end{bmatrix} + \begin{bmatrix} x_p(t) \\ y_p(t) \end{bmatrix}$$

For a linear model, from our discussions above, we know how to find a complete solution.

8.2.3.17 Worked Out Example

Example 8.2.6

$$\begin{bmatrix} x'(t) \\ y'(t) \end{bmatrix} = \begin{bmatrix} 2 & -5 \\ 4 & -7 \end{bmatrix} \begin{bmatrix} x(t) \\ y(t) \end{bmatrix} + \begin{bmatrix} 2 \\ 1 \end{bmatrix}$$
$$\begin{bmatrix} x(0) \\ y(0) \end{bmatrix} = \begin{bmatrix} 1 \\ 5 \end{bmatrix}$$

Solution *It is straightforward to find the eigenvalues here are* $r_1 = -3$ *and* $r_2 = -2$ *with corresponding eigenvectors*

$$\boldsymbol{E}_1 = \begin{bmatrix} 1 \\ 1 \end{bmatrix}, \quad \boldsymbol{E}_2 = \begin{bmatrix} 1.0 \\ 0.8 \end{bmatrix}$$

The particular solution is

$$\begin{bmatrix} x_p \\ y_p \end{bmatrix} = -\begin{bmatrix} 2 & -5 \\ 4 & -7 \end{bmatrix}^{-1} \begin{bmatrix} 2 \\ 1 \end{bmatrix} = -\frac{1}{(-14+20)} \begin{bmatrix} -7 & 5 \\ -4 & 2 \end{bmatrix} \begin{bmatrix} 2 \\ 1 \end{bmatrix} = \begin{bmatrix} \frac{9}{6} \\ \frac{6}{6} \end{bmatrix}$$

Note, this particular solution is biologically reasonable if x and y are proteins and the dynamics represent their interaction since x_p *and* y_p *are positive.*

The homogeneous solution is

$$\begin{bmatrix} x_h(t) \\ y_h(t) \end{bmatrix} = a\boldsymbol{E}_1 e^{-3t} + b\boldsymbol{E}_2 e^{-2t}$$

and the general solution is thus

$$\begin{bmatrix} x(t) \\ y(t) \end{bmatrix} = a \begin{bmatrix} 1 \\ 1 \end{bmatrix} e^{-3t} + b \begin{bmatrix} 1 \\ 0.8 \end{bmatrix} e^{-2t} + \begin{bmatrix} 9/6 \\ 1 \end{bmatrix}$$

Finally, to solve the IVP, we know a and b must satisfy

$$\begin{bmatrix} 1 \\ 5 \end{bmatrix} = a \begin{bmatrix} 1 \\ 1 \end{bmatrix} + b \begin{bmatrix} 1 \\ 0.8 \end{bmatrix} + \begin{bmatrix} \frac{3}{2} \\ 1 \end{bmatrix}$$

which is two equations in two unknowns which we solve in the usual way. We find

$$1 = a + b + 3/2$$
$$5 = a + 0.8b + 1$$

or

$$\begin{aligned} -1/2 &= a + b \\ 4 &= a + 0.8b \end{aligned}$$

which implies $a = 22$ and $b = -45/2$. The solution is then

$$\begin{bmatrix} x(t) \\ y(t) \end{bmatrix} = 22 \begin{bmatrix} 1 \\ 1 \end{bmatrix} e^{-3t} - 45/2 \begin{bmatrix} 1 \\ 0.8 \end{bmatrix} e^{-2t} + \begin{bmatrix} 9/6 \\ 1 \end{bmatrix}$$

8.2.3.18 Homework

Solve these IVP problems showing all details.

Exercise 8.2.14

$$\begin{aligned} \begin{bmatrix} x'(t) \\ y'(t) \end{bmatrix} &= \begin{bmatrix} 1 & 3 \\ 3 & 1 \end{bmatrix} \begin{bmatrix} x(t) \\ y(t) \end{bmatrix} + \begin{bmatrix} 3 \\ 5 \end{bmatrix} \\ \begin{bmatrix} x(0) \\ y(0) \end{bmatrix} &= \begin{bmatrix} -3 \\ 1 \end{bmatrix}. \end{aligned}$$

Exercise 8.2.15

$$\begin{aligned} \begin{bmatrix} x'(t) \\ y'(t) \end{bmatrix} &= \begin{bmatrix} 3 & 12 \\ 2 & 1 \end{bmatrix} \begin{bmatrix} x(t) \\ y(t) \end{bmatrix} + \begin{bmatrix} 6 \\ 8 \end{bmatrix} \\ \begin{bmatrix} x(0) \\ y(0) \end{bmatrix} &= \begin{bmatrix} 6 \\ 1 \end{bmatrix}. \end{aligned}$$

Exercise 8.2.16

$$\begin{aligned} \begin{bmatrix} x'(t) \\ y'(t) \end{bmatrix} &= \begin{bmatrix} -1 & 1 \\ -2 & -4 \end{bmatrix} \begin{bmatrix} x(t) \\ y(t) \end{bmatrix} + \begin{bmatrix} 1 \\ 10 \end{bmatrix} \\ \begin{bmatrix} x(0) \\ y(0) \end{bmatrix} &= \begin{bmatrix} 3 \\ 8 \end{bmatrix}. \end{aligned}$$

Exercise 8.2.17

$$\begin{aligned} \begin{bmatrix} x'(t) \\ y'(t) \end{bmatrix} &= \begin{bmatrix} 3 & 4 \\ -7 & -8 \end{bmatrix} \begin{bmatrix} x(t) \\ y(t) \end{bmatrix} + \begin{bmatrix} 0.3 \\ 6 \end{bmatrix} \\ \begin{bmatrix} x(0) \\ y(0) \end{bmatrix} &= \begin{bmatrix} -2 \\ 4 \end{bmatrix}. \end{aligned}$$

Exercise 8.2.18

$$\begin{bmatrix} x'(t) \\ y'(t) \end{bmatrix} = \begin{bmatrix} -1 & 1 \\ -3 & -5 \end{bmatrix} \begin{bmatrix} x(t) \\ y(t) \end{bmatrix} + \begin{bmatrix} 7 \\ 0.5 \end{bmatrix}$$
$$\begin{bmatrix} x(0) \\ y(0) \end{bmatrix} = \begin{bmatrix} 2 \\ -4 \end{bmatrix}.$$

Exercise 8.2.19

$$\begin{bmatrix} x'(t) \\ y'(t) \end{bmatrix} = \begin{bmatrix} -5 & 2 \\ -4 & 1 \end{bmatrix} \begin{bmatrix} x(t) \\ y(t) \end{bmatrix} + \begin{bmatrix} 23 \\ 11 \end{bmatrix}$$
$$\begin{bmatrix} x(0) \\ y(0) \end{bmatrix} = \begin{bmatrix} 21 \\ 5 \end{bmatrix}.$$

8.3 Repeated Eigenvalues

In this case, the two roots of the characteristic equation are both the same real value. For us, there are two cases where this can happen.

8.3.1 The Repeated Eigenvalue Has Two Linearly Independent Eigenvectors

Let the matrix $\boldsymbol{A}$ be a multiple of the identity. For example, we have the linear system

$$\begin{aligned} x'(t) &= -2\,x(t) + 0\,y(t) \\ y'(t) &= 0\,x(t) - 2\,y(t) \\ x(0) &= 2 \\ y(0) &= 6 \end{aligned}$$

which can clearly be written more succinctly as

$$\begin{aligned} x' &= -2x; \quad x(0) = 2 \Longrightarrow x(t) = 2e^{-2t} \\ y' &= -2y; \quad y(0) = 6 \Longrightarrow y(t) = 6e^{-2t} \end{aligned}$$

The characteristic equation is then $(r+2)^2 = 0$ with repeated roots $r = -2$. When we solve the eigenvalue–eigenvector equation, we find when we substitute in the eigenvalue $r = -2$ that we must solve

$$\begin{bmatrix} 0 & 0 \\ 0 & 0 \end{bmatrix} \begin{bmatrix} V_1 \\ V_2 \end{bmatrix} = \begin{bmatrix} 0 \\ 0 \end{bmatrix}$$

which is a strange looking system of equations which we have not seen before. The way to interpret this is to go back to looking at it as two equations in the unknowns V_1 and V_2. We have

$$\begin{aligned} 0\,V_1 + 0\,V_2 &= 0 \\ 0\,V_1 + 0\,V_2 &= 0 \end{aligned}$$

As usual both rows of the eigenvalue–eigenvector equation are the same and so choosing the top row to work with, the equation says there are **no** constraints on the values of V_1 and V_2. Hence, any nonzero vector $[a, b]^T$ will work. This gives us the two parameter family

$$\begin{bmatrix} V_1 \\ V_2 \end{bmatrix} = \begin{bmatrix} a \\ b \end{bmatrix} \; a \begin{bmatrix} 1 \\ 0 \end{bmatrix} + b \begin{bmatrix} 0 \\ 1 \end{bmatrix}.$$

We choose as the first eigenvector $E_1 = [1, 0]^T$ and as the second eigenvector $E_2 = [0, 1]^T$. Hence, the two linearly independent solutions to this system are $E_1 e^{-2t}$ and $E_2 e^{-2t}$ with the general solution

$$\begin{bmatrix} x(t) \\ y(t) \end{bmatrix} = A \begin{bmatrix} 1 \\ 0 \end{bmatrix} e^{-2t} + B \begin{bmatrix} 0 \\ 1 \end{bmatrix} e^{-2t} = \begin{bmatrix} Ae^{-2t} \\ Be^{-2t} \end{bmatrix}$$

Using the initial conditions, we get the solution we had before

$$\begin{bmatrix} x(t) \\ y(t) \end{bmatrix} = \begin{bmatrix} Ae^{-2t} \\ Be^{-2t} \end{bmatrix}$$

8.3.2 The Repeated Eigenvalue Has only One Eigenvector

Let's start with an example. Consider the model

$$\begin{aligned} x'(t) &= 2\,x(t) - y(t) \\ y'(t) &= x(t) + 4\,y(t) \\ x(0) &= 2 \\ y(0) &= 5 \end{aligned}$$

The characteristic is $r^2 - 6r + 9 = 0$ which gives the repeated eigenvalue 3. The eigenvalue equation for this root leads to this system to solve for nonzero $\boldsymbol{V}$.

$$\begin{bmatrix} 3-2 & 1 \\ -1 & 3-4 \end{bmatrix} \begin{bmatrix} V_1 \\ V_2 \end{bmatrix} = \begin{bmatrix} 0 \\ 0 \end{bmatrix}$$

This reduces to the system

$$\begin{bmatrix} 1 & 1 \\ -1 & -1 \end{bmatrix} \begin{bmatrix} V_1 \\ V_2 \end{bmatrix} = \begin{bmatrix} 0 \\ 0 \end{bmatrix}$$

Clearly, these two rows are equivalent and hence we only need to choose one to solve for V_2 in terms of V_1. The first row gives $V_1 + V_2 = 0$. Letting $V_1 = a$, we find $V_2 = -a$. Hence, the eigenvectors have the form

$$\boldsymbol{V} = a \begin{bmatrix} 1 \\ -1 \end{bmatrix}$$

Hence, we choose as our eigenvector

$$\boldsymbol{E} = \begin{bmatrix} 1 \\ -1 \end{bmatrix}.$$

Our first solution is thus

$$\begin{bmatrix} x_1(t) \\ y_1(t) \end{bmatrix} = \boldsymbol{E}\, e^{3t}.$$

However, the eigenvalue–eigenvector equation we solve here does not allow us to find two linearly independent eigenvectors corresponding to the eigenvalue $r = 3$. However, we know there is another linearly independent solution. The first solution is $\boldsymbol{E}e^{3t}$ and from our experiences with repeated root second order models, we suspect the second solution should have te^{3t} in it. Let's try as the second solution,

$$\begin{bmatrix} x_2(t) \\ y_2(t) \end{bmatrix} = \left(\boldsymbol{F} + \boldsymbol{G}t\right) e^{3t}.$$

and find conditions on the vectors $\boldsymbol{F}$ and $\boldsymbol{G}$ that make this true. If this is a solution, then we must have

$$\begin{bmatrix} x_2'(t) \\ y_2'(t) \end{bmatrix} = \boldsymbol{A} \begin{bmatrix} x_2(t) \\ y_2(t) \end{bmatrix}$$

Hence, by direct calculation

$$\boldsymbol{A}\left(\boldsymbol{F} + \boldsymbol{G}t\right) e^{3t} = \left(3\boldsymbol{F} + \boldsymbol{G}\right) e^{3t} + 3t\boldsymbol{G}\, e^{3t}$$

Rewriting, we obtain

$$\boldsymbol{A}\,\boldsymbol{F}\, e^{3t} + \boldsymbol{A}\,\boldsymbol{G}\left(t\, e^{3t}\right) = \left(3\,\boldsymbol{F} + \boldsymbol{G}\right) e^{3t} + 3t\boldsymbol{G}\, e^{3t}$$

Equating the coefficients of the terms e^{3t} and te^{3t}, we find

$$\begin{aligned} \boldsymbol{A}\,\boldsymbol{G} &= 3\,\boldsymbol{G} \\ \boldsymbol{A}\,\boldsymbol{F} &= 3\,\boldsymbol{F} + \boldsymbol{G}. \end{aligned}$$

The solution to the first equation is simply $\boldsymbol{G} = \boldsymbol{E}$ as it is a restatement of the eigenvalue–eigenvector equation for the eigenvalue 3. We can then rewrite the second equation as

$$\left(3\boldsymbol{I} - \boldsymbol{A}\right)\boldsymbol{F} = -\boldsymbol{E}$$

which can easily be solved for $\boldsymbol{F}$. The next linearly independent solution therefore has the form $(\boldsymbol{F} + \boldsymbol{E}t)\,e^{3t}$ where $\boldsymbol{F}$ solves the system

$$\begin{bmatrix} 1 & 1 \\ -1 & -1 \end{bmatrix}\begin{bmatrix} F_1 \\ F_2 \end{bmatrix} = -\boldsymbol{E} = \begin{bmatrix} -1 \\ 1 \end{bmatrix}$$

From the first row, we find $F_1 + F_2 = -1$ so letting $F_1 = b$, we have $F_2 = -1 - b$. Hence the most general solution is

$$\begin{bmatrix} b \\ -1-b \end{bmatrix} = \begin{bmatrix} 0 \\ -1 \end{bmatrix} + b\begin{bmatrix} 1 \\ -1 \end{bmatrix} = \begin{bmatrix} 0 \\ -1 \end{bmatrix} + b\,\boldsymbol{E}.$$

The contribution to the general solution from $b\boldsymbol{E}$ can be lumped in with the first solution $\boldsymbol{E}e^{3t}$ as we have discussed, so we only need choose

$$\boldsymbol{F} = \begin{bmatrix} 0 \\ -1 \end{bmatrix}.$$

The general real solution is therefore

$$\begin{aligned} \begin{bmatrix} x(t) \\ y(t) \end{bmatrix} &= a\begin{bmatrix} 1 \\ -1 \end{bmatrix}e^{3t} + b\left(\begin{bmatrix} 0 \\ -1 \end{bmatrix} + t\begin{bmatrix} 1 \\ -1 \end{bmatrix}\right)e^{3t} \\ &= \begin{bmatrix} a + bt \\ -a - b - bt \end{bmatrix}e^{3t} \end{aligned}$$

Now apply the initial conditions to obtain

$$\begin{bmatrix} 2 \\ 5 \end{bmatrix} = \begin{bmatrix} a \\ -a - b \end{bmatrix}$$

Thus, $a = 2$ and $b = -7$. The solution is therefore

$$\begin{bmatrix} x(t) \\ y(t) \end{bmatrix} = \begin{bmatrix} 2 - 7t \\ 5 + 7t \end{bmatrix}e^{3t}$$

Note for large t, the (x, y) trajectory is essentially parallel to the eigenvector line for $\boldsymbol{E}$ which has slope -1. Hence, if we scale by e^{3t} to obtain

$$\begin{bmatrix} x(t)/e^{3t} \\ y(t)/e^{3t} \end{bmatrix} = \begin{bmatrix} 2 - 7t \\ 5 + 7t \end{bmatrix}$$

the phase plane trajectories will look like a series of parallel lines. If we keep in the e^{3t} factor, the phase plane trajectories will exhibit some curvature. Without scaling, a typical phase plane plot for multiple trajectories looks like what we see in Fig. 8.15. If we do the scaling, we see the parallel lines as we mentioned. This is shown in Fig. 8.16.

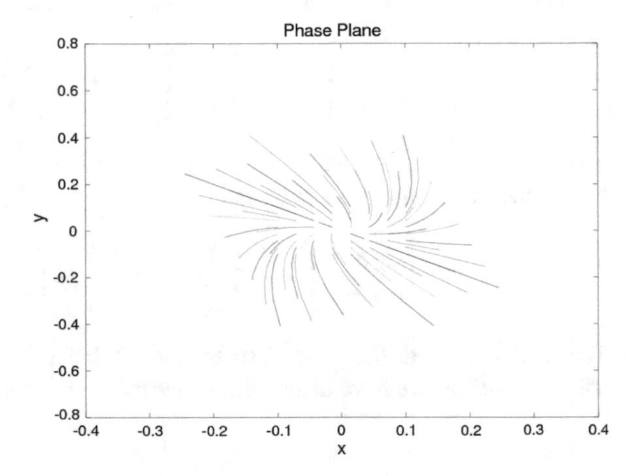

Fig. 8.15 A phase plane plot

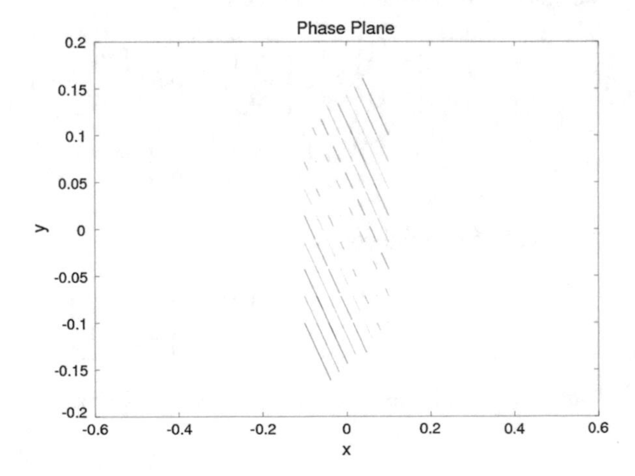

Fig. 8.16 A scaled phase plane plot

From our discussions for this case of repeated eigenvalues, we can see the general rule that if $r = \alpha$ is a repeated eigenvalue with only one eigenvector $\boldsymbol{E}$, the two linearly independent solutions are

$$\begin{bmatrix} x_1(t) \\ y_1(t) \end{bmatrix} = \boldsymbol{E}\, e^{\alpha t}$$
$$\begin{bmatrix} x_2(t) \\ y_2(t) \end{bmatrix} = \left(\boldsymbol{F} + \boldsymbol{E} t\right) e^{\alpha t}.$$

where $\boldsymbol{F}$ is a solution to the system

$$\left(\alpha\, \boldsymbol{I} - \boldsymbol{A}\right) \boldsymbol{F} = -\boldsymbol{E}.$$

8.3.2.1 Homework

We can now do some problems. Find the solution to the following models.

Exercise 8.3.1

$$\begin{aligned} x'(t) &= 3x(t) + y(t) \\ y'(t) &= -x(t) + y(t) \\ x(0) &= 2 \\ y(0) &= 3. \end{aligned}$$

Exercise 8.3.2

$$\begin{aligned} x'(t) &= 5\, x(t) + 4\, y(t) \\ y'(t) &= -16\, x(t) - 11\, y(t) \\ x(0) &= 5 \\ y(0) &= -6. \end{aligned}$$

Exercise 8.3.3

$$\begin{aligned} x'(t) &= 5\, x(t) - 9\, y(t) \\ y'(t) &= 4\, x(t) - 7\, y(t) \\ x(0) &= 10 \\ y(0) &= -20. \end{aligned}$$

Exercise 8.3.4

$$\begin{aligned}x'(t) &= 2\,x(t) + y(t)\\ y'(t) &= -4\,x(t) + 6\,y(t)\\ x(0) &= -4\\ y(0) &= -12.\end{aligned}$$

We still have to determine the direction of motion in these trajectories. If we need this, we do the usual nullcline analysis to get the algebraic sign pairs for (x', y') as usual.

8.4 Complex Eigenvalues

Let's begin with a theoretical analysis for a change of pace. If the real valued matrix $\boldsymbol{A}$ has a complex eigenvalue $r = \alpha + i\beta$, then there is a nonzero vector $\boldsymbol{G}$ so that

$$\boldsymbol{A}\boldsymbol{G} = (\alpha + i\beta)\boldsymbol{G}.$$

Now take the complex conjugate of both sides to find

$$\overline{\boldsymbol{A}}\;\overline{\boldsymbol{G}} = \overline{(\alpha + i\beta)}\;\overline{\boldsymbol{G}}.$$

However, since $\boldsymbol{A}$ has real entries, its complex conjugate is simply $\boldsymbol{A}$ back. Thus, after taking complex conjugates, we find

$$\boldsymbol{A}\;\overline{\boldsymbol{G}} = (\alpha - i\beta)\;\overline{\boldsymbol{G}}$$

and we conclude that if $\alpha + i\beta$ is an eigenvalue of $\boldsymbol{A}$ with eigenvector $\boldsymbol{G}$, then the eigenvalue $\alpha - i\beta$ has eigenvector $\overline{\boldsymbol{G}}$. Hence, letting $\boldsymbol{E}$ be the real part of $\boldsymbol{G}$ and $\boldsymbol{F}$ be the imaginary part, we see $\boldsymbol{E} + i\boldsymbol{F}$ is the eigenvector for $\alpha + i\beta$ and $\boldsymbol{E} - i\boldsymbol{F}$ is the eigenvector for $\alpha - i\beta$.

8.4.1 The General Real and Complex Solution

We can write down the general complex solution immediately.

$$\begin{bmatrix}\phi(t)\\ \psi(t)\end{bmatrix} = c_1\,(\boldsymbol{E} + i\boldsymbol{F})\,e^{(\alpha+i\beta)t} + c_2\,(\boldsymbol{E} - i\boldsymbol{F})\,e^{(\alpha-i\beta)t}$$

for arbitrary complex numbers c_1 and c_2. We can reorganize this solution into a more convenient form as follows.

$$\begin{aligned}\begin{bmatrix}\phi(t)\\ \psi(t)\end{bmatrix} &= e^{\alpha t}\left(c_1\,(\boldsymbol{E}+i\boldsymbol{F})\,e^{(i\beta)t} + c_2\,(\boldsymbol{E}-i\boldsymbol{F})\,e^{(-i\beta)t}\right)\\ &= e^{\alpha t}\left(\left(c_1 e^{(i\beta)t} + c_2 e^{(-i\beta)t}\right)\boldsymbol{E} + i\left(c_1 e^{(i\beta)t} - c_2 e^{(-i\beta)t}\right)\boldsymbol{F}\right).\end{aligned}$$

The first real solution is found by choosing $c_1 = 1/2$ and $c_2 = 1/2$. This give

$$\begin{bmatrix}x_1(t)\\ y_1(t)\end{bmatrix} = e^{\alpha t}\left(\left((1/2)\left(e^{(i\beta)t} + e^{(-i\beta)t}\right)\right)\boldsymbol{E} + i\left((1/2)\left(e^{(i\beta)t} - e^{(-i\beta)t}\right)\right)\boldsymbol{F}\right).$$

However, we know that $(1/2)\left(e^{(i\beta)t}+e^{(-i\beta)t}\right) = \cos(\beta t)$ and $(1/2)\left(e^{(i\beta)t}-e^{(-i\beta)t}\right) = i\ \sin(\beta t)$. Thus, we have

$$\begin{bmatrix}x_1(t)\\ y_1(t)\end{bmatrix} = e^{\alpha t}\left(\boldsymbol{E}\ \cos(\beta t) - \boldsymbol{F}\ \sin(\beta t)\right).$$

The second real solution is found by setting $c_1 = 1/2i$ and $c_2 = -1/2i$ which gives

$$\begin{aligned}\begin{bmatrix}x_2(t)\\ y_2(t)\end{bmatrix} &= e^{\alpha t}\left(\left((1/2i)\left(e^{(i\beta)t} - e^{(-i\beta)t}\right)\right)\boldsymbol{E} + i\left((1/2i)\left(e^{(i\beta)t} + e^{(-i\beta)t}\right)\right)\boldsymbol{F}\right)\\ &= e^{\alpha t}\left(\boldsymbol{E}\sin(\beta t) + \boldsymbol{F}\cos(\beta t)\right).\end{aligned}$$

The general real solution is therefore

$$\begin{bmatrix}x(t)\\ y(t)\end{bmatrix} = e^{\alpha t}\left(a\left(\boldsymbol{E}\ \cos(\beta t) - \boldsymbol{F}\ \sin(\beta t)\right) + b\left(\boldsymbol{E}\sin(\beta t) + \boldsymbol{F}\cos(\beta t)\right)\right)$$

for arbitrary real numbers a and b.

Example 8.4.1

$$\begin{aligned}x'(t) &= 2\,x(t) + 5\,y(t)\\ y'(t) &= -x(t) + 4\,y(t)\\ x(0) &= 6\\ y(0) &= -1\end{aligned}$$

Solution *The characteristic is* $r^2 - 6r + 13 = 0$ *which gives the eigenvalues* $3 \pm 2i$. *The eigenvalue equation for the first root,* $3 + 2i$ *leads to this system to solve for nonzero* $\boldsymbol{V}$.

$$\begin{bmatrix} (3+2i)-2 & -5 \\ 1 & (3+2i)-4 \end{bmatrix} \begin{bmatrix} V_1 \\ V_2 \end{bmatrix} = \begin{bmatrix} 0 \\ 0 \end{bmatrix}$$

This reduces to the system

$$\begin{bmatrix} 1+2i & -5 \\ 1 & -1+2i \end{bmatrix} \begin{bmatrix} V_1 \\ V_2 \end{bmatrix} = \begin{bmatrix} 0 \\ 0 \end{bmatrix}$$

Although it is not immediately apparent, the second row is a multiple of row one. Multiply row one by $-1-2i$. *This gives the row* $[-1-2i, 5]$. *Now multiple this new row by* -1 *to get* $[1+2i, -5]$ *which is row one. So even though it is harder to see, these two rows are equivalent and hence we only need to choose one to solve for* V_2 *in terms of* V_1. *The first row gives* $(1+2i)V_1 + 5V_2 = 0$. *Letting* $V_1 = a$, *we find* $V_2 = (-1-2i)/5\,a$. *Hence, the eigenvectors have the form*

$$\boldsymbol{G} = a \begin{bmatrix} 1 \\ -\frac{1+2i}{5} \end{bmatrix} = \begin{bmatrix} 1 \\ -\frac{1}{5} \end{bmatrix} + i \begin{bmatrix} 0 \\ -\frac{2}{5} \end{bmatrix}$$

Hence,

$$\boldsymbol{E} = \begin{bmatrix} 1 \\ -\frac{1}{5} \end{bmatrix} \quad \text{and} \quad \boldsymbol{F} = \begin{bmatrix} 0 \\ -\frac{2}{5} \end{bmatrix}$$

The general real solution is therefore

$$\begin{aligned} \begin{bmatrix} x(t) \\ y(t) \end{bmatrix} &= e^{3t}\left(a\left(\boldsymbol{E}\cos(2t) - \boldsymbol{F}\sin(2t)\right) + b\left(\boldsymbol{E}\sin(2t) + \boldsymbol{F}\cos(2t)\right)\right) \\ &= e^{3t}\left(a\left(\begin{bmatrix} 1 \\ -\frac{1}{5} \end{bmatrix}\cos(2t) - \begin{bmatrix} 0 \\ -\frac{2}{5} \end{bmatrix}\sin(2t)\right) + b\left(\begin{bmatrix} 1 \\ -\frac{1}{5} \end{bmatrix}\sin(2t) + \begin{bmatrix} 0 \\ -\frac{2}{5} \end{bmatrix}\cos(2t)\right)\right) \\ &= e^{3t}\begin{bmatrix} a\cos(2t) + b\sin(2t) \\ \left(-\frac{1}{5}a - \frac{2}{5}b\right)\cos(2t)\left(\frac{2}{5}a + \frac{1}{5}b\right)\sin(2t) \end{bmatrix} \end{aligned}$$

Now apply the initial conditions to obtain

$$\begin{bmatrix} 6 \\ -1 \end{bmatrix} = e^{3t}\begin{bmatrix} a \\ \left(-\frac{1}{5}a - \frac{2}{5}b\right) \end{bmatrix}$$

Thus, $a = 6$ *and* $b = -1/2$. *The solution is therefore*

$$\begin{bmatrix} x(t) \\ y(t) \end{bmatrix} = e^{3t}\begin{bmatrix} 6\cos(2t) - \frac{1}{2}\sin(2t) \\ -\cos(2t) - \frac{23}{10}\sin(2t) \end{bmatrix}$$

8.4.1.1 Homework

We can now do some problems. Find the solution to the following models.

Exercise 8.4.1

$$
\begin{aligned}
x'(t) &= x(t) - 3\,y(t)\\
y'(t) &= 6\,x(t) - 5\,y(t)\\
x(0) &= 2\\
y(0) &= -1.
\end{aligned}
$$

Exercise 8.4.2

$$
\begin{aligned}
x'(t) &= 2\,x(t) - 5\,y(t)\\
y'(t) &= 5\,x(t) - 6\,y(t)\\
x(0) &= 15\\
y(0) &= -2.
\end{aligned}
$$

Exercise 8.4.3

$$
\begin{aligned}
x'(t) &= 4\,x(t) + y(t)\\
y'(t) &= -41\,x(t) - 6\,y(t)\\
x(0) &= 1\\
y(0) &= -2.
\end{aligned}
$$

Exercise 8.4.4

$$
\begin{aligned}
x'(t) &= 3\,x(t) + 13\,y(t)\\
y'(t) &= -2\,x(t) + y(t)\\
x(0) &= 4\\
y(0) &= 2.
\end{aligned}
$$

8.4.2 Rewriting the Real Solution

This is then rewritten as

$$
\begin{bmatrix} x(t) \\ y(t) \end{bmatrix} = e^{\alpha t}\left(\left(a\,\boldsymbol{E} + b\,\boldsymbol{F}\right)\cos(\beta t) + \left(b\,\boldsymbol{E} - a\,\boldsymbol{F}\right)\sin(\beta t)\right)
$$

Now rewrite again in terms of the components of $\boldsymbol{E}$ and $\boldsymbol{F}$ to obtain

$$\begin{aligned}\begin{bmatrix} x(t) \\ y(t) \end{bmatrix} &= e^{\alpha t} \left(\left(a \begin{bmatrix} E_1 \\ E_2 \end{bmatrix} + b \begin{bmatrix} F_1 \\ F_2 \end{bmatrix} \right) \cos(\beta t) + \left(b \begin{bmatrix} E_1 \\ E_2 \end{bmatrix} - a \begin{bmatrix} F_1 \\ F_2 \end{bmatrix} \right) \sin(\beta t) \right) \\ &= e^{\alpha t} \begin{bmatrix} aE_1 + bF_1 & bE_1 - aF_1 \\ aE_2 + bF_2 & bE_2 - aF_2 \end{bmatrix} \begin{bmatrix} \cos(\beta t) \\ \sin(\beta t) \end{bmatrix}.\end{aligned}$$

Finally, we can move back to the vector form and write

$$\begin{bmatrix} x(t) \\ y(t) \end{bmatrix} = e^{\alpha t} \left[a\boldsymbol{E} + b\boldsymbol{F},\ b\boldsymbol{E} - a\boldsymbol{F} \right] \begin{bmatrix} \cos(\beta t) \\ \sin(\beta t) \end{bmatrix}.$$

8.4.3 The Representation of A

We know

$$\begin{bmatrix} x'(t) \\ y'(t) \end{bmatrix} = \boldsymbol{A} \begin{bmatrix} x(t) \\ y(t) \end{bmatrix}.$$

We can also calculate the derivative directly to find

$$\begin{aligned}\begin{bmatrix} x'(t) \\ y'(t) \end{bmatrix} &= \alpha\, e^{\alpha t} \left[a\boldsymbol{E} + b\boldsymbol{F},\ b\boldsymbol{E} - a\boldsymbol{F} \right] \begin{bmatrix} \cos(\beta t) \\ \sin(\beta t) \end{bmatrix} \\ &\quad + e^{\alpha t} \left[a\boldsymbol{E} + b\boldsymbol{F},\ b\boldsymbol{E} - a\boldsymbol{F} \right] \begin{bmatrix} -\beta \sin(\beta t) \\ \beta \cos(\beta t) \end{bmatrix}.\end{aligned}$$

To make these equations a bit shorter, define the matrix $\boldsymbol{\Lambda_{ab}}$ by

$$\boldsymbol{\Lambda_{ab}} = \left[a\boldsymbol{E} + b\boldsymbol{F},\ b\boldsymbol{E} - a\boldsymbol{F} \right].$$

Then the derivative can be written more compactly as

$$\begin{bmatrix} x'(t) \\ y'(t) \end{bmatrix} = \alpha\, e^{\alpha t}\, \boldsymbol{\Lambda_{ab}} \begin{bmatrix} \cos(\beta t) \\ \sin(\beta t) \end{bmatrix} + e^{\alpha t}\, \boldsymbol{\Lambda_{ab}} \begin{bmatrix} -\beta \sin(\beta t) \\ \beta \cos(\beta t) \end{bmatrix}.$$

Whew! But wait, we can do more! With a bit more factoring, we have

$$\begin{aligned}\begin{bmatrix} x'(t) \\ y'(t) \end{bmatrix} &= e^{\alpha t}\, \boldsymbol{\Lambda_{ab}} \begin{bmatrix} \alpha \cos(\beta t) - \beta \sin(\beta t) \\ \beta \cos(\beta t) + \alpha \sin(\beta t) \end{bmatrix} \\ &= e^{\alpha t}\, \boldsymbol{\Lambda_{ab}} \begin{bmatrix} \alpha & -\beta \\ \beta & \alpha \end{bmatrix} \begin{bmatrix} \cos(\beta t) \\ \sin(\beta t) \end{bmatrix}.\end{aligned}$$

Hence, since

$$\begin{bmatrix} x(t) \\ y(t) \end{bmatrix} = e^{\alpha t} \, \mathbf{\Lambda_{ab}} \begin{bmatrix} \cos(\beta t) \\ \sin(\beta t) \end{bmatrix}.$$

we can equate our two expressions for the derivative to find

$$\boldsymbol{A} \, e^{\alpha t} \, \mathbf{\Lambda_{ab}} \begin{bmatrix} \cos(\beta t) \\ \sin(\beta t) \end{bmatrix} = e^{\alpha t} \, \mathbf{\Lambda_{ab}} \begin{bmatrix} \alpha & -\beta \\ \beta & \alpha \end{bmatrix} \begin{bmatrix} \cos(\beta t) \\ \sin(\beta t) \end{bmatrix}.$$

Since $e^{\alpha t} > 0$ always, we have

$$\left(\boldsymbol{A} \, \mathbf{\Lambda_{ab}} - \mathbf{\Lambda_{ab}} \begin{bmatrix} \alpha & -\beta \\ \beta & \alpha \end{bmatrix} \right) \begin{bmatrix} \cos(\beta t) \\ \sin(\beta t) \end{bmatrix} = 0.$$

Therefore, after a fair bit of work, we have found the identity

$$\boldsymbol{A} \, \mathbf{\Lambda_{ab}} = \mathbf{\Lambda_{ab}} \begin{bmatrix} \alpha & -\beta \\ \beta & \alpha \end{bmatrix}$$

8.4.4 The Transformed Model

It is straightforward, albeit messy, to show $det(\mathbf{\Lambda_{ab}}) \neq 0$ and so $\mathbf{\Lambda_{ab}}$ is invertible as long as at least one of a and b is not zero. Thus, if we define the change of variable

$$\begin{bmatrix} u(t) \\ v(t) \end{bmatrix} = \left(\mathbf{\Lambda_{ab}} \right)^{-1} \begin{bmatrix} x(t) \\ y(t) \end{bmatrix}$$

we see

$$\begin{aligned} \begin{bmatrix} u'(t) \\ v'(t) \end{bmatrix} &= \left(\mathbf{\Lambda_{ab}} \right)^{-1} \begin{bmatrix} x'(t) \\ y'(t) \end{bmatrix} \\ &= \left(\mathbf{\Lambda_{ab}} \right)^{-1} \boldsymbol{A} \begin{bmatrix} x(t) \\ y(t) \end{bmatrix}. \end{aligned}$$

But, we know

$$\left(\mathbf{\Lambda_{ab}} \right)^{-1} \boldsymbol{A} = \begin{bmatrix} \alpha & -\beta \\ \beta & \alpha \end{bmatrix} \left(\mathbf{\Lambda_{ab}} \right)^{-1}.$$

Thus, we have

$$\begin{bmatrix} u'(t) \\ v'(t) \end{bmatrix} = \begin{bmatrix} \alpha & -\beta \\ \beta & \alpha \end{bmatrix} \left(\boldsymbol{\Lambda_{ab}} \right)^{-1} \begin{bmatrix} x(t) \\ y(t) \end{bmatrix}$$
$$= \begin{bmatrix} \alpha & -\beta \\ \beta & \alpha \end{bmatrix} \begin{bmatrix} u(t) \\ v(t) \end{bmatrix}.$$

This transformed system in the variables u and v also has the eigenvalues $\alpha + i\beta$ but it is simpler to solve.

8.4.5 The Canonical Solution

We can solve the canonical model

$$\begin{bmatrix} u'(t) \\ v'(t) \end{bmatrix} = \begin{bmatrix} \alpha & -\beta \\ \beta & \alpha \end{bmatrix} \begin{bmatrix} u(t) \\ v(t) \end{bmatrix}.$$

as usual. The characteristic equation is $(r-\alpha)^2 + \beta^2 = 0$ giving eigenvalues $\alpha \pm \beta i$ as usual. We find the eigenvector for $\alpha + i\beta$ satisfies

$$\begin{bmatrix} i\beta & \beta \\ -\beta & i\beta \end{bmatrix} \begin{bmatrix} V_1 \\ V_2 \end{bmatrix} = \begin{bmatrix} 0 \\ 0 \end{bmatrix}$$

This gives the equation $i\beta V_1 + \beta V_2 = 0$. Letting $V_1 = 1$, we have $V_2 = -i\beta$. Thus,

$$\begin{bmatrix} V_1 \\ V_2 \end{bmatrix} = \begin{bmatrix} 1 \\ -i\beta \end{bmatrix} = \begin{bmatrix} 1 \\ 0 \end{bmatrix} + i \begin{bmatrix} 0 \\ -1 \end{bmatrix}$$

Thus,

$$\boldsymbol{E} = \begin{bmatrix} 1 \\ 0 \end{bmatrix} \quad \text{and} \quad \boldsymbol{F} = \begin{bmatrix} 0 \\ -1 \end{bmatrix}$$

The general real solution is then

$$\begin{bmatrix} u(t) \\ v(t) \end{bmatrix} = e^{\alpha t} \begin{bmatrix} aE_1 + bF_1 & bE_1 - aF_1 \\ aE_2 + bF_2 & bE_2 - aF_2 \end{bmatrix} \begin{bmatrix} \cos(\beta t) \\ \sin(\beta t) \end{bmatrix} = e^{\alpha t} \begin{bmatrix} a & b \\ -b & a \end{bmatrix} \begin{bmatrix} \cos(\beta t) \\ \sin(\beta t) \end{bmatrix}$$
$$= e^{\alpha t} \begin{bmatrix} a\cos(\beta t) + b\sin(\beta t) \\ -b\cos(\beta t) + a\sin(\beta t) \end{bmatrix}$$

We can rewrite this as follows. Let $R = \sqrt{a^2 + b^2}$. Then

$$\begin{bmatrix} u(t) \\ v(t) \end{bmatrix} = R\, e^{\alpha t} \begin{bmatrix} \frac{a}{R} \cos(\beta t) + \frac{b}{R} \sin(\beta t) \\ -\frac{b}{R} \cos(\beta t) + \frac{a}{R} \sin(\beta t) \end{bmatrix}$$

Let the angle δ be defined by $\tan(\delta) = b/a$. This tells us $\cos(\delta) = a/R$ and $\sin(\delta) = b/R$. Then, $\cos(\pi/2 - \delta) = b/R$ and $\sin(\pi/2 - \delta) = a/R$. Plugging these values into our expression, we find

$$\begin{aligned} \begin{bmatrix} u(t) \\ v(t) \end{bmatrix} &= R\, e^{\alpha t} \begin{bmatrix} \cos(\delta)\cos(\beta t) + \sin(\delta)\sin(\beta t) \\ -\cos(\pi/2 - \delta)\cos(\beta t) + \sin(\pi/2 - \delta)\sin(\beta t) \end{bmatrix} \\ &= R\, e^{\alpha t} \begin{bmatrix} \cos(\delta)\cos(\beta t) + \sin(\delta)\sin(\beta t) \\ -\sin(\delta)\cos(\beta t) + \cos(\delta)\sin(\beta t) \end{bmatrix} \end{aligned}$$

Then using the standard cos addition formulae, we obtain

$$\begin{bmatrix} u(t) \\ v(t) \end{bmatrix} = R\, e^{\alpha t} \begin{bmatrix} \cos(\beta t - \delta) \\ \sin(\beta t - \delta) \end{bmatrix}$$

Hence, $u^2(t) + v^2(t) = e^{2\alpha t}\big(R^2 \cos^2(\beta t - \delta) + R^2 \sin^2(\beta t - \delta)\big)$. This simplifies to $u^2(t) + v^2(t) = e^{2\alpha t} R^2$ and hence in this canonical case, the phase plane trajectory is spiral in if $\alpha < 0$, a circle of radius $R = \sqrt{a^2 + b^2}$ if $\alpha = 0$ and a spiral out if $\alpha > 0$. Of course, the values of a and b are determined by the initial conditions.

8.4.5.1 Homework

Solve these canonical problems.

Exercise 8.4.5

$$\begin{aligned} x'(t) &= 2\,x(t) - 3\,y(t) \\ y'(t) &= 3\,x(t) + 2\,y(t) \\ x(0) &= -6 \\ y(0) &= 8. \end{aligned}$$

Exercise 8.4.6

$$\begin{aligned} x'(t) &= -4\,x(t) - 3\,y(t) \\ y'(t) &= 3\,x(t) - 4\,y(t) \\ x(0) &= 6 \\ y(0) &= 5. \end{aligned}$$

Exercise 8.4.7

$$\begin{aligned} x'(t) &= 5\,x(t) - 6\,y(t) \\ y'(t) &= 6\,x(t) + 5\,y(t) \\ x(0) &= -2 \\ y(0) &= 4. \end{aligned}$$

Exercise 8.4.8

$$\begin{aligned} x'(t) &= x(t) - 2\,y(t) \\ y'(t) &= 2\,x(t) + y(t) \\ x(0) &= 1 \\ y(0) &= -1. \end{aligned}$$

8.4.6 The General Model Solution

In general, we know the real solution is given by

$$\begin{bmatrix} x(t) \\ y(t) \end{bmatrix} = e^{\alpha t}\ \boldsymbol{\Lambda_{ab}} \begin{bmatrix} \cos(\beta t) \\ \sin(\beta t) \end{bmatrix}.$$

Let $\boldsymbol{\Lambda_{ab}}$ have components

$$\boldsymbol{\Lambda_{ab}} = \begin{bmatrix} \lambda_{11} & \lambda_{12} \\ \lambda_{21} & \lambda_{22} \end{bmatrix}$$

Then we have

$$\begin{bmatrix} x(t) \\ y(t) \end{bmatrix} = e^{\alpha t} \begin{bmatrix} \lambda_{11}\ \cos(\beta t) + \lambda_{12}\ \sin(\beta t) \\ \lambda_{21}\ \cos(\beta t) + \lambda_{22}\ \sin(\beta t) \end{bmatrix}$$

Let $R_1 = \sqrt{\lambda_{11}^2 + \lambda_{12}^2}$ and $R_2 = \sqrt{\lambda_{21}^2 + \lambda_{22}^2}$. Then

$$\begin{bmatrix} x(t) \\ y(t) \end{bmatrix} = e^{\alpha t} \begin{bmatrix} R_1\left(\frac{\lambda_{11}}{R_1}\ \cos(\beta t) + \frac{\lambda_{12}}{R_1}\ \sin(\beta t)\right) \\ R_2\left(\frac{\lambda_{21}}{R_2}\ \cos(\beta t) + \frac{\lambda_{22}}{R_2}\ \sin(\beta t)\right) \end{bmatrix}$$

Let the angles δ_1 and δ_2 be defined by $\tan(\delta_1) = \lambda_{12}/\lambda_{11}$ and $\tan(\delta_2) = \lambda_{22}/\lambda_{21}$ where in the case where we divide by zero, these angles are assigned the value of $\pm\pi/2$ as needed. For convenience of exposition, we will assume here that all the entries of $\mathbf{\Lambda}_{ab}$ are nonzero although, of course, what we really know is that this matrix is invertible and so it is possible for some entries to be zero. But that is just a messy complication and easy enough to fix with a little thought. So once we have our angles δ_1 and δ_2, we can rewrite the solution as

$$\begin{bmatrix} x(t) \\ y(t) \end{bmatrix} = e^{\alpha t} \begin{bmatrix} R_1 \left(\cos(\delta_1)\ \cos(\beta t) + \sin(\delta_1)\ \sin(\beta t) \right) \\ R_2 \left(\cos(\delta_2)\ \cos(\beta t) + \sin(\delta_2)\ \sin(\beta t) \right) \end{bmatrix} = e^{\alpha t} \begin{bmatrix} R_1 \cos(\beta t - \delta_1) \\ R_2 \cos(\beta t - \delta_2) \end{bmatrix}$$

This is a little confusing as is, so let's do an example with numbers. Suppose our solution was

$$\begin{bmatrix} x(t) \\ y(t) \end{bmatrix} = e^{\alpha t} \begin{bmatrix} 2\cos(\beta t - \pi/6) \\ 3\cos(\beta t - \pi/3) \end{bmatrix}$$

Then this solution is clearly periodic since the cos functions are periodic. Note that x hits its maximum and minimum value of 2 and -2 at the times t_1 where $\beta t_1 - \pi/6 = 0$ and t_2 with $\beta t_2 - \pi/6 = \pi$. This gives $t_1 = \pi/(6\beta)$ and $t2 = 7\pi/(6\beta)$. At these values of time, the y values are $3\cos(\beta t_1 - \pi/3) = 3\cos(\pi/6 - \pi/3)$ or $y = 3\cos(-\pi/6) = 3\sqrt{3}/2$ and $3\cos(\beta t_2 - \pi/3) = 3\cos(\pi/3 - \pi/6)$ or $y = 3\cos(\pi/6) = -3\sqrt{3}/2$. Thus, the points $(2, 3\sqrt{3}/2)$ and $(-2, -3\sqrt{3}/2)$ are on this trajectory. The trajectory reaches the point $(2, 3\sqrt{3}/2)$ in time $\pi/(6\beta)$ and then hits the point $(-2, -3\sqrt{3}/2)$ at time $\pi/\beta + \pi/(6\beta)$.

The extremal y values occur at ± 3. The maximum y of 3 is obtained at $\beta t_3 - \pi/3 = 0$ or $t_3 = \pi/(3\beta)$. The corresponding x value is then $2\cos(\beta t_3 - \pi/6) = 2\cos(\pi/3 - \pi/6) = 2\sqrt{3}/2$. So at time t_3, the trajectory passes through $(2\sqrt{3}/2, 3)$. Finally, the minimum y value of -3 is achieved at $\beta t_4 - \pi/3 = \pi$ or $\beta t_4 = 8\pi/6$. At this time, the corresponding x value is $2\cos(8\pi/6 - \pi/6) = -2\sqrt{3}/2$.

This is probably not very helpful; however, what we have shown here is that this trajectory is an ellipse which is rotated. Take a sheet of paper and mark off the x and y axes. Choose a small angle to represent $\pi/(6\beta)$. Draw a line through the origin at angle $\pi/(6\beta)$ and on this line mark the two points $(2, 3\sqrt{3}/2)$ (angle is $\pi/(6\beta)$ and $(-3\sqrt{3}/2, -2)$ (angle $7\pi/(6\beta)$). This line is the *horizontal* axis of the ellipse. Now draw another line through the origin at angle $\pi/(3\beta)$ which is double the first angle. This line is the *vertical* axis of the ellipse. On this line plot the two points $(2\sqrt{3}/2, 3)$ (angle is $\pi/(3\beta)$ and $(-2\sqrt{3}/2, -3)$ (angle $4\pi/(3\beta)$). This is a phase shifted ellipse. At time $\pi/(6\beta)$, we start at the farthest positive x value of the ellipse on the *horizontal* axis. Then in $\pi/(6\beta)$ additional time, we hit the largest y value of the *vertical* axis. Next, in an additional $5\pi/(6\beta)$ we reach the most negative x

value of the ellipse on the *horizontal* axis. After another $\pi/(6\beta)$ we arrive at the most negative y value on the *vertical* axis. Finally, we arrive back at the start point after another $5\pi/(6\beta)$. Try drawing it!

Maybe a little Matlab/Octave will help. Consider the following quick plot. We won't bother to label axes and so forth as we just want to double check all of the complicated arithmetic above.

Listing 8.1: Checking our arithmetic!

```
beta = 3;
x = @(t) 2*cos(beta*t - pi/6);
y = @(t) 3*cos(beta*t - pi/3);
t = linspace(0,1,21);
u = x(t);
v = y(t);
plot(u,v);
```

This generates part of this ellipse as we can see in Fig. 8.17. We can close the plot by plotting for a longer time.

Listing 8.2: Plotting for more time

```
beta = 3;
x = @(t) 2*cos(beta*t - pi/6);
y = @(t) 3*cos(beta*t - pi/3);
t = linspace(0,3,42);
u = x(t);
v = y(t);
plot(u,v);
```

This fills in the rest of the ellipse as we can see in Fig. 8.18.
We can plot the *axes* of this *ellipse* easily as well using the MatLab/Octave session below.

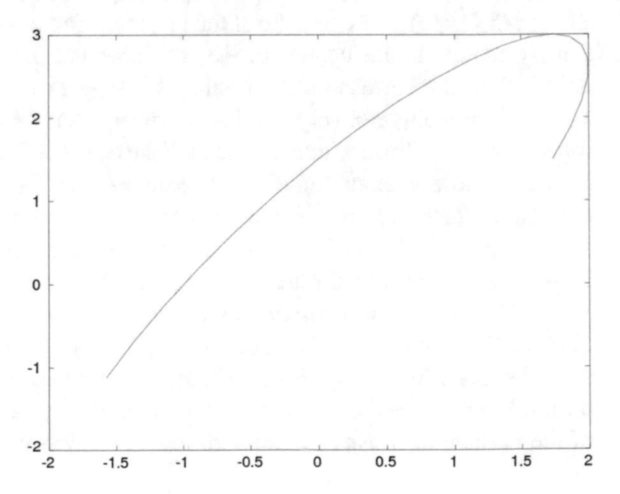

Fig. 8.17 The partial ellipse for the phase plane portrait

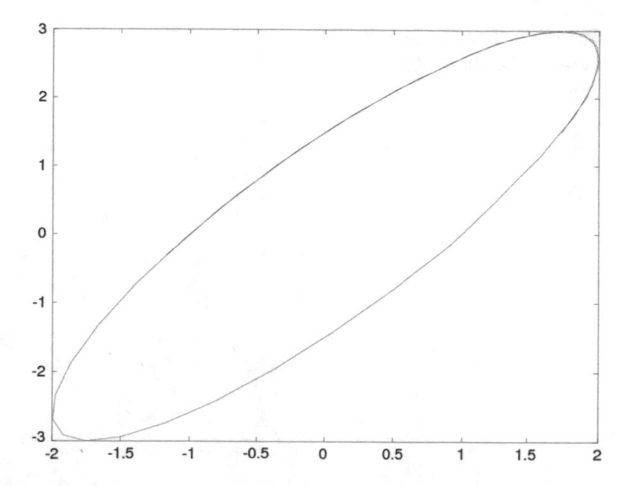

Fig. 8.18 The complete ellipse for the phase plane portrait

Listing 8.3: Plotting the ellipse and its axis

```
beta = 3;
x = @(t) 2*cos(beta*t - pi/6);
y = @(t) 3*cos(beta*t - pi/3);
A = [-2;2];
B = [-3*sqrt(3)/2;3*sqrt(3)/2];
C = [-2*sqrt(3)/2;2*sqrt(3)/2)];
D = [-3;3];
time = linspace(0,3,42);
u = x(time);
v = y(time);
hold on
plot(u,v);
plot(A,B);
plot(C,D);
hold off;
```

We can now see the axes lines we were talking about earlier in Fig. 8.19. Note these axes are not perpendicular like we would usually see in an ellipse! Now, this picture does not include the exponential decay or growth we would get by multiplying by the scaling factor $e^{\alpha t}$ for various α's. A typical spiral out would look like Fig. 8.20. To determine the direction of motion in these trajectories, we do the usual nullcline analysis to get the algebraic sign pairs for (x', y') as usual.

8.4.6.1 Homework

Write MatLab/Octave code to graph the following solutions and their axes. We are using the same models we have solved before but with different initial conditions.

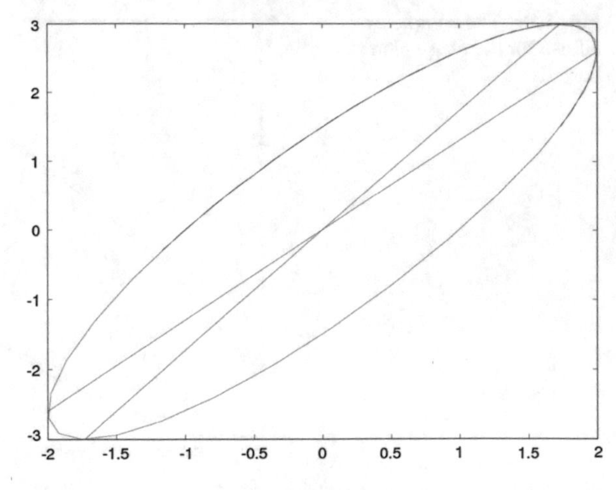

Fig. 8.19 The ellipse with axes for the phase plane portrait

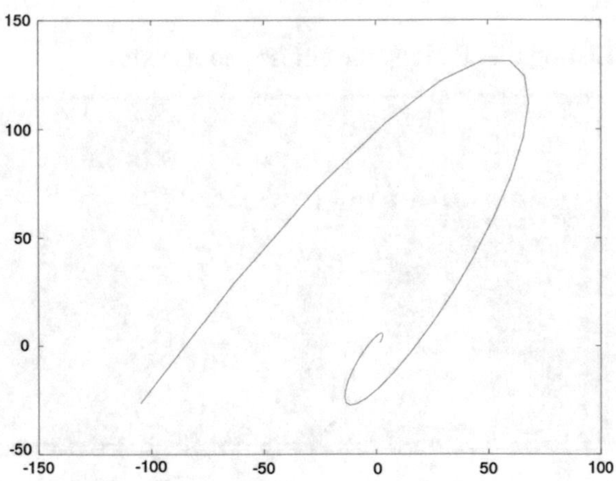

Fig. 8.20 The spiral out phase plane portrait

Exercise 8.4.9

$$\begin{aligned} x'(t) &= x(t) - 3\,y(t) \\ y'(t) &= 6\,x(t) - 5\,y(t) \\ x(0) &= 6 \\ y(0) &= 8. \end{aligned}$$

Exercise 8.4.10

$$\begin{aligned} x'(t) &= 2\,x(t) - 5\,y(t) \\ y'(t) &= 5\,x(t) - 6\,y(t) \\ x(0) &= 9 \\ y(0) &= -20. \end{aligned}$$

Exercise 8.4.11

$$\begin{aligned}
x'(t) &= 4\,x(t) + y(t)\\
y'(t) &= -41\,x(t) - 6\,y(t)\\
x(0) &= 10\\
y(0) &= 8.
\end{aligned}$$

Exercise 8.4.12

$$\begin{aligned}
x'(t) &= 3\,x(t) + 13\,y(t)\\
y'(t) &= -2\,x(t) + y(t)\\
x(0) &= -5\\
y(0) &= -2.
\end{aligned}$$

Chapter 9
Numerical Methods Systems of ODEs

We now want to learn how to solve systems of differential equations. A typical system is the following

$$x'(t) = f(t, x(t), y(t)) \tag{9.1}$$
$$y'(t) = g(t, x(t), y(t)) \tag{9.2}$$
$$x(t_0) = x_0; \quad y(t_0) = y_0 \tag{9.3}$$

where x and y are our variables of interest which might represent populations to two competing species or other quantities of biological interest. The system starts at time t_0 (which for us is usually 0) and we specify the values the variables x and y start at as x_0 and y_0 respectively. The functions f and g are "nice" meaning that as functions of three arguments (t, x, y) they do not have jumps and corners. The exact nature of "nice" here is a bit beyond our ability to discuss in this introductory course, so we will leave it at that. For example, we could be asked to solve the system

$$x'(t) = 3x(t) + 2x(t)y(t) - 3y^2(t), \tag{9.4}$$
$$y'(t) = -2x(t) + 2x^2(t)y(t) + 5y(t), \tag{9.5}$$
$$x(0) = 2; \quad y(0) = 3. \tag{9.6}$$

In the system given by Eqs. 9.4 and 9.5, the function f is

$$f(t, x, y) = 3x + 2xy - 3y^2$$

and the function g is

$$g(t, x, y) = -2x + 2x^2y + 5y.$$

J.K. Peterson, *Calculus for Cognitive Scientists*,
Cognitive Science and Technology, DOI 10.1007/978-981-287-877-9_9

or the system

$$x'(t) = 3x(t) + 2x(t)y(t) + 10\sin(t^2 + 5), \tag{9.7}$$

$$y'(t) = -2x(t) + 2x(t)y^2(t) + 20t^3e^{-2t^2}, \tag{9.8}$$

$$x(0) = 12; \;\; y(0) = -5. \tag{9.9}$$

The functions f and g in these examples are not linear in the variables x and y; hence the system Eqs. 9.4 and 9.5 and system Eqs. 9.7 and 9.8 are what are called a *nonlinear systems*. In general, arbitrary functions of time, ϕ and ψ to the model giving

$$x'(t) = 3x(t) + 2x(t)y(t) - 3y^2(t) + \phi(t), \tag{9.10}$$

$$y'(t) = -2x(t) + 2x^2(t)y(t) + 5y(t) + \psi(t), \tag{9.11}$$

$$x(0) = 2; \;\; y(0) = 3. \tag{9.12}$$

The functions ϕ and ψ are what we could call *data* functions. For example, if $\phi(t) = \sin(t)$ and $\psi(t) = te^{-t}$, the system Eqs. 9.10 and 9.11 would become

$$\begin{aligned} x'(t) &= 3x(t) + 2x(t)y(t) - 3y^2(t) + \sin(t), \\ y'(t) &= -2x(t) + 2x^2(t)y(t) + 5y(t) + te^{-t}, \\ x(0) &= 2; \;\; y(0) = 3. \end{aligned}$$

The functions f and g would then become

$$\begin{aligned} f(t, x, y) &= 3x + 2xy - 3y^2 + \sin(t) \\ g(t, x, y) &= -2x + 2x^2y + 5y + te^{-t} \end{aligned}$$

How do we solve such a system of differential equations? There are some things we can do if the functions f and g are linear in x and y, but many times we will be forced to look at the solutions using numerical techniques. We explored how to solve first order equations in Chap. 3 and we will now adapt the tools developed in that chapter to systems of differential equations. First, we will show you how to write a system in terms of matrices and vectors and then we will solve some particular systems.

9.1 Setting Up the Matrix and the Vector Functions

It is easiest to see how to convert problems into a matrix–vector form by looking at some examples.

Example 9.1.1 Convert to a matrix–vector system

$$
\begin{aligned}
2.0\,x''(t) \;-\; 3.0\,x'(t)\; 5.0\,x(t) &= 0 \\
x(0) &= 2.0 \\
x'(0) &= -4.0
\end{aligned}
$$

Solution *We let the vector* $\boldsymbol{x}$ *be given by*

$$
\boldsymbol{x}(t) = \begin{bmatrix} \boldsymbol{x}_1(t) \\ \boldsymbol{x}_2(t) \end{bmatrix} = \begin{bmatrix} u(t) \\ u'(t) \end{bmatrix}
$$

Then,

$$
\begin{aligned}
\boldsymbol{x}_1'(t) &= u'(t) \;=\; x_2(t), \\
\boldsymbol{x}_2'(t) &= u''(t) \\
&= -(5/2)u(t) \;+\; (3/2)u'(t) \\
&= -(5/2)\boldsymbol{x}_1(t) \;+\; (3/2)\boldsymbol{x}_2(t).
\end{aligned}
$$

We then convert the above into the matrix–vector system

$$
\boldsymbol{x}'(t) = \begin{bmatrix} \boldsymbol{x}_1'(t) \\ \boldsymbol{x}_2'(t) \end{bmatrix} = \begin{bmatrix} 0 & 1 \\ -(5/2) & (3/2) \end{bmatrix} \begin{bmatrix} \boldsymbol{x}_1(t) \\ \boldsymbol{x}_2(t) \end{bmatrix}
$$

Also, note that

$$
\boldsymbol{x}(0) = \begin{bmatrix} \boldsymbol{x}_1(0) \\ \boldsymbol{x}_2(0) \end{bmatrix} = \begin{bmatrix} u(0) \\ u'(0) \end{bmatrix} = \begin{bmatrix} 2 \\ -4 \end{bmatrix} = \boldsymbol{x}_0.
$$

Let the matrix above be called A. Then we have converted the original system into the matrix–vector equation $\boldsymbol{x}'(t) \;=\; A\,\boldsymbol{x}(t),\ \boldsymbol{x}(0) = \boldsymbol{x}_0$.

Example 9.1.2 Convert to a matrix–vector system

$$
\begin{aligned}
x''(t) \;+ 4.0\,x'(t) \;-\; 5.0\,x(t) &= 0 \\
x(0) &= -1.0 \\
x'(0) &= 1.0
\end{aligned}
$$

Solution *Again, we let the vector* $\boldsymbol{x}$ *be given by*

$$
\boldsymbol{x}(t) = \begin{bmatrix} \boldsymbol{x}_1(t) \\ \boldsymbol{x}_2(t) \end{bmatrix} = \begin{bmatrix} u(t) \\ u'(t) \end{bmatrix}
$$

Then, similar to what we did in the previous example, we find

$$\begin{aligned} \boldsymbol{x}_1'(t) &= u'(t) \;=\; x_2(t), \\ \boldsymbol{x}_2'(t) &= u''(t) \\ &= 5u(t) \;-\; (4)u'(t) \\ &= 5\boldsymbol{x}_1(t) \;-\; 4\boldsymbol{x}_2(t). \end{aligned}$$

We then convert the above into the matrix–vector system

$$\boldsymbol{x}'(t) = \begin{bmatrix} \boldsymbol{x}_1'(t) \\ \boldsymbol{x}_2'(t) \end{bmatrix} = \begin{bmatrix} 0 & 1 \\ 5 & -4 \end{bmatrix} \begin{bmatrix} \boldsymbol{x}_1(t) \\ \boldsymbol{x}_2(t) \end{bmatrix}$$

Also, note that

$$\boldsymbol{x}(0) = \begin{bmatrix} \boldsymbol{x}_1(0) \\ \boldsymbol{x}_2(0) \end{bmatrix} = \begin{bmatrix} u(0) \\ u'(0) \end{bmatrix} = \begin{bmatrix} -1 \\ 1 \end{bmatrix} = \boldsymbol{x}(0).$$

Let the matrix above be called A. Then we have converted the original system into the matrix–vector equation $\boldsymbol{x}'(t) \;=\; A\,\boldsymbol{x}(t),\; x(0),\; \boldsymbol{x}(0) = \boldsymbol{x}_0$.

Example 9.1.3 Convert to a matrix–vector system

$$\begin{aligned} x'(t) \;+2.0\,y'(t) \;-2.0\,x(t)\;3.0\,y(t) &= 0 \\ 4.0\,x'(t) \;-1.0\,y'(t) \;+6.0\,x(t) \;-8.0\,y(t) &= 0 \\ x(0) &= -1.0 \\ y(0) &= 2.0 \end{aligned}$$

Solution *Let the vector* $\boldsymbol{u}$ *be given by*

$$\boldsymbol{u}(t) = \begin{bmatrix} \boldsymbol{u}_1(t) \\ \boldsymbol{u}_2(t) \end{bmatrix} = \begin{bmatrix} x(t) \\ y(t) \end{bmatrix}.$$

It is then easy to see that if we define the matrices A and B by

$$A = \begin{bmatrix} 1 & 2 \\ 4 & -1 \end{bmatrix}, \quad \textit{and} \;\; B = \begin{bmatrix} -2 & 3 \\ 6 & -8 \end{bmatrix},$$

we can convert the original system into

$$\begin{aligned} A\,\boldsymbol{u}'(t) \;+\; B\,\boldsymbol{u}(t) &= \begin{bmatrix} 0 \\ 0 \end{bmatrix}, \\ \boldsymbol{u}(0) &= \begin{bmatrix} -1 \\ 2 \end{bmatrix} \end{aligned}$$

9.1.1 Homework

Exercise 9.1.1 *Convert to a matrix–vector system* $X' = AX$

$$\begin{aligned} 6.0\,x''(t) \ + 14.0\,x'(t) \ + 1.0\,x(t) &= 0 \\ x(0) &= 1.0 \\ x'(0) &= -1.0 \end{aligned}$$

Exercise 9.1.2 *Convert to a matrix–vector system* $AX' + BX = 0$

$$\begin{aligned} -2.0\,x'(t) \ + 1.0\,y'(t) \ + 4.0\,x(t) \ + 1.0\,y(t) &= 0 \\ 6.0\,x'(t) \ + 9.0\,y'(t) \ + 1.0\,x(t) \ - 4.0\,y(t) &= 0 \\ x(0) &= 10.0 \\ y(0) &= 20.0 \end{aligned}$$

Exercise 9.1.3 *Convert to a matrix–vector system* $AX' + BX = 0$

$$\begin{aligned} 1.0\,x'(t) \ + 0.0\,y'(t) \ + 2.0\,x(t) \ + 3.0\,y(t) &= 0 \\ 4.0\,x'(t) \ - 5.0\,y'(t) \ + 6.0\,x(t) \ - 2.0\,y(t) &= 0 \\ x(0) &= 1.0 \\ y(0) &= -2.0 \end{aligned}$$

9.2 Linear Second Order Problems as Systems

Now let's consider how to adapt our previous code to handle these systems of differential equations. We will begin with the linear second order problems because we know how to do those already. First, let's consider a general second order linear problem.

$$\begin{aligned} a\,u''(t) \ + \ b\,u'(t) \ + \ c\,u(t) &= g(t) \\ u(0) &= e \\ u'(0) &= f \end{aligned}$$

where we assume a is not zero, so that we really do have a second order problem! As usual, we let the vector $\boldsymbol{x}$ be given by

$$\boldsymbol{x}(t) = \begin{bmatrix} \boldsymbol{x}_1(t) \\ \boldsymbol{x}_2(t) \end{bmatrix} = \begin{bmatrix} u(t) \\ u'(t) \end{bmatrix}$$

Then,

$$\begin{aligned} \boldsymbol{x}_1'(t) &= u'(t) \;=\; x_2(t), \\ \boldsymbol{x}_2'(t) &= u''(t) \\ &= -(c/a)u(t) \;-\; (b/a)u'(t) + (1/a)\,g(t) \end{aligned}$$

We then convert the above into the matrix–vector system

$$\boldsymbol{x}'(t) = \begin{bmatrix} \boldsymbol{x}_1'(t) \\ \boldsymbol{x}_2'(t) \end{bmatrix} = \begin{bmatrix} 0 & 1 \\ -(c/a) & -(b/a) \end{bmatrix} \begin{bmatrix} \boldsymbol{x}_1(t) \\ \boldsymbol{x}_2(t) \end{bmatrix} + \begin{bmatrix} 0 \\ (1/a)\,g(t) \end{bmatrix}$$

Also, note that

$$\boldsymbol{x}(0) = \begin{bmatrix} \boldsymbol{x}_1(0) \\ \boldsymbol{x}_2(0) \end{bmatrix} = \begin{bmatrix} u(0) \\ u'(0) \end{bmatrix} = \begin{bmatrix} e \\ f \end{bmatrix} = \boldsymbol{x}_0.$$

Let the matrix above be called A. Then we have converted the original system into the matrix–vector equation $\boldsymbol{x}'(t) \;=\; A\,\boldsymbol{x}(t),\ \boldsymbol{x}(0) = \boldsymbol{x}_0$. For our purposes of using MatLab, we need to write this in terms of vectors. We have

$$\begin{bmatrix} \boldsymbol{x}_1'(t) \\ \boldsymbol{x}_2'(t) \end{bmatrix} = \begin{bmatrix} \boldsymbol{x}_2 \\ -(c/a)\,\boldsymbol{x}_1 \;-\; (b/a)\,\boldsymbol{x}_2 \;+\; (1/a)\,g(t) \end{bmatrix}$$

Now, let the dynamics vector f be defined by

$$f = \begin{bmatrix} f_1 \\ f_2 \end{bmatrix} = \begin{bmatrix} \boldsymbol{x}_2 \\ -(c/a)\,\boldsymbol{x}_1 \;-\; (b/a)\,\boldsymbol{x}_2 \;+\; (1/a)\,g(t) \end{bmatrix}$$

For example, given the model

$$\begin{aligned} x'' + 4x' - 5x &= te^{-0.03t} \\ x(0) &= -1.0 \\ x'(0) &= 1.0 \end{aligned}$$

we write this in a MatLab session as is shown in Listing 9.1.

Listing 9.1: Linear Second Order Dynamics

```
% for second order model
% given the model  y'' + 4y - 5y = t e^{-.03t}
a = 1;
b = 4;
c = -5;
B = -(b/a);
A = -(c/a);
C = 1/a;
g = @(t) t.*exp(-.03*t);
f = @(t,y) [y(2);A*y(1) + B*y(2) + C*g(t)];
```

9.2.1 A Practice Problem

Consider the problem

$$
\begin{aligned}
y''(t) \; + 4.0 \, y'(t) \; - 5.0 \, y(t) &= 0 \\
y(0) &= -1.0 \\
y'(0) &= 1.0
\end{aligned}
$$

This has characteristic equation $r^2 + 4r - 5 = 0$ with roots $r_1 = -5$ and $r_2 = 1$. Hence, the general solution is $x(t) = Ae^{-5t} + Be^{t}$. The initial conditions give

$$
\begin{aligned}
A + B &= -1 \\
-5A + B &= 1
\end{aligned}
$$

It is straightforward to see that $A = -1/3$ and $B = -2/3$. Then we solve the system using MatLab with this session:

Listing 9.2: Solving $x'' + 4x' - 5x = 0, \;\; x(0) = -1, x'(0) = 1$

```
% define the dynamics for  y'' + 4y' - 5y = 0
a = 1; b = 4; c = -5;
B = -(b/a); A = -(c/a); C = 1/a;
g = @(t) 0.0;
f = @(t,y) [y(2);A*y(1) + B*y(2) + C*g(t)];
% define the true solution
 true= @(t) [-(1.0/3.0)*exp(-5*t)-(2.0/3.0)*exp(t);...
                    (5.0/3.0)*exp(-5*t)-(2.0/3.0)*exp(t)];
y0 = [-1;1];
h = .2;
T = 3;
time = linspace(0,T,101);
N = ceil(T/h);
[htime1,rkapprox1] = FixedRK(f,0,y0,h,1,N);
yhat1 = rkapprox1(1,:);
[htime2,rkapprox2] = FixedRK(f,0,y0,h,2,N);
yhat2 = rkapprox2(1,:);
[htime3,rkapprox3] = FixedRK(f,0,y0,h,3,N);
yhat3 = rkapprox3(1,:);
[htime4,rkapprox4] = FixedRK(f,0,y0,h,4,N);
yhat4 = rkapprox4(1,:);
ytrue = true(time);
plot(time,ytrue(1,:),htime1,yhat1,'o',htime2,yhat2,'*',...
         htime3,yhat3,'+',htime4,yhat4,'-');
xlabel('Time');
ylabel('y');
title('Solution to  x'''' + 4x'' - 5x = 0, x(0) = -1, x''(0) = 1 on [1.3]');
legend('True','RK1','RK2','RK3','RK4','Location','Best');
```

One comment about this code. The function **`true`** has vector values; the first component is the true solution and the second component is the true solution's derivative. Since we want to plot only the true solution, we need to extract it from **`true`**. We do this with the command **`ytrue=true(time)`** which saves the vector of **`true`** values into the new variable **`ytrue`**. Then in the plot command, we plot only the solution by using **`ytrue(1,:)`**. This generates a plot as shown in Fig. 9.1.

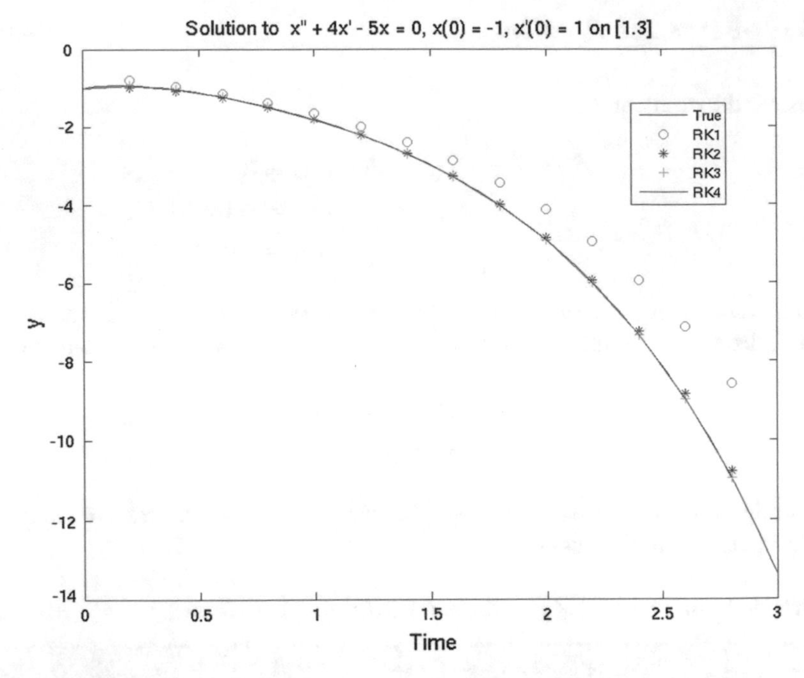

Fig. 9.1 Solution to $x'' + 4x' - 5x = 0, x(0) = -1, x'(0) = 1$ on [1.3]

9.2.2 *What If We Don't Know the True Solution?*

Now let's add an external input, $g(t) = 10\sin(5 * t)e^{-0.03t}$. In a more advanced class, we could find the true solution for this external input, but it if we changed to $g(t) = 10\sin(5 * t)e^{-0.03t^2}$ we would not be able to do that. So in general, there are many models we can not find the true solution to. However, the Runge–Kutta methods work quite well. Still, we always have the question in the back of our minds: is this plot accurate?

Listing 9.3: Solving $x'' + 4x' - 5x = 10\sin(5 * t)e^{-0.03t^2}$, $x(0) = -1, x'(0) = 1$

```
% define the dynamics for  y'' + 4y - 5y = 10 sin(5t) e^{-.03t}
a = 1; b = 4; c = -5;
B = -(b/a); A = -(c/a); C = 1/a;
g = @(t) 10*sin(5*t).*exp(-.03*t);
5 f = @(t,y) [y(2);A*y(1) + B*y(2) + C*g(t)];
y0 = [-1;1];
h = .2;
T = 3;
N = ceil(T/h);
10 [htime1,rkapprox1] = FixedRK(f,0,y0,h,1,N);
yhat1 = rkapprox1(1,:);
[htime2,rkapprox2] = FixedRK(f,0,y0,h,2,N);
yhat2 = rkapprox2(1,:);
[htime3,rkapprox3] = FixedRK(f,0,y0,h,3,N);
```

```
15 yhat3 = rkapprox3(1,:);
   [htime4,rkapprox4] = FixedRK(f,0,y0,h,4,N);
   yhat4 = rkapprox4(1,:);
   plot(htime1,yhat1,'o',htime2,yhat2,'*',...
              htime3,yhat3,'+',htime4,yhat4,'-');
20 xlabel('Time');
   ylabel('Approx y');
   title('Solution to  x'''' + 4x'' - 5x = 10 sin(5t) e^{-.03t}, x(0) = -1, x''(0) =
      1 on [0,3]');
   legend('RK1','RK2','RK3','RK4','Location','Best');
```

This generates a plot as shown in Fig. 9.2.

9.2.3 Homework: No External Input

On all of these problems, choose an appropriate stepsize h, time interval $[0, T]$ for some positive T and

- find the true solution and write this as MatLab code.
- find the Runge–Kutta order 1 through 4 solutions.
- Write this up with attached plots.

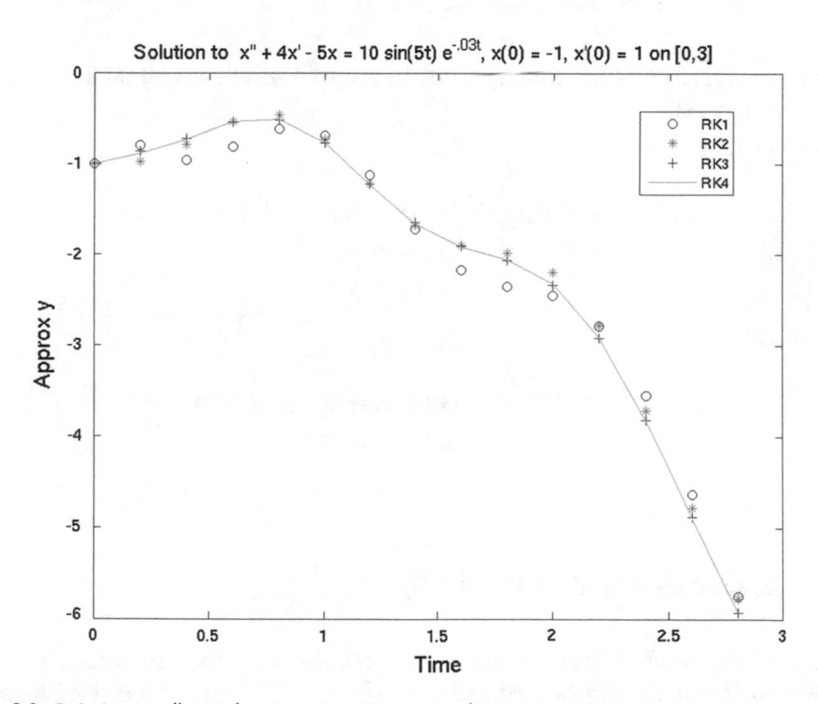

Fig. 9.2 Solution to $x'' + 4x' - 5x = 0, x(0) = -1, x'(0) = 1$ on [1.3]

Exercise 9.2.1

$$\begin{aligned} u''(t) + u'(t) - 2\,u(t) &= 0 \\ u(0) &= 1 \\ u'(0) &= -2 \end{aligned}$$

Exercise 9.2.2

$$\begin{aligned} x''(t) + 6\,x'(t) + 9\,x(t) &= 0 \\ x(0) &= 1 \\ x'(0) &= 2 \end{aligned}$$

Exercise 9.2.3

$$\begin{aligned} y''(t) + 4\,y'(t) + 13\,y(t) &= 0 \\ y(0) &= 1 \\ y'(0) &= 2 \end{aligned}$$

9.2.4 Homework: External Inputs

For these models, find the Runge–Kutta 1 through 4 solutions and do the write up with plot as usual.

Exercise 9.2.4

$$\begin{aligned} 2\,u''(t) + 4\,u'(t) - 3\,u(t) &= \exp(-2t)\,\cos(3t+5) \\ u(0) &= -2 \\ u'(0) &= 3 \end{aligned}$$

Exercise 9.2.5

$$\begin{aligned} u''(t) - 2\,u'(t) + 13\,u(t) &= \exp(-5t)\,\sin(3t^2+5) \\ u(0) &= -12 \\ u'(0) &= 6 \end{aligned}$$

9.3 Linear Systems Numerically

We now turn our attention to solving systems of linear ODEs numerically. We will show you how to do it in two worked out problems. We then generate the plot of y versus x, the two lines representing the eigenvectors of the problem, the $x' = 0$ and the $y' = 0$ lines on the same plot. A typical MatLab session would look like this:

Listing 9.4: Solving $x' = -3x + 4y, \;\; y' = -x + 2y; \;\; x(0) = -1; \, y(0) = 1$

```
f = @(t,x) [-3*x(1)+4*x(2);-x(1)+2*x(2)];
E1 = @(x) 0.25*x;
E2 = @(x) x;
xp = @(x) (3/4)*x;
yp = @(x) 0.5*x;
T = 1.4;
h = .03;
x0 = [-1;1];
N = ceil(T/h);
[ht,rk] = FixedRK(f,0,x0,h,4,N);
X = rk(1,:);
Y = rk(2,:);
xmin = min(X);
xmax = max(X);
xtop = max( abs(xmin), abs(xmax) );
ymin = min(Y);
ymax = max(Y);
ytop = max( abs(ymin), abs(ymax) );
D = max(xtop,ytop);
s = linspace(-D,D,201);
plot(s,E1(s),'-r' ,s,E2(s),'-m' ,s,xp(s),'-b' ,s,yp(s),'-g' ,X,Y,'-k' );
xlabel('x' );
ylabel('y' );
title('Phase Plane for Linear System x'' = -3x+4y, y''=-x+2y, x(0) = -1, y(0) = 1'
    );
legend('E1' ,'E2' ,'x''=0' ,'y''=0' ,'y vs x' ,'Location' ,'Best' );
```

This generates Fig. 9.3. This matches the kind of qualitative analysis we have done by hand, although when we do the plots by hand we get a more complete picture. Here, we see only one plot instead of the many trajectories we would normally sketch.

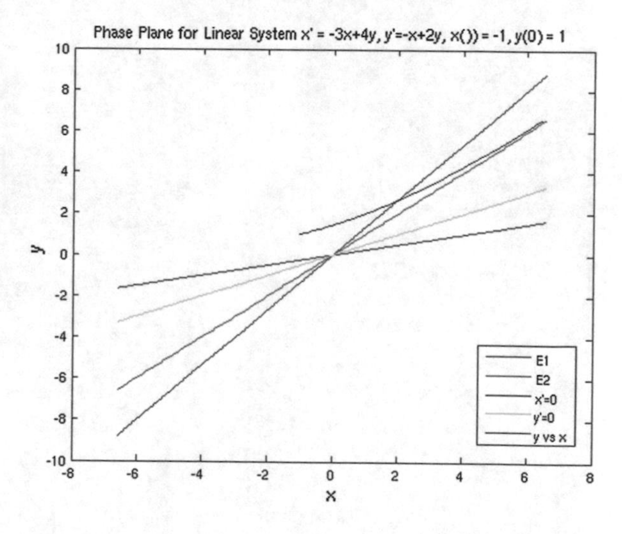

Fig. 9.3 Solving $x' = -3x + 4xy, \; y' = -x + 2y; \;\; x(0) = -1; \, y(0) = 1$

Now let's annotate the code.

Listing 9.5: Solving $x' = -3x + 4y,\ \ y' = -x + 2y;\ \ x(0) = -1; y(0) = 1$

```
% Set up the dynamics for the model
f = @(t,x) [-3*x(1)+4*x(2);-x(1)+2*x(2)];
% Eigenvector 1 is [1;.25] so slope is .25
% set up a straight line using this slope
E1 = @(x) 0.25*x;
% Eigenvector 2 is [1;.1] so slope is 1
% set up a straight line using this slope
E2 = @(x) x;
% the x'=0 nullcline is -3x+4y = 0 or y=3/4 x
% set up a line with this slope
xp = @(x) (3/4)*x;
% the y'=0 nullcline is -x+2y = 0 or y = .5 x
% so set up a line with this slope
yp = @(x) 0.5*x;
% set the final time
T = 1.4;
% set the step size
h = .03;
% set the IC
y0 = [-1;1];
% find how many steps we'll take
N = ceil(T/h);
% Find N approximate RK 4 values
% store the times in ht and the values in rk
% rk(1,:) is the  first column which is x
% rk(2,:) is the second column which is y
[ht,rk] = FixedRK(f,0,y0,h,4,N);
% rk(1,:) is the  first row which is set to X
% rk(2,:) is the second row which is set to Y
X = rk(1,:);
Y = rk(2,:);
% the x values range from xmin to xman
xmin = min(X);
xmax = max(X);
% find max of their absolute values
% example: xmin = -7, xmax = 4 ==> xtop = 7
xtop = max( abs(xmin), abs(xmax) );
% the x values range from xmin to xman
ymin = min(Y);
ymax = max(Y);
% find max of their absolute values
% example: ymin = -3, ymax = 10 ==> ytop = 10
ytop = max( abs(ymin), abs(ymax) );
% find max of xtop and ytop; our example gives D = 10
D = max(xtop,ytop);
% The plot lives in the box [-D,D] x [-D,D]
% set up a linspace of [-D,D]
s = linspace(-D,D,201);
% plot the eigenvector lines, the nullclines and Y vs X
% -r is red, -m is magenta, -b is blue, -g is green and -k is black
plot(s,E1(s),'-r' ,s,E2(s),'-m' ,s,xp(s),'-b' ,s,yp(s),'-g' ,X,Y, '-k' );
% set x and y labels
xlabel('x' );
ylabel('y' );
% set title
title('Phase Plane for Linear System x'' = -3x+4y, y''=-x+2y, x(0) = -1, y(0) = 1'
    );
% set legend
legend('E1' ,'E2' ,'x''=0'  ,'y''=0'  ,'y vs x' ,'Location'  ,'Best'  );
```

Let's do another example.

Example 9.3.1 For the system below

$$\begin{bmatrix} x'(t) \\ y'(t) \end{bmatrix} = \begin{bmatrix} 4 & 9 \\ -1 & -6 \end{bmatrix} \begin{bmatrix} x(t) \\ y(t) \end{bmatrix}$$
$$\begin{bmatrix} x(0) \\ y(0) \end{bmatrix} = \begin{bmatrix} 4 \\ -2 \end{bmatrix}$$

- Find the characteristic equation
- Find the general solution
- Solve the IVP
- Solve the System Numerically

Solution *The eigenvalue and eigenvector portion of this solution has already been done in Example 2.10.2 and so we only have to copy the results here. The characteristic equation is*

$$det \left(r \begin{bmatrix} 1 & 0 \\ 0 & 1 \end{bmatrix} - \begin{bmatrix} 4 & 9 \\ -1 & -6 \end{bmatrix} \right) = 0$$

with **eigenvalues** $r_1 = -5$ *and* $r_2 = 3$.

1. *For eigenvalue* $r_1 = -5$, *we found the eigenvector to be*

$$\begin{bmatrix} 1 \\ -1 \end{bmatrix}.$$

2. *For eigenvalue* $r_2 = 3$, *we found the eigenvector to be*

$$\begin{bmatrix} 1 \\ -1/9 \end{bmatrix}$$

The general solution to our system is thus

$$\begin{bmatrix} x(t) \\ y(t) \end{bmatrix} = A \begin{bmatrix} 1 \\ -1 \end{bmatrix} e^{-5t} + B \begin{bmatrix} 1 \\ -1/9 \end{bmatrix} e^{3t}$$

We solve the IVP by finding the A and B that will give the desired initial conditions. This gives

$$\begin{bmatrix} 4 \\ -2 \end{bmatrix} = A \begin{bmatrix} 1 \\ -1 \end{bmatrix} + B \begin{bmatrix} 1 \\ -1/9 \end{bmatrix}$$

or

$$4 = A + B$$
$$-2 = -A + \frac{-1}{9} B$$

This is easily solved using elimination to give $A = 7/4$ and $B = 9/4$. The solution to the IVP is therefore

$$\begin{bmatrix} x(t) \\ y(t) \end{bmatrix} = \frac{7}{4} \begin{bmatrix} 1 \\ -1 \end{bmatrix} e^{-5t} + \frac{9}{4} \begin{bmatrix} 1 \\ \frac{-1}{9} \end{bmatrix} e^{3t}$$

$$= \begin{bmatrix} 7/4\, e^{-5t} \ + 9/4\, e^{3t} \\ -7/4\, e^{-5t} \ - 1/4\, e^{3t} \end{bmatrix}$$

We now have all the information needed to solve this numerically. The MatLab session for this problem is then

Listing 9.6: Phase Plane for $x' = 4x + 9y$, $y' = -x - 6y$, $x(0) = 4$, $y(0) = -2$

```
f = @(t,x) [4*x(1)+9*x(2);-x(1)-6*x(2)];
E1 = @(x) -1*x;
E2 = @(x) -(1/9)*x;
xp = @(x) -(4/9)*x;
yp = @(x) -(1/6)*x;
T = 0.4;
h = .01;
x0 = [4;-2];
N = ceil(T/h);
[ht,rk] = FixedRK(f,0,x0,h,4,N);
X = rk(1,:);
Y = rk(2,:);
xmin = min(X);
xmax = max(X);
xtop = max( abs(xmin), abs(xmax) );
ymin = min(Y);
ymax = max(Y);
ytop = max( abs(ymin), abs(ymax) );
D = max(xtop,ytop);
x = linspace(-D,D,201);
plot(x,E1(x),'-r',x,E2(x),'-m',x,xp(x),'-b',x,yp(x),'-c',X,Y,'-k');
xlabel('x');
ylabel('y');
title('Phase Plane for Linear System x'' = 4x+9y, y''=-x-6y, x(0) = 4, y(0) = -2')
    ;
legend('E1','E2','x''=0','y''=0','y vs x','Location','Best');
```

This generates Fig. 9.4. Again this matches the kind of qualitative analysis we have done by hand, but for only one plot instead of the many trajectories we would normally sketch. We had to choose the final time T and the step size h by trial and error to generate the plot you see. If T is too large, the growth term in the solution generates x and y values that are too big and the trajectory just looks like it lies on top of the dominant eigenvector line.

9.3.1 Homework

For these models,

- Find the characteristic equation
- Find the general solution

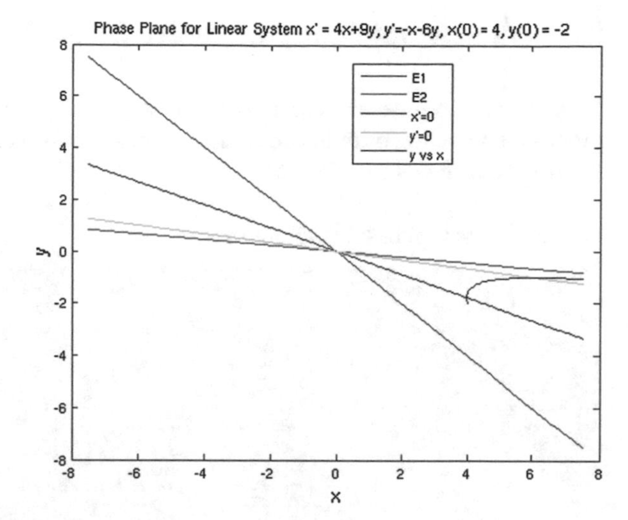

Fig. 9.4 Solving $x' = 4x + 9xy, \;\; y' = -x - 6y; x(0) = 4; y(0) = -2$

- Solve the IVP
- Solve the System Numerically

Exercise 9.3.1

$$\begin{bmatrix} x'(t) \\ y'(t) \end{bmatrix} = \begin{bmatrix} 1 & 1 \\ -1 & -3/2 \end{bmatrix} \begin{bmatrix} x(t) \\ y(t) \end{bmatrix}$$
$$\begin{bmatrix} x(0) \\ y(0) \end{bmatrix} = \begin{bmatrix} -5 \\ -2 \end{bmatrix}$$

Exercise 9.3.2

$$\begin{bmatrix} x'(t) \\ y'(t) \end{bmatrix} = \begin{bmatrix} 4 & 2 \\ -9 & -5 \end{bmatrix} \begin{bmatrix} x(t) \\ y(t) \end{bmatrix}$$
$$\begin{bmatrix} x(0) \\ y(0) \end{bmatrix} = \begin{bmatrix} -1 \\ 2 \end{bmatrix}$$

Exercise 9.3.3

$$\begin{bmatrix} x'(t) \\ y'(t) \end{bmatrix} = \begin{bmatrix} 3 & 7 \\ 4 & 6 \end{bmatrix} \begin{bmatrix} x(t) \\ y(t) \end{bmatrix}$$
$$\begin{bmatrix} x(0) \\ y(0) \end{bmatrix} = \begin{bmatrix} -2 \\ -3 \end{bmatrix}$$

9.4 An Attempt at an Automated Phase Plane Plot

Now let's try to generate a real phase plane portrait by automating the phase plane plots for a selection of initial conditions. Consider the code below which is saved in the file **AutoPhasePlanePlot.m**.

Listing 9.7: AutoPhasePlanePlot.m

```
function AutoPhasePlanePlot(fname,stepsize,tinit,tfinal,rkorder,xboxsize,yboxsize,
    xmin,xmax,ymin,ymax)
  % fname is the name of the model dynamics
  % stepsize is the chosen step size
  % tinit is the initial time
5 % tfinal is the final time
  % rkorder is the RK order
  % we will use initial conditions chosen from the box
  % [xmin, xmax] x [ymin,ymax]
  %   This is done using the linspace command
10 %   so xboxsize is the number of points in the interval [xmin,xmax]
  %        yboxsize is the number of points in the interval [ymin,ymax]
  % u and v are the vectors we use to compute our ICs
  n = ceil((tfinal-tinit)/stepsize);
  u = linspace(xmin,xmax,xboxsize);
15 v = linspace(ymin,ymax,yboxsize);
  % hold plot and cycle line colors
  xlabel('x');
  ylabel('y');
  newplot;
20 hold all;
  for i=1:xboxsize
    for j=1:yboxsize
      x0 = [u(i);v(j)];
      [htime,rk,frk] = FixedRK(fname,tinit,x0,stepsize,rkorder,n);
25    X = rk(1,:);
      Y = rk(2,:);
      plot(X,Y);
    end
  end
30 title('Phase Plane Plot');
  hold off;
```

There are some new elements here. We set up vectors u and v to construct our initial conditions from. Each initial condition is of the form (u_i, v_j) and we use that to set the initial condition $x0$ we pass into **FixedRK()** as usual. We start by telling MatLab the plot we are going to build is a *new* one; so the previous plot should be erased. The command **hold all** then tells MatLab to keep all the plots we generate as well as the line colors and so forth until a **hold off** is encountered. So here we generate a bunch of plots and we then see them on the same plot at the end! A typical session usually requires a lot of trial and error. In fact, you should find the analysis by hand is actually more informative! As discussed, the **`AutoPhasePlanePlot.m`** script is used by filling in values for the inputs it needs. Again the script has these inputs

Listing 9.8: AutoPhasePlanePlot Arguments

```
AutoPhasePlanePlot(fname,stepsize,tinit,tfinal,rkorder,xboxsize,yboxsize,xmin,
    xmax,ymin,ymax)
  % fname is the name of our dynamics function
  % stepsize is our call, here .01 seems good
  % tinit is the starting time, here always 0
  % tfinal is hard to pick, here small seems best; .2 - .6
  %          because one of our eigenvalues is positive and the solution
  %          grows too fast
  % rkorder is the Runge-Kutta order, here 4
  % xboxsize is how many different x initial values we want, here 4
  % yboxsize is how mnay different y initial values we want, here 4
  % xmin and xmax give the interval we pick the initial x values from;
  %          here, they come from [-.3,.3] in the last attempt.
  % ymin and ymax give the interval we pick the initial x values from
  %          here, they come from [-.3,.3] in the last attempt.
```

Now some attempts for the model

$$\begin{bmatrix} x'(t) \\ y'(t) \end{bmatrix} = \begin{bmatrix} 4 & -1 \\ 8 & -5 \end{bmatrix} \begin{bmatrix} x(t) \\ y(t) \end{bmatrix}$$

This is encoded in Matlab as **`vecfunc = @(t,y) [4*y(1)-y(2);8*y(1)-5*y(2)];`**. We can then try a few phase plane plots.

Listing 9.9: Trying Some Phase Plane Plots

```
AutoPhasePlanePlot(vecfunc,.1,0,1,4,4,4,-1,1,-1,1);
AutoPhasePlanePlot(vecfunc,.01,0,.1,4,4,4,-1,1,-1,1);
AutoPhasePlanePlot(vecfunc,.01,0,.4,4,4,4,-1,1,-1,1);
AutoPhasePlanePlot(vecfunc,.01,0,.6,4,4,4,-1,1,-1,1);
AutoPhasePlanePlot(vecfunc,.01,0,.6,4,4,4,-1.5,1.5,-1.5,1.5);
AutoPhasePlanePlot(vecfunc,.01,0,.6,4,4,4,-.5,.5,-.5,.5);
AutoPhasePlanePlot(vecfunc,.01,0,.8,4,4,4,-.5,.5,-.5,.5);
AutoPhasePlanePlot(vecfunc,.01,0,.8,4,4,4,-.3,.3,-.3,.3);
AutoPhasePlanePlot(vecfunc,.01,0,.8,4,4,4,-.2,.2,-.2,.2);
AutoPhasePlanePlot(vecfunc,.01,0,.4,4,4,4,-.2,.2,-.2,.2);
AutoPhasePlanePlot(vecfunc,.01,0,.4,4,4,4,-.3,.3,-.3,.3);
xlabel('x axis');
ylabel('y axis');
title('Phase Plane Plot');
```

After a while, we get a plot we like as shown in Fig. 9.5.

Not bad! But we don't show the nullclines and the eigenvector lines on top of the plots! That would be nice to do and we can do that using tools in MatLab.

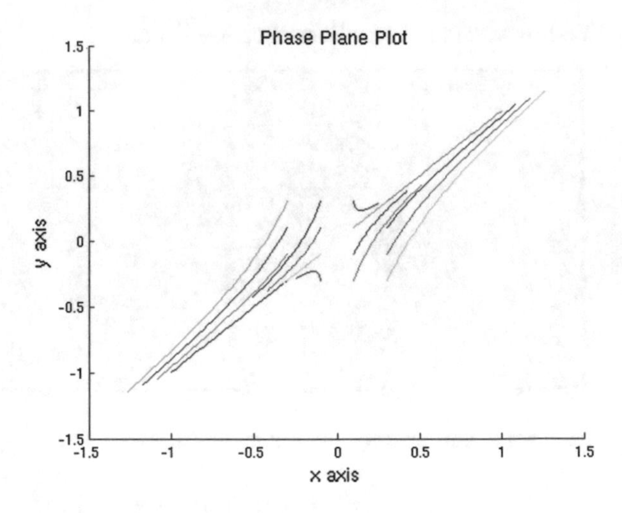

Fig. 9.5 Phase plane $x' = 4x - y,\ y' = 8x - 5y$

9.5 Further Automation!

So let's start using some more MatLab tools. As always, with power comes an increased need for responsible behavior!

9.5.1 Eigenvalues in MatLab

We will now discuss certain ways to compute eigenvalues and eigenvectors for a square matrix in MatLab. For a given $\boldsymbol{A}$, we can compute its eigenvalues as follows:

Listing 9.10: Eigenvalues in Matlab: eig

```
A = [1 2 3; 4 5 6; 7 8 -1]

A =

     1     2     3
     4     5     6
     7     8    -1

E = eig(A)

E =

   -0.3954
   11.8161
   -6.4206
```

So we have found the eigenvalues of this small 3×3 matrix. Note, in general they are not returned in any sorted order like small to large. Bummer! To get the eigenvectors, we do this:

Listing 9.11: Eigenvectors and Eigenvalues in Matlab: eig

```
[V,D] = eig(A)

V =

    0.7530   -0.3054   -0.2580
   -0.6525   -0.7238   -0.3770
    0.0847   -0.6187    0.8896


D =

   -0.3954         0         0
         0   11.8161         0
         0         0   -6.4206
```

The eigenvalue/eigenvector pairs are thus

$$\lambda_1 = -0.3954$$
$$V_1 = \begin{bmatrix} 0.7530 \\ -0.6525 \\ 0.0847 \end{bmatrix}$$

$$\lambda_2 = 11.8161$$
$$V_2 = \begin{bmatrix} -0.3054 \\ -0.7238 \\ -0.6187 \end{bmatrix}$$

$$\lambda_3 = -6.4206$$
$$V_3 = \begin{bmatrix} -0.2580 \\ -0.3770 \\ 0.8896 \end{bmatrix}$$

Now let's try a nice 5×5 array that is symmetric:

Listing 9.12: Example 5 × 5 Eigenvalue and Eigenvector Calculation in Matlab

```
B = [1 2 3 4 5;
     2 5 6 7 9;
     3 6 1 2 3;
     4 7 2 8 9;
     5 9 3 9 6]

B =

     1     2     3     4     5
     2     5     6     7     9
     3     6     1     2     3
     4     7     2     8     9
     5     9     3     9     6

[W,Z] = eig(B)

W =

    0.8757    0.0181   -0.0389    0.4023    0.2637
   -0.4289   -0.4216   -0.0846    0.6134    0.5049
    0.1804   -0.6752    0.4567   -0.4866    0.2571
   -0.1283    0.5964    0.5736   -0.0489    0.5445
    0.0163    0.1019   -0.6736   -0.4720    0.5594


Z =

    0.1454         0         0         0         0
         0    2.4465         0         0         0
         0         0   -2.2795         0         0
         0         0         0   -5.9321         0
         0         0         0         0   26.6197
```

It is possible to show that the eigenvalues of a symmetric matrix will be real and eigenvectors corresponding to distinct eigenvalues will be 90° apart. Such vectors are called **orthogonal** and recall this means their inner product is 0. Let's check it out. The eigenvectors of our matrix are the columns of W above. So their dot product should be 0!

Listing 9.13: Inner Products in Matlab: dot

```
C = dot(W(1:5,1),W(1:5,2))

C =

  1.3336e-16
```

Well, the dot product is not actually 0 because we are dealing with floating point numbers here, but as you can see it is close to machine zero (the smallest number our computer chip can *detect*). Welcome to the world of computing!

9.5.2 Linear System Models in MatLab Again

We have already solves linear models using MatLab tools. Now we will learn to do a bit more. We begin with a sample problem. Note to analyze a linear systems model, we can do everything by hand, we can then try to emulate the hand work, using computational tools. A sketch of the process is thus:

1. For the system below, first do the work by hand.

$$\begin{bmatrix} x'(t) \\ y'(t) \end{bmatrix} = \begin{bmatrix} 4 & -1 \\ 8 & -5 \end{bmatrix} \begin{bmatrix} x(t) \\ y(t) \end{bmatrix}$$
$$\begin{bmatrix} x(0) \\ y(0) \end{bmatrix} = \begin{bmatrix} -3 \\ 1 \end{bmatrix}$$

   - Find the characteristic equation
   - Find the general solution
   - Solve the IVP
   - On the same x–y graph,
   (a) draw the $x' = 0$ line
   (b) draw the $y' = 0$ line
   (c) draw the eigenvector one line
   (d) draw the eigenvector two line
   (e) divide the $x - y$ into four regions corresponding to the algebraic signs of x' and y'
   (f) draw the trajectories of enough solutions for various initial conditions to create the phase plane portrait

2. Now do the graphical work with MatLab. Use the MatLab scripts below. Make sure you have the relevant codes in your directory. For the problem above, a typical MatLab session would be

Listing 9.14: Sample Linear Model: $x' = 4x - y$, $y' = 8x - 5y$

```
vecfunc = @(t,x) [4*x(1)-x(2);8*x(1)-5*x(2)];
% E1 is [1;8] so slope is 8
E1 = @(x) 8*x;
% E2 is [1;1] so slope is 1
E2 = @(x) x;
% x' = 0 is 4x-y=0 or y = 4x
xp = @(x) 4*x;
%y'=0 is 8x-5y=0 so 5y = 8x or y = 8/5 x
yp = @(x) (8/5)*x;
T = 0.4;
h = .01;
x0 = [-3;1];
N = ceil(T/h);
[ht,rk,fk] = FixedRK(vecfunc,0,x0,h,4,N);
X = rk(1,:);
Y = rk(2,:);
xmin = min(X);
xmax = max(X);
xtop = max( abs(xmin), abs(xmax) );
ymin = min(Y);
ymax = max(Y);
ytop = max( abs(ymin), abs(ymax) );
```

```
   D = max(xtop,ytop);
   x = linspace(-D,D,201);
25 plot(x,E1(x),'-r',x,E2(x),'-m',x,xp(x),'-b',x,yp(x),'-c',X,Y,'-k');
   xlabel('x');
   ylabel('y');
   title('Phase Plane for Linear System x'' = 4x-y, y''=8x-5y, x(0) = -3, y(0) =
       1');
   legend('E1','E2','x''=0','y''=0','y vs x','Location','Best');
```

which generates the plot seen in Fig. 9.6.

3. Now find eigenvectors and eigenvalues with MatLab for this problem. In MatLab this is done like this

Listing 9.15: Find the Eigenvalues and Eigenvectors in Matlab

```
1     A = [4,-1;8,-5]

      A =

           4    -1
6          8    -5

      [V,D] = eigs(A)

      V =
11
         0.1240    0.7071
         0.9923    0.7071


16    D =

          -4    0
           0    3
```

you should be able to see that the eigenvectors we got before by hand are the same as these except they are written as vectors of length one.

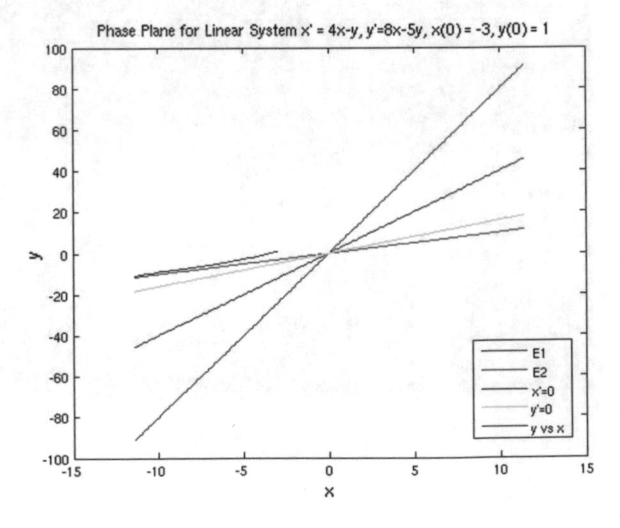

Fig. 9.6 Sample plot

4. Now plot many trajectories at the same time. We discussed this a bit earlier. It is very important to note that the hand analysis is in many ways easier. A typical MatLab session usually requires a lot of trial and error! So try not to get too frustrated. The **AutoPhasePlanePlot.m** script is used by filling in values for the inputs it needs.
   Now some attempts.

Listing 9.16: Session for $x' = 4x + 9y, \quad y' = -x - 6y$ Phase Plane Plots

```
AutoPhasePlanePlot(vecfunc,.1,0,1,4,4,4,-1,1,-1,1);
AutoPhasePlanePlot(vecfunc,.01,0,.1,4,4,4,-1,1,-1,1);
AutoPhasePlanePlot(vecfunc,.01,0,.4,4,4,4,-1,1,-1,1);
AutoPhasePlanePlot(vecfunc,.01,0,.6,4,4,4,-1,1,-1,1);
AutoPhasePlanePlot(vecfunc,.01,0,.6,4,4,4,-1.5,1.5,-1.5,1.5);
AutoPhasePlanePlot(vecfunc,.01,0,.6,4,4,4,-.5,.5,-.5,.5);
AutoPhasePlanePlot(vecfunc,.01,0,.8,4,4,4,-.5,.5,-.5,.5);
AutoPhasePlanePlot(vecfunc,.01,0,.8,4,4,4,-.3,.3,-.3,.3);
AutoPhasePlanePlot(vecfunc,.01,0,.8,4,4,4,-.2,.2,-.2,.2);
AutoPhasePlanePlot(vecfunc,.01,0,.4,4,4,4,-.2,.2,-.2,.2);
AutoPhasePlanePlot(vecfunc,.01,0,.4,4,4,4,-.3,.3,-.3,.3);
xlabel('x axis');
ylabel('y axis');
title('Phase Plane Plot');
```

which generates the plot seen in Fig. 9.7.

9.5.3 Project

For this project, follow the outline just discussed above to solve any one of these models. Then

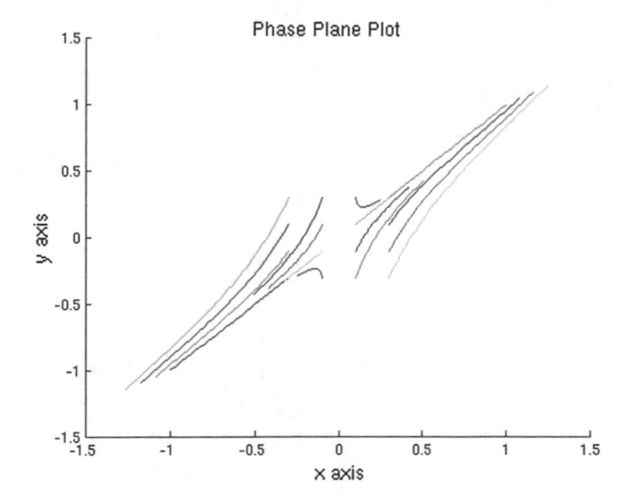

Fig. 9.7 Phase plane plot

Solve the Model By Hand: Do this and attach to your project report.

Plot One Trajectory Using MatLab: Follow the outline above. This part of the report is done in a word processor with appropriate comments, discussion etc. Make sure you document all of your MatLab work thoroughly! Show your MatLab code and sessions as well as plots.

Find the Eigenvalues and Eigenvectors in MatLab: Explain how the MatLab work connects to the calculations we do by hand.

Plot Many Trajectories Simultaneously Using MatLab: This part of the report is also done in a word processor with appropriate comments, discussion etc. Show your MatLab code and sessions as well as plots with appropriate documentation.

Exercise 9.5.1

$$\begin{bmatrix} x'(t) \\ y'(t) \end{bmatrix} = \begin{bmatrix} 1 & 5 \\ 5 & 1 \end{bmatrix} \begin{bmatrix} x(t) \\ y(t) \end{bmatrix}$$
$$\begin{bmatrix} x(0) \\ y(0) \end{bmatrix} = \begin{bmatrix} 4 \\ 10 \end{bmatrix}$$

Exercise 9.5.2

$$\begin{bmatrix} x'(t) \\ y'(t) \end{bmatrix} = \begin{bmatrix} 1 & 1 \\ -1 & -3/2 \end{bmatrix} \begin{bmatrix} x(t) \\ y(t) \end{bmatrix}$$
$$\begin{bmatrix} x(0) \\ y(0) \end{bmatrix} = \begin{bmatrix} -4 \\ -6 \end{bmatrix}$$

Exercise 9.5.3

$$\begin{bmatrix} x'(t) \\ y'(t) \end{bmatrix} = \begin{bmatrix} 4 & 2 \\ -9 & -5 \end{bmatrix} \begin{bmatrix} x(t) \\ y(t) \end{bmatrix}$$
$$\begin{bmatrix} x(0) \\ y(0) \end{bmatrix} = \begin{bmatrix} 2 \\ -5 \end{bmatrix}$$

Exercise 9.5.4

$$\begin{bmatrix} x'(t) \\ y'(t) \end{bmatrix} = \begin{bmatrix} 3 & 7 \\ 4 & 6 \end{bmatrix} \begin{bmatrix} x(t) \\ y(t) \end{bmatrix}$$
$$\begin{bmatrix} x(0) \\ y(0) \end{bmatrix} = \begin{bmatrix} -2 \\ 5 \end{bmatrix}$$

$$\begin{bmatrix} x'(t) \\ y'(t) \end{bmatrix} = \begin{bmatrix} 1 & 5 \\ 5 & 1 \end{bmatrix} \begin{bmatrix} x(t) \\ y(t) \end{bmatrix}$$
$$\begin{bmatrix} x(0) \\ y(0) \end{bmatrix} = \begin{bmatrix} 2 \\ 3 \end{bmatrix}$$

9.6 AutoPhasePlanePlot Again

Here is an enhanced version of the automatic phase plane plot tool. It would be nice to automate the plotting of the eigenvector lines, the nullclines and the trajectories so that we didn't have to do so much work by hand. Consider the new function in Listing 9.17. In this function, we pass in **vecfunc** into the argument **fname**. We evaluate this function using the command **feval(fname,0,[a time; an x]);**. The linear model has a coefficient matrix A of the form

$$\begin{bmatrix} a & b \\ c & d \end{bmatrix}$$

and so **u feval(fname,0,[1;0]);** returns column one of A and **u feval (fname,0,[0;1]);** returns column two of A. This is how we can extract a, b, c and d for our linear model without having to type them in. With them we can define our A using **A = [a,b;c,d];** and then grab the eigenvalues and eigenvectors.

Listing 9.17: AutoPhasePlanePlotLinearModel

```
function Approx = AutoPhasePlanePlotLinearModel(fname,stepsize,tinit,tfinal,
    rkorder, ...
                                   xboxsize,yboxsize,xmin,xmax,ymin,ymax,mag)
% fname is the name of the model dynamics
% stepsize is the chosen step size
5 % tinit is the initial time
% tfinal is the final time
% rkorder is the RK order
% we will use initial conditions chosen from the box
% [xmin, xmax] x [ymin,ymax]
10 %   This is done using the linspace command
%   so xboxsize is the number of points in the interval [xmin,xmax]
%        yboxsize is the number of points in the interval [ymin,ymax]
% mag is related to the zoom in level of our plot.
%
15 n = ceil((tfinal-tinit)/stepsize);
% extract a, b, c, and d
u = feval(fname,0,[1;0]);
a = u(1);
c = u(2);
20 v = feval(fname,0,[0;1]);
b = v(1);
d = v(2);
% construct A
A = [a,b;c,d];
25 % get eigenvalues and eigenvectors
[V,D] = eig(A);
evals = diag(D);
% the first column of V is E1, second is E2
E1 = V(:,1);
30 E2 = V(:,2);
% The rise over run of E1 is the slope
% The rise over run of E2 is the slope
E1slope = E1(2)/E1(1);
E2slope = E2(2)/E2(1);
35 % define the eigenvector lines
E1line = @(x) E1slope*x;
E2line = @(x) E2slope*x;
% setup the nullcline lines
```

```
xp = @(x) -(a/b)*x;
40 yp = @(x) -(c/d)*x;
% clear out any old pictures
clf
% setup x and y initial condition box
xic = linspace(xmin,xmax,xboxsize);
45 yic = linspace(ymin,ymax,yboxsize);
% find all the trajectories and store them
Approx = {};
for i=1:xboxsize
  for j=1:yboxsize
50     x0 = [xic(i);yic(j)];
     [ht,rk] = FixedRK(fname,0,x0,stepsize,4,n);
     Approx{i,j} = rk;
     U = Approx{i,j};
     X = U(1,:);
55     Y = U(2,:);
     % get the plotting square for each trajectory
     umin = min(X);
     umax = max(X);
     utop = max( abs(umin), abs(umax) );
60     vmin = min(Y);
     vmax = max(Y);
     vtop = max( abs(vmin), abs(vmax) );
     D(i,j) = max(utop,vtop);
    end
65 end
% get the largest square to put all the plots into
E = max(max(D))
% setup the x linspace for the plot
x = linspace(-E,E,201);
70 % start the hold
hold on
% plot the eigenvector lines and then the nullclines
plot(x,E1line(x),'-r',x,E2line(x),'-m',x,xp(x),'-b',x,yp(x),'-c');
% loop over all the ICS and get all trajectories
75 for i=1:xboxsize
  for j=1:yboxsize
     U = Approx{i,j};
     X = U(1,:);
     Y = U(2,:);
80     plot(X,Y,'-k');
    end
  end
% set labels and so forth
xlabel('x');
85 ylabel('y');
title('Phase Plane');
legend('E1','E2','x''=0','y''=0','y vs x','Location','BestOutside');
% set zoom for plot using mag
axis([-E*mag E*mag -E*mag E*mag]);
90 % finish the hold
hold off;
end
```

In use, for **`f = @(t,x)[4*x(1)+9*x(2);-x(1)-6*x(2)]`** we can generate nice looking plots and *zero* in on a nice view with an appropriate use of the parameter **`max`**.

Listing 9.18: Session for $x' = 4x + 9y,\ \ y' = -x - 6y$ Enhanced Phase Plane Plots

```
Approx = AutoPhasePlanePlotLinearModelE(f,.01,0,.45,4,8,8,-.5,.5,-.5,.5,.2);
E =
    4.1421
Approx = AutoPhasePlanePlotLinearModelE(f,.01,0,.65,4,8,8,-.5,.5,-.5,.5,.1);
E =
    7.6481
Approx = AutoPhasePlanePlotLinearModelE(f,.01,0,.65,4,12,12,-1.5,1.5,-1.5,1.5,.1);
E =
   22.9443
Approx = AutoPhasePlanePlotLinearModelE(f
   ,.01,0,1.65,4,12,12,-1.5,1.5,-1.5,1.5,.01);
E =
  462.3833
Approx = AutoPhasePlanePlotLinearModelE(f
   ,.01,0,1.65,4,12,12,-1.5,1.5,-1.5,1.5,.005);
E =
  462.3833
Approx = AutoPhasePlanePlotLinearModelE(f,.01,0,1.65,4,6,6,-1.5,1.5,-1.5,1.5,.005)
   ;
E =
  462.3833
Approx = AutoPhasePlanePlotLinearModelE(f,.01,0,1.65,4,6,6,-1.5,1.5,-1.5,1.5,.008)
   ;
E =
  462.3833
```

The last command generates the plot in Fig. 9.8.

Let's look at all the pieces of this code in detail. It will help you learn a bit more about how to write code to solve your models. First, we find the number of steps needed as we have done before; **`n = ceil((tfinal-tinit)/stepsize);`** Now we don't want to have to hand calculate the eigenvalues and eigenvectors anymore. To use the **`eig`** command, we need the matrix $\boldsymbol{A}$ from the dynamics

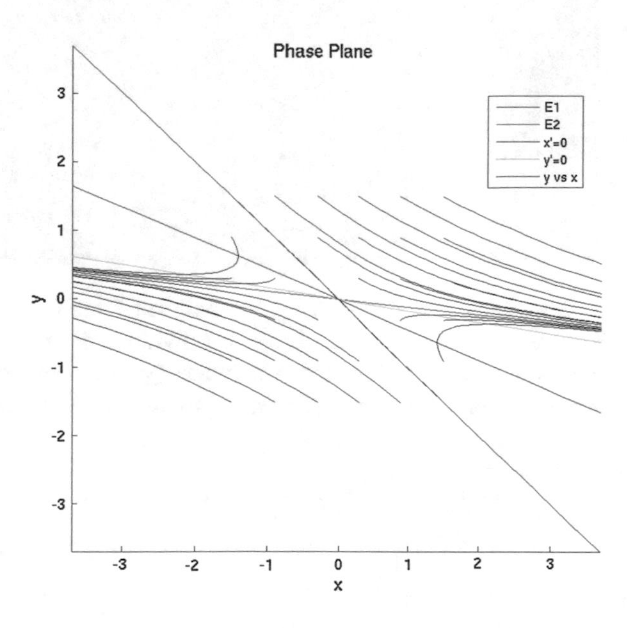

Fig. 9.8 Phase plane $x' = 4x + 9y,\ \ y' = -x - 6y$

fname. To extract A, we use the following lines of code. Note, for a linear model, **f = @(t,x) [a*x(1) + b*x(2); c*x(1)+d*x(2)];**. Hence, we can do evaluations to find the coefficients: **f(0,[1;0]) = [a;c]** and **f(0,[0;1]) = [b;d]**.

Listing 9.19: Extracting A

```
% extract [a;c]
u = feval(fname,0,[1;0]);
% extract a and c
a = u(1);
c = u(2);
% extract [b;d]
v = feval(fname,0,[0;1]);
% extract b and d
b = v(1);
d = v(2);
% set up A
A = [a,b;c,d];
% get the eigenvalues and eigenvectors of A
[V,D] = eigs(A);
% the diag(D) command gets the diagonal of D
% which gives the two eigenvalues
evals = diag(D);
```

Next, we set the first column of **V** to be the eigenvector **E1** and the second column of **V** to be the eigenvector **E1**. Then we setup the lines we need to plot the eigenvectors and the nullclines. The $x' = 0$ nullcline is $ax + by = 0$ or $y = -(a/b)x$ and the $y' = 0$ nullcline is $cx + dy = 0$ or $y = -(c/d)x$.

Listing 9.20: Setting up the eigenvector and nullcline lines

```
% extract eigenvector 1 and eigenvector 2
E1 = V(:,1);
E2 = V(:,2);
% find the slope of eigenvector 1 and
% the slope of eigenvector 2
E1slope = E1(2)/E1(1);
E2slope = E2(2)/E2(1);
% set the eigenvector lines
E1line = @(x) E1slope*x;
E2line = @(x) E2slope*x;
% set the nullcline lines
xp = @(x) -(a/b)*x;
yp = @(x) -(c/d)*x;
```

Then clear any previous figures with **clf** before we get started with the plot. The initial conditions are chosen by dividing the interval **[xmin, xmax]** into **xboxsize** points. Similarly, we divide **[ymin, ymax]** into **yboxsize** points. We use these $xboxsize \times yboxsize$ possible pairs as our initial conditions.

Listing 9.21: Set up Initial conditions, find trajectories and the bounding boxes

```
% set up possible x coordinates of the ICs
xic = linspace(xmin,xmax,xboxsize);
% set up possible y coordinates of the ICs
yic = linspace(ymin,ymax,yboxsize);
% set up a data structure call a cell to store
% each trajectory
Approx = {};
% loop over all possible initial condition
for i=1:xboxsize
  for j=1:yboxsize
     % set the IC
     x0 = [xic(i);yic(j)];
     % solve the model usign RK4.
     % return the approximate values in rk
     [ht,rk] = FixedRK(fname,0,x0,stepsize,4,n);
     % store rk as the i,j th entry in the cell Approx
     Approx{i,j} = rk;
     % set U to be the current Approx cell entry
     % which is the same as the returned rk
     U = Approx{i,j};
     % set the first row of U to be X
     % set the second row of U to be Y
     X = U(1,:);
     Y = U(2,:);
     % find the square the trajectory fits inside
     umin = min(X);
     umax = max(X);
     utop = max( abs(umin), abs(umax) );
     vmin = min(Y);
     vmax = max(Y);
     vtop = max( abs(vmin), abs(vmax) );
     % store the size of the square that fits the trajectory
     % for this IC
     D(i,j) = max(utop,vtop);
   end
end
```

Now that we have all these squares for the possible trajectories, we find the biggest one possible and set up the linspace command for this box. All of our trajectories will be drawn inside the square $[-E, E] \times [-E, E]$.

Listing 9.22: Set up the bounding box for all the trajectories

```
E = max(max(D))
x = linspace(-E,E,201);
```

Next, plot all the trajectories and set labels and so forth. The last thing we do is to set the axis as **`axis([-E*mag E*mag -E*mag E*mag]);`** which zeros or zooms in on the interval $[-E * mag, E * mag] \times [-E * magE * mag]$.

Listing 9.23: Plot the trajectories at the chosen zoom level

```
hold on
plot(x,E1line(x),'-r',x,E2line(x),'-m',x,xp(x),'-b',x,yp(x),'-c');
for i=1:xboxsize
  for j=1:yboxsize
    U = Approx{i,j};
    X = U(1,:);
    Y = U(2,:);
    plot(X,Y,'-k');
   end
  end
xlabel('x');
ylabel('y');
title('Phase Plane');
legend('E1','E2','x''=0','y''=0','y vs x','Location','BestOutside');
axis([-E*mag E*mag -E*mag E*mag]);
hold off;
```

9.6.1 Project

Here is another project which uses **`AutoPhasePlanePlotLinearModel`**. For a given model

Plot Many Trajectories Simultaneously Using MatLab: This part of the report is done in a word processor with appropriate comments, discussion etc. Show your MatLab code and sessions as well as plots with appropriate documentation.

Choose Initial Conditions wisely: Choose a useful set of initial conditions to plot trajectories for by choosing **`xmin, xmax`**, **`ymin, ymax`**, and **`xboxsize, yboxsize`** appropriately.

Choose the final time and step size: You'll also have to find the right final time and step size to use.

Generate a nice phase plane plot: Work hard on generating a really nice phase plane plot by an appropriate use of the **`mag`** factor.

Exercise 9.6.1

$$\begin{bmatrix} x'(t) \\ y'(t) \end{bmatrix} = \begin{bmatrix} 1 & 1 \\ -1 & -3/2 \end{bmatrix} \begin{bmatrix} x(t) \\ y(t) \end{bmatrix}$$

Exercise 9.6.2

$$\begin{bmatrix} x'(t) \\ y'(t) \end{bmatrix} = \begin{bmatrix} 4 & 2 \\ -9 & -5 \end{bmatrix} \begin{bmatrix} x(t) \\ y(t) \end{bmatrix}$$

Exercise 9.6.3

$$\begin{bmatrix} x'(t) \\ y'(t) \end{bmatrix} = \begin{bmatrix} 3 & 7 \\ 4 & 6 \end{bmatrix} \begin{bmatrix} x(t) \\ y(t) \end{bmatrix}$$

Exercise 9.6.4

$$\begin{bmatrix} x'(t) \\ y'(t) \end{bmatrix} = \begin{bmatrix} 1 & 5 \\ 5 & 1 \end{bmatrix} \begin{bmatrix} x(t) \\ y(t) \end{bmatrix}$$

9.7 A Two Protein Model

We begin with a description of what you need to do by using a sample problem. Consider the following two protein system.

$$\begin{bmatrix} x'(t) \\ y'(t) \end{bmatrix} = \begin{bmatrix} -0.001 & -0.001 \\ -0.004 & -0.008 \end{bmatrix} \begin{bmatrix} x(t) \\ y(t) \end{bmatrix} + \begin{bmatrix} 0.1 \\ 0.5 \end{bmatrix}$$
$$\begin{bmatrix} x(0) \\ y(0) \end{bmatrix} = \begin{bmatrix} 0 \\ 0 \end{bmatrix}$$

Note we can group terms like this:

$$x' = (-0.001x + 0.1) - 0.001y$$
$$y' = (-0.008y + 0.5) - 0.004x.$$

Hence, we can interpret the x dynamics as protein x follows a standard protein synthesis model ($P' = -\alpha P + \beta$) but as protein y is created it binds with the promoter for the gene controlling x to shut off x production at the rate -0.001. Similarly, the y dynamics reflect a standard protein synthesis model for y with the production of x curtailing the production of y through binding with y's promoter. We know this sort of linear system with constant inputs has a solution which we can find using our standard tools. But we do want to make sure this model has a solution where the protein levels are always positive. Otherwise, it is not a biologically realistic two protein model!

9.7.1 Protein Values Should be Positive!

In general, although it is easy to write down a two protein system, it is a bit harder to make sure you get positive values for the long term protein concentration levels. So as we were making up problems, we used a MatLab script to help get a good coefficient matrix. This is an interesting use of Matlab and for those of you who would like to dig more deeply into this aspect of modeling, we encourage you to study closely what we do. The terms that inhibit or enhance the other protein's production in general are a $c_1 y$ in the x dynamics and a $c_2 x$ in the y dynamics. Note, we can always think of a two dimensional vector like this

$$\begin{bmatrix} c_1 \\ c_2 \end{bmatrix} = c_1 \begin{bmatrix} 1 \\ c_2/c_1 \end{bmatrix}$$

and letting $t = c_2/c_1$, we have for $c_1 = \gamma$, that

$$\begin{bmatrix} c_1 \\ c_2 \end{bmatrix} = \gamma \begin{bmatrix} 1 \\ t \end{bmatrix} = \begin{bmatrix} \gamma \\ t\gamma \end{bmatrix}.$$

We can do a similar thing for the constant production terms d_1 for x and d_2 for y to write

$$\begin{bmatrix} d_1 \\ d_2 \end{bmatrix} = d_1 \begin{bmatrix} 1 \\ d_2/d_1 \end{bmatrix} = d_1 \begin{bmatrix} 1 \\ s \end{bmatrix} = \begin{bmatrix} \beta \\ s\beta \end{bmatrix}$$

for $d_1 = \beta$ and $s = d_1/d_2$. Finally, we can handle the two decay rates for x and y similarly. If these two rates are $-\alpha_1$ for x and $-\alpha_2$ for y, we can model that as

$$\begin{bmatrix} \alpha_1 \\ \alpha_2 \end{bmatrix} = \alpha_1 \begin{bmatrix} 1 \\ \alpha_2/\alpha_1 \end{bmatrix} = \alpha_1 \begin{bmatrix} 1 \\ u \end{bmatrix} = \begin{bmatrix} \alpha \\ u\alpha \end{bmatrix}$$

for $\alpha_1 = \alpha$ and $u = \alpha_1/\alpha_2$. So, if u is a lot more than 1, y will decay fast and we would expect x to take a lot longer to reach equilibrium. Our general two protein model will then have the form

$$\begin{aligned} x' &= -\alpha\, x - \gamma\, y + \beta \\ y' &= -u\alpha\, y + s\beta - t\gamma\, x \end{aligned}$$

and we want to choose these parameters so we get positive protein levels. The coefficient matrix $\boldsymbol{A}$ here is

$$\boldsymbol{A} = \begin{bmatrix} -\alpha & -\gamma \\ -t\,\gamma & -u\,\alpha \end{bmatrix}$$

with characteristic equation

$$r^2 + (\alpha + u\alpha)r + u\alpha^2 - t\alpha\gamma = 0.$$

So the roots are

$$r = \frac{1}{2}\left(-(\alpha + u\alpha) \pm \sqrt{\alpha^2(1-u)^2 + 4t\alpha\gamma}\right)$$

We want both roots negative so we have a decay situation. Hence,

$$-(\alpha + u\alpha) + \sqrt{\alpha^2(1-u)^2 + 4t\alpha\gamma} < 0.$$

This implies

$$\begin{aligned}(\alpha + u\alpha) &> \sqrt{\alpha^2(1-u)^2 + 4t\gamma^2}\\ \alpha^2(1+u)^2 &> \alpha^2(1-u)^2 + 4t\gamma^2\end{aligned}$$

After simplifying, we find we need to choose t so that

$$t < \frac{u\alpha^2}{\gamma^2}.$$

Let's choose $t = (1/2)\alpha u/\gamma^2$. Hence, a good choice for $\boldsymbol{A}$ is

$$\boldsymbol{A} = \begin{bmatrix} -\alpha & \gamma \\ -t\gamma = -(1/2)\alpha^2 u/\gamma & -u\alpha \end{bmatrix}$$

Next, we set up the growth levels. We want the particular solution to have positive components. If we let the particular solution be X_p, we know $X_p = -A^{-1}\, F$ where F is the vector of constant external inputs. Hence, we want $-A^{-1}\, F$ to have positive components. Thus,

$$-(1/det(A)) \begin{bmatrix} -u * \alpha & \gamma \\ t * \gamma & -\alpha \end{bmatrix} \begin{bmatrix} \beta \\ s\beta \end{bmatrix} > \begin{bmatrix} 0 \\ 0 \end{bmatrix}$$

where we interpret the inequality componentwise. Now $det(\boldsymbol{A}) = u\alpha^2 - t\gamma^2$. Plugging in $t = 0.5\, u\, \alpha^2/\gamma^2$, we find $det(\boldsymbol{A}) = 0.5\alpha\gamma^2 > 0$. Hence, we satisfy our component inequalities if

$$\begin{aligned}-u\,\alpha\,\beta + s\,\beta\,\gamma &< 0\\ t\,\beta\,\gamma - s\,\alpha\,\beta &< 0.\end{aligned}$$

Cancelling β, we find

$$\begin{aligned}-u\,\alpha + s\,\gamma &< 0 \Rightarrow s\,\gamma < u\,\alpha\\ t\,\gamma - s\,\alpha &< 0 \Rightarrow t\gamma < s\alpha.\end{aligned}$$

Plugging in t, we want

$$\begin{aligned}s\,\gamma &< u\,\alpha\\ \left(0.5\,u\,\frac{\alpha^2}{\gamma^2}\gamma\right) &< s\alpha.\end{aligned}$$

Simplifying, we find we want s to satisfy

$$0.5u\frac{\alpha}{\gamma} < s < u\frac{\alpha}{\gamma}.$$

Hence, any s on the line between $0.5u\alpha$ and $u\alpha/\gamma$ will work. Parameterize this line as $s = u(\alpha/\gamma)(1 - 0.5z)$ which at $z = 0$ gives $u\alpha/\gamma$ and at $z = 1$ gives $0.5u\alpha\gamma$. We will choose $z = 1/3$. The commands in Matlab are then

Listing 9.24: Choose s and set the external inputs

```
% choose s
z = 1/3;
s = (1 - 0.5*z)*alpha*u/gamma;
% set external input
F = [beta;s*beta]
```

We can easily write MatLab code to do all of this in a function we'll call **twoproteins**. It will return a good $\boldsymbol{A}$ and $\boldsymbol{F}$ for our choices of α, γ, u and β.

Listing 9.25: Finding the A and F for a two protein model

```
function [A,F] = twoprotsGetAF(alpha,gamma,u,beta)
%
% -alpha is decay rate for protein x
% -gamma is decay rate for protein y influencing protein x
% -t*gamma is decay rate for protein x influencing protein y
% -u*alpha is decay rate for protein y
% beta is the growth rate for protein x
% s*beta is the growth rate for protein y
% A is the dynamics
%
% now setup A
%  The roots satisfy
% 2 r = -alpha*(1+u) pm sqrt(D)
% where D = alpha^2(1-u)^2 + 4t gamma^2
% We want both roots negative, so we want
%
%  -alpha*(1+u) + sqrt(D) < 0
%   alpha*(1+u) > sqrt(D)
%   alpha^2(1+u)^2 > alpha^2(1-u)^2 + 4t gamma^2
%   alpha^2(1+2u+u^2) > alpha^2(1-2u+u^2) + 4t gamma^2
%   4 alpha^2 u > 4 t gamma^2
%     t < u alpha^2 /gamma^2
%
% so set t to satisfy this requirement
t = 0.5*alpha^2*u/(gamma^2)
A = [-alpha,-gamma;-t*gamma,-u*alpha];
% set up growth levels
% We want XP > 0 so we want
% - Ainv*F > 0
% - (1/det(A)) [-u*alpha,gamma; t*gamma,-alpha]*[1;s]*beta > 0
% Now det(A) = u alpha^2 - t gamma^2
%            = u alpha^2 - (0.5*alpha^2*u/(gamma^2)) gamma^2
%            = u alpha^2 - 0.5 alpha^2 u = 0.5 alpha^2
% so det(A) > 0
% So we want
%  (1/det(A)) [-u*alpha,gamma; t*gamma,-alpha]*[1;s]*beta < 0
% or
% [-u*alpha,gamma; t*gamma,-alpha]*[1;s] < 0
% or
```

```
40 %  -u alpha + s gamma < 0 ==> gamma s < u alpha
   %                        ==> s < u alpha/gamma
   %  t gamma - alpha s   < 0 ==> alpha s > t gamma
   %                        ==> s > (0.5*alpha^2*u/(gamma^2)) (gamma/alpha)
   %                        ==> s > 0.5 alpha u/gamma
45 % So
   %  0.5 alpha u/gamma < s < u alpha/gamma
   %  s = (alpha u/gamma) (0.5 z + (1-z))
   %    = (alpha u/gamma) * (1 - 0.5 z)
   %
50 z = 1/3;
   s = (1 - 0.5*z)*alpha*u/gamma
   F = [beta;s*beta];
   %
```

Note without the comments, this code is short:

Listing 9.26: Finding the A and F for a two protein model: uncommented

```
   function [A,F] = twoprotsGetAF(alpha,gamma,u,beta)
 2 %
   % -alpha is decay rate for protein x
   % -gamma is decay rate for protein y influencing protein x
   % -t*gamma is decay rate for protein x influencing protein y
   % -u*alpha is decay rate for protein y
 7 % beta is the growth rate for protein x
   % s*beta is the growth rate for protein y
   % A is the dynamics
   %
   t = 0.5*alpha^2*u/(gamma^2)
12 A = [-alpha,-gamma;-t*gamma,-u*alpha];
   % set up growth levels
   z = 1/3;
   s = (1 - 0.5*z)*alpha*u/gamma
   F = [beta;s*beta];
```

9.7.2 *Solving the Two Protein Model*

The section above shows us how to design the two protein system so the protein levels are always positive so let's try it out. Let $\alpha = \gamma = a$ for any positive a and $u = 8$. Then for our nice choice of t, we have could do the work by hand and find

$$\boldsymbol{A} = \begin{bmatrix} -a & -a \\ -4a & -8a \end{bmatrix}$$

But let's be smarter and let MatLab do this for us! We can do this for any positive a and generate a reasonable two protein model to play with. In Matlab. we begin by setting up the matrix $\boldsymbol{A}$ with **`twoprotsGetAF`**. We will choose do this for $\beta = 10$.

Listing 9.27: Setting up the coefficient matrix A

```
a = 10^(-3);
a =  0.0010000
[A,F] = twoprotsGetAF(a,a,8,10);
t =  4
s =  6.6667
A
A =

   -0.0010000   -0.0010000
   -0.0040000   -0.0080000
F
F =

   10.000
   66.667
```

Then we find the inverse of A using the formulae we developed in class for the inverse of a 2×2 matrix of numbers.

Listing 9.28: Finding the inverse of A

```
AInv = (1/det(A))*[A(2,2),-A(1,2);-A(2,1),A(1,1)];
AInv
AInv =

  -2000     250
   1000    -250
```

Next, we find the particular solution, XP.

Listing 9.29: Find the particular solution

```
XP = -AInv*F
XP =

   3333.67
   6666.67
```

Now we find the eigenvalues and eigenvectors of the matrix A using the **`eig`** command as we have done before. Recall, this commands returns the eigenvectors as columns of a matrix V and returns the eigenvalues as the diagonal entries of the matrix D.

Listing 9.30: Find eigenvalues and eigenvectors of A

```
[V,D] = eig(A)
V =

   0.88316    0.13163
  -0.46907    0.99130

D =

Diagonal Matrix

  -4.6887e-04            0
            0  -8.5311e-03
```

So column one of V is the eigenvector associated with the eigenvalue $-4.6887e{-}04$ and column two of V is the eigenvector associated with the eigenvalue $-8.5311e{-}03$. Note the first eigenvalue is the dominant one as MatLab doesn't necessarily follow our conventions on labeling the eigenvalues in a small to large order! The general solution is known to be

$$\begin{bmatrix} x(t) \\ y(t) \end{bmatrix} = C_1 \begin{bmatrix} 0.88316 \\ -0.46907 \end{bmatrix} e^{-4.6887e-04t} + C_2 \begin{bmatrix} 0.13163 \\ 0.99130 \end{bmatrix} e^{-8.5311e-03t} + \begin{bmatrix} 3333.67 \\ 6666.67 \end{bmatrix}$$

Notice that since eigenvector one and eigenvector two are the columns of the matrix **V**, we can rewrite this as in matrix–vector form as

$$\begin{bmatrix} x(t) \\ y(t) \end{bmatrix} = V \begin{bmatrix} C_1 \\ C_2 \end{bmatrix} + \boldsymbol{XP}.$$

Our protein models start with zero levels of proteins, so to satisfy the initial conditions, we have to solve

$$\begin{bmatrix} 0 \\ 0 \end{bmatrix} = C_1 \begin{bmatrix} 0.88316 \\ -0.46907 \end{bmatrix} + C_2 \begin{bmatrix} 0.13163 \\ 0.99130 \end{bmatrix} + \begin{bmatrix} 3333.67 \\ 6666.67 \end{bmatrix}$$

Hence, the initial data gives us

$$V \begin{bmatrix} C_1 \\ C_2 \end{bmatrix} = -\boldsymbol{XP}.$$

and the solution is then

$$\begin{bmatrix} C_1 \\ C_2 \end{bmatrix} = \boldsymbol{V}^{-1} (-\boldsymbol{XP}) = -\boldsymbol{V}^{-1} (\boldsymbol{XP}).$$

In Matlab, we then have

Listing 9.31: Solving the IVP

```
VInv = (1/det(V))*[V(2,2),-V(1,2);-V(2,1),V(1,1)];
C =  VInv XP
C =

   -2589.4
   -7950.4
```

Then we can construct the solutions $x(t)$ and $y(t)$ so we can plot the protein concentrations as a function of time. First, store the eigenvalues in a more convenient form by grabbing the diagonal entries of `D` using `Ev = diag(D)`. Then, we construct the solutions like this:

Listing 9.32: Constructing the solutions x and y and plotting

```
Ev = diag(D);
x = @(t) ( C(1)*V(1,1)*exp(Ev(1)*t)+C(2)*V(1,2)*exp(Ev(2)*t)+XP(1));
y = @(t) ( C(1)*V(2,1)*exp(Ev(1)*t)+C(2)*V(2,2)*exp(Ev(2)*t)+XP(2));
T = linspace(0,12000,1000);
plot(T,x(T),T,y(T).'-+');
xlabel('Time');
ylabel('Protein Levels');
title('Two Protein Model: alpha = gamma = 1.-e-3, u = 8, beta = 10');
legend('Protein 1','Protein 2');
```

This generates Fig. 9.9.
Note we can also do a quick phase plane plot: plotting $y(t)$ versus $x(t)$ as shown in Fig. 9.10. We see the protein trajectory converges to the particular solution as $t \rightarrow \infty$. Of course, you have to plot the solutions over long enough time to see this!

Listing 9.33: The protein phase plane plot

```
T = linspace(0,12000,1000);
plot(x(T),y(T));
xlabel('Protein One');
ylabel('Protein Two');
title('Protein Two vs Protein One');
```

In summary, to solve the two protein problem, you would just type a few lines in MatLab as follows:

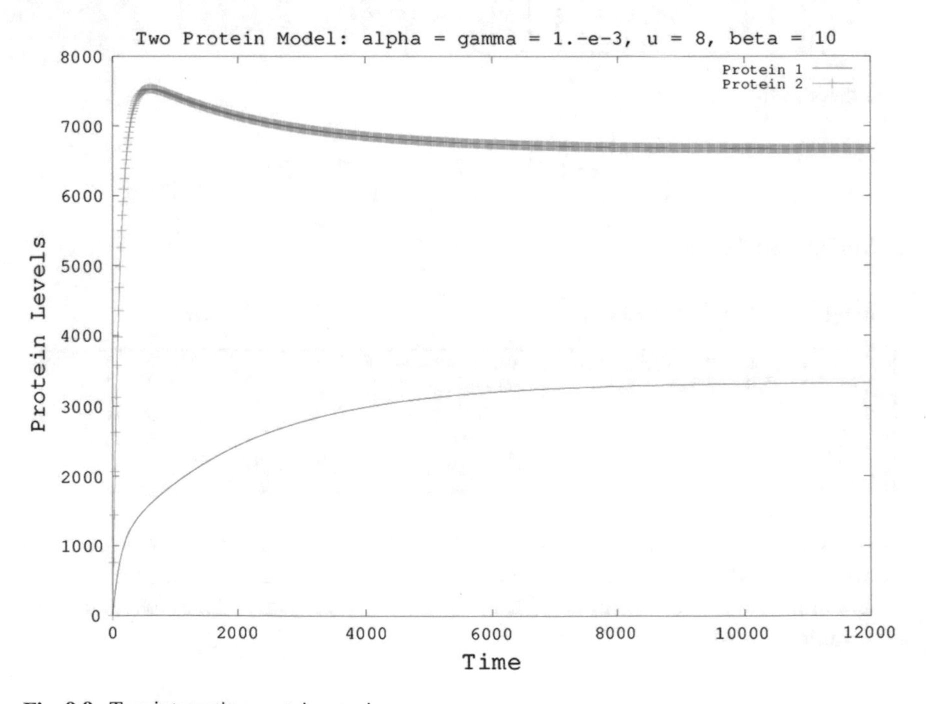

Fig. 9.9 Two interacting proteins again

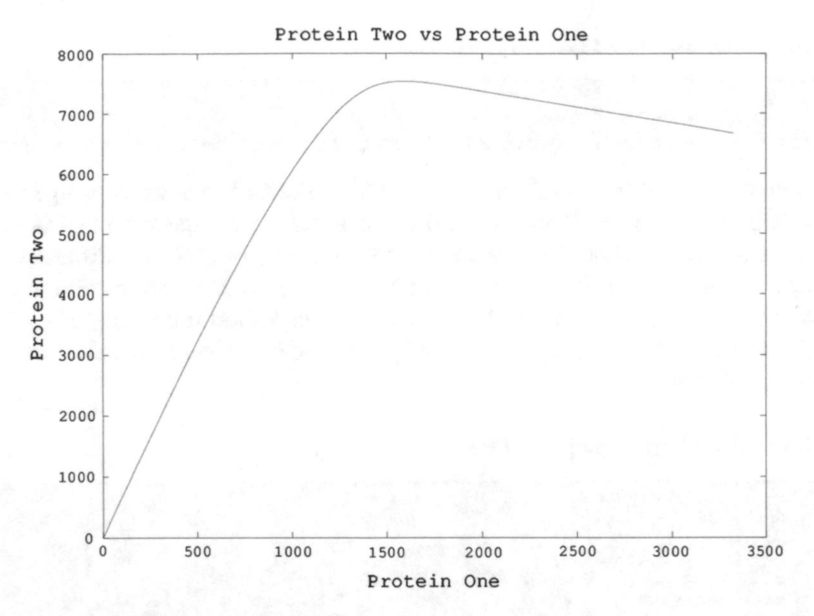

Fig. 9.10 Phase plane for two interacting proteins again

Listing 9.34: Solving the problem

```
a = 10^(-3);
[A,F] = twoprotsGetAF(a,a,8,10);
% find eigenvalues and eigenvectors
[V,D] = eig(A);
Ev = diag(D)
VInv = (1/det(V))*[V(2,2),-V(1,2);-V(2,1),V(1,1)];
AInv = (1/det(A))*[A(2,2),-A(1,2);-A(2,1),A(1,1)];
% find particular solution
XP = -AInv*F
% Use initial conditions
C = -VInv*XP
% set up x and y
x = @(t) ( C(1)*V(1,1)*exp(Ev(1)*t)+C(2)*V(1,2)*exp(Ev(2)*t)+XP(1));
y = @(t) ( C(1)*V(2,1)*exp(Ev(1)*t)+C(2)*V(2,2)*exp(Ev(2)*t)+XP(2));
```

The last thing is to think about response times. Here x grows a lot slower by design as we set $u = 8$ and so we could temporarily think of x as ≈ 0 giving the y dynamics, $y' \approx -u\alpha y + s\beta$. Thus, the response time for y is about $t_r^y = \ln(2)/u\alpha$ or in MatLab `tRx = log(2)/(-A(2,2))`. To approximate the response time of x, start with y achieving its approximate steady state value from our y dynamics approximation above. This is $y_\infty \approx s\beta/(u\alpha)$. Using this in the x dynamics, we have $x' \approx -\alpha x + \beta - \gamma y_\infty$ which has a response time of $t_r^y \approx \ln(2)/(\alpha)$ which in Matlab is `tRx = log(2)/(-A(1,1))`. Hence, in our problem, one protein grows very slowly and so we typically see two time scales. On the first, one protein changes rapidly while the other is essentially constant. But on the longer time scale, the protein which grows slower initially eventually overtakes the first protein on its way to its higher equilibrium value. As discussed above, we estimate these time scales as

Listing 9.35: Estimated Protein Response Times

```
tRy = log(2)/(-A(2,2))
tRx = log(2)/(-A(1,1))
```

Note the large difference in time scales! The protein levels we converge to are $x = 3333.67$ and $y = 6666.67$, but the response time for the x protein is much longer than the response time for the y protein. So if we plot over different time scales, we should see primarily y protein until we reach a suitable fraction of the x protein response time. Figure 9.11 shows a plot over a short time scale, $10t_r^y$. Over 10 short time scales, protein y is essentially saturated but protein x is only about $1/3$ of its equilibrium value.

Listing 9.36: Short Time Scale Plot

```
T = linspace(0,10*tRy,100);
plot(T,x(T),T,y(T),'+');
xlabel('Time');
ylabel('Protein Levels');
title('Two Protein Model over Short Time Scale');
legend('Protein X','Protein Y');
```

Figure 9.12 shows a plot over a long time scale, $10t_r^x$ which is about $80t_r^y$. This allows x to get close to its asymptotic value. Note the long term value of y drops some from its earlier peak. Our *response times* here are just approximations and the

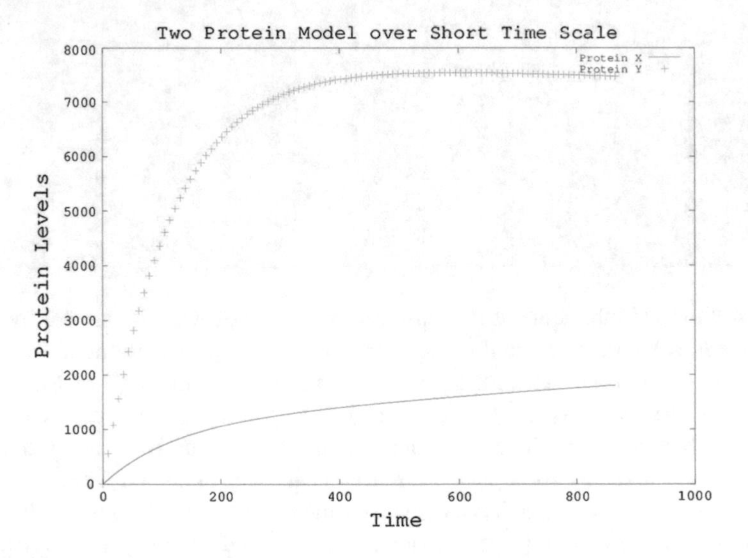

Fig. 9.11 Phase plane for two interacting proteins on a short time scale

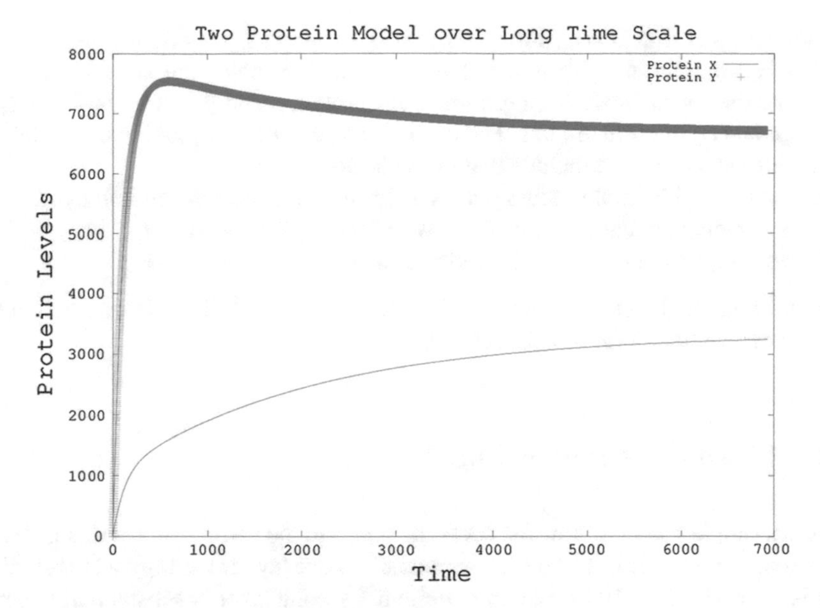

Fig. 9.12 Phase plane for two interacting proteins on a long time scale

protein interactions of the model eventually take effect the the equilibrium value of y drops to its final particular solution value.

Listing 9.37: Long Time Scale Plot

```
T = linspace(0,10*tRx,10000);
plot(T,x(T),T,y(T),'+');
xlabel('Time');
ylabel('Protein Levels');
title('Two Protein Model over Long Time Scale');
legend('Protein X','Protein Y');
```

9.7.3 Project

Now we are ready for the project. For the model choice of $\alpha = 0.003$, $\gamma = 0.005$ and $u = 8$ and $\beta = 0.01$, we expect x to grow a lot slower. Your job is the follow the discussion above and generate a nice report. The report has this format.

Introduction (5 Points) Explain what we are trying to do here with this model. Contrast this model with two proteins to our earlier model for one protein.

Description of Model (10 Points) Describe carefully what each term in the model represents in terms of protein transcription. Since these models are abstractions of reality, explain how these models are a simplified version of the protein transcription process.

Annotated Solution Discussion (27 Points) In this part, you solve the model using MatLab as I did in the example above. As usual, explain your steps nicely. Calculate the approximate response times for protein x and protein y and use them to generate appropriate plots. For each plot you generate, provide useful labels, legends and titles and include them in your document.

Conclusion (5 Points) Discuss what you have done here. Do you think you could do something similar for more than two proteins using MatLab?

References (3 Points) Put any reference material you use in here.

Now do it all again, but this time set $\beta = 2$ with $u = 0.1$. This should switch the short term and long term protein behavior.

9.8 A Two Box Climate Model

This is a simple model of how the ocean responds to the various inputs that give rise to greenhouse warming. The ocean is modeled as two layers: the top layer is shallow and is approximately 100 m in depth while the second layer is substantially deeper as it is on average 4000 m in depth. The top layer is called the **mixed layer** as it interacts with the atmosphere and so there is a transfer of energy back and forth from the atmosphere to the top layer. The bottom layer is the **deep layer** and it exchanges energy only with the mixed layer. The model we use is as follows:

$$C_m\, T_m' = F \;-\; \lambda_1 T_m \;-\; \lambda_2(T_m - T_d)$$
$$C_d\, T_d' = \lambda_2(T_m - T_d)$$

There are a lot of parameters here and they all have a physical meaning.

t: This is time.

T_m: This is the temperature of the mixed layer. We assume it is measured as the deviation from the mixed layer's equilibrium temperature. Hence, $T_m = 1$ would mean the temperature has risen 1° from its equilibrium temperature. We therefore set $T_m = 0$ initially and when we solve our model, we will know how much T_m has gone up.

T_d: This is the temperature of the deep layer. Again, this is measured as the difference from the equilibrium temperature of the deep layer. We set $T_d = 0$ initially also.

F: This is the external input which represents all the outside things that contribute to global warming such as CO_2 release and so forth. It does not have to be a constant but in our project we will use a constant value for F.

C_m: This is the heat capacity of the mixed layer.

C_d: This is the heat capacity of the deep layer.

λ_1: This is the exchange coefficient which determines the rate at which heat is transferred from the mixed layer to the atmosphere.

λ_2: This is the exchange coefficient which determines the rate at which heat is transferred from the mixed layer to the deep layer.

Looking at the mixed layer dynamics, we see there are two *loss* terms for the mixed layer temperature. The first is based on the usual exponential decay model using the exchange coefficient λ_1 which specifies how fast heat is transferred from the mixed layer to the atmosphere. The second loss term models how heat is transferred from the mixed layer to the deep layer. We assume this rate is proportional to the temperature difference between the layers which is why this loss is in terms of $T_m - T_d$. The deep layer temperature dynamics is all growth as the deep layer is picking up energy from the mixed layer above it. Note this type of modeling—a loss in T_m and the loss written as a gain for T_d—is exactly what we will do in the SIR disease model that comes later. Our reference for this model is the nice book on climate modeling (Vallis 2012). It is a book all of you can read with profit as it uses mathematics you now know very well. We can rewrite the dynamics into the standard form with a little manipulation.

$$\begin{aligned} C_m\, T_m' &= -(\lambda_1 + \lambda_2)T_m \,+\, \lambda_2 T_d \,+\, F \\ C_d\, T_d' &= \lambda_2 T_m - \lambda_2 T_d \end{aligned}$$

In matrix form we then have

$$\begin{bmatrix} C_m\, T_m' \\ C_d\, T_d' \end{bmatrix} = \begin{bmatrix} -(\lambda_1 + \lambda_2) & \lambda_2 \\ \lambda_2 & -\lambda_2 \end{bmatrix} \begin{bmatrix} T_m \\ T_d \end{bmatrix} + \begin{bmatrix} F \\ 0 \end{bmatrix}$$

Then dividing by the heat capacities, we find

$$\begin{bmatrix} T_m' \\ T_d' \end{bmatrix} = \begin{bmatrix} -(\lambda_1 + \lambda_2)/C_m & \lambda_2/C_m \\ \lambda_2/C_d & -\lambda_2/C_d \end{bmatrix} \begin{bmatrix} T_m \\ T_d \end{bmatrix} + \begin{bmatrix} F/C_m \\ 0 \end{bmatrix}$$

Hence, this model is another of our standard linear models with an external input such as we solved in the two protein model. The $\boldsymbol{A}$ matrix here is

$$\boldsymbol{A} = \begin{bmatrix} -(\lambda_1 + \lambda_2)/C_m & \lambda_2/C_m \\ \lambda_2/C_d & -\lambda_2/C_d \end{bmatrix}$$

and letting $\boldsymbol{F}$ denote the external input vector and $\boldsymbol{T}$ denote the vector of layer derivatives, we see the two box climate model is represented by our usual dynamics $\boldsymbol{T}' = \boldsymbol{A}\boldsymbol{T} + \boldsymbol{F}$ which we know how to solve.

Note the particular solution here is $\boldsymbol{TP} = -\boldsymbol{A}^{-1}\boldsymbol{F}$.

$$\boldsymbol{TP} = -\frac{1}{det(\boldsymbol{A})} \begin{bmatrix} -\lambda_2/C_d & -\lambda_2/C_m \\ -\lambda_2/C_d & -(\lambda_1 + \lambda_2)/C_m \end{bmatrix} \begin{bmatrix} F/C_m \\ 0 \end{bmatrix}$$

We find $det(\boldsymbol{A}) = (\lambda_1\lambda_2)/(C_m C_d)$ and so

$$\begin{aligned} \boldsymbol{TP} &= -(C_m C_d)/(\lambda_1\lambda_2) \begin{bmatrix} -\lambda_2 F/(C_m C_d) \\ -\lambda_2 F/(C_m C_d) \end{bmatrix} \\ &= \frac{F}{\lambda_1} \begin{bmatrix} 1 \\ 1 \end{bmatrix} \end{aligned}$$

So, if both eigenvalues of this model were negative, we would have the long term equilibrium of both the mixed and deep layer would be the same $T_d^\infty = T_m^\infty = F/\lambda_1$. Next, we show the eigenvalues are indeed negative here.

For convenience of exposition, let $\alpha = \lambda_1/C_m$, $\beta = \lambda_2/C_m$ and $\gamma = \lambda_2/C_d$. Also, it is helpful to express C_d as a multiple of C_m, so we write $C_d = \rho C_m$ where ρ is our multiplier. With these changes, the model can be rewritten. We find

$$\begin{bmatrix} T_m' \\ T_d' \end{bmatrix} = \begin{bmatrix} -(\alpha+\beta) & \beta \\ \gamma & -\gamma \end{bmatrix} \begin{bmatrix} T_m \\ T_d \end{bmatrix} + \begin{bmatrix} F/C_m \\ 0 \end{bmatrix}$$

But since $\gamma = \lambda_2/C_d$ and $C_d = \rho C_m$ we have $\gamma = \beta/\rho$. This gives the new form

$$\begin{bmatrix} T_m' \\ T_d' \end{bmatrix} = \begin{bmatrix} -(\alpha+\beta) & \beta \\ \beta/\rho & -\beta/\rho \end{bmatrix} \begin{bmatrix} T_m \\ T_d \end{bmatrix} + \begin{bmatrix} F/C_m \\ 0 \end{bmatrix}$$

Hence, any two box climate model can be represented by the dynamics $\boldsymbol{T}' = \boldsymbol{AT} + \boldsymbol{F}$ where

$$\boldsymbol{A} = \begin{bmatrix} -\alpha-\beta & \beta \\ \beta/\rho & -\beta/\rho \end{bmatrix}$$

Let's look at the eigenvalues of $\boldsymbol{A}$. The characteristic equation is $(r+\alpha+\beta)(r+\beta/\rho) - \beta^2/\rho = 0$. This simplifies to $r^2 + (\alpha+\beta+\beta/\rho)r + \alpha\beta/\rho = 0$. We suspect both eigenvalues are negative for our climate model as the physics here implies the temperatures stabilize. From the quadratic formula

$$r = (1/2)\left(-(\alpha+\beta+\beta/\rho) \pm \sqrt{(\alpha+\beta+\beta/\rho)^2 - 4\alpha\beta/\rho}\right)$$

The eigenvalue corresponding to the minus is negative. Next, look at the other root. The discriminant D is the term inside the square root. Let's show it is positive and that will help us show the other root is negative also. We have

$$\begin{aligned}
D &= (\alpha+\beta+\beta/\rho)^2 - 4\alpha\beta/\rho \\
&= (\alpha+\beta)^2 + 2(\alpha+\beta)(\beta/\rho) + \beta^2/\rho^2 - 4\alpha\beta/\rho \\
&= (\alpha+\beta)^2 - 2\alpha\beta/\rho + 2\beta^2/\rho + \beta^2/\rho^2 \\
&= 2\alpha\beta + \beta^2 + 2\beta^2/\rho + \left(\beta^2/\rho^2 - 2\alpha\beta/\rho + \alpha^2\right)
\end{aligned}$$

So we see

$$D = 2\alpha\beta + \beta^2 + 2\beta^2/\rho + (\beta/\rho - \alpha)^2 > 0$$

as ρ, α and β are positive. We now know $D = (\alpha+\beta+\beta/\rho)^2 - 4\alpha\beta/\rho > 0$. Hence, if we drop the $-4\alpha\beta/\rho$ term, the result is still positive and we have $0 < D < (\alpha+\beta+\beta/\rho)^2$. The plus root then is

$$\begin{aligned}
r &= (1/2)\left(-(\alpha+\beta+\beta/\rho) + \sqrt{(\alpha+\beta+\beta/\rho)^2 - 4\alpha\beta/\rho}\right) \\
&< (1/2)\left(-(\alpha+\beta+\beta/\rho) + \sqrt{(\alpha+\beta+\beta/\rho)^2}\right) \\
&= (1/2)\left(-(\alpha+\beta+\beta/\rho) + (\alpha+\beta+\beta/\rho)\right) = 0.
\end{aligned}$$

Hence, the eigenvalues r_1 and r_2 are both negative and the general solution is

$$\begin{bmatrix} T_m \\ T_d \end{bmatrix} = a\boldsymbol{E_1}e^{r_1 t} + b\boldsymbol{E_2}e^{r_2 t} + \begin{bmatrix} F/\lambda_1 \\ F/\lambda_1 \end{bmatrix}$$

where $\boldsymbol{E_1}$ and $\boldsymbol{E_2}$ are the eigenvectors of the two eigenvalues.

The values of these parameters in real climate modeling are fairly well known. We know that C_d is much larger than C_m and that the mixed layer takes a short time to reach its temporary equilibrium and that in the long term both T_m and T_d reach an equilibrium value that is larger. To estimate how this works, assume that the value of T_d stays close to zero at first. Then the T_m dynamics becomes simpler as $T_d = 0$:

$$T_m' = F/C_m \; - \; (\lambda_1 + \lambda_2)/C_m T_m$$

This is our familiar protein synthesis model with steady state value of $\hat{T}_m^{\infty} = F/(\lambda_1 + \lambda_2)$ and response time $t_R^m = C_m \ln(2)/(\lambda_1 + \lambda_2)$. After sufficient response times have passed for the mixed layer temperature to reach quasi equilibrium, we will have $T_m' \approx 0$ as T_m is no longer changing by much. Hence, setting $T_m' = 0$, we find a relationship between T_m and T_d once this new equilibrium is achieved. We have

$$-(\lambda_1 + \lambda_2)T_m \; + \; F \; + \; \lambda_2 T_d = 0$$

which tells us that $T_m = (\lambda_2 T_d + F)/(\lambda_1 + \lambda_2)$. Substitute this value of T_m into the T_d dynamics and we will get the final equilibrium value of T_d for time much larger that t_R^m. The deep layer temperature dynamics now become

$$
\begin{aligned}
T_d' &= \lambda_2 T_m / C_d - \lambda_2 / C_d T_d \\
&= \lambda_2 \frac{\lambda_2 T_d + F}{\lambda_1 + \lambda_2} - \lambda_2 / C_d T_d \\
&= -\frac{\lambda_1 \lambda_2}{C_d(\lambda_1 + \lambda_2)} T_d + \frac{\lambda_2}{C_d(\lambda_1 + \lambda_2)} F
\end{aligned}
$$

This is also a standard protein synthesis problem with a steady state value of

$$
\begin{aligned}
\hat{T}_d^{\infty} &= \left(\frac{\lambda_2}{C_d(\lambda_1 + \lambda_2)} F\right) / \left(\frac{\lambda_1 \lambda_2}{C_d(\lambda_1 + \lambda_2)}\right) \\
&= \frac{F}{\lambda_1}
\end{aligned}
$$

and a response time of

$$
t_R^d = \frac{\ln(2) C_d(\lambda_1 + \lambda_2)}{\lambda_1 \lambda_2}
$$

and as t gets large, both temperatures approach the common steady state value of F/λ_1.

So with a little reasoning and a lot of algebra we find the two response times of the climate model are t_R^m and t_R^d as given above. To see these results in a specific problem, we wrote a MatLab script to solve a typical climate problem. For the values $\lambda_1 = 0.1$, $\lambda_2 = 0.15000$, $C_m = 1.0820$, $C_d = 4.328 = 4C_m$ so that $\rho = 4$ and an external input of $F = 0.5$, we can compute the solutions to the climate model as follows

Listing 9.38: A Sample Two Box Climate Model Solution

```
% Set up A
lambda1 = .1
lambda2 = 0.15
Cm = 1.0820
Cd = 4.328
rho = Cd/Cm
alpha = lambda1/Cm;
beta = lambda2/Cm;
A = [-alpha-beta, beta; beta/rho, -beta/rho];
% Find A inverse
Ainv = (1/det(A))*[A(2,2),-A(1,2);-A(2,1),A(1,1)];
% set up external input
F = 0.5
% Find particular solution
TP = -Ainv*[F/Cm;0]
% Find solution for initial conditions Tm = 0 and Td = 0
[V,D] = eig(A);
VInv = (1/det(V))*[V(2,2),-V(1,2);-V(2,1),V(1,1)];
C = -VInv*TP;
Ev = diag(D);
% Find response times using equations above
tRmixed = Cm*log(2)/(lambda1+lambda2)
tRmixed = 2.9999
```

```
tRdeep = rho*Cm*(log(2)/(lambda1*lambda2))*(lambda1+lambda2)
tRdeep =   49.999
% Quasi Equilibrium Estimate of Tm
Tmshort = F/(lambda1+lambda2)
% Long term equilibrium Tm and Td
Tmlong = F/lambda1
Tdlong = F/lambda1
% Set up Tm and Td solutions
Tm = @(t) ( C(1)*V(1,1)*exp(Ev(1)*t)+C(2)*V(1,2)*exp(Ev(2)*t)+TP(1));
Td = @(t) ( C(1)*V(2,1)*exp(Ev(1)*t)+C(2)*V(2,2)*exp(Ev(2)*t)+TP(2));
% Plot Tm and Td over a few Tm response times
time = linspace(0,4*tRmixed,100);
figure
plot(time,Tm(time),time,Td(time),'+');
xlabel('Time')
ylabel('T_m');
title('Ocean Layer Temperatures');
legend('Mixed Layer','Deep Layer');
% plot Tm and Td until we get close to Tm steady state
time = linspace(0,8*tRmixed,300);
figure
plot(time,Tm(time),time,Td(time),'+');
xlabel('Time in Years')
ylabel('Temperature Degrees C');
title('Ocean Layer Temperatures');
legend('Mixed Layer','Deep Layer');
% plot Tm and Td over Td response times
time = linspace(0,6*tRdeep,500);
figure
plot(time,Tm(time),time,Td(time),'+');
xlabel('Time in Years')
ylabel('Temperature Degrees C');
title('Ocean Layer Temperatures');
legend('Mixed Layer','Deep Layer');
```

We see the approximate response time of the mixed layer is only 3 years but the approximate response time for the deep layer is about 50 years. This MatLab session generates three plots as shown in Figs. 9.13, 9.14 and 9.15.

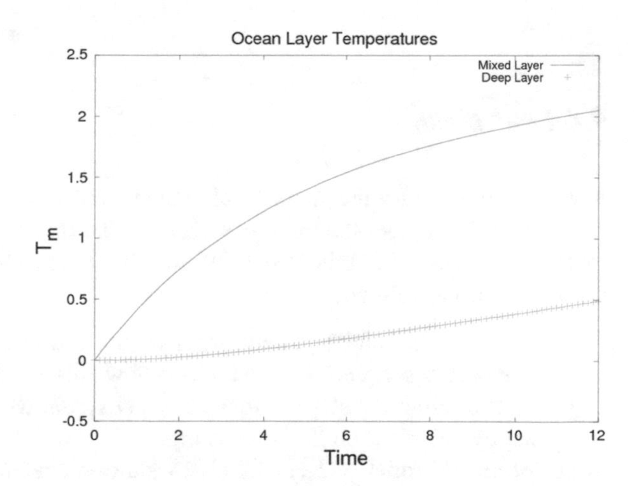

Fig. 9.13 Simple two box climate model: short time scale

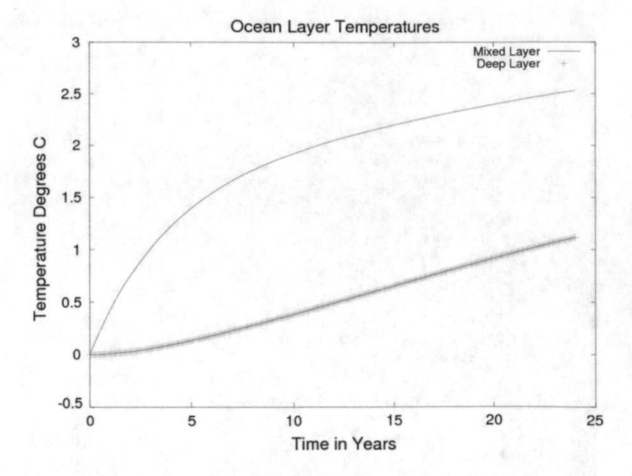

Fig. 9.14 Simple two box climate model: mid time scale

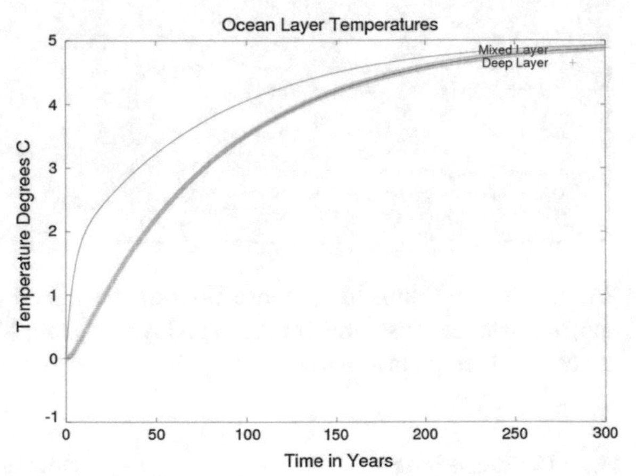

Fig. 9.15 Simple two box climate model: long time scale

9.8.1 Project

Now we are ready for the project. Solve the two box climate model with $\lambda_1 = 0.01$, $\lambda_2 = 0.02000$, $C_m = 0.34625$, $\rho = 2.2222$ (so that $C_d = \rho C_m$) and an external input of $F = 0.06$. For this model, follow what I did in the example and generate a report in word as follows:

Introduction (8 Points) Explain what we are trying to do here with this model. The project description here tells you how this model works. Make sure you understand all of the steps so that you can explain what this model really means for policy decisions on global warning.

Description of Model (12 Points) Find additional references and use them to build a two page description of this kind of simple climate model. Pay particular attention to a discussion of what the parameters in the model mean and what the external input F might represent. Our description was deliberately vague as we want you to find out more.

Annotated Solution Discussion (22 Points) In this part, you solve the model using MatLab as I did in the example above. As usual, explain your steps nicely. Calculate the approximate response times for the temperatures in each ocean layer and also state the quasi mixed ocean layer equilibrium. We want to see plots for short term, midterm and long term time scales with full discussion of what the plots mean.

Conclusion (5 Points) Discuss what you have done here and how you might use it for social policy decisions. How could you convince a skeptical audience that the model results are valid enough to warrant action?

References (3 Points) Put any reference material you use in here.

9.8.2 Control of Carbon Loading

There is talk currently about strategies to reduce carbon loading via a genetically engineered bacteria or a nano machine. Such a strategy is called **carbon sequestering** and it is not clear if this is possible. If you have read or watched much science fiction, you'll recognize carbon sequestering as a type of **terraforming**. This is certainly both ambitious and no doubt a plan having a lot of risk that needs to be thought about carefully. However, if we assume such a control strategy for carbon loading is implemented say 25 years in the future, what does our simple model say? The analysis is straightforward. At 25 years, our model reach a value T_m^{25} and T_d^{25} which we know are far from the true equilibrium values of $\approx 5°$ of global warming in our example. If the sequestering strategy were instantly successful this would correspond to setting $T = 0$ in our external data. This gives a straight exponential decay system

$$\begin{bmatrix} T_m' \\ T_d' \end{bmatrix} = \begin{bmatrix} -(\lambda_1 + \lambda_2)/C_m & \lambda_2/C_m \\ \lambda_2/C_d & -\lambda_2/C_d \end{bmatrix} \begin{bmatrix} T_m \\ T_d \end{bmatrix}, \quad \begin{bmatrix} T_m \\ T_d \end{bmatrix} = \begin{bmatrix} T_m^{25} \\ T_d^{25} \end{bmatrix}.$$

The eigenvalue and eigenvectors for this system are the same as before and we know half life of about 50 years.

$$t_R^d \approx \frac{\ln(2) C_d (\lambda_1 + \lambda_2)}{\lambda_1 \lambda_2} \approx 50.$$

It doesn't really matter at what point carbon loading stops. So we can do this analysis whether carbon loading stops 25 years in the future or 50 years. The deep layer still takes a long time to return to a temperature of 0 which for us represents current temperature and thus no global warming. Note the half life $t_R^d \approx 50$ is what drives this. The constants λ_1 and λ_2 represent exchange rates between the mixed layer and the atmosphere and the deep layer and the mixed layer. The constant C_d is the heat capacity of the deep layer. If we also assumed carbon sequestering continued even after carbon loading was shut off, perhaps this would correspond to increasing the two critical parameters λ_1 and λ_2 and decreasing C_d. Since

$$t_R^d \approx \ln(2)\, C_d \left(\frac{1}{\lambda_1} + \frac{1}{\lambda_2} \right)$$

such changes would decrease t_R^d and bring the deep layer temperature to 0 faster. It is conceivable administering nano machines and gene engineered bacteria might do this but it is a strategy that would alter fundamental balances that we have had in place for millions of years. Hence, it is hard to say if this is wise; certainly further study is needed. But the bottom line is the carbon sequestering strategies will not easily return the deep layer to a state of no global increase in temperature quickly. Note we could also assume the carbon sequestering control strategy alters carbon loading as a simple exponential decay $Fe^{-\epsilon t}$ for some positive epsilon. Then our model becomes

$$\begin{bmatrix} T'_m \\ T'_d \end{bmatrix} = \begin{bmatrix} -(\lambda_1 + \lambda_2)/C_m & \lambda_2/C_m \\ \lambda_2/C_d & -\lambda_2/C_d \end{bmatrix} \begin{bmatrix} T_m \\ T_d \end{bmatrix} + \begin{bmatrix} Fe^{-\epsilon t} \\ 0 \end{bmatrix} \begin{bmatrix} T_m \\ T_d \end{bmatrix} = \begin{bmatrix} T_m^{25} \\ T_d^{25} \end{bmatrix}.$$

which we could solve numerically. Here, the carbon loading would not instantaneously drop to zero at some time like 25; instead is decays gracefully. However, this still does not change the fact that the response time is determined by the coefficient matrix $\boldsymbol{A}$ and so the return to a no global warming state will be slow. Finally, it is sobering to think about how all of this analysis plays out in the backdrop of political structures that remain in authority for about 4 years or so. We can see how hard it is to elicit change when the results won't show up for 10–40 administrations!

Reference

G. Vallis (ed.), *Climate and the Oceans*, Princeton Primers on Climate (Princeton University Press, Princeton, 2012)

Part IV
Interesting Models

Chapter 10
Predator–Prey Models

In the 1920s, the Italian biologist Umberto D'Ancona studied population variations of various species of fish that interact with one another. He came across the data shown in Table 10.1.

Here, we interpret the percentage we see in column two of Table 10.1 as *predator* fish, such as sharks, skates and so forth. Also, the catches used to calculate these percentages were reported from all over the Mediterranean. The tonnage from all the different catches for the entire year were then added and used to calculate the percentages in the table. Thus, we can also calculate the percentage of catch that was *food* by subtracting the predator percentages from the *predator* ones. This leads to what we see in Table 10.2.

D'Ancona noted the time period coinciding with World War One, when fishing was drastically cut back due to military actions, had puzzling data. Let's highlight this in Table 10.3. D'Ancona expected both food fish and predator fish to increase when the rate of fishing was cut back. But in these war years, there is a substantial increase in the percentage of predators caught at the same time the percentage of food fish went down. Note, we are looking at percentages here. Of course, the raw

Table 10.1 The percent of the total fish catch in the Mediterranean Sea which was considered not food fish

Year	Percent not food fish
1914	11.9
1915	21.4
1916	22.1
1917	21.2
1918	36.4
1919	27.3
1920	16.0
1921	15.9
1922	14.8
1923	10.7

J.K. Peterson, *Calculus for Cognitive Scientists*,
Cognitive Science and Technology, DOI 10.1007/978-981-287-877-9_10

Table 10.2 The percent of the total fish catch in the Mediterranean Sea considered predator and considered food

Year	Percent food	Percent predator
1914	88.1	11.9
1915	78.6	21.4
1916	77.9	22.1
1917	78.8	21.2
1918	63.6	36.4
1919	72.7	27.3
1920	84.0	16.0
1921	84.1	15.9
1922	85.2	14.8
1923	89.3	10.7

Table 10.3 During World War One, fishing is drastically curtailed, yet the predator percentage went up while the food percentage went down

Year	Percent food	Percent predator
1915	78.6	21.4
1916	77.9	22.1
1917	78.8	21.2
1918	63.6	36.4
1919	72.7	27.3

tonnage of fish caught went down during the war years, but the expectation was that since there is reduced fishing, there should be a higher percentage of food fish because they have not been harvested. D'Ancona could not understand this, so he asked the mathematician Vito Volterra for help.

10.1 Theory

Volterra approached the modeling this way. He let the variable $x(t)$ denote the population of food fish and $y(t)$, the population of predator fish at time t. He was constructing what you might call a *coarse* model. The food fish are divided into categories like halibut, mackerel with a separate variable for each and the predators are also not divided into different classes like sharks, squids and so forth. Hence, instead of dozens of variables for both the food and predator population, everything was lumped together. Following Volterra, we make the following assumptions:

1. The food population grows exponentially. Letting x'_g denote the growth rate of the food fish, we must have

$$x'_g = a\,x$$

for some positive constant a.

2. The number of contacts per unit time between predators and prey is proportional to the product of their populations. We assume the food fish are eaten by the predators at a rate proportional to this contact rate. Letting the decay rate of the food be denoted by x'_d, we see

$$x'_d = -\, b\, x\, y$$

for some positive constant b.

Thus, the net rate of change of food is $x' = x'_g + x'_d$ giving

$$x' = a\, x - b\, x\, y.$$

for some positive constants a and b. He made assumptions about the predators as well.

1. Predators naturally die following an exponential decay; letting this decay rate be given by y'_d, we have

$$y'_d = -c\, y$$

for some positive constant c.
2. We assume the predator fish can grow proportional to how much they eat. In turn, how much they eat is assumed to be proportional to the rate of contact between food and predator fish. We model the contact rate just like before and let y'_g be the growth rate of the predators. We find

$$y'_g = d\, x\, y$$

for some positive constant d.

Thus, the net rate of change of predators is $y' = y'_g + y'_d$ giving

$$y' = -c\, y + d\, x\, y.$$

for some positive constants c and d. The full Volterra model is thus

$$\begin{aligned} x' &= \ \ a\, x - b\, x\, y && (10.1)\\ y' &= -c\, y + d\, x\, y && (10.2)\\ x(0) &= x_0 && (10.3)\\ y(0) &= y_0 && (10.4) \end{aligned}$$

Equations 10.1 and 10.2 give the dynamics of this system. Note these are nonlinear dynamics, the first we have seen since the logistics model. Equations 10.3 and 10.4

are the initial conditions for the system. Together, these four equations are called a **Predator–Prey** system. Since Volterra's work, this model has been applied in many other places. A famous example is the wolf–moose predator–prey system which has been extensively modeled for Island Royale in Lake Superior. We are now going to analyze this model. We have been inspired by the analysis given Braun (1978), but Braun can use a bit more mathematics in his explanations and we will try to use only calculus ideas.

10.2 The Nullcline Analysis

Once we obtain a solution (x, y) to the Predator–Prey problem, we have two nice curves $x(t)$ and $y(t)$ defined for all non negative time t. As we did in Chap. 8, if we graph in the x–y plane the ordered pairs $(x(t), y(t))$, we will draw a curve $\mathscr{C}$ where any point on $\mathscr{C}$ corresponds to an ordered pair $(x(t), y(t))$ for some time t. At $t = 0$, we are at the point (x_0, y_0) on $\mathscr{C}$. Hence, the initial conditions for the Predator–Prey problem determine the starting point on $\mathscr{C}$. As time increases, the pairs $(x(t), y(t))$ move in the direction of the tangent line to $\mathscr{C}$. If we knew the algebraic sign of the derivatives x' and y' at any point on $\mathscr{C}$, we could decide the direction in which we are moving along the curve $\mathscr{C}$. So we begin our analysis by looking at the curves in the x–y plane where x' and y' become 0. From these curves, we will be able to find out the different regions in the plane where each is positive or negative. From that, we will be able to decide in which direction a point moves along the curve.

10.2.1 The $x' = 0$ Analysis

Looking at the predator–prey equations, we see that if t^* is a time point when $x'(t^*)$ is zero, the food dynamic of the predator–prey system reduce to

$$0 = a\, x(t^*) - b\, x(t^*)\, y(t^*)$$

or

$$0 = x(t^*)\Big(a - b\, y(t^*)\Big)$$

Thus, the (x, y) pairs in the x–y plane where

$$0 = x\Big(a - b\, y\Big)$$

are the ones where the rate of change of the food fish will be zero. Now these pairs can correspond to many different time values t^* so what we really need to do is to find all the (x, y) pairs where this happens. Since this is a product, there are two possibilities: $x = 0$; the y axis and $y = \frac{a}{b}$; a horizontal line.

10.2.2 The $y' = 0$ Analysis

In a similar way, the pairs (x, y) where y' becomes zero satisfy the equation

$$0 = y\left(-c + d\,x\right).$$

Again, there are two possibilities: $y = 0$; the x axis and $x = \frac{c}{d}$; a vertical line.

10.2.2.1 An Example

Here's an example: consider the predator–prey model

$$\begin{aligned} x'(t) &= 2\,x(t) - 5\,x(t)\,y(t) \\ y'(t) &= -6\,y(t) + 3\,x(t)\,y(t) \end{aligned}$$

The x' nullclines satisfy $2x - 5xy = 0$ or $x(2 - 5y) = 0$. Hence $x = 0$ or $y = 2/5$. Draw the lines $x = 0$, the y axis, and the horizontal line $y = 2/5$ in the xy plane. Then note the factor $2 - 5y$ is positive when $2 - 5y > 0$ or when $y < 2/5$. Hence, the factor is negative when $y > 2/5$. The factor x is positive when $x > 0$ and negative when $x < 0$. So the combination $x(2 - 5y)$ has a sign that can be determined easily as shown in Fig. 10.1.

Next, for our predator–prey model

$$\begin{aligned} x'(t) &= 2\,x(t) - 5\,x(t)\,y(t) \\ y'(t) &= -6\,y(t) + 3\,x(t)\,y(t) \end{aligned}$$

the y' nullclines satisfy $-6y + 3xy = 0$ or $y(-6 + 3x) = 0$. Hence $y = 0$ or $x = 2$. Draw the lines $y = 0$, the x axis, and the vertical line $x = 2$ in the xy plane. The factor $-6 + 3x$ is positive when $-6 + 3x > 0$ or when $x > 2$. Hence, the factor is negative when $x < 2$. The factor y is positive when $y > 0$ and negative when $y < 0$. So the combination $y(-6 + 3x)$ has a sign that can be determined easily as shown in Fig. 10.2.

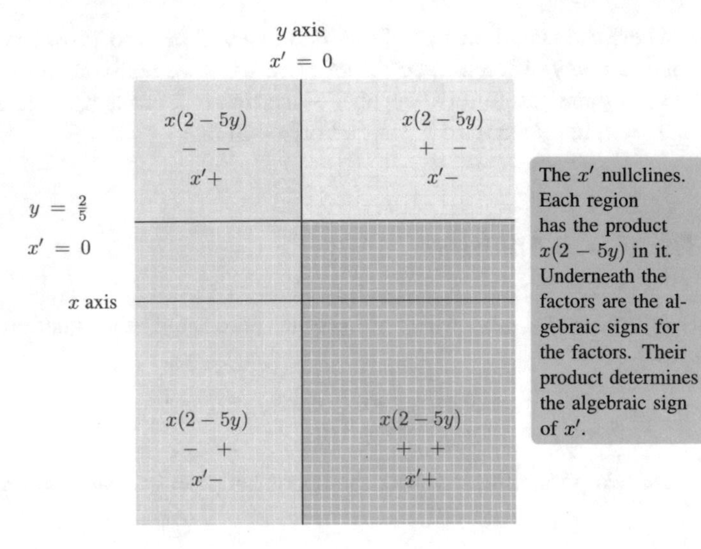

Fig. 10.1 $x' = 0$ nullcline for $x' = 2x + 5y$

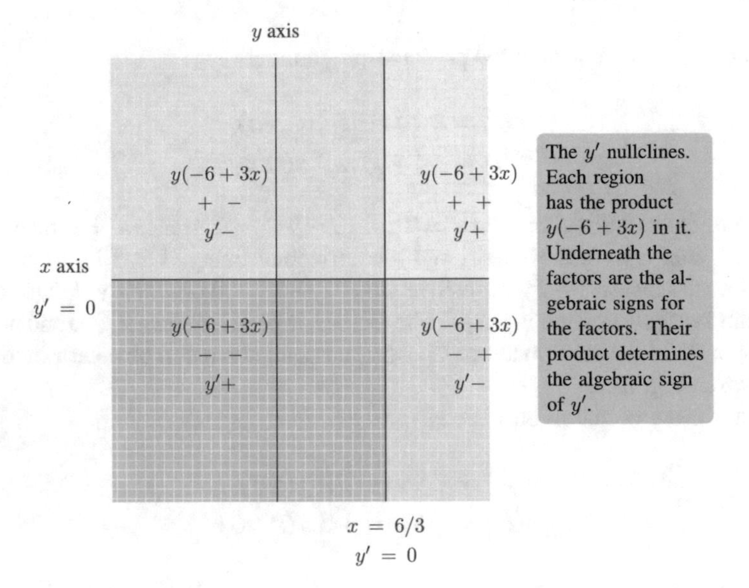

Fig. 10.2 $y' = 0$ nullcline for $y' = 6x + 3y$

10.2.3 *The Nullcline Plane*

Just like we did in Chap. 8, we find the parts of the x–y plane where the algebraic signs of x' and y' are $(+, +)$, $(+, -)$, $(-, +)$ and $(-, -)$. As usual, the set of (x, y) pairs where $x' = 0$ is called the **nullcline** for x; similarly, the points where $y' = 0$ is the **nullcline** for y. The $x' = 0$ equation gives us the y axis and the horizontal line $y = \frac{a}{b}$ while the $y' = 0$ gives the x axis and the vertical line $x = \frac{c}{d}$. The x' dynamics thus divide the plane into three pieces: the part where $x' > 0$; the part where $x' = 0$; and, the part where $x' < 0$.

10.2.3.1 Back to Our Example

We go back to the model

$$x'(t) = 2\,x(t) - 5\,x(t)\,y(t)$$
$$y'(t) = -6\,y(t) + 3\,x(t)\,y(t)$$

We have already determined the nullclines for this model as shown in Figs. 10.1 and 10.2. We combinc the x' and y' nullcline information to create a map of how x' and y' change sign in the xy plane.

- Regions I, II, III and IV divide Quadrant 1.
- We will show there are trajectories moving down the positive y axis and out along the positive x axis.
- Thus, a trajectory that starts in Quadrant 1 with positive initial conditions can't cross the trajectory on the positive x axis or the trajectory on the positive y axis.
- Thus, the Predator–Prey trajectories that start in Quadrant 1 with positive initial conditions will stay in Quadrant 1.

This is shown in Fig. 10.3.

10.2.4 *The General Results*

For the general predator–prey model

$$x'(t) = a\,x(t) - b\,x(t)\,y(t)$$
$$y'(t) = -c\,y(t) + d\,x(t)\,y(t)$$

The x' nullclines satisfy $ax - bxy = 0$ or $x(a - by) = 0$. Hence $x = 0$ or $y = a/b$. Draw the lines $x = 0$ (the y axis and the horizontal line $y = a/b$ in the xy plane.

- The factor $a - by$ is positive when $a - by > 0$ or when $y < a/b$. Hence, the factor is negative when $y > a/b$.

y axis
$x' = 0$

V (+,−)	II (−,−)	I (−,+)
VI (−,−)	III (+,−)	IV (+,+)
VII (−,+)	VIII (+,+)	IX (+,−)

$y = 2/5$
$x' = 0$

x axis
$y' = 0$

$x = 6/3$
$y' = 0$

The $x' = 0$ and the $y' = 0$ equations determine regions in the xy plane. In each region, the algebraic sign of x' and y' are shown as an ordered pair. For example, in region I, x' is negative and y' is positive and so this is denoted by $(-, +)$.

Fig. 10.3 The combined $x' = 0$ nullcline for $x' = 2x + 5y$ and $y' = 0$ nullcline for $y' = 6x + 3y$ information

- The factor x is positive when $x > 0$ and negative when $x < 0$.
- So the combination $x(a - by)$ has a sign that can be determined easily.

In Fig. 10.4, we show the part of the x–y plane where $x' > 0$ with one shading and the part where it is negative with another. Next, the y' nullclines satisfy $-cy + dxy = 0$ or $y(-c + dx) = 0$. Hence $y = 0$ or $x = c/d$. Draw the lines $y = 0$ (the x axis and the vertical line $x = c/d$ in the xy plane.

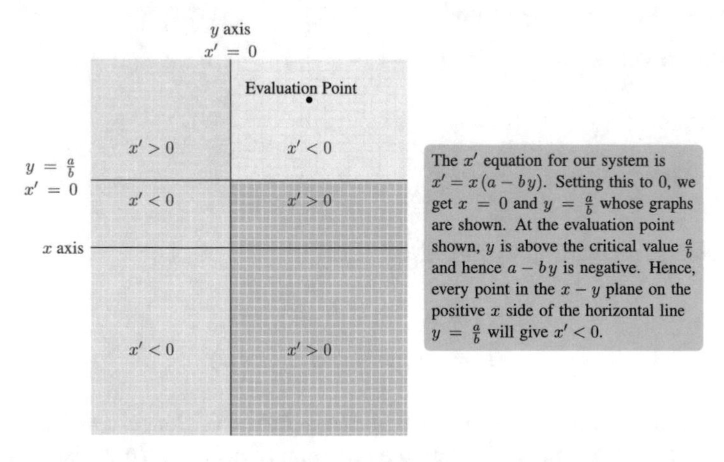

Fig. 10.4 Finding where $x' < 0$ and $x' > 0$ for the predator–prey model

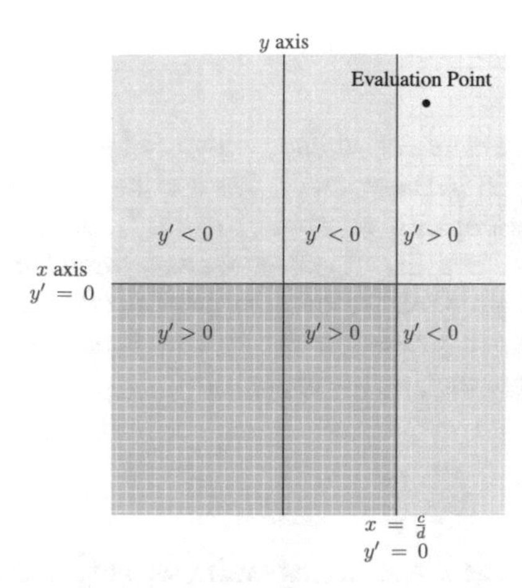

The y' equation for our system is $y' = y\,(-c + d\,y)$. Setting this to 0, we get $y = 0$ and $x = \frac{c}{d}$ whose graphs are shown. At the evaluation point shown, x is to the right of the critical value $\frac{c}{d}$ and hence $-c + d\,x$ is positive. Hence, every point in the $x - y$ plane on the positive x side of the vertical line $x = \frac{c}{d}$ will give $y' > 0$.

Fig. 10.5 Finding where $y' < 0$ and $y' > 0$ for the predator–prey model

- The factor $-c + dx$ is positive when $-c + dx > 0$ or when $x > c/d$. Hence, the factor is negative when $x < c/d$.
- The factor y is positive when $y > 0$ and negative when $y < 0$.
- So the combination $y(-c + dx)$ has a sign that can also be determined easily.

In Fig. 10.5, we show how the y' nullcline divides the x–y plane into three pieces as well.

The shaded areas shown in Figs. 10.4 and 10.5 can be combined into Fig. 10.6. In each region, x' and y' are either positive or negative. Hence, each region can be marked with an ordered pair, $(x'\pm,\ y'\pm)$.

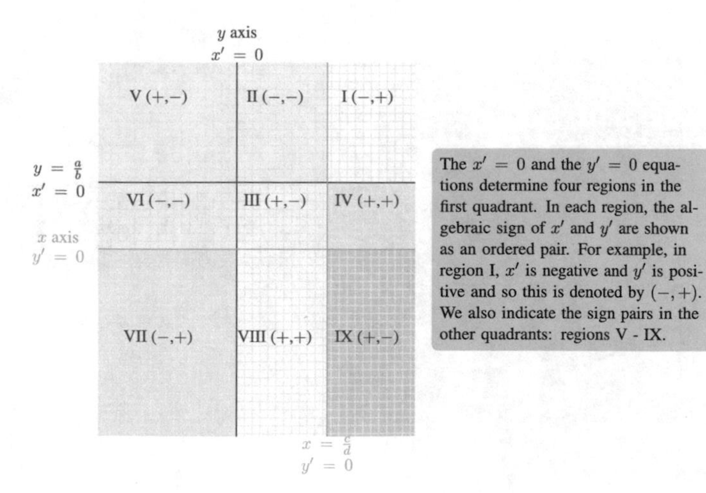

Fig. 10.6 Combining the x' and y' algebraic sign regions

10.2.5 Drawing Trajectories

We drew trajectories for the linear system models already without a lot of background discussion. Now we'll go over it again in more detail. We use the algebraic signs of x' and y' to determine this. For example, if we are in Region I, the sign of x' is negative and the sign of y' is positive. Thus, the variable x decreases and the variable y increases in this region. So if we graphed the ordered pairs $(x(t), y(t))$ in the x–y plane for all $t > 0$, we would plot a y versus x curve. That is, we would have $y = f(x)$ for some function of x. Note that, by the chain rule

$$\frac{dy}{dt} = f'(x)\,\frac{dx}{dt}.$$

Hence, as long as x' is not zero (and this is true in Region I!), we have at each time t, that the slope of the curve $y = f(x)$ is given by

$$\frac{df}{dx}(t) = \frac{y'(t)}{x'(t)}.$$

Since our pair (x, y) is the solution to a differential equation, we expect that x and y both are continuously differentiable with respect to t. So if we draw the curve for y versus x in the x–y plane, we do not expect to see a corner in it (as a corner means the derivative fails to exist). So we can see three possibilities:

- a straight line as x' equals y' at each t meaning the slope is always the same,

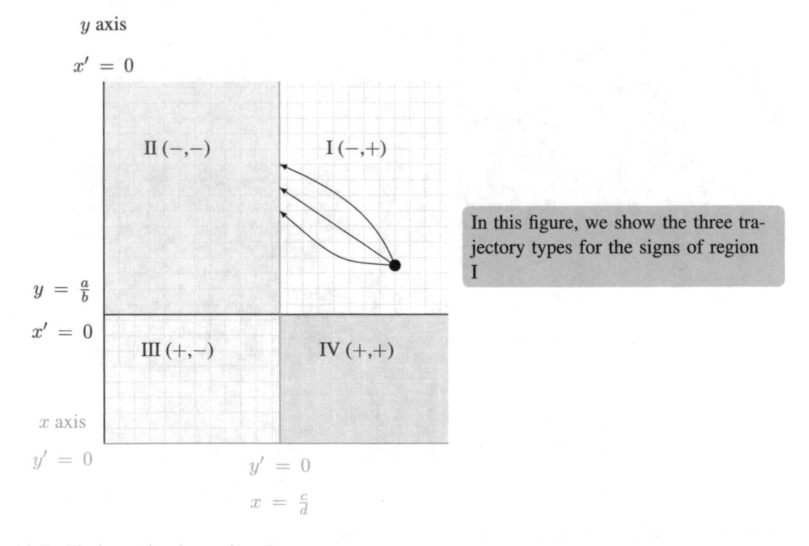

Fig. 10.7 Trajectories in region I

- a curve that is concave up or
- a curve that is concave down.

We illustrate this three possibilities in Fig. 10.7.

When we combine trajectories from one region with another, we must attach them so that we do not get corners in the curves. This is how we can determine whether or not we should use concave up or down or straight in a given region. We can do this for all the different regions shown in Fig. 10.7.

10.3 Only Quadrant One Is Biologically Relevant

To analyze this nonlinear model, we need a fact from more advanced courses. For these kinds of nonlinear models, trajectories that start at different initial conditions can not cross.

Assumption 10.3.1 (*Trajectories do not cross in the Predator–Prey model*)
We can show, in a more advanced course, that two distinct trajectories to the Predator–Prey model

$$\begin{aligned} x' &= \;\; a\,x - b\,x\,y && (10.5)\\ y' &= -c\,y + d\,x\,y && (10.6)\\ x(0) &= x_0 && (10.7)\\ y(0) &= y_0 && (10.8) \end{aligned}$$

can not cross.

10.3.1 Trajectories on the y^+ Axis

Let's begin by looking at a trajectory that starts on the positive y axis. We therefore need to solve the system

$$\begin{aligned} x' &= \;\; a\,x - b\,x\,y && (10.9)\\ y' &= -c\,y + d\,x\,y && (10.10)\\ x(0) &= 0 && (10.11)\\ y(0) &= y_0 > 0 && (10.12) \end{aligned}$$

It is easy to guess the solution is the pair $(x(t), y(t))$ with $x(t) = 0$ always and $y(t)$ satisfying $y' = -c\, y(t)$. Hence,

$$y(t) = y_0\, e^{-ct}$$

and y decays nicely down to 0 as t increases.

10.3.2 Trajectories on the x^+ Axis

If we start on the positive x axis, we want to solve

$$x' = \ a\, x - b\, x\, y \tag{10.13}$$

$$y' = -c\, y + d\, x\, y \tag{10.14}$$

$$x(0) = x_0 > 0 \tag{10.15}$$

$$y(0) = 0 \tag{10.16}$$

Again, it is easy to guess the solution is the pair $(x(t), y(t))$ with $y(t) = 0$ always and $x(t)$ satisfying $x' = a\, x(t)$. Hence,

$$x(t) = x_0\, e^{at}$$

and the trajectory moves along the positive x axis always increasing as t increases. Since trajectories can't cross other trajectories, this tells us a trajectory that begins in Quadrant 1 with a positive (x_0, y_0) can't hit the x axis or the y axis in a finite amount of time because it did, we would have two trajectories crossing.

10.3.3 What Does This Mean Biologically?

This is good news for our biological model. Since we are trying to model food and predator interactions in a real biological system, we always start with initial conditions (x_0, y_0) that are in Quadrant One. It is very comforting to know that these solutions will always remain positive and, therefore, biologically realistic. In fact, it doesn't seem biologically possible for the food or predators to become negative, so if our model permitted that, it would tell us our model is seriously flawed! Hence, for our modeling purposes, we need not consider initial conditions that start in Regions V–IX. Indeed, if you look at Fig. 10.7, you can see that a solution trajectory could only hit the y axis from Region II. But that can't happen as if it did, two trajectories would cross! Also, a trajectory could only hit the x axis from a start in Region III. Again, since trajectories can't cross, this is not possible either. So, a **trajectory that starts in Quadrant 1, stays in Quadrant 1**—kind of has a *Las Vegas feel doesn't it?*.

10.3.4 Homework

For the following problems, find the x' and y' nullclines and sketch using multiple colors, the algebraic sign pairs (x', y') the nullclines determine in the x–y plane.

Exercise 10.3.1

$$\begin{aligned} x'(t) &= 100\, x(t) - 25\, x(t)\, y(t) \\ y'(t) &= -200\, y(t) + 40\, x(t)\, y(t) \end{aligned}$$

Exercise 10.3.2

$$\begin{aligned} x'(t) &= 1000\, x(t) - 250\, x(t)\, y(t) \\ y'(t) &= -2000\, y(t) + 40\, x(t)\, y(t) \end{aligned}$$

Exercise 10.3.3

$$\begin{aligned} x'(t) &= 900\, x(t) - 45\, x(t)\, y(t) \\ y'(t) &= -100\, y(t) + 50\, x(t)\, y(t) \end{aligned}$$

Exercise 10.3.4

$$\begin{aligned} x'(t) &= 10\, x(t) - 25\, x(t)\, y(t) \\ y'(t) &= -20\, y(t) + 40\, x(t)\, y(t) \end{aligned}$$

Exercise 10.3.5

$$\begin{aligned} x'(t) &= 90\, x(t) - 2.5\, x(t)\, y(t) \\ y'(t) &= -200\, y(t) + 4.5\, x(t)\, y(t) \end{aligned}$$

10.4 The Nonlinear Conservation Law

So we can assume that for a start in Quadrant 1, the solution pair is always positive. Let's see how far we can get with a preliminary mathematical analysis. We can analyze these trajectories like this. For convenience, assume we start in Region II and the resulting trajectory hits the $y = \frac{a}{b}$ line at some time t^*. At that time, we will have $x'(t^*) = 0$ and $y(t^*) < 0$. We show this situation in Fig. 10.8.

Look at the Predator–Prey model dynamics for $0 \leq t < t^*$. Since all variables are positive and their derivatives are not zero for these times, we can look at the fraction $y'(t)/x'(t)$.

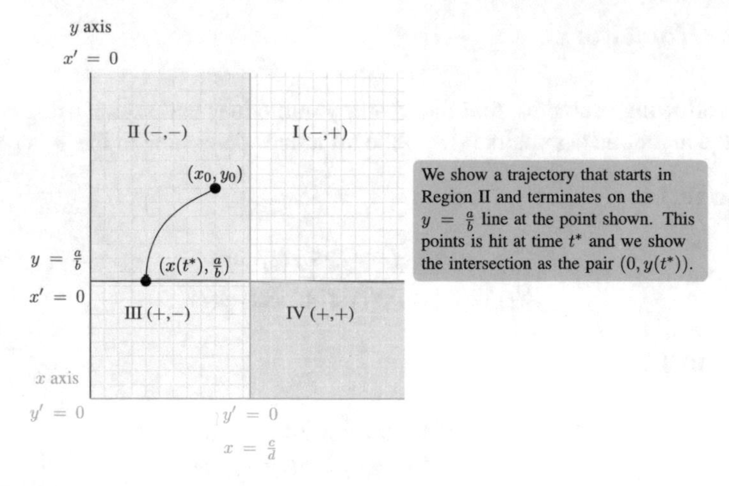

Fig. 10.8 Trajectories in region II

$$\frac{y'(t)}{x'(t)} = \frac{y(t)\,(-c + d\,x(t))}{x(t)\,(a - b\,y(t))}.$$

10.4.1 Example

To make this easier to understand, let's do a specific example.

$$\begin{aligned} x'(t) &= 2\,x(t) - 5\,x(t)\,y(t) \\ y'(t) &= -6\,y(t) + 3\,x(t)\,y(t) \end{aligned}$$

- Rewrite as y'/x':

$$\begin{aligned} \frac{y'(t)}{x'(t)} &= \frac{-6y(t) + 3x(t)y(t)}{2x(t) - 5x(t)y(t)} \\ &= \frac{y(t)(-6 + 3x(t))}{x(t)(2 - 5y(t))}. \end{aligned}$$

- Put all the y stuff on the left and all the x stuff on the right:

$$y'(t)\frac{2 - 5y(t)}{y(t)} = x'(t)\frac{-6 + 3x(t)}{x(t)}.$$

- Rewrite as separate pieces:

$$2\,\frac{y'(t)}{y(t)} - 5y'(t) = -6\,\frac{x'(t)}{x(t)} + 3x'(t).$$

- Integrate both sides from 0 to t:

$$2\int_0^t \frac{y'(s)}{y(s)}\,ds - 5\int_0^t y'(s)\,ds = -6\int_0^t \frac{x'(s)}{x(s)}\,ds + 3\int_0^t x'(s)\,ds.$$

- Do the integrations: everything is positive so we don't need absolute values in the ln's

$$2\,\ln(y(s))\Big|_0^t - 5y(s)\Big|_0^t = -6\ln(x(s))\Big|_0^t + 3x(s)\Big|_0^t.$$

- Evaluate:

$$2\,\ln\left(\frac{y(t)}{y_0}\right) - 5(y(t) - y_0) = -6\ln\left(\frac{x(t)}{x_0}\right) + 3(x(t) - x_0).$$

- Put ln's on the left and other terms on the right:

$$6\ln\left(\frac{x(t)}{x_0}\right) + 2\,\ln\left(\frac{y(t)}{y_0}\right) = 3(x(t) - x_0) + 5(y(t) - y_0).$$

- Combine ln terms:

$$\ln\left(\left(\frac{x(t)}{x_0}\right)^6\right) + \ln\left(\left(\frac{y(t)}{y_0}\right)^2\right) = 3(x(t) - x_0) + 5(y(t) - y_0).$$

- Combine ln terms again:

$$\ln\left(\left(\frac{x(t)}{x_0}\right)^6\left(\frac{y(t)}{y_0}\right)^2\right) = 3(x(t) - x_0) + 5(y(t) - y_0).$$

- Exponentiate both sides:

$$\left(\frac{x(t)}{x_0}\right)^6\left(\frac{y(t)}{y_0}\right)^2 = e^{3(x(t)-x_0)+5(y(t)-y_0)}.$$

- Simplify the exponential term:

$$\left(\frac{x(t)}{x_0}\right)^6\left(\frac{y(t)}{y_0}\right)^2 = \frac{e^{3x(t)}}{e^{3x_0}}\,\frac{e^{5y(t)}}{e^{5y_0}}$$

- Put all function terms on the left and all constant terms on the right:

$$\frac{(x(t))^6}{e^{3x(t)}} \frac{(y(t))^2}{e^{5y(t)}} = \frac{(x_0)^6}{e^{3x_0}} \frac{(y_0)^2}{e^{5y_0}}$$

- Define the functions f and g by
  - $f(x) = x^6/e^{3x}$.
  - $g(y) = y^2/e^{5y}$.
- Then we can rewrite our result as

$$f(x(t))\, g(y(t)) = f(x_0)\, g(y_0).$$

- We did this analysis for Region II, but it works in all the regions. So for the entire trajectory, we know

$$f(x(t))\, g(y(t)) = f(x_0)\, g(y_0).$$

- The equation

$$f(x(t))\, g(y(t)) = f(x_0)\, g(y_0).$$

for $f(x) = x^6/e^{3x}$ and $g(y) = y^2/e^{5y}$, is called the **Nonlinear Conservation Law** or **NLCL** for the Predator–Prey model

$$\begin{aligned} x'(t) &= 2\, x(t) - 5\, x(t)\, y(t) \\ y'(t) &= -6\, y(t) + 3\, x(t)\, y(t) \end{aligned}$$

10.4.2 The General Derivation

Recall, we were looking at the Predator–Prey model dynamics for $0 \le t < t^*$. Since all variables are positive and their derivatives are not zero for these times, we can look at the fraction $y'(t)/x'(t)$.

$$\frac{y'(t)}{x'(t)} = \frac{y(t)\,(-c + d\, x(t))}{x(t)\,(a - b\, y(t))}.$$

The equation above will not hold at t^*, however, because at that point $x'(t^*) = 0$. But for t below that critical value, it is ok to look at this fraction.

- Rearranging a bit, we find

$$\frac{(a - b\,y(t))\,y'(t)}{y(t)} = \frac{(-c + d\,x(t))\,x'(t)}{x(t)}.$$

- switching to the variable s for $0 \leq s < t$, for any value t strictly less than our special value t^*, we have

$$a\,\frac{y'(s)}{y(s)} - b\,y'(s) = -c\,\frac{x'(s)}{x(s)} + d\,x'(s).$$

- Now integrate from $s = 0$ to $s = t$ to obtain

$$\int_0^t \left(a\,\frac{y'(s)}{y(s)} - b\,y'(s) \right) ds = \int_0^t \left(-c\,\frac{x'(s)}{x(s)} + d\,x'(s) \right) ds.$$

- These integrals can be split into separate pieces giving

$$a \int_0^t \frac{y'(s)}{y(s)}\,ds - b \int_0^t y'(s)\,ds = -c \int_0^t \frac{x'(s)}{x(s)}\,ds + d \int_0^t x'(s)\,ds.$$

- These can be integrated easily (yes, it's true!) and we get

$$a\,\ln y(s)\Big|_0^t - b\,y(s)\Big|_0^t = -c\,\ln x(s)\Big|_0^t + d\,x(s)\Big|_0^t.$$

- Evaluating these expressions at $s = t$ and $s = 0$, using the initial condition $x(0) = x_0$ and $y(0) = y_0$, we find

$$a\left(\ln y(t) - \ln y_0\right) - b\left(y(t) - y_0\right) = -c\left(\ln x(t) - \ln x_0\right) + d\left(x(t) - x_0\right).$$

- Now we simplify a lot (remember x_0 and y_0 are positive so absolute values are not needed around them). First, we use a standard logarithm property:

$$a\,\ln\left(\frac{y(t)}{y_0}\right) - b\left(y(T) - y(0)\right) = -c\,\ln\left(\frac{x(t)}{x_0}\right) + d\left(x(T) - x_0\right).$$

  Then, put all the logarithm terms on the left side and pull the powers a and c inside the logarithms:

$$\ln\left(\frac{y(t)}{y_0}\right)^a + \ln\left(\frac{x(t)}{x_0}\right)^c = b\left(y(t) - y_0\right) + d\left(x(t) - x_0\right).$$

Then using properties of the logarithm again,

$$\ln\left(\left(\frac{y(t)}{y_0}\right)^a \left(\frac{x(t)}{x_0}\right)^c\right) = b\left(y(t) - y_0\right) + d\left(x(t) - x_0\right).$$

- Now exponentiate both sides and use the properties of the exponential function to simplify. We find

$$\left(\frac{y(t)}{y_0}\right)^a \left(\frac{x(t)}{x_0}\right)^c = \frac{e^{b\,y(t)}}{e^{b\,y_0}} \frac{e^{d\,x(t)}}{e^{d\,x_0}}.$$

- We can rearrange this as follows:

$$\left(\frac{(x(t))^c}{e^{d\,x(t)}}\right)\left(\frac{(y(t))^a}{e^{b\,y(t)}}\right) = \frac{x_0^c}{e^{d\,x_0}} \frac{y_0^a}{e^{b\,y_0}}. \tag{10.17}$$

The equation

$$f(x(t))\,g(y(t)) = f(x_0)\,g(y_0).$$

for $f(x) = x^c/e^{dx}$ and $g(y) = y^a/e^{by}$, is called the **Nonlinear Conservation Law** or **NLCL** for the general Predator–Prey model

$$\begin{aligned} x'(t) &= a\,x(t) - b\,x(t)\,y(t) \\ y'(t) &= -c\,y(t) + d\,x(t)\,y(t) \end{aligned}$$

10.4.2.1 Approaching t^*

Now the right hand side is a positive number which for convenience we will call α. Hence, we have the equation

$$\left(\frac{(y(t))^a}{e^{b\,y(t)}}\right)\left(\frac{(x(t))^c}{e^{d\,x(t)}}\right) = \alpha$$

holds for all time t strictly less than t^*. Thus, as we allow t to approach t^* from below, the continuity of our solutions $x(t)$ and $y(t)$ allows us to say

$$\lim_{t->t^*}\left(\frac{(y(t))^a}{e^{b\,y(t)}}\right) \lim_{t->t^*}\left(\frac{(x(t))^c}{e^{d\,x(t)}}\right) = \left(\frac{(y(t^*))^a}{e^{b\,y(t^*)}}\right)\left(\frac{(x(t^*))^c}{e^{d\,x(t^*)}}\right) = \alpha.$$

Thus, Eq. 10.17 holds at t^* also.

10.4.2.2 The Other Regions

We can do a similar analysis for a trajectory that starts in Region IV and moves up until it hits the $y = \frac{a}{b}$ line where $x' = 0$. This one will start at an initial point (x_0, y_0) in Region IV and terminate on the $y = \frac{a}{b}$ line at the point $(x(t^*), \frac{a}{b})$ for some time t^*. In this case, we continue the analysis as before. For any time $t < t^*$, the variables $x(t)$ and $y(t)$ are positive and their derivatives non zero. Hence, we can manipulate the Predator–Prey Equations just like before to end up with

$$a \int_0^t \frac{y'(s)}{y(s)}\, ds - b \int_0^t y'(s)\, ds = -c \int_0^t \frac{x'(s)}{x(s)}\, ds + d \int_0^t x'(s)\, ds.$$

We integrate in the same way and apply the initial conditions to obtain Eq. 10.17 again.

$$\left(\frac{(y(t))^a}{e^{b\,y(t)}}\right)\left(\frac{(x(t))^c}{e^{d\,x(t)}}\right) = \frac{y_0^a}{e^{b\,y_0}}\,\frac{x_0^c}{e^{d\,x_0}}.$$

Then, taking the limit at t goes to t^*, we see this equation holds at t^* also. Again, label the right hand side as the positive constant α. We then have

$$\left(\frac{(y(t))^a}{e^{b\,y(t)}}\right)\left(\frac{(x(t))^c}{e^{d\,x(t)}}\right) = \alpha.$$

Letting t approach t^*, as we did earlier, we find

$$\left(\frac{(y(t^*))^a}{e^{b\,y(t^*)}}\right)\left(\frac{(x(t^*))^c}{e^{d\,x(t^*)}}\right) = \alpha.$$

We conclude Eq. 10.17 holds for trajectories that start in regions that terminate on the $x' = 0$ line $y = \frac{a}{b}$. Since trajectories that start in region I and III never have x' become 0, all of the analysis we did above works perfectly. Hence, we can conclude that Eq. 10.17 holds for all trajectories starting at a positive initial point (x_0, y_0) in Quadrant 1.

We know the pairs $(x(t), y(t))$ are on the trajectory that corresponds to the initial start of (x_0, y_0). Hence, we can drop the time dependence (t) above and write Eq. 10.18 which holds for any (x, y) pair that is on the trajectory.

$$\left(\frac{y^a}{e^{b\,y}}\right)\left(\frac{x^c}{e^{d\,x}}\right) = \frac{y_0^a}{e^{b\,y_0}}\,\frac{x_0^c}{e^{d\,x_0}}. \tag{10.18}$$

Equation 10.18 is called the **Nonlinear Conservation Law** associated with the Predator–Prey model.

10.4.3 Can a Trajectory Hit the y Axis Redux?

Although we have assumed trajectories can't cross and therefore a trajectory starting in Region II can't hit the y axis for that reason, we can also see this using the nonlinear conservation law. We can do the same derivation for a trajectory starting in Region II with a positive x_0 and y_0 and this time assume the trajectory hits the y axis at a time t^* at the point $(0, y_1)$ with $y_1 > 0$. We can repeat all of the integration steps to obtain

$$\left(\frac{(y(t))^a}{e^{b\,y(t)}}\right)\left(\frac{(x(t))^c}{e^{d\,x(t)}}\right) = \frac{y_0^a}{e^{b\,y_0}}\,\frac{x_0^c}{e^{d\,x_0}}.$$

This equation holds for all t before t^*. Taking the limit as t goes to t^*, we obtain

$$\left(\frac{(y(t^*))^a}{e^{b\,y(t^*)}}\right)\left(\frac{(x(t^*))^c}{e^{d\,x(t^*)}}\right) = \left(\frac{(y_1)^a}{e^{b\,y_1}}\right)\left(\frac{(0)^c}{e^{d\,0}}\right) = 0 = \frac{y_0^a}{e^{b\,y_0}}\,\frac{x_0^c}{e^{d\,x_0}}.$$

This is not possible, so we have another way to seeing that a trajectory can't hit the y axis. A similar argument shows a trajectory in Region III can't hit the x axis. We will leave the details of that argument to you!

10.4.4 Homework

For the following Predator–Prey models, derive the nonlinear conservation law. Since our discussions have shown us the times when $x' = 0$ in the fraction y'/x' do not give us any trouble, you can derive this law by integrating

$$\frac{y'(t)}{x'(t)} = \frac{y(t)\,(-c + d\,x(t))}{x(t)\,(a - b\,y(t))}.$$

in the way we have described in this section for the particular values of a, b, c and d in the given model. So you should derive the equation

$$\frac{y^a}{e^{by}}\,\frac{x^c}{e^{dx}} = \frac{(y(0))^a}{e^{by(0)}}\,\frac{(x(0))^c}{e^{dx(0)}}$$

Exercise 10.4.1

$$\begin{aligned} x'(t) &= 100\,x(t) - 25\,x(t)\,y(t) \\ y'(t) &= -200\,y(t) + 40\,x(t)\,y(t) \end{aligned}$$

Exercise 10.4.2

$$x'(t) = 1000\, x(t) - 250\, x(t)\, y(t)$$
$$y'(t) = -2000\, y(t) + 40\, x(t)\, y(t)$$

Exercise 10.4.3

$$x'(t) = 900\, x(t) - 45\, x(t)\, y(t)$$
$$y'(t) = -100\, y(t) + 50\, x(t)\, y(t)$$

Exercise 10.4.4

$$x'(t) = 10\, x(t) - 25\, x(t)\, y(t)$$
$$y'(t) = -20\, y(t) + 40\, x(t)\, y(t)$$

Exercise 10.4.5

$$x'(t) = 90\, x(t) - 2.5\, x(t)\, y(t)$$
$$y'(t) = -200\, y(t) + 4.5\, x(t)\, y(t)$$

10.5 Qualitative Analysis

From the discussions above, we now know that given an initial start (x_0, y_0) in Quadrant 1 of the x–y plane, the solution to the Predator–Prey system will not leave Quadrant 1. If we piece the various trajectories together for Regions I, II, III and IV, the solution trajectories will either periodic, spiraling in to some center or spiraling out to give unbounded motion. These possibilities are shown in Fig. 10.9 (periodic), Fig. 10.10 (spiraling out) and Fig. 10.11 (spiraling in). We want to find out which of these possible trajectories is possible.

10.5.1 The Predator–Prey Growth Functions

Recall the Predator–Prey nonlinear conservation law is given by

$$\left(\frac{y^a}{e^{b\,y}}\right)\left(\frac{x^c}{e^{d\,x}}\right) = \frac{y_0^a}{e^{b\,y_0}}\,\frac{x_0^c}{e^{d\,x_0}}.$$

We have already defined the functions f and g by for all non negative real numbers by

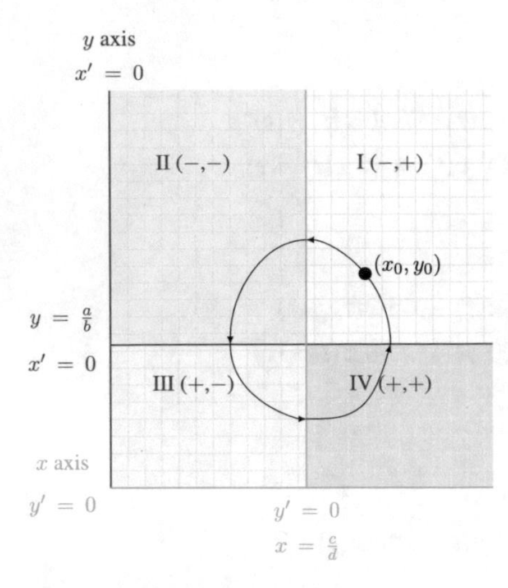

We show a possible periodic trajectories from a given start (x_0, y_0) in Quadrant 1. Note that there is a time value, call it T, so that $x(0) = x(T) = x_0$ and $y(0) = y(T) = y_0$. T is called the **period** of the trajectory. Note that horizontal or vertical lines intersect the trajectory at most twice.

Fig. 10.9 A periodic trajectory

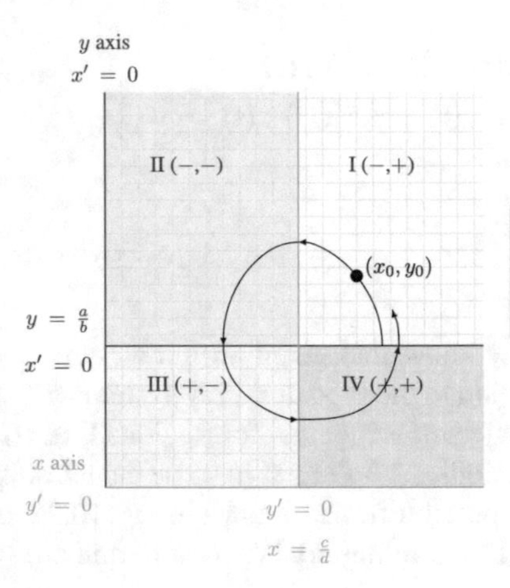

We show a trajectory starting from (x_0, y_0) in Quadrant 1 that spirals out. Note that horizontal or vertical lines intersect the trajectory more than two times.

Fig. 10.10 A spiraling out trajectory

$$f(x) = \frac{x^c}{e^{dx}}$$

$$g(y) = \frac{y^a}{e^{by}}.$$

These functions have a very specific look. We can figure this out using a bit of common sense and some first semester calculus.

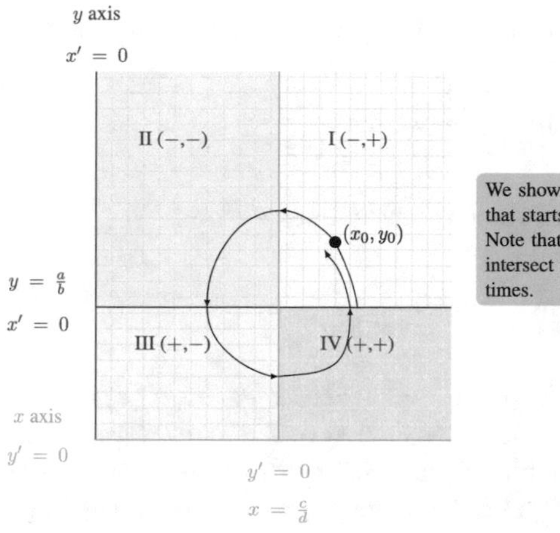

Fig. 10.11 A spiraling in trajectory

10.5.1.1 An Example

Consider a typical Predator–Prey Model

$$x'(t) = 8\,x(t) - 6\,x(t)\,y(t)$$
$$y'(t) = -7\,y(t) + 3\,x(t)\,y(t)$$

We know for any $(x_0 >, y_0 > 0)$, the trajectory $(x(t), y(t))$ satisfies the NLCL

$$f(x(t))g(y(t)) = f(x_0)g(y_0)$$

where

$$f(x) = \frac{x^7}{e^{3x}}$$
$$g(y) = \frac{y^8}{e^{6y}}$$

What do f and g look like? Let's look at f first. Recall L'Hôpital's rule.

$$\lim_{x\to\infty} \frac{x^7}{e^{3x}} = \frac{\infty}{\infty}$$

and so

$$\lim_{x\to\infty} \frac{x^7}{e^{3x}} = \lim_{x\to\infty} \frac{(x^7)'}{(e^{3x})'} = \lim_{x\to\infty} \frac{7x^6}{3e^{3x}}.$$

But this limit is also ∞/∞ and so we can apply L'Hôpital's rule again.

$$\lim_{x \to \infty} \frac{7x^6}{3e^{3x}} = \lim_{x \to \infty} \frac{42x^5}{(9e^{3x}}$$

After we have taken the derivative of x^7 seven times we find

$$\lim_{x \to \infty} \frac{x^7}{e^{3x}} = \lim_{x \to \infty} \frac{7!}{3^7 e^{3x}} = 0$$

as $1/e^{3x}$ goes to 0 as x grows large. So we know as $x \to \infty$, $f(x) \to 0$ from above as $f(x)$ is always positive. We also know $f(0) = 0$. These two facts tell us that since f is continuous and differentiable, f must rise up to a maximum at some positive point c. We find c by setting $f'(x) = 0$ and finding the critical point that is positive. We have

$$f'(x) = \frac{(7x^6)\, e^{3x} - x^7\, (3e^{3x})}{(e^{3x})^2} = \frac{7x^6 - 3x^7}{e^{3x}} = (7 - 3x)\frac{x^6}{e^{3x}}.$$

Since e^{3x} is never zero, $f'(x) = 0$ when $x = 0$ or when $x = 7/3$. Note this is c/d for our Predator–Prey model. A similar analysis holds for $g(y) = y^8/e^{6y}$. We find $y \to \infty$, $g(y) \to 0$ from above as $g(y)$ is always positive and since $g(0) = 0$, we know g has a maximum. We use calculus to show the maximum occurs at $y = 8/6$ which is a/b for our Predator–Prey model.

10.5.2 General Results

We can do the same sort of analysis for a general Predator–Prey model. From our specific example, it is easy to infer that f and g have the same generic form which are shown in Figs. 10.12 and 10.13.

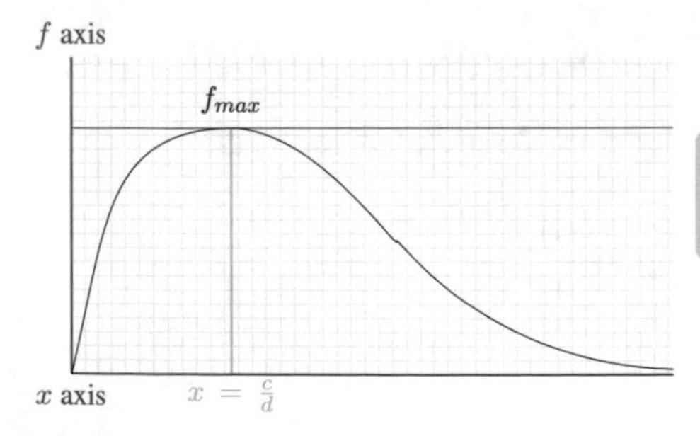

The Predator - Prey model f - growth function has the form $f(x) = x^c/e^{dx}$ for non negative x.

Fig. 10.12 The predator–prey f growth graph

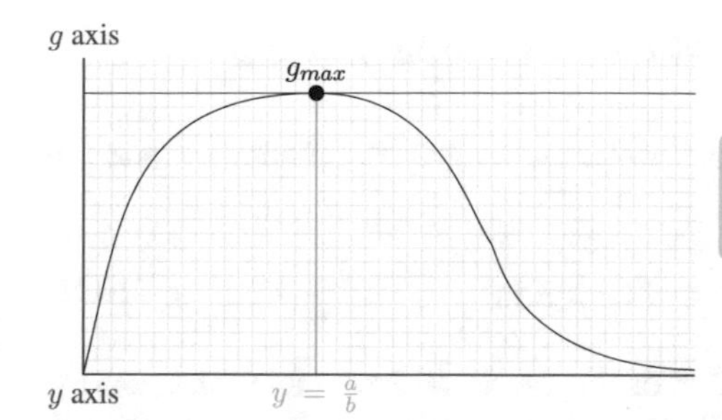

The Predator - Prey model g - growth function has the form $g(y) = y^a/e^{by}$ for non negative y.

Fig. 10.13 The predator–prey g growth graph

10.5.2.1 Homework

For the following Predator–Prey models, state what the f and g growth functions are, use calculus to derive where there maximum occurs (you can do either f or g as the derivation is the same for both) and sketch their graphs nicely.

Exercise 10.5.1

$$x'(t) = 10\, x(t) - 25\, x(t)\, y(t)$$
$$y'(t) = -20\, y(t) + 40\, x(t)\, y(t)$$

Exercise 10.5.2

$$x'(t) = 100\, x(t) - 25\, x(t)\, y(t)$$
$$y'(t) = -200\, y(t) + 40\, x(t)\, y(t)$$

Exercise 10.5.3

$$x'(t) = 90\, x(t) - 45\, x(t)\, y(t)$$
$$y'(t) = -10\, y(t) + 5\, x(t)\, y(t)$$

Exercise 10.5.4

$$x'(t) = 10\, x(t) - 2.5\, x(t)\, y(t)$$
$$y'(t) = -20\, y(t) + 4\, x(t)\, y(t)$$

Exercise 10.5.5

$$x'(t) = 9\, x(t) - 3\, x(t)\, y(t)$$
$$y'(t) = -300\, y(t) + 50\, x(t)\, y(t)$$

10.5.3 The Nonlinear Conservation Law Using f and g

We can write the nonlinear conservation law using the growth functions f and g in the form of Eq. 10.19:

$$f(x)\, g(y) = f(x_0)\, g(y_0). \tag{10.19}$$

The trajectories formed by the solutions of the Predator–Prey model that start in Quadrant 1 are powerfully shaped by these growth functions. It is easy to see that if we choose $(x_0 = \frac{c}{d}, y_0 = \frac{a}{b})$, i.e. we start at the places where f and g have their maximums, the resulting trajectory is very simple. It is the single point $(x(t) = \frac{c}{d}, y(t) = \frac{a}{b})$ for all time t. The solution to this Predator–Prey model with this initial condition is thus to simply stay at the point where we start. If $x_0 = c/d$ and $y_0 = a/b$, then the NLCL says

$$f(x(t))\, g(y(t)) = f(x_0)\, g(y_0) = f(c/d)\, g(a/b) = f_{max}\, g_{max}.$$

But $f_{max}\, g_{max}$ is just a number so this trajectory is the constant $x(t) = c/d$ and $y(t) = a/b$ for all time. Otherwise, there are two cases: if $x_0 \neq c/d$, then $f(x_0) < f_{max}$. So we can write $f(x_0) = r_1 f_{max}$ where $r_1 < 1$. We don't know where y_0 is, but we do know $g(y_0) \leq g_{max}$. So we can write $g(y_0) = r_2 g_{max}$ with $r_2 \leq 1$. So in this case, the NLCL gives

$$f(x(t))\, g(y(t)) = f(x_0)\, g(y_0) = r_1\, r_2 f_{max}\, g_{max}.$$

Let $\mu = r_1\, r_2$. Then we can say

$$f(x(t))\, g(y(t)) = f(x_0)\, g(y_0) = \mu f_{max}\, g_{max}.$$

where $\mu < 1$. Finally, if $y_0 \neq a/b$, then $g(y_0) < g_{max}$. We can thus write $g(y_0) = r_2 f_{max}$ where $r_2 < 1$. Although we don't know where x_0 is, we do know $f(x_0) \leq f_{max}$. So we can write $f(x_0) = r_1 f_{max}$ with $r_1 \leq 1$. So in this case, the NLCL again gives

$$f(x(t))\, g(y(t)) = f(x_0)\, g(y_0) = r_1\, r_2 f_{max}\, g_{max}.$$

Letting $\mu = r_1\, r_2$, we can say

$$f(x(t))\, g(y(t)) = f(x_0)\, g(y_0) = \mu f_{max}\, g_{max}.$$

where $\mu < 1$. We conclude all trajectories with $x_0 > 0$ and $y_0 > 0$ have an associated $\mu \leq 1$ so that the NCLC that can be written

$$f(x(t))\; g(y(t)) = f(x_0, y_0) = \mu f_{max}\; g_{max}.$$

and $\mu < 1$ for any trajectory with ICs different from the pair$(c/d, a/b)$.

The next step is to examine what happens if we choose a value of $\mu < 1$.

10.5.4 Trajectories are Bounded in x Example

Let's do this for the specific Predator–Prey model

$$x'(t) = 8\,x(t) - 6\,x(t)\;y(t)$$
$$y'(t) = -7\;y(t) + 3\,x(t)\;y(t)$$

Let's assume an IC corresponding to $\mu = 0.7$. The arguments work for any μ but it is nice to be able to pick a number and work off of it.

- Step 1: draw the f curve and the horizontal line of value $0.7 f_{max}$. The horizontal line will cross the f curve twice giving two corresponding x values. Label these x_1 and x_2 as shown. Also label the point $c/d = 7/3$ and draw the vertical lines from these x values to the f curve itself.
- Also pick a point $x_1^* < x_1$ and a point $x_2^* > x_2$ and draw them in along with their vertical lines that go up to the f curve.
- We show all this in Fig. 10.14.
- Step 2: Is it possible for the trajectory to contain the point x_1^*? If so, there is a corresponding y value so the NLCL holds: $f(x_1^*)g(y) = 0.7 f_{max} g_{max}$. But this implies that

$$g(y) = \frac{0.7 f_{max}}{f(x_1^*)}\; g_{max} > g_{max}$$

  as the bottom of the fraction $0.7 f_{max}/f(x_1^*)$ is smaller than the top making the fraction larger than 1. But no y can have a value larger than g_{max}. Hence, the point x_1^* is not on the trajectory.
- Step 3: Is it possible for x_1 to be on the trajectory? If so, there is a y value so the NLCL holds giving $f(x_1)g(y) = 0.7 f_{max} g_{max}$. But $f(x_1) = 0.7 f_{max}$, so cancelling, we find $g(y) = g_{max}$. Thus $y = a/b = 8/6$ and $(x_1, a/b = 8/6)$ is on the trajectory.
- Step 4: Is it possible for the trajectory to contain the point x_2^*? If so, there is a corresponding y value so the NLCL holds: $f(x_2^*)g(y) = 0.7 f_{max} g_{max}$. But this implies that

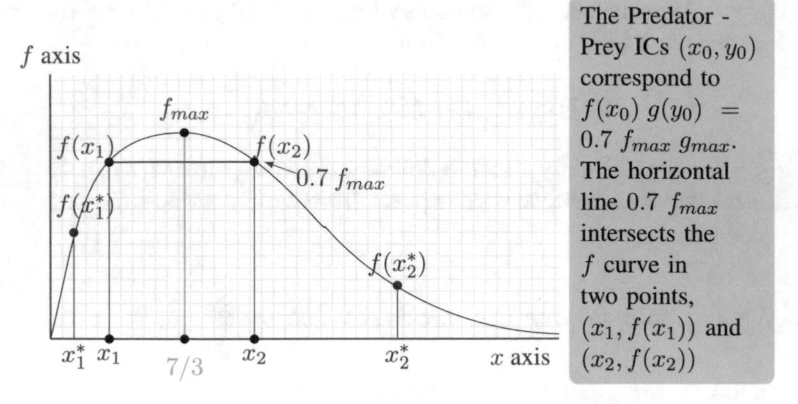

Fig. 10.14 The conservation law $f(x)\ g(y) = 0.7\ f_{max}\ g_{max}$ implies there are two critical points x_1 and x_2 of interest

$$g(y) = \frac{0.7 f_{max}}{f(x_2^*)}\ g_{max} > g_{max}$$

as the bottom of the fraction $0.7 f_{max}/f(x_2^*)$ is smaller than the top making the fraction larger than 1. But no y can have a value larger than g_{max}. Hence, the point x_2^* is not on the trajectory.

- Step 5: Is it possible for x_2 to be on the trajectory? If so, there is a y value so the NLCL holds giving $f(x_2)g(y) = 0.7 f_{max} g_{max}$. But $f(x_2) = 0.7 f_{max}$, so cancelling, we find $g(y) = g_{max}$. Thus $y = a/b = 8/6$ and $(x_2, a/b = 8/6)$ is on the trajectory. We conclude if $(x(t), y(t))$ is on the trajectory, then $x_1 \leq x(t) \leq x_2$. We show this in Fig. 10.15.

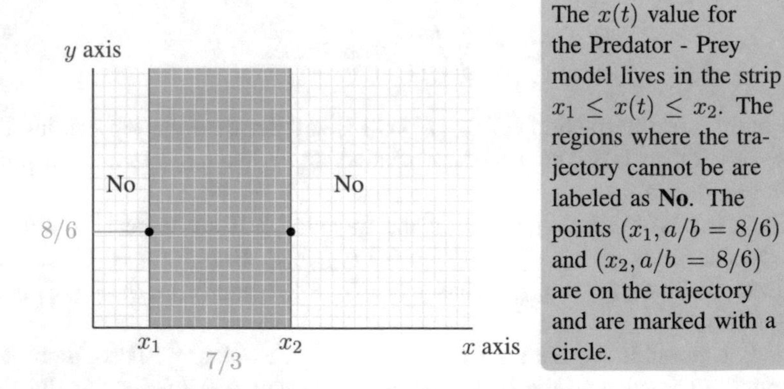

Fig. 10.15 Predator–prey trajectories with initial conditions from Quadrant 1 are bounded in x

10.5.5 Trajectories are Bounded in y Example

Again we start with our specific Predator–Prey model

$$x'(t) = 8\,x(t) - 6\,x(t)\,y(t)$$
$$y'(t) = -7\,y(t) + 3\,x(t)\,y(t)$$

Our IC corresponds to $\mu = 0.7$.

- Step 1: draw the g curve and the horizontal line of value $0.7g_{max}$. The horizontal line will cross the g curve twice giving two corresponding y values. Label these y_1 and y_2 as shown. Also label the point $a/b = 8/6$ and draw the vertical lines from these y values to the g curve itself.
- Also pick a point $y_1^* < y_1$ and a point $y_2^* > y_2$ and draw them in along with their vertical lines that go up to the g curve.
- We show all this in Fig. 10.16.
- Step 2: Is it possible for the trajectory to contain the point y_1^*? If so, there is a corresponding x value so the NLCL holds: $f(x)g(y_1^*) = 0.7 f_{max} g_{max}$. But this implies that

$$f(x) = \frac{0.7 g_{max}}{g(y_1^*)}\, f_{max} > f_{max}$$

  as the bottom of the fraction $0.7g_{max}/g(y_1^*)$ is smaller than the top making the fraction larger than 1. But no x can have a value larger than f_{max}. Hence, the point y_1^* is not on the trajectory.
- Step 3: Is it possible for y_1 to be on the trajectory? If so, there is a x value so the NLCL holds giving $f(x)g(y_1) = 0.7 f_{max} g_{max}$. But $g(y_1) = 0.7g_{max}$, so

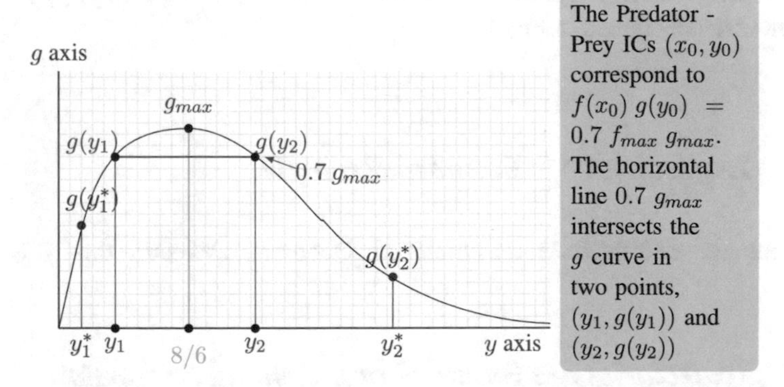

Fig. 10.16 The conservation law $f(x)\ g(y) = 0.7\ f_{max}\ g_{max}$ implies there are two critical points y_1 and y_2 of interest

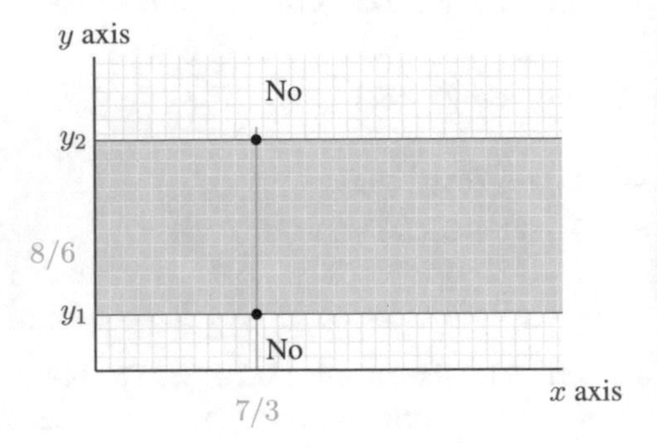

The $y(t)$ value for the Predator - Prey model lives in the strip $y_1 \leq y(t) \leq y_2$. The regions where the trajectory cannot be are labeled as **No**. The points $(c/d = 7/3, y_1)$ and $(c/d = 7/3, y_2)$ are on the trajectory and are marked with a circle.

Fig. 10.17 Predator–prey trajectories with initial conditions from Quadrant 1 are bounded in y

cancelling, we find $f(x) = f_{max}$. Thus $x = c/d = 7/3$ and $(c/d = 7/3, y_1)$ is on the trajectory.

- Step 4: Is it possible for the trajectory to contain the point y_2^*? If so, there is a corresponding x value so the NLCL holds: $f(xg(y_2^*) = 0.7 f_{max} g_{max}$. But this implies that

$$f(x) = \frac{0.7 g_{max}}{g(y_2^*)} f_{max} > f_{max}$$

as the bottom of the fraction $0.7 g_{max}/g(y_2^*)$ is smaller than the top making the fraction larger than 1. But no x can have a value larger than f_{max}. Hence, the point y_2^* is not on the trajectory.

- Step 5: Is it possible for y_2 to be on the trajectory? If so, there is a x value so the NLCL holds giving $f(x)g(y_2) = 0.7 f_{max} g_{max}$. But $g(y_2) = 0.7 g_{max}$, so cancelling, we find $f(x) = f_{max}$. Thus $x = c/d = 7/3$ and $(c/d = 7/3, y_2)$ is on the trajectory. We conclude if $(x(t), y(t))$ is on the trajectory, then $y_1 \leq y(t) \leq y_2$. We show these bounds in Fig. 10.17.

10.5.6 Trajectories are Bounded Example

Combining, we see trajectories are bounded in Quadrant 1. We show this in Fig. 10.18.

10.5.7 Trajectories are Bounded in x General Argument

We can do the analysis we did for the specific Predator–Prey model for the general one. Some people like to see these arguments with the parameters a, b, c and d and

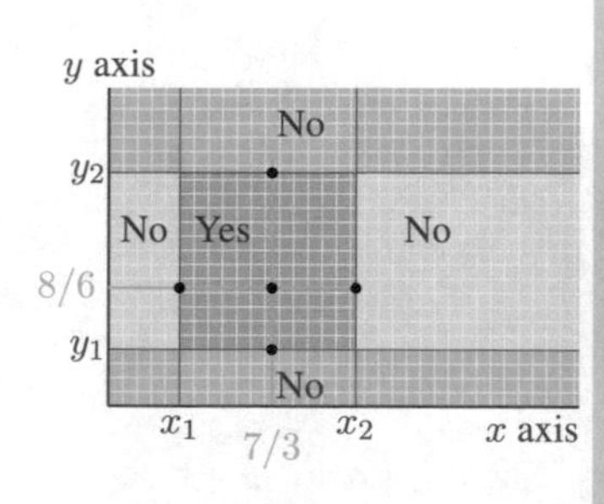

The $x(t)$ value for the Predator - Prey model lives in the strip $x_1 \leq x(t) \leq x_2$ and the $y(t)$ value lives in the strip $y_1 \leq y(t) \leq y_2$. Hence, the trajectory lives in a box. The regions where the trajectory cannot be are labeled as **No**. The points $(x_1, a/b = 8/6)$, $(x_2, a/b = 8/6)$, $(c/d = 7/3, y_1)$ and $(c/d = 7/3, y_2)$ are on the trajectory and are marked with a circle.

Fig. 10.18 Predator–prey trajectories with initial conditions from Quadrant 1 are bounded in x and y

others like to see the argument with numbers. However, learning how to see things abstractly is a skill that is honed by thinking with general terms and not specific numbers. So as we redo the arguments from the specific example in this general way, reflect on how similar there are even though you don't see numbers! We now look at the general model

$$x'(t) = a\,x(t) - b\,x(t)\,y(t)$$
$$y'(t) = -c\,y(t) + d\,x(t)\,y(t)$$

Let's again assume an IC corresponding to $\mu = 0.7$. Again, the arguments work for any μ but it is nice to be able to pick a number and work off of it. So our argument is a bit of a hybrid: general parameter values and a specific mu value.

- Step 1: draw the f curve and the horizontal line of value $0.7 f_{max}$. The horizontal line will cross the f curve twice giving two corresponding x values. Label these x_1 and x_2 as shown. Also label the point c/d and draw the vertical lines from these x values to the f curve itself.
- Also pick a point $x_1^* < x_1$ and a point $x_2^* > x_2$ and draw them in along with their vertical lines that go up to the f curve.
- We show all this in Fig. 10.19.
- Step 2: Is it possible for the trajectory to contain the point x_1^*? If so, there is a corresponding y value so the NLCL holds: $f(x_1^*)g(y) = 0.7 f_{max} g_{max}$. But this implies that

$$g(y) = \frac{0.7 f_{max}}{f(x_1^*)}\, g_{max} > g_{max}$$

  as the bottom of the fraction $0.7 f_{max}/f(x_1^*)$ is smaller than the top making the fraction larger than 1. But no y can have a value larger than g_{max}. Hence, the point x_1^* is not on the trajectory.

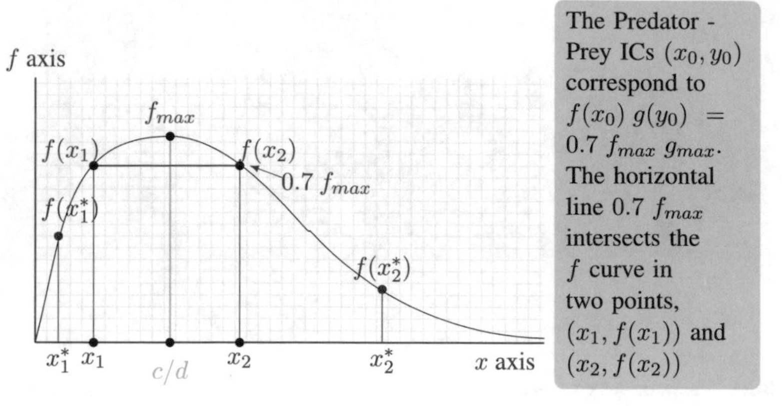

Fig. 10.19 The conservation law $f(x)\,g(y) = 0.7\, f_{max}\, g_{max}$ implies there are two critical points x_1 and x_2 of interest

- Step 3: Is it possible for x_1 to be on the trajectory? If so, there is a y value so the NLCL holds giving $f(x_1)g(y) = 0.7 f_{max} g_{max}$. But $f(x_1) = 0.7 f_{max}$, so cancelling, we find $g(y) = g_{max}$. Thus $y = a/b$ and $(x_1, a/b)$ is on the trajectory.
- Step 4: Is it possible for the trajectory to contain the point x_2^*? If so, there is a corresponding y value so the NLCL holds: $f(x_2^*)g(y) = 0.7 f_{max} g_{max}$. But this implies that

$$g(y) = \frac{0.7 f_{max}}{f(x_2^*)}\, g_{max} > g_{max}$$

as the bottom of the fraction $0.7 f_{max}/f(x_2^*)$ is smaller than the top making the fraction larger than 1. But no y can have a value larger than g_{max}. Hence, the point x_2^* is not on the trajectory.
- Step 5: Is it possible for x_2 to be on the trajectory? If so, there is a y value so the NLCL holds giving $f(x_2)g(y) = 0.7 f_{max} g_{max}$. But $f(x_2) = 0.7 f_{max}$, so cancelling, we find $g(y) = g_{max}$. Thus $y = a/b$ and $(x_2, a/b)$ is on the trajectory. We conclude if $(x(t), y(t))$ is on the trajectory, then $x_1 \le x(t) \le x_2$. We show this in Fig. 10.20.

10.5.8 Trajectories are Bounded in y General Argument

Again we look at the general Predator–Prey model

$$\begin{aligned} x'(t) &= a\, x(t) - b\, x(t)\, y(t) \\ y'(t) &= -c\, y(t) + d\, x(t)\, y(t) \end{aligned}$$

with an IC corresponding to $\mu = 0.7$.

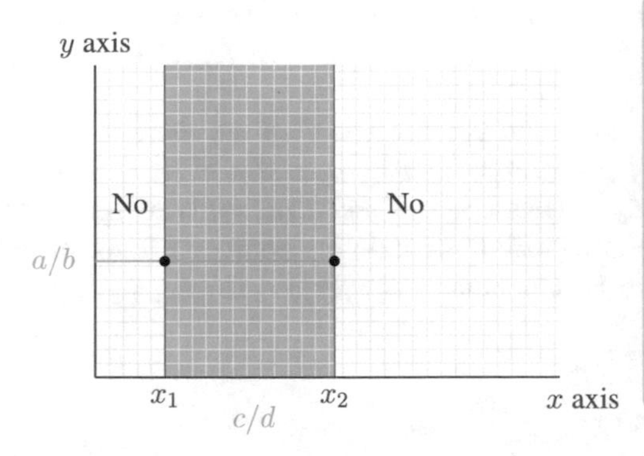

The $x(t)$ value for the Predator - Prey model lives in the strip $x_1 \leq x(t) \leq x_2$. The regions where the trajectory cannot be are labeled as **No**. The points $(x_1, a/b)$ and $(x_2, a/b)$ are on the trajectory and are marked with a circle.

Fig. 10.20 Predator–Prey trajectories with initial conditions from Quadrant 1 are bounded in x

- Step 1: draw the g curve and the horizontal line of value $0.7g_{max}$. The horizontal line will cross the g curve twice giving two corresponding y values. Label these y_1 and y_2 as shown. Also label the point a/b and draw the vertical lines from these y values to the g curve itself.
- Also pick a point $y_1^* < y_1$ and a point $y_2^* > y_2$ and draw them in along with their vertical lines that go up to the g curve.
- We show all this in Fig. 10.21.
- Step 2: Is it possible for the trajectory to contain the point y_1^*? If so, there is a corresponding x value so the NLCL holds: $f(x)g(y_1^*) = 0.7f_{max}g_{max}$. But this implies that

$$f(x) = \frac{0.7g_{max}}{g(y_1^*)} f_{max} > f_{max}$$

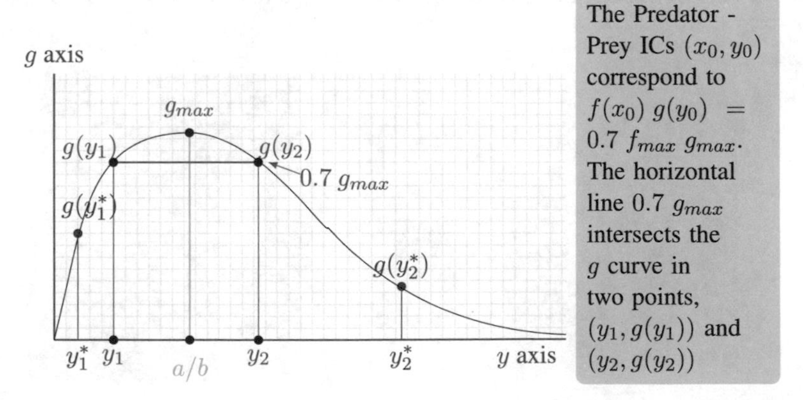

The Predator - Prey ICs (x_0, y_0) correspond to $f(x_0)\ g(y_0) = 0.7\ f_{max}\ g_{max}$. The horizontal line $0.7\ g_{max}$ intersects the g curve in two points, $(y_1, g(y_1))$ and $(y_2, g(y_2))$

Fig. 10.21 The conservation law $f(x)\ g(y) = 0.7\ f_{max}\ g_{max}$ implies there are two critical points y_1 and y_2 of interest

as the bottom of the fraction $0.7g_{max}/g(y_1^*)$ is smaller than the top making the fraction larger than 1. But no x can have a value larger than f_{max}. Hence, the point y_1^* is not on the trajectory.

- Step 3: Is it possible for y_1 to be on the trajectory? If so, there is a x value so the NLCL holds giving $f(x)g(y_1) = 0.7 f_{max} g_{max}$. But $g(y_1) = 0.7 g_{max}$, so cancelling, we find $f(x) = f_{max}$. Thus $x = c/d$ and $(c/d, y_1)$ is on the trajectory.
- Step 4: Is it possible for the trajectory to contain the point y_2^*? If so, there is a corresponding x value so the NLCL holds: $f(xg(y_2^*) = 0.7 f_{max} g_{max}$. But this implies that

$$f(x) = \frac{0.7 g_{max}}{g(y_2^*)} f_{max} > f_{max}$$

as the bottom of the fraction $0.7g_{max}/g(y_2^*)$ is smaller than the top making the fraction larger than 1. But no x can have a value larger than f_{max}. Hence, the point y_2^* is not on the trajectory.

- Step 5: Is it possible for y_2 to be on the trajectory? If so, there is a x value so the NLCL holds giving $f(x)g(y_2) = 0.7 f_{max} g_{max}$. But $g(y_2) = 0.7 g_{max}$, so cancelling, we find $f(x) = f_{max}$. Thus $x = c/d$ and $(c/d, y_2)$ is on the trajectory. We conclude if $(x(t), y(t))$ is on the trajectory, then $y_1 \leq y(t) \leq y_2$. We show these bounds in Fig. 10.22.

10.5.9 Trajectories are Bounded General Argument

Combining, we see trajectories are bounded in Quadrant 1. We show this in Fig. 10.23.

Now that we have discussed these two cases, note that we could have just done the x variable case and said that a similar thing happens for the y variable. In many texts, it is very common to do this. Since you are beginners at this kind of reasoning,

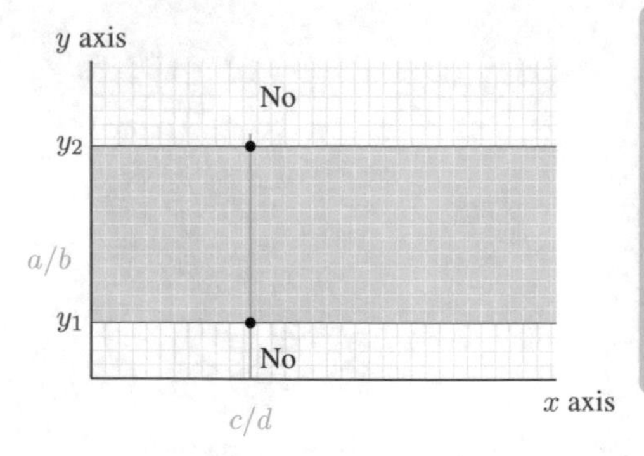

The $y(t)$ value for the Predator - Prey model lives in the strip $y_1 \leq y(t) \leq y_2$. The regions where the trajectory cannot be are labeled as **No**. The points $(c/d, y_1)$ and $(c/d, y_2)$ are on the trajectory and are marked with a circle.

Fig. 10.22 Predator–prey trajectories with initial conditions from Quadrant 1 are bounded in y

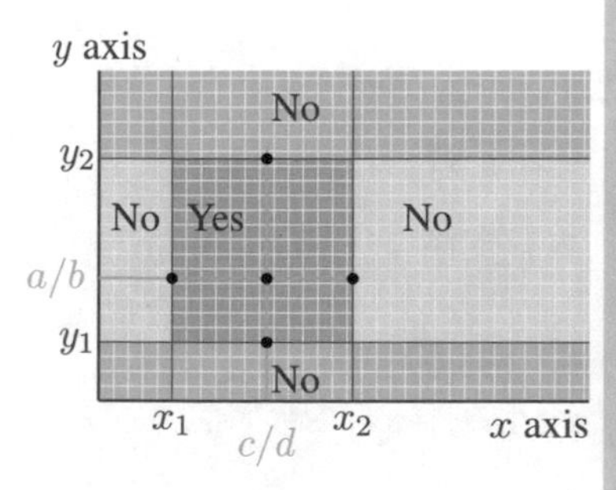

The $x(t)$ value for the Predator - Prey model lives in the strip $x_1 \leq x(t) \leq x_2$ and the $y(t)$ value lives in the strip $y_1 \leq y(t) \leq y_2$. Hence, the trajectory lives in a box. The regions where the trajectory cannot be are labeled as **No**. The points $(x_1, a/b)$, $(x_2, a/b)$, $(c/d, y_1)$ and $(c/d, y_2)$ are on the trajectory and are marked with a circle.

Fig. 10.23 Predator–prey trajectories with initial conditions from Quadrant 1 are bounded in x and y

we have presented both cases in detail. But you should start training your mind to see that presenting one case is actually enough!

10.5.10 Homework

For these Predator–Prey models, follow the analysis of the section above to show that the trajectories must bounded.

Exercise 10.5.6

$$\begin{aligned} x'(t) &= 10\,x(t) - 25\,x(t)\,y(t) \\ y'(t) &= -20\,y(t) + 40\,x(t)\,y(t) \end{aligned}$$

Exercise 10.5.7

$$\begin{aligned} x'(t) &= 100\,x(t) - 25\,x(t)\,y(t) \\ y'(t) &= -20\,y(t) + 4\,x(t)\,y(t) \end{aligned}$$

Exercise 10.5.8

$$\begin{aligned} x'(t) &= 80\,x(t) - 4\,x(t)\,y(t) \\ y'(t) &= -10\,y(t) + 5\,x(t)\,y(t) \end{aligned}$$

Exercise 10.5.9

$$x'(t) = 10\,x(t) - 2\,x(t)\,y(t)$$
$$y'(t) = -25\,y(t) + 10\,x(t)\,y(t)$$

Exercise 10.5.10

$$x'(t) = 12\,x(t) - 4\,x(t)\,y(t)$$
$$y'(t) = -60\,y(t) + 15\,x(t)\,y(t)$$

10.6 The Trajectory Must Be Periodic

If the trajectory was not periodic, then there would be horizontal and vertical lines that would intersect the trajectory in more than two places. We will show that we can have at most two intersections which tells us the trajectory must be periodic. We go back to our specific example

$$x'(t) = 8\,x(t) - 6\,x(t)\,y(t)$$
$$y'(t) = -7\,y(t) + 3\,x(t)\,y(t)$$

This time we will show the argument for this specific case only and not do a general example. But it is easy to infer from this specific example how to handle this argument for any other Predator–Prey model!

- We draw the same figure as before for f but we don't need the points x_1^* and x_2^*. This time we add a point x^* between x_1 and x_2. We'll draw it so that it is between $c/d = 7/3$ and x_2 but just remember it could have been chosen on the other side. We show this in Fig. 10.24.

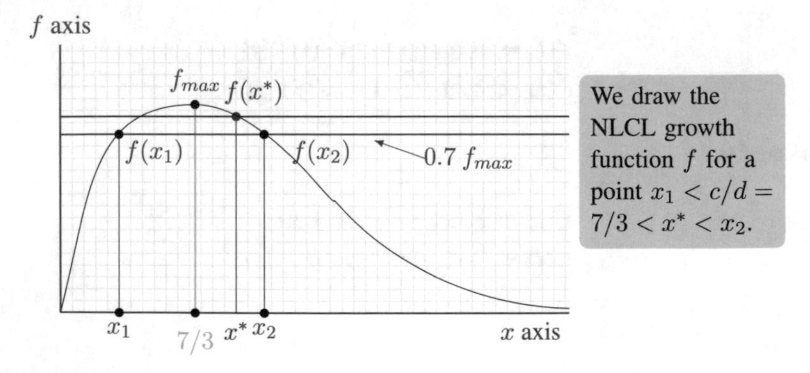

Fig. 10.24 The f curve with the point $x_1 < c/d = 7/3 < x^* < x_2$ added

- At the point x^*, the NLCL says the corresponding y values satisfy $f(x^*)g(y) = 0.7 f_{max}\, g_{max}$. This tells us

$$g(y) = \frac{0.7 f_{max}}{f(x^*)}\, g_{max}$$

- The biggest the ratio $0.7 f_{max}/f(x^*)$ can be is when the bottom $f(x^*)$ is the smallest. This occurs when x^* is chosen to be x_1 or x_2. Then the ratio is $0.7 f_{max}/(0.7 f_{max}) = 1$.
- The smallest the ratio $0.7 f_{max}/f(x^*)$ can be is when the bottom $f(x^*)$ is the largest. This occurs when x^* is chosen to be $c/d = 7/3$. Then the ratio is $0.7 f_{max}/f_{max}) = 0.7$.
- So the ratio $0.7 f_{max}/f(x^*)$ is between 0.7 and 1.
- Draw the g now adding in a horizontal line for $r g_{max}$ where $r = 0.7 f_{max}/f(x^*)$. The lowest this line can be is on the line $0.7 g_{max}$ and the highest it can be is the line of value g_{max}. This is shown in Fig. 10.25.
- The figure above shows that there are at most two intersections with the g curve.
- The case of spiral in or spiral out trajectories implies there are points x^* with more than two corresponding y values. Hence, spiral in and spiral out trajectories are not possible and the only possibility is that the trajectory is periodic.
- So there is a smallest positive number T called the period of the trajectory which means

$$\begin{aligned}(x(0), y(0)) &= (x(T), y(T)) = (x(2T), y(2T)) \\ &= (x(3T), y(3T)) = \ldots\end{aligned}$$

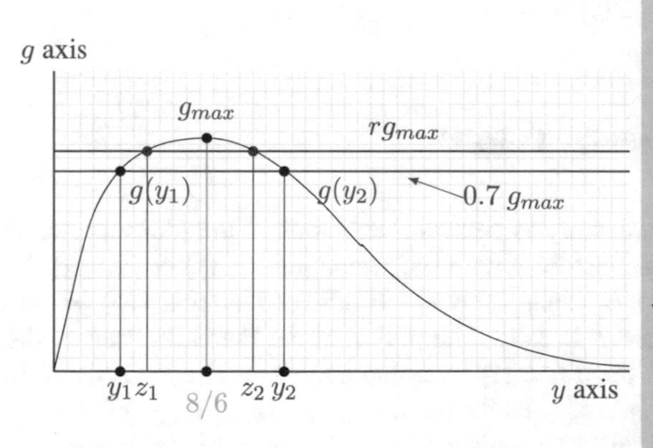

We draw the NLCL growth function g along with the line $r g_{max}$. This line crosses the g two times only except when $r = 1$ when it just touches g at its maximum point. However, no more than two intersections.

Fig. 10.25 The g curve with the $r g_{max}$ line added showing the y values for the chosen x^*

10.6.1 Homework

For the following problems, show the details of the periodic nature of the Predator–Prey trajectories by mimicking the analysis in the section above.

Exercise 10.6.1

$$x'(t) = 8\,x(t) - 25\,x(t)\,y(t)$$
$$y'(t) = -10\,y(t) + 50\,x(t)\,y(t)$$

Exercise 10.6.2

$$x'(t) = 30\,x(t) - 3\,x(t)\,y(t)$$
$$y'(t) = -45\,y(t) + 9\,x(t)\,y(t)$$

Exercise 10.6.3

$$x'(t) = 50\,x(t) - 12.5\,x(t)\,y(t)$$
$$y'(t) = -100\,y(t) + 50\,x(t)\,y(t)$$

Exercise 10.6.4

$$x'(t) = 10\,x(t) - 2.5\,x(t)\,y(t)$$
$$y'(t) = -2\,y(t) + 1\,x(t)\,y(t)$$

Exercise 10.6.5

$$x'(t) = 7\,x(t) - 2.5\,x(t)\,y(t)$$
$$y'(t) = -13\,y(t) + 2\,x(t)\,y(t)$$

10.7 Plotting Trajectory Points!

Now that we know the trajectory is periodic, let's look at the plot more carefully. We know the trajectories must lie within the rectangle $[x_1, x_2] \times [y_1, y_2]$. Mathematically, this means there is a smallest positive number T so that $x(0) = x(T)$ and $y(0) = y(T)$. This number T is called the period of the Predator–Prey model. We can see the periodicity of the trajectory by doing a more careful analysis of the trajectories. We know the trajectory hits the points $(x_1, \frac{a}{b})$, $(x_2, \frac{a}{b})$, $(\frac{c}{d}, y_1)$ and $(\frac{c}{d}, y_2)$. What happens when we look at x points u with $x_1 < u < x_2$? For convenience, let's look at the case $x_1 < u < \frac{c}{d}$ and the case $u = \frac{c}{d}$ separately.

10.7.1 Case 1: $u = \frac{c}{d}$

In this case, the nonlinear conservation law gives

$$f(u)\, g(v) = \mu\, f_{max}\, g_{max}.$$

However, we also know $f(\frac{c}{d}) = f_{max}$ and so we must have

$$f_{max}\, g(v) = \mu\, f_{max}\, g_{max}.$$

or

$$g(v) = \mu\, g_{max}.$$

Since μ is less than 1, we draw the $\mu\, g_{max}$ horizontal line on the g graph as usual to obtain the figure we previously drew as Fig. 10.21. Hence, there are two values of v that give the value $\mu\, g_{max}$; namely, $v = y_1$ and $v = y_2$. We conclude there are two possible points on the trajectory, $(\frac{c}{d}, v = y_1)$ and $(\frac{c}{d}, v = y_2)$. This gives the usual points shown as the vertical points in Fig. 10.23.

10.7.2 Case 2: $x_1 < U < \frac{c}{d}$

The analysis is very similar to the one we just did for $u = \frac{c}{d}$. First, for this choice of u, we can draw a new graph as shown in Fig. 10.26.

Here, the conservation law gives

$$f(u)\, g(v) = \mu\, f_{max}\, g_{max}.$$

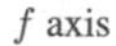

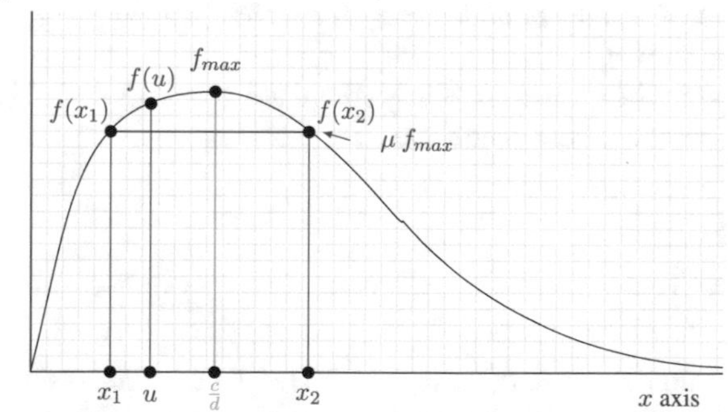

The horizontal line $\mu\, f_{max}$ intersects the f curve in two points, $(x_1, f(x_1))$ and $(x_2, f(x_2))$. The choice $x_1 < u < \frac{c}{d}$ gives the vertical line shown which intersects the f curve in the point $(u, f(u))$ with $f(x_1) < f(u) < f(\frac{c}{d})$.

Fig. 10.26 The predator–prey f growth graph trajectory analysis for $x_1 < u < \frac{c}{d}$

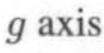

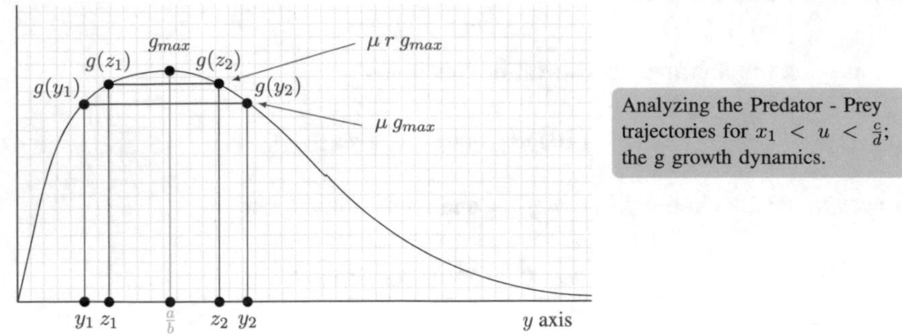

Fig. 10.27 The predator–prey g growth analysis for one point $x_1 < u < \frac{c}{d}$

Dividing through by $f(u)$, we seek v values satisfying

$$g(v) = \mu \frac{f_{max}}{f(u)} g_{max}.$$

Here the ratio $f_{max}/f(u)$ is larger than 1 (just look at Fig. 10.26 to see this). Call this ratio r. Hence, $\mu < \mu\,(f_{max}/f(u))$ and so $\mu g_{max} < \mu\,(f_{max}/f(u))\,g_{max}$. Also from Fig. 10.26, we see the ratio $\mu\, f_{max} < f(u)$ which tells us $(\mu\, f_{max})/f(u)\, g_{max} < g_{max}$. Now look at Fig. 10.27. The inequalities above show us we must draw the horizontal line $\mu\, r\, g_{max}$ above the line $\mu\, g_{max}$ and below the line g_{max}. So we seek v values that satisfy

$$\mu\, g_{max} < g(v) = \mu \frac{f_{max}}{f(u)} g_{max} = \mu\, r\, g_{max} < g_{max}.$$

We already know the values of v that satisfy $g(v) = \mu\, g_{max}$ which are labeled in Fig. 10.26 as $v = y_1$ and $v = y_2$. Since the number $\mu\, r$ is larger than μ, we see from Fig. 10.27 there are two values of v, $v = z_1$ and $v = z_2$ for which $g(v) = \mu\, r\, g_{max}$ and $y_1 < z_1 < \frac{a}{b} < z_2 < y_2$ as shown.

From the above, we see that in the case $x_1 < u < \frac{c}{d}$, there are always 2 and only 2 possible v values on the trajectory. These points are (u, z_1) and (u, z_2).

10.7.3 $x_1 < u_1 < u_2 < \frac{c}{d}$

What happens if we pick two points, $x_1 < u_1 < u_2 < \frac{c}{d}$? The f curve analysis is essentially the same but now there are two vertical lines that we draw as shown in Fig. 10.28.

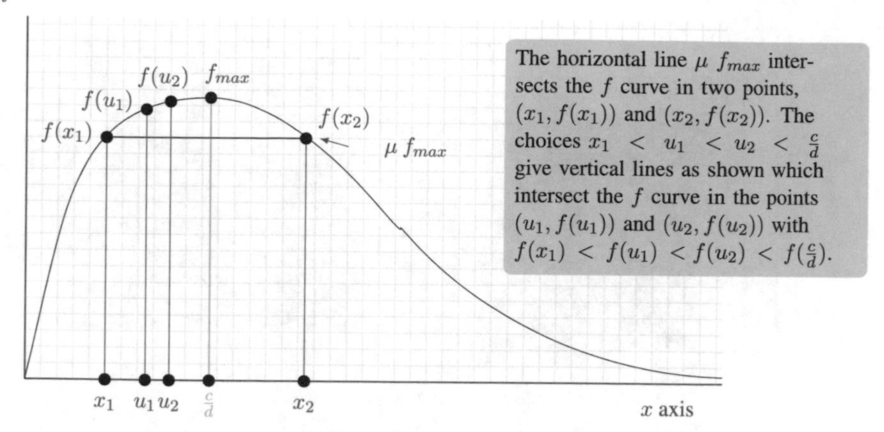

Fig. 10.28 The predator–prey f growth graph trajectory analysis for $x_1 < u_1 < u_2 < \frac{c}{d}$ points

Now, applying conservation law gives two equations

$$
\begin{aligned}
f(u_1)\, g(v) &= \mu\, f_{max}\, g_{max} \\
f(u_2)\, g(v) &= \mu\, f_{max}\, g_{max}
\end{aligned}
$$

This implies we are searching for v values in the following two cases:

$$g(v) = \mu\, \frac{f_{max}}{f(u_1)}\, g_{max}$$

and

$$g(v) = \mu\, \frac{f_{max}}{f(u_2)}\, g_{max}.$$

Since $f(u_1)$ is smaller than $f(u_2)$, we see the ratio $f_{max}/f(u_1)$ is larger than $f_{max}/f(u_2)$ and both ratios are larger than 1 (just look at Fig. 10.28 to see this). Call these ratios r_1 (the one for u_1) and r_2 (for u_2). It is easy to see $r_2 < r_1$ from the figure. We also still have (as in our analysis of the case of one point u) that both $\mu\, r_1$ and $\mu\, r_2$ are less than 1. We conclude

$$
\begin{aligned}
\mu &< \mu\, \frac{f_{max}}{f(u_2)} = \mu\, r_1 \\
&< \mu\, \frac{f_{max}}{f(u_1)} = \mu\, r_2 \\
&< 1.
\end{aligned}
$$

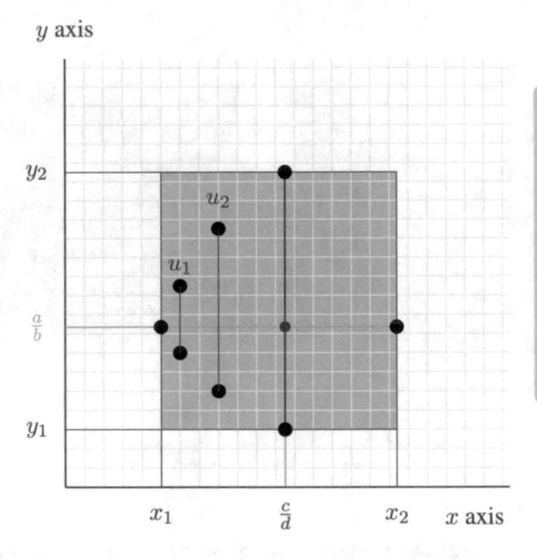

Fig. 10.29 The spread of the trajectory through fixed lines on the x axis gets smaller as we move away from the center point $\frac{c}{d}$

Now look at Fig. 10.29. The inequalities above show us we must draw the horizontal line $\mu\, r_1\, g_{max}$ above the line $\mu\, r_2\, g_{max}$ which is above the line $\mu\, g_{max}$. We already know the values of v that satisfy $g(v) = \mu\, g_{max}$ which are labeled in Fig. 10.27 as $v = y_1$ and $v = y_2$. Since the number $\mu\, r_2$ is larger than μ, we see from Fig. 10.29 there are two values of v, $v = z_{21}$ and $v = z_{22}$ for which $g(v) = \mu\, r_2\, g_{max}$ and $y_1 < z_{21} < \frac{a}{b} < z_{22} < y_2$ as shown. But we can also do this for the line $\mu\, r_1\, g_{max}$ to find two more points z_{11} and z_{12} satisfying

$$y_1 < z_{21} < z_{11} < \frac{a}{b} < z_{12} < z_{22} < y_2$$

as seen in Fig. 10.29 also.

We also see that the largest spread in the y direction is at $x = \frac{c}{d}$ giving the two points $(\frac{c}{d}, y_1)$ and $(\frac{c}{d}, y_2)$ which corresponds to the line segment $[y_1, y_2]$ drawn at the $x = \frac{c}{d}$ location. If we pick the point $x_1 < u_2 < \frac{c}{d}$, the two points on the trajectory give a line segment $[z_{21}, z_{22}]$ drawn at the $x = u_2$ location. Note this line segment is *smaller* and contained in the largest one $[y_1, y_2]$. The corresponding line segment for the point u_1 is $[z_{11}, z_{12}]$ which is smaller yet.

10.7.4 Three u Points

If you think about it a bit, if we picked three points as follows, $x_1 < u_1 < u_2 < u_3 < \frac{c}{d}$ and three more points $\frac{c}{d} < u_4 < u_5 < u_6 < x_2$, we would find line segments as follows:

Point	Spread
x_1	One point $(x_1, \frac{a}{b})$
u_1	$[z_{11}, z_{12}]$
u_2	$[z_{21}, z_{22}]$ contains $[z_{11}, z_{12}]$
u_3	$[z_{31}, z_{32}]$ contains $[z_{21}, z_{22}]$
$\frac{c}{d}$	$[y_1, y_2]$ contains $[z_{21}, z_{22}]$
u_4	$[z_{41}, z_{42}]$ inside $[y_1, y_2]$
u_2	$[z_{51}, z_{52}]$ inside $[z_{41}, z_{42}]$
u_1	$[z_{61}, z_{62}]$ inside $[z_{51}, z_{52}]$
x_2	One point $(x_2, \frac{a}{b})$ inside $[z_{51}, z_{52}]$

We draw these line segments in Fig. 10.30. We know the Predator–Prey trajectory must go through these points. Every time the trajectory hits the x value $\frac{c}{d}$, the corresponding y spread is $[y_1, y_2]$. If the trajectory was spiraling inwards, then the first time we hit $\frac{c}{d}$, the spread would be $[y_1, y_2]$ and the next time, the spread would have to be less so that the trajectory moved inwards. This can't happen as the second time we hit $\frac{c}{d}$, the spread is exactly the same. The points shown in Fig. 10.30 are always the same. Again, note since the trajectory is periodic is there is a smallest positive number T so that

$$x(t + T) = x(t) \text{ and } y(t + T) = y(t)$$

for all values of t. This is the behavior we are seeing in Fig. 10.30. Note the value of this period is really determined by the initial values (x_0, y_0) as they determine the bounding box $[x_1, x_2] \times [y_1, y_2]$ since the initial condition determines μ.

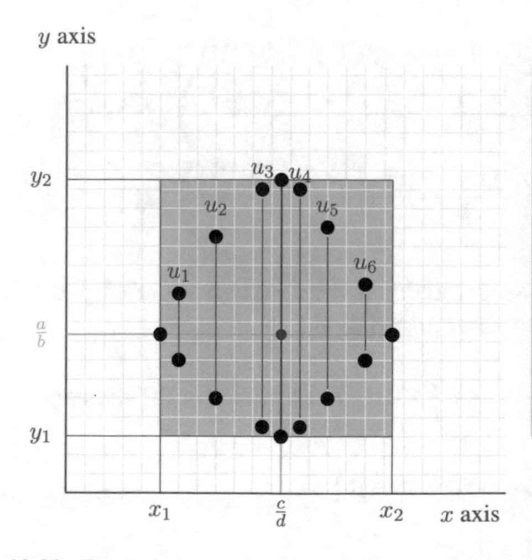

The Predator - Prey model trajectory is shown for the points (u_1, z_{11}), (u_1, z_{12}), (u_2, z_{21}), (u_2, z_{22}), (u_3, z_{31}), (u_4, z_{42}) and (u_5, z_{51}), (u_6, z_{62}) which all live in the bounding box $[x_1, x_2] \times [y_1, y_2]$. Note the length of the line segments in the vertical direction is decreasing as we move away from the center line through $x = \frac{c}{d}$.

Fig. 10.30 The trajectory must be periodic

10.8 The Average Value of a Predator–Prey Solution

If we had a positive function h defined on an interval $[a, b]$, we can define the average value of h over $[a, b]$ by the integral

$$\bar{h} = \frac{1}{b-a} \int_a^b h(t)\, dt. \tag{10.20}$$

To motivate this definition, let's look at the Riemann sums of some nice function f on the interval $[1, 3]$. Take a uniform partition $[1, 3]$ with 5 points. $P_4 = \{1, 1+h, 1+2h, 1+3h, 3\}$ where $h = (3-1)/4 = 0.5$. The evaluation set is the left hand endpoints: $E_4 = \{1, 1+h, 1+2h, 1+3h = 3\}$. Note $4h = 3 - 1 = 2$ which is the length of $[1, 3]$. The Riemann sum is

$$\begin{aligned} RS &= \Big(f(1) + f(1+h) + f(1+2h) + f(1+3h) \Big)\, h \\ &= \left(\frac{f(1) + f(1+h) + f(1+2h) + f(1+3h)}{4} \right) (4h) \\ &= \left(\frac{f(1) + f(1+h) + f(1+2h) + f(1+3h)}{4} \right) (2) \end{aligned}$$

Note $(\sum_{j=0}^{3} f(1+jh))/4$ is an estimate of the **average** value of f on $[1, 3]$ using 4 values of the function.

Now cut h in half. The new partition is $\{1, 1+h/2, 1+2(h/2), 1+3(h/2), \ldots, 3\}$ which has 8 subintervals with 9 points. The evaluation set is left hand endpoints again. Note $8(h/2) = 3 - 1 = 2$ which is the length of $[1, 3]$

$$\begin{aligned} RS &= (f(1) + f(1+(h/2)) + \cdots + f(1+7(h/2)))\,(h/2) \\ &= \left(\frac{f(1) + f(1+(h/2)) + \cdots + f(1+7(h/2))}{8} \right) (8(h/2)) \\ &= \left(\frac{f(1) + f(1+(h/2)) + \cdots + f(1+7(h/2))}{8} \right) (2) \end{aligned}$$

Now $(\sum_{j=0}^{7} f(1+j(h/2)))/8$ is an estimate of the **average** value of f on $[1, 3]$ using 8 values of the function. Do this again and again

- For $h/4$, $\left(\frac{f(1)+f(1+(h/4))+\cdots+f(1+15(h/4))}{16} \right)$ (2) is an estimate for the **average value** of f on $[1, 3]$ using 16 values.
- For $h/8 = h/2^3$, $\left(\frac{f(1)+f(1+(h/8))+\cdots+f(1+31(h/8))}{32} \right)$ (2) is an estimate for the **average value** of f on $[1, 3]$ using 32 values.

- For $h/64 = h/2^6$, $\left(\frac{f(1)+f(1+(h/64))+\cdots+f(1+255\,(h/64))}{256}\right)$ (2) is an estimate for the **average value** of f on $[1, 3]$ using 256 values.
- These all have the form **Riemann sum** is **average value of f for the partition** times **the length of** $[1, 3]$.

Letting the number of points in the partition go to ∞, the RS's converge to $\int_1^3 f(t)\,dt$. So we have $\boldsymbol{\int_1^3 f(t)dt}$ is **average value of f on** $[1, 3]$ times **the length of** $[1, 3]$. This leads to the definition we used at the start of this section: **Average value of f on** $\boldsymbol{[a, b] = 1/(b-a) \int_a^b f(t)dt}$. So for a Predator–Prey model, since there is a period T for the model, we can define

$$\bar{x} = \textbf{Average Value of } x \textbf{ on } [0, T] = \frac{1}{T}\int_0^T x(t)\,dt$$

$$\bar{y} = \textbf{Average Value of } y \textbf{ on } [0, T] = \frac{1}{T}\int_0^T y(t)\,dt$$

Here are the details. Now, recall the Predator–Prey model is given by

$$\begin{aligned} x' &= x\left(a - b\,y\right) \\ y' &= y\left(-c + d\,x\right) \\ x(0) &= x_0 \\ y(0) &= y_0 \end{aligned}$$

Rearrange the x' equation like this:

$$\frac{x'(s)}{x(s)} = a - b\,y(s)$$

for all $0 \leq s \leq T$ where T is the period for this trajectory. Now integrate from $s = 0$ to $s = T$ to get

$$\int_0^T \frac{x'(s)}{x(s)}\,ds = \int_0^T \left(a - b\,y(s)\right)ds.$$

Hence, we have

$$\ln\left(x(s)\right)\Big|_0^T = a\,T - b\int_0^T y(s)\,ds.$$

Simplifying, we find

$$\ln\left(\frac{x(T)}{x_0}\right) = a\,T - b\int_0^T y(s)\,ds.$$

However, since T is the period for this trajectory, we know $x(T)$ must equal $x(0)$. Hence, $\ln(x(T)/x_0) = \ln(1) = 0$. Rearranging, we conclude

$$0 = a\,T - b\int_0^T y(s)\,ds,$$

$$b\int_0^T y(s)\,ds = a\,T,$$

$$\frac{1}{T}\int_0^T y(s)\,ds = \frac{a}{b}.$$

The term on the left hand side is the *average value* of the solution y over the one period of time, $[0, T]$. Using the usual average notation, we will call this $\bar{y}$. Thus, we have

$$\bar{y} = \frac{1}{T}\int_0^T y(s)\,ds = \frac{a}{b}. \tag{10.21}$$

We can do a similar analysis for the average value of the x component of the solution. We find

$$\frac{y'(s)}{y(s)} = -c + d\,x(s),\ \ 0 \le s \le T,$$

$$\int_0^T \frac{y'(s)}{y(s)}\,ds = \int_0^T \left(-c + d\,x(s)\right) ds,$$

$$\ln\left(y(s)\right)\Big|_0^T = -c\,T + d\int_0^T x(s)\,ds,$$

$$\ln\left(\frac{y(T)}{y_0}\right) = -c\,T + d\int_0^T x(s)\,ds.$$

However, since T is the period for this trajectory, we know $y(T)$ must equal $y(0)$. Hence, $\ln(y(T)/y_0) = \ln(1) = 0$. Rearranging, we conclude

$$0 = -c\,T + d\int_0^T x(s)\,ds,$$

$$d\int_0^T x(s)\,ds = c\,T,$$

$$\frac{1}{T}\int_0^T x(s)\,ds = \frac{c}{d}.$$

The term on the left hand side is the *average value* of the solution x over the one period of time, $[0, T]$. Using the usual average notation, we will call this $\bar{x}$. Thus, we have

$$\bar{x} = \frac{1}{T} \int_0^T x(s)\, ds = \frac{c}{d}. \tag{10.22}$$

The point $(\bar{x} = \frac{c}{d}, \bar{y} = \frac{a}{b})$ has an important interpretation. It is the average value of the solution over the period of the trajectory.

10.8.1 Homework

For the following Predator–Prey models, derive the average x and y equations.

Exercise 10.8.1

$$\begin{aligned} x'(t) &= 100\, x(t) - 25\, x(t)\, y(t) \\ y'(t) &= -200\, y(t) + 40\, x(t)\, y(t) \end{aligned}$$

Exercise 10.8.2

$$\begin{aligned} x'(t) &= 1000\, x(t) - 250\, x(t)\, y(t) \\ y'(t) &= -2000\, y(t) + 40\, x(t)\, y(t) \end{aligned}$$

Exercise 10.8.3

$$\begin{aligned} x'(t) &= 900\, x(t) - 45\, x(t)\, y(t) \\ y'(t) &= -100\, y(t) + 50\, x(t)\, y(t) \end{aligned}$$

Exercise 10.8.4

$$\begin{aligned} x'(t) &= 10\, x(t) - 25\, x(t)\, y(t) \\ y'(t) &= -20\, y(t) + 40\, x(t)\, y(t) \end{aligned}$$

Exercise 10.8.5

$$\begin{aligned} x'(t) &= 90\, x(t) - 2.5\, x(t)\, y(t) \\ y'(t) &= -200\, y(t) + 4.5\, x(t)\, y(t) \end{aligned}$$

10.9 A Sample Predator–Prey Model

Example 10.9.1 Consider the following Predator–Prey model:

$$x'(t) = 2\,x(t) - 10\,x(t)\,y(t)$$
$$y'(t) = -3\,y(t) + 18\,x(t)\,y(t)$$

Solution *For any choice of initial conditions* (x_0, y_0), *we can solve this as discussed in the previous sections. We find* $a = 2$, $b = 10$ *so that* $\frac{a}{b} = 0.2$ *and* $c = 3$, $d = 18$ *so that* $\frac{c}{d} = 0.1666\bar{6}$. *We know a lot about these solutions now.*

1. *The solution* $(x(t), y(t))$ *has average value x value* $\bar{x} = \frac{c}{d}$ *which is* $0.1666\bar{6}$ *and an average y value,* $\bar{y} = \frac{a}{b} = 0.2$.
2. *The initial condition* (x_0, y_0) *is some point on the curve.*
3. *For each choice of initial condition* (x_0, y_0), *there is a corresponding period T so that* $(x(t), y(t)) = (x(t+T), y(t+T))$ *for all time t.*
4. *Looking at Fig. 10.30, we can connect the* dots *so to speak to generate the trajectory shown in Fig. 10.31.*

Now that you know how to analyze the predator–prey models, you can look in the literature and see how they are used. We will leave it up to you to find the many references on how this model is used to study the wolve–moose population on Isle Royal in Lake Superior and instead point you to a different one. In Axelsen et al. (2001), the predators are Atlantic Puffins and the prey are juvenile herring and the research is trying to understand the shapes that schools of herring take in the wild under predation. This study is primarily descriptive with no mathematics at all but the references point to other papers where simulations are carried out. You have enough training now to follow this paper trail and see how the simulation papers and the descriptive papers work hand in hand. But we leave the details and hard work to do this to you. Happy hunting! If you look at the references in this paper, you'll note that one the papers there is by Hamilton—the same biologist whose work on altruism we studied in Peterson (2015).

10.9.1 Homework

Do these analyses for some specific value of μ; for example, $mu = 0.8$ or something similar. Of course, the specific value doesn't matter that much, but it is easier to see the graphical analysis is $\mu\ f_{max}$ is not too close to the peak of the f. This time also draw the bounding boxes that we get for different values of μ. You will see that when μ is close to 1, the bounding box is very small and as μ get close to 0, the bounding boxes get very large. Also, you will see they are nested inside each other.

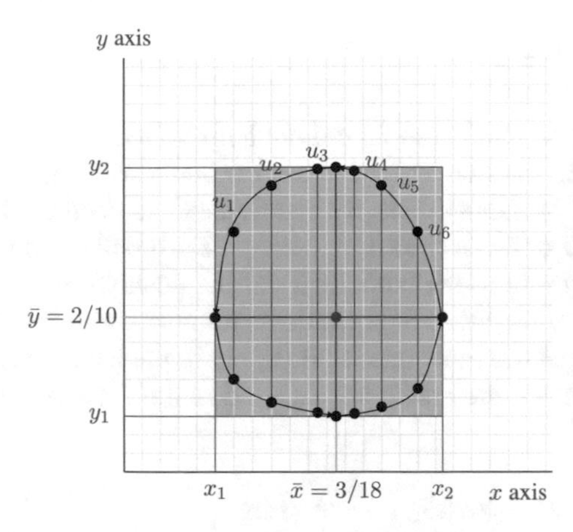

Our example The Predator - Prey model trajectory has average x value of $3/18 = .1666\bar{6}$ and average y value of $2/10 = .2$. We shown the spreads for 6 different u values which all lie inside the bounding box $[x_1, x_2] \times [y_1, y_2]$. Note the length of the line segments in the vertical direction is decreasing as we move away from the center line through $x = .166\bar{6}$.

Fig. 10.31 The theoretical trajectory for $x' = 2x - 10xy$; $y' = -3y + 18xy$. We do not know the actual trajectory as we can not solve for x and y explicitly as functions of time. However, our analysis tells us the trajectory has the qualitative features shown

Exercise 10.9.1 *Provide a complete analysis for the Predator–Prey Model*

$$
\begin{aligned}
x'(t) &= 4\,x(t) - 7\,x(t)\,y(t) \\
y'(t) &= -9\,y(t) + 7\,x(t)\,y(t)
\end{aligned}
$$

Exercise 10.9.2 *Provide a complete analysis for the Predator–Prey Model*

$$
\begin{aligned}
x'(t) &= 90\,x(t) - 45\,x(t)\,y(t) \\
y'(t) &= -180\,y(t) + 20\,x(t)\,y(t)
\end{aligned}
$$

Exercise 10.9.3 *Provide a complete analysis for the Predator–Prey Model*

$$
\begin{aligned}
x'(t) &= 80\,x(t) - 4\,x(t)\,y(t) \\
y'(t) &= -100\,y(t) + 5\,x(t)\,y(t)
\end{aligned}
$$

Exercise 10.9.4 *Provide a complete analysis for the Predator–Prey Model*

$$
\begin{aligned}
x'(t) &= 9\,x(t) - 27\,x(t)\,y(t) \\
y'(t) &= -18\,y(t) + 6\,x(t)\,y(t)
\end{aligned}
$$

Exercise 10.9.5 *Provide a complete analysis for the Predator–Prey Model*

$$
\begin{aligned}
x'(t) &= 4\,x(t) - 18\,x(t)\,y(t) \\
y'(t) &= -3\,y(t) + 21\,x(t)\,y(t)
\end{aligned}
$$

10.10 Adding Fishing Rates

The Predator–Prey model we have looked at so far did not help Volterra explain the food and predator fish data seen in the Mediterranean sea during World War I. The model must also handle changes in fishing rates. War activities had decreased the rate of fishing from 1915 to 1919 or so as shown in Table 10.2. To understand this data, Volterra added a new decay rate to the model. He let the positive constant r represent the rate of fishing and assumed that $-r\,x$ would be removed from food fish due to fishing and also assumed that the same rate would apply to predator removal. Hence, $-r\,y$ would be removed from the predators. This led to the *Predator–Prey with fishing* given by

$$x'(t) = a\,x(t) - b\,x(t)\,y(t) - r\,x(t) \tag{10.23}$$
$$y'(t) = -c\,y(t) + d\,x(t)\,y(t) - r\,y(t). \tag{10.24}$$

We don't have to work to hard to understand what adding the fishing does to our model results. We can rewrite the model as

$$x'(t) = (a - r)\,x(t) - b\,x(t)\,y(t) \tag{10.25}$$
$$y'(t) = -(c + r)\,y(t) + d\,x(t)\,y(t). \tag{10.26}$$

We see immediately that it doesn't make sense for the fishing rate to exceed a as we want $a - r$ to be positive. We also know the new averages are

$$\bar{x}_r = \frac{c + r}{d}$$
$$\bar{y}_r = \frac{a - r}{b}.$$

where we label the new averages with a subscript r to denote their dependence on the fishing rate r. What happens if we halve the fishing rate r? The new model is

$$x'(t) = a\,x(t) - b\,x(t)\,y(t) - r/2\,x(t)$$
$$y'(t) = -c\,y(t) + d\,x(t)\,y(t) - r/2\,y(t).$$

which can be reorganized as

$$x'(t) = (a - r/2)\,x(t) - b\,x(t)\,y(t) \tag{10.27}$$
$$y'(t) = -(c + r/2)\,y(t) + d\,x(t)\,y(t). \tag{10.28}$$

leading to the new averages

$$\bar{x}_{r/2} = \frac{c + r/2}{d}$$

$$\bar{y}_{r/2} = \frac{a - r/2}{b}.$$

Note that as long as we use a *feasible* r value (i.e. $r < a$), we have the following inequality relationships:

$$\bar{x}_{r/2} = \frac{c + r/2}{d} < \bar{x}_r = \frac{c + r}{d}$$
$$\bar{x}_{r/2} = \frac{a - r/2}{b} > \bar{y}_r = \frac{a - r}{b}.$$

Hence, if we decrease the fishing rate r, the predator percentage goes up and the food percentage goes down. Now look at Table 10.2 rewritten with the percentages listed as fractions and interpreted as $\bar{x}$ and $\bar{y}$. We show this in Table 10.4.

Note that Volterra's Predator–Prey model with fishing rates added has now explained this data. During the war years, predator amounts went up and food fish amounts went down. A wonderful use of modeling, don't you think? Insight was gained from the modeling that had not been able to be achieved using other types of analysis.

Let's do an example to set this in place. Consider the following Predator–Prey model with fishing added.

Example 10.10.1

$$x'(t) = 4\,x(t) - 18\,x(t)\,y(t) - 2\,x(t)$$
$$y'(t) = -3\,y(t) + 21\,x(t)\,y(t) - 2\,y(t)$$

Solution *The averages are as follows:*

$$\bar{x}_{r=0} = 3/21 = 0.1429, \;\; \bar{y}_{r=0} = 4/18\; 0.2222$$

Table 10.4 The average food and predator fish caught in the Mediterranean Sea

Year	$\bar{x}$	$\bar{y}$	Fishing rate change	$(\Delta\,\bar{x}, \Delta\,\bar{y})$
1914	0.881	0.119	Starting value	No change yet
1915	0.786	0.214	Down relative to 1914	$(-, +)$
1916	0.779	0.221	Down relative to 1914	$(-, +)$
1917	0.788	0.212	Down relative to 1914	$(-, +)$
1918	0.636	0.364	Down relative to 1914	$(-, +)$
1919	0.727	0.273	Increased relative to 1918	$(+, -)$
1920	0.840	0.160	Increased relative to 1918	$(+, -)$
1921	0.841	0.159	Increased relative to 1918	$(+, -)$
1922	0.852	0.148	Increased relative to 1918	$(+, -)$
1923	0.893	0.107	Back to normal 1914 rate	Back to normal

$$\bar{x}_{r=2} = 5/21 = 0.2381, \quad \bar{y}_{r=2} = 2/18\, 0.1111$$
$$\bar{x}_{r=1} = 4/21 = 0.1905, \quad \bar{y}_{r=1} = 3/18\, 0.1667.$$

We see that halving the fishing rate, decreases the food fish amounts (0.2381 *down to* 0.1905) *and increases the predator amounts* (0.1111 *up to* 0.1667)*. We could also shown this graphically by drawing all three average pairs on the same x–y plane but we will leave that to you in the exercises.*

10.10.1 Homework

For the following problems, add fishing to the model at some rate r which is given. Find the new average solutions $(\bar{x}, \bar{y})$ and explain what happens if we half the fishing rate and how this relates to the way Volterra explained the Mediterranean Sea fishing data from World War I. Draw a simple picture shown these three averages on the same x–y graph: show original $(\bar{x}, \bar{y})$, the $(\bar{x}, \bar{y})$ when the fishing is added and the $(\bar{x}, \bar{y})$ when the fishing is halved. You should clearly see that adding halving the fishing rate leads to the average predator value going up with the average food fish value going down.

Exercise 10.10.1

$$x'(t) = 4\, x(t) - 18\, x(t)\, y(t) - 2\, x(t)$$
$$y'(t) = -3\, y(t) + 21\, x(t)\, y(t) - 2\, y(t)$$

Exercise 10.10.2

$$x'(t) = 3\, x(t) - 10\, x(t)\, y(t) - 1\, x(t)$$
$$y'(t) = -4\, y(t) + 20\, x(t)\, y(t) - 1\, y(t)$$

Exercise 10.10.3

$$x'(t) = 1\, x(t) - 2\, x(t)\, y(t) - 0.5\, x(t)$$
$$y'(t) = -4\, y(t) + 8\, x(t)\, y(t) - 0.5\, y(t)$$

Exercise 10.10.4

$$x'(t) = 40\, x(t) - 18\, x(t)\, y(t) - 0.2\, x(t)$$
$$y'(t) = -30\, y(t) + 20\, x(t)\, y(t) - 0.2\, y(t)$$

Exercise 10.10.5

$$x'(t) = 7\, x(t) - 8\, x(t)\, y(t) - 0.2\, x(t)$$
$$y'(t) = -3\, y(t) + 4\, x(t)\, y(t) - 0.2\, y(t)$$

10.11 Numerical Solutions

Let's try to solve a typical predator–prey system such as the one given below numerically.

$$x'(t) = a\,x(t) - b\,x(t)\,y(t)$$
$$y'(t) = -c\,y(t) + d\,x(t)\,y(t)$$

As we have done in Chap. 9, we convert our model to a matrix-vector system. The right hand side of our system is now a column vector: we identify x with the component $x(1)$ and y with the component $x(2)$. This gives the vector function

$$f(t,x) = \begin{bmatrix} a\,x(1) & -b\,x(1)\,x(2) \\ -c\,x(2) + d\,x(1)\,x(2) \end{bmatrix}$$

and we can no longer find the true solution, although our theoretical investigations have told us a lot about the behavior that the true solution must have.

Let's solve a Predator–Prey Model with Runge–Kutta Order 4.

$$x'(t) = 12\,x(t) - 5\,x(t)\,y(t)$$
$$y'(t) = -6\,y(t) + 3\,x(t)\,y(t)$$

We can do this with the following Matlab session.

Listing 10.1: Phase Plane $x' = 12x - 5xy$, $y' = -6y + 3xy$, $x(0) = 0.2$, $y(0) = 8.6$

```
f = @(t,x) [12*x(1) - 5*x(1).*x(2);-6*x(2)+3*x(1).*x(2)];
T = 1.7;
h = .01;
% xbar = 6/3 = 2.0; ybar = 12/5 = 2.2
x0 = [0.2;8.6];
N = ceil(T/h);
[ht,rk] = FixedRK(f,0,x0,h,4,N);
X = rk(1,:);
Y = rk(2,:);
x1 = min(X)
x2 = max(X);
y1 = min(Y)
y2 = max(Y);
clf
hold on
plot([x1 x1],[y1 y2],'-b');
plot([x2 x2],[y1 y2],'-g');
plot([x1 x2],[y1 y1],'-r');
plot([x1 x2],[y2 y2],'-m');
plot(X,Y,'-k');
xlabel('x');
ylabel('y');
title(' Phase Plane for Predator - Prey model x''=12x - 5xy, y''=-6y+3xy, x(0) =
    0.2, y(0) = 8.6');
legend('x1','x2','y1','y2','y vs x','Location','Best');
hold off
```

This gives the plot of Fig. 10.32. Let's annotate this code.

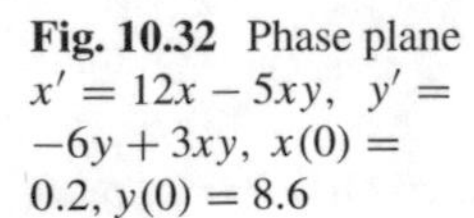

Fig. 10.32 Phase plane $x' = 12x - 5xy$, $y' = -6y + 3xy$, $x(0) = 0.2$, $y(0) = 8.6$

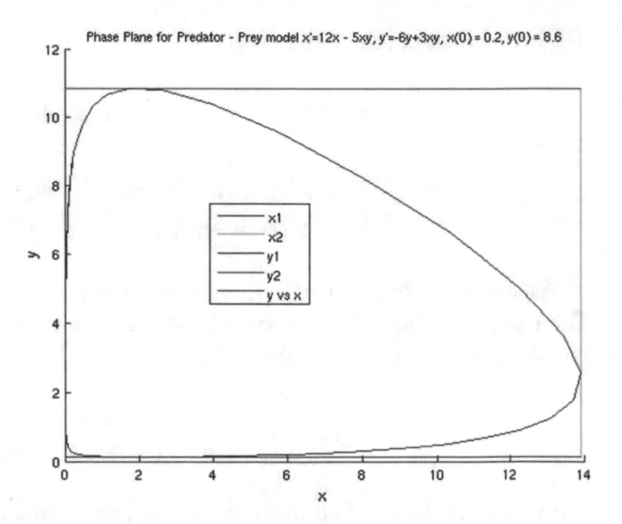

Listing 10.2: Annotated Predator–Prey Phase Plane Code

```
% set up Predator - Prey dynamics
f = @(t,x) [12*x(1) - 5*x(1).*x(2);-6*x(2)+3*x(1).*x(2)];
% set up final time: too small and trajectory does not close
% too big and trajectory wraps around multiple times
T = 1.7;
% set stepsize: too small and trajectory looks jagged
h = .01;
% find averages xbar = 6/3 = 2.0; ybar = 12/5 = 2.2
% pick IC so mu is really small
x0 = [0.2;8.6];
% find number of steps
N = ceil(T/h);
% find RK 4 approximate solution
[ht,rk] = FixedRK(f,0,x0,h,4,N);
% set X and Y
X = rk(1,:);
Y = rk(2,:);
% get x1 and x2 from our bounding box argument
% using min and max of X
x1 = min(X)
x2 = max(X);
% get y1 and y2 from our bounding box argument
% using min and max of Y
y1 = min(Y)
y2 = max(Y);
% clear previous graph
clf
% we will do multiple plots so set a hold on
% plot x1 vertical line in blue
plot([x1 x1],[y1 y2],'-b');
% plot x2 vertical line in green
plot([x2 x2],[y1 y2],'-g');
% plot y1 horizontal line in red
plot([x1 x2],[y1 y1],'-r');
% plot y1 horizontal line in magenta
plot([x1 x2],[y2 y2],'-m');
% plot Y vs X in black
plot(X,Y,'-k');
```

```
% set x and y labels
xlabel('x');
ylabel('y');
% set title
title(' Phase Plane for Predator - Prey model x''=12x - 5xy, y''=-6y+3xy, x(0) =
    0.2, y(0) = 8.6');
% set legend
legend('x1','x2','y1','y2','y vs x','Location','Best');
% cancel hold
hold off
```

We can use this type of session to do interesting things.

- For a given Predator–Prey model with IC, set the final time T so low the trajectory does not close. Then increase T slowly until trajectory just touches. This gives a good estimate of the period T. The step size h needs to be adjusted to make sure the graph is smooth to get a good value of T.
- For different IC's you get different bounding boxes, so with a little care you can plot a sequence of nested bounding boxes. You'll find the smaller μ value bounding box is inside the larger μ value box.

10.11.1 Estimating the Period T Numerically

We'll estimate the period for our sample problem. We start with a small final time T and move it up until the trajectory is almost closed.

Listing 10.3: First Estimate of the period T

```
f = @(t,x) [12*x(1) - 5*x(1).*x(2);-6*x(2)+3*x(1).*x(2)];
T = 1.01;
h = .01;
% xbar = 6/3 = 2.0; ybar = 12/5 = 2.2
x0 = [1.2;8.6];
N = ceil(T/h);
[ht,rk] = FixedRK(f,0,x0,h,4,N);
X = rk(1,:);
Y = rk(2,:);
x1 = min(X)
x2 = max(X);
y1 = min(Y)
y2 = max(Y);
clf
hold on
plot([x1 x1],[y1 y2],'-b');
plot([x2 x2],[y1 y2],'-g');
plot([x1 x2],[y1 y1],'-r');
plot([x1 x2],[y2 y2],'-m');
plot(X,Y,'-k');
xlabel('x');
ylabel('y');
title(' Phase Plane for Predator - Prey model x''=12x - 5xy, y''=-6y+3xy, x(0) =
    0.2, y(0) = 8.6');
legend('x1','x2','y1','y2','y vs x','Location','Best');
hold off
```

This gives us Fig. 10.33 and we can see the period $T > 1.01$.

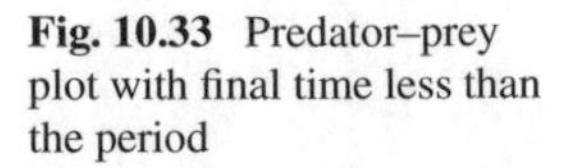

Fig. 10.33 Predator–prey plot with final time less than the period

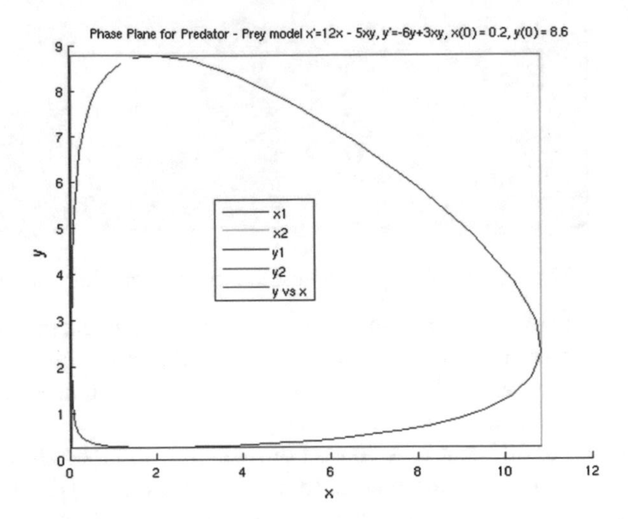

Now we increase the final time a bit.

Listing 10.4: Second Estimate of the period T

```
f = @(t,x) [12*x(1) - 5*x(1).*x(2);-6*x(2)+3*x(1).*x(2)];
T = 1.02;
h = .01;
% xbar = 6/3 = 2.0; ybar = 12/5 = 2.2
x0 = [1.2;8.6];
N = ceil(T/h);
[ht,rk] = FixedRK(f,0,x0,h,4,N);
X = rk(1,:);
Y = rk(2,:);
x1 = min(X)
x2 = max(X);
y1 = min(Y)
y2 = max(Y);
clf
hold on
plot([x1 x1],[y1 y2],'-b');
plot([x2 x2],[y1 y2],'-g');
plot([x1 x2],[y1 y1],'-r');
plot([x1 x2],[y2 y2],'-m');
plot(X,Y,'-k');
xlabel('x');
ylabel('y');
title(' Phase Plane for Predator - Prey model x''=12x - 5xy, y''=-6y+3xy, x(0) =
    0.2, y(0) = 8.6');
legend('x1','x2','y1','y2','y vs x','Location','Best');
hold off
```

This gives us Fig. 10.34 and we can see the trajectory is now closed. So the period $T \leq 1.02$. Hence, we know $1.01 < T \leq 1.02$.

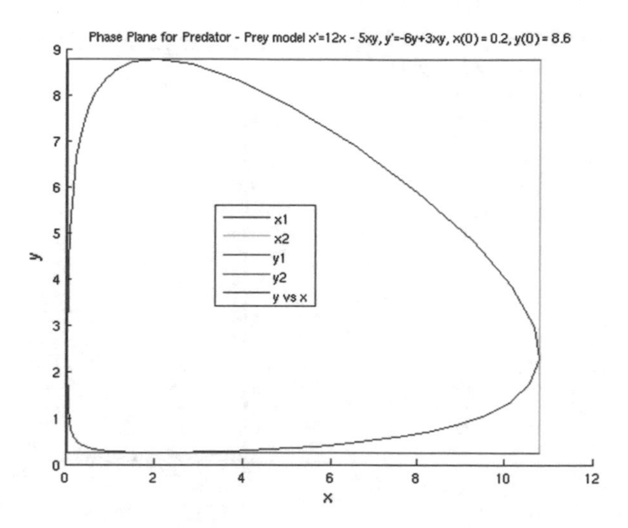

Fig. 10.34 Predator–prey plot with final time about the period

10.11.2 Plotting Predator and Prey Versus Time

We can also write code to generate x versus t plots and y versus t plots From these, we can also estimate the period T. The x versus t code is shown below and right after it is the one line modification need to generate the plot of y versus t. The codes below shows a little bit more than one period for x and y.

Listing 10.5: x versus time code

```
f = @(t,x) [12*x(1) - 5*x(1).*x(2);-6*x(2)+3*x(1).*x(2)];
T = 1.04;
h = .01;
% xbar = 6/3 = 2.0; ybar = 12/5 = 2.2
x0 = [1.2;8.6];
N = ceil(T/h);
[ht,rk] = FixedRK(f,0,x0,h,4,N);
X = rk(1,:);
plot(ht,X,'-k');
xlabel('t');
ylabel('y');
title(' x vs time for Predator - Prey model x''=12x - 5xy, y''=-6y+3xy, x(0) =
    0.2, y(0) = 8.6');
```

The x versus t plot is shown in Fig. 10.35.

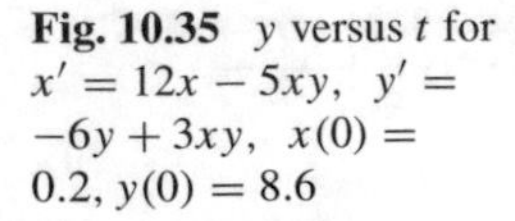

Fig. 10.35 y versus t for $x' = 12x - 5xy,\ y' = -6y + 3xy,\ x(0) = 0.2, y(0) = 8.6$

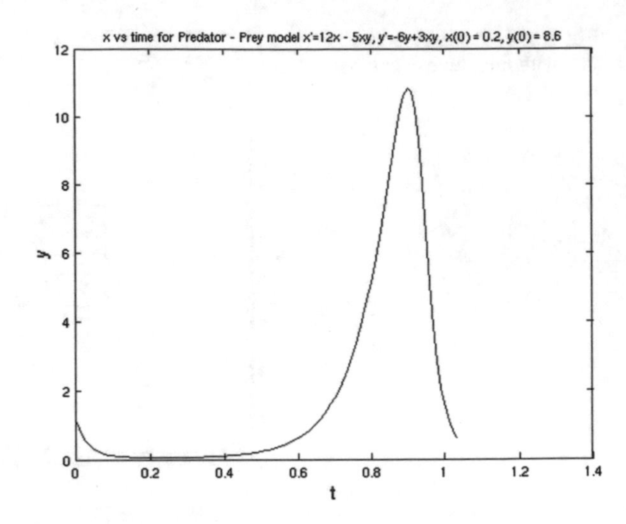

Listing 10.6: y versus time code

```
f = @(t,x) [12*x(1) - 5*x(1).*x(2);-6*x(2)+3*x(1).*x(2)];
T = 1.04;
h = .01;
% xbar = 6/3 = 2.0; ybar = 12/5 = 2.2
x0 = [1.2;8.6];
N = ceil(T/h);
[ht,rk] = FixedRK(f,0,x0,h,4,N);
Y = rk(2,:);
plot(ht,Y,'-k');
xlabel('t');
ylabel('y');
title(' y vs time for Predator - Prey model x''=12x - 5xy, y''=-6y+3xy, x(0) =
    0.2, y(0) = 8.6');
```

The y versus t plot is shown in Fig. 10.36.

10.11.3 Plotting Using a Function

Although it is nice to do things interactively, sometimes it gets a bit tedious. Let's rewrite this as a function.

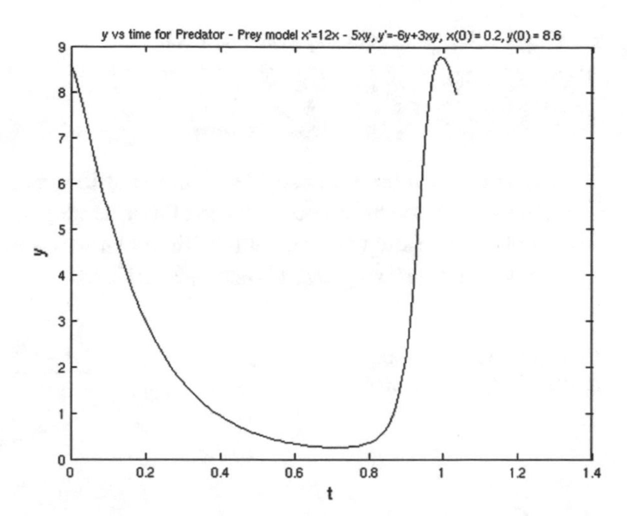

Fig. 10.36 y versus t for $x' = 12x - 5xy,\ y' = -6y + 3xy,\ x(0) = 0.2, y(0) = 8.6$

Listing 10.7: AutoSystemFixedRK.m

```
function AutoSystemFixedRK(fname, stepsize, tinit, tfinal, yinit, rkorder)
  % fname is the name of the model dynamics
  % stepsize is our step size choice
  % tinit is the initial time
  % tfinal is the final time
  % rkorder is the RK order
  % yinit is the initial data entered as [ number 1 ; number 2]
  n = ceil((tfinal-tinit)/stepsize);
  [htime,rk] = FixedRK(fname,tinit,yinit,stepsize,rkorder,n);
  X = rk(1,:);
  Y = rk(2,:);
  x1 = min(X)
  x2 = max(X);
  y1 = min(Y)
  y2 = max(Y);
  clf
  hold on
  plot([x1 x1],[y1 y2],'-b');
  plot([x2 x2],[y1 y2],'-g');
  plot([x1 x2],[y1 y1],'-r');
  plot([x1 x2],[y2 y2],'-m');
  plot(X,Y,'-k');
  xlabel('x');
  ylabel('y');
  title('Phase Plane Plot of y vs x');
  legend('x1','x2','y1','y2','y vs x','Location','Best');
  hold off
end
```

This is saved in the file **`AutoSystemFixedRK.m`** and we use it in MatLab like this:

Listing 10.8: Using AutoSystemFixedRK

```
f = @(t,x) [12*x(1) - 5*x(1).*x(2);-6*x(2)+3*x(1).*x(2)];
AutoSystemFixedRK(f,0.02,0,5,[2;4],4);
```

This is much more compact! We can use it to generate our graphs much faster. We can use this to show you how the choice of step size is crucial to generating a decent plot. We show what happens with too large a step size in Fig. 10.37 and what we see with a better step size choice in Fig. 10.38.

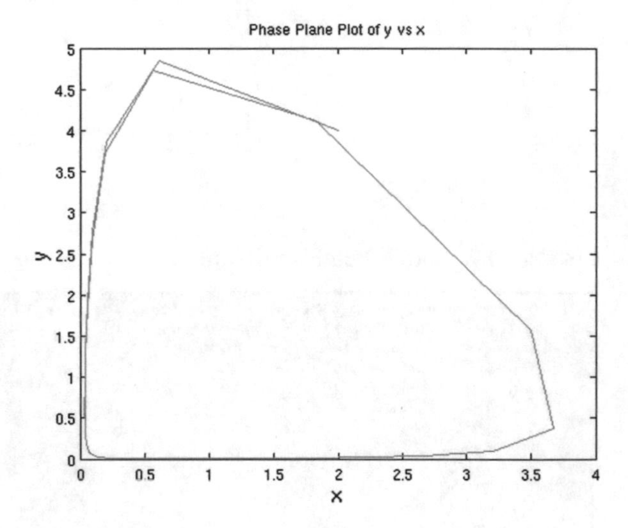

Fig. 10.37 Predator–prey plot with step size too large

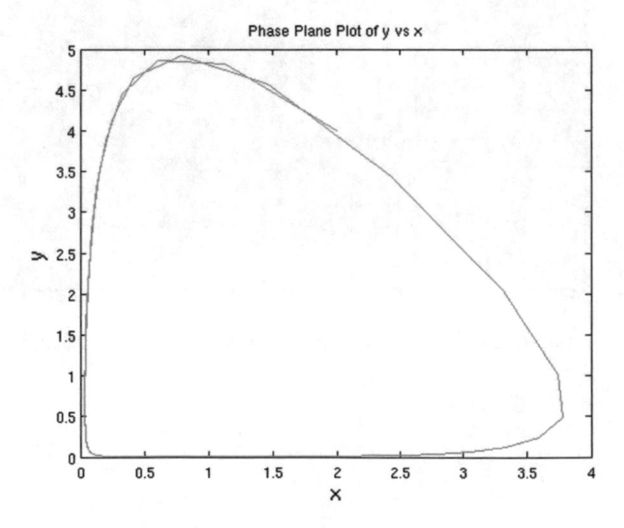

Fig. 10.38 Predator–prey plot with step size better!

10.11.4 Automated Phase Plane Plots

Next, we can generate a real phase plane portrait by automating the phase plane plots for a selection of initial conditions. This uses the code **AutoPhasePlanePlot.m** which we discussed in Sect. 9.4.

We generate a very nice phase plane plot as shown in Fig. 10.39 for the model

$$
\begin{aligned}
x'(t) &= 6\,x(t) - 5\,x(t)\,y(t) \\
y'(t) &= -7\,y(t) + 4\,x(t)\,y(t)
\end{aligned}
$$

for initial conditions from the box $[0.1, 4.5] \times [0.1, 4.5]$ using a fairly small step size of 0.2.

Listing 10.9: Automated Phase Plane Plot for $x' = 6x - 5xy$, $y' = -7y + 4xy$

```
f = @(t,x) [6*x(1)-5*x(1)*x(2);-7*x(2)+4*x(1)*x(2)];
AutoPhasePlanePlot('PredPrey',0.02,0.0,16.5,4,5,5,.1,4.5,.1,4.5);
```

10.11.5 Homework

Here are some problems on using MatLab!

Exercise 10.11.1

$$
\begin{aligned}
x'(t) &= 4\,x(t) - 7\,x(t)\,y(t) \\
y'(t) &= -9\,y(t) + 7\,x(t)\,y(t)
\end{aligned}
$$

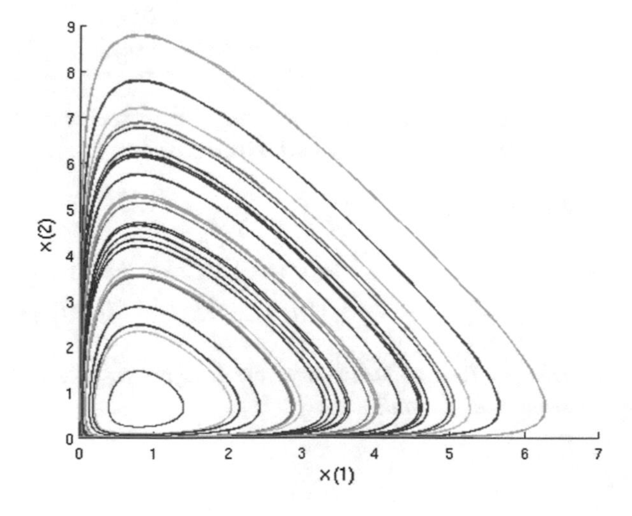

Fig. 10.39 Predator–prey plot for multiple initial conditions!

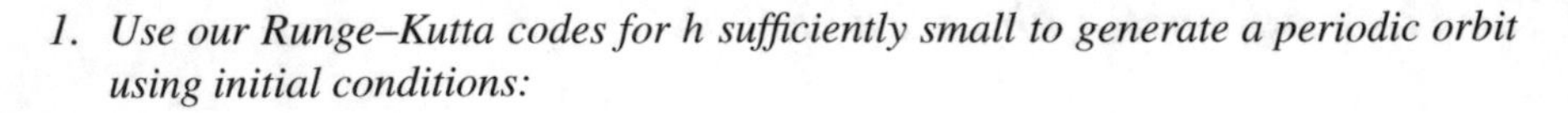

1. *Use our Runge–Kutta codes for h sufficiently small to generate a periodic orbit using initial conditions:*

   (a)

$$\begin{bmatrix} 2 \\ 1 \end{bmatrix}$$

   (b)

$$\begin{bmatrix} 5 \\ 2 \end{bmatrix}$$

2. *Use our MatLab codes to estimate the period in each case*
3. *Generate plots of the xand y trajectories.*

Exercise 10.11.2

$$\begin{aligned} x'(t) &= 90\,x(t) - 45\,x(t)\,y(t) \\ y'(t) &= -180\,y(t) + 20\,x(t)\,y(t) \end{aligned}$$

1. *Use our Runge–Kutta codes for h sufficiently small to generate a periodic orbit using initial conditions:*

   (a)

$$\begin{bmatrix} 4 \\ 12 \end{bmatrix}$$

   (b)

$$\begin{bmatrix} 5 \\ 20 \end{bmatrix}$$

2. *Use our MatLab codes to estimate the period in each case*
3. *Generate plots of the x and y trajectories.*

Exercise 10.11.3

$$\begin{aligned} x'(t) &= 10\,x(t) - 5\,x(t)\,y(t) \\ y'(t) &= -4\,y(t) + 20\,x(t)\,y(t) \end{aligned}$$

1. *Use our Runge–Kutta codes for h sufficiently small to generate a periodic orbit using initial conditions:*

(a)

$$\begin{bmatrix} 40 \\ 2 \end{bmatrix}$$

(b)

$$\begin{bmatrix} 5 \\ 25 \end{bmatrix}$$

2. *Use our MatLab codes to estimate the period in each case*
3. *Generate plots of the xand y trajectories.*

Exercise 10.11.4

$$\begin{aligned} x'(t) &= 7\,x(t) - 14\,x(t)\,y(t) \\ y'(t) &= -6\,y(t) + 3\,x(t)\,y(t) \end{aligned}$$

1. *Use our Runge–Kutta codes for h sufficiently small to generate a periodic orbit using initial conditions:*

   (a)

$$\begin{bmatrix} 7 \\ 12 \end{bmatrix}$$

   (b)

$$\begin{bmatrix} 0.2 \\ 2 \end{bmatrix}$$

2. *Use our MatLab codes to estimate the period in each case*
3. *Generate plots of the xand y trajectories.*

Exercise 10.11.5

$$\begin{aligned} x'(t) &= 8\,x(t) - 4\,x(t)\,y(t) \\ y'(t) &= -10\,y(t) + 2\,x(t)\,y(t) \end{aligned}$$

1. *Use our the Runge–Kutta codes for h sufficiently small to generate a periodic orbit using initial conditions:*

   (a)

$$\begin{bmatrix} 0.1 \\ 18 \end{bmatrix}$$

(b)

$$\begin{bmatrix} 6 \\ 0.1 \end{bmatrix}$$

2. *Use our MatLab codes to estimate the period in each case*
3. *Generate plots of the $x$and y trajectories.*

10.12 A Sketch of the Predator–Prey Solution Process

We now have quite a few tools for analyzing Predator–Prey models. Let's look at a sample problem. We can analyze by hand or with computational tools. Here is a sketch of the process on a sample problem.

1. For the system below, first do the work by hand. For the model

$$x'(t) = 56\,x(t) - 25\,x(t)\,y(t)$$
$$y'(t) = -207\,y(t) + 40\,x(t)\,y(t)$$

   (a) Derive the algebraic sign regions in Q1 determined by the $x' = 0$ and $y' = 0$ lines. Do the $x' = 0$ analysis first, then the $y' = 0$ analysis. Then assemble the results into one graph. Use colors. In each of the four regions you get, draw appropriate trajectory pieces and explain why at this point the full trajectory could be spiral in, spiral out or periodic.
   (b) Derive the nonlinear conservation law.
   (c) The nonlinear conservation law has the form $f(x)\,g(y) = f(x_0)\,g(y_0)$ for a very particular f and g. State what f and g are for the following problems and derive their properties. Give a nice graph of your results.
   (d) Derive the trajectories that start on the positive x or positive y axis. Draw nice pictures.
   (e) Prove the trajectories that start in Q1 with positive x_0 and y_0 must be bounded. Do this very nicely with lots of colors and explanation. Explain why this result tells us the trajectories can't spiral out or spiral in.
   (f) Draw typical trajectories for various initial conditions showing how the bounding boxes nest.

2. Now do the graphical work with MatLab. A typical generated phase plane plot for this initial condition would be the plot seen in Fig. 10.40. In this plot,we commented out the code that generates the bounding box.
3. Now plot the x versus time and y versus time for this Predator–Prey model. This generates the plots seen in Figs. 10.35 and 10.36.

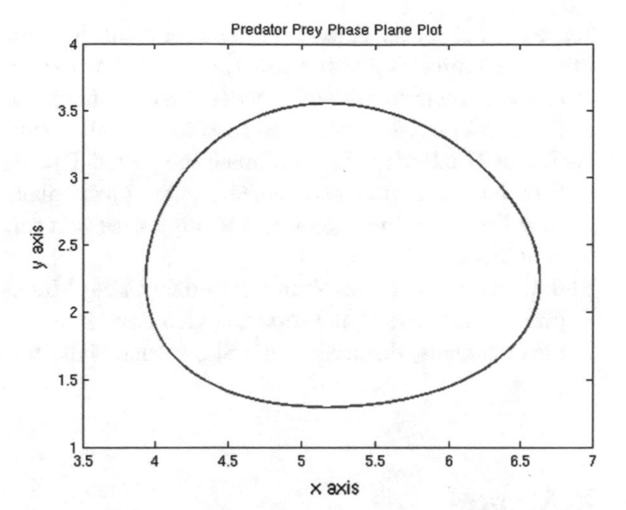

Fig. 10.40 Predator–prey phase plot

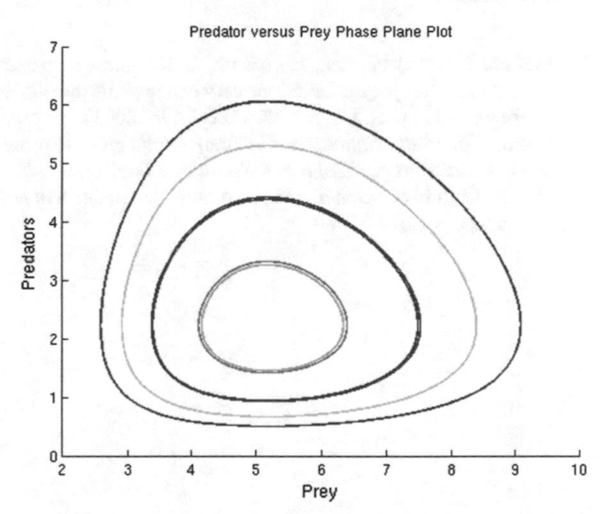

Fig. 10.41 Predator–prey phase plane plot

4. Now plot many trajectories at the same time. A typical session usually requires a lot of trial and error. The **`AutoPhasePlanePlot.m`** script is used by filling in values for the inputs it needs. We generate the plot seen in Fig. 10.41.

10.12.1 Project

Now your actual project. For the model

$$x'(t) = 10\,x(t) - 5\,x(t)\,y(t)$$
$$y'(t) = -40\,y(t) + 9\,x(t)\,y(t)$$

Solve the Model By Hand: Do this and attach to your project report.

Plot One Trajectory Using MatLab: Follow the outline above. This part of the report is done in a word processor with appropriate comments, discussion etc. Show your MatLab code and sessions as well as plots.

Estimate The Period T: Estimate the period T using the x versus time plot and then fine tune your estimate using the phase plane plot—keep increasing the final time until the trajectories touch for the first time. Pick an interesting initial condition, of course!

Plot Many Trajectories Simultaneously Using MatLab: Follow the outline above. This part of the report is also done in a word processor with appropriate comments, discussion etc. Show your MatLab code and sessions as well as plots.

References

B. Axelsen, T. Anker-Nilssen, P. Fossum, C. Kvamme, L. Nettestad, Pretty patterns but a simple strategy: predator–prey interactions between juvenile herring and Atlantic puffins observed with multibeam sonar. Can. J. Zool. **79**, 1586–1596 (2001)

M. Braun, *Differential Equations and Their Applications* (Springer, New York, 1978)

J. Peterson, *Calculus for Cognitive Scientists: Derivatives, Integration and Modeling*, Springer Series on Cognitive Science and Technology (Springer Science+Business Media Singapore Pte Ltd, Singapore, 2015 In press)

Chapter 11
Predator–Prey Models with Self Interaction

Many biologists of Volterra's time criticized his Predator–Prey model because it did not include self-interaction terms. These are terms that model how food fish interactions with other food fish and sharks interactions with other predators effect their populations. We can model these effects by assuming their magnitude is proportional to the interaction. Mathematically, we assume these are both *decay* terms giving us the *Predator–Prey Self Interaction* model

$$\begin{aligned} x'_{self} &= -\, e\, x\, x \\ y'_{self} &= -\, f\, y\, y. \end{aligned}$$

for positive constants e and f. We are thus led to the new self-interaction model given below:

$$\begin{aligned} x'(t) &= a\, x(t) - b\, x(t)\, y(t) - e\, x(t)^2 \\ y'(t) &= -c\, y(t) + d\, x(t)\, y(t) - f\, y(t)^2 \end{aligned}$$

The nullclines for the self-interaction model are a bit more complicated, but still straightforward to work with. First, we can factor the dynamics to obtain

$$\begin{aligned} x' &= x\, (a - b\, y - e\, x), \\ y' &= y\, (-c + d\, x - f\, y). \end{aligned}$$

11.1 The Nullcline Analysis

11.1.1 The $x' = 0$ Analysis

Looking at the predator–prey self interaction dynamics, equations, we see the (x, y) pairs in the x–y plane where

J.K. Peterson, *Calculus for Cognitive Scientists*,
Cognitive Science and Technology, DOI 10.1007/978-981-287-877-9_11

$$0 = x\left(a - b\,y - e\,x\right)$$

are the ones where the rate of change of the food fish will be zero. Now these pairs can correspond to many different time values so what we really need to do is to find all the (x, y) pairs where this happens. Since this is a product, there are two possibilities:

- $x = 0$; the y axis and
- $y = \frac{a}{b} - \frac{e}{b}\,x$.

11.1.2 The $y' = 0$ Analysis

In a similar way, the pairs (x, y) where y' becomes zero satisfy the equation

$$0 = y\left(-c + d\,x - f\,y\right).$$

Again, there are two possibilities:

- $y = 0$; the x axis and
- $y = -\frac{c}{f} + \frac{d}{f}\,x$.

11.1.3 The Nullcline Plane

Just like we did in Chap. 8, we find the parts of the x–y plane where the algebraic signs of x' and y' are $(+, +)$, $(+, -)$, $(-, +)$ and $(-, -)$. As usual, the set of (x, y) pairs where $x' = 0$ is called the **nullcline** for x; similarly, the points where $y' = 0$ is the **nullcline** for y. The $x' = 0$ equation gives us the y axis and the line $y = \frac{a}{b} - \frac{e}{b}x$ while the $y' = 0$ gives the x axis and the line $y = -\frac{c}{f} + \frac{d}{f}x$. The x' and y' nullclines thus divide the plane into the usual three pieces: the part where the derivative is positive, zero or negative. In Fig. 11.1, we show the part of the x–y plane where $x' > 0$ with one shading and the part where it is negative with another. In Fig. 11.2, we show how the y' nullcline divides the x–y plane into three pieces as well. For x', in each region of interest, we know the term x' has the two factors x and $a - by - ex$. The second factor is positive when

$$\begin{aligned} a - by - ex &> 0 \\ \frac{a}{b} - \frac{e}{b}x &> y \end{aligned}$$

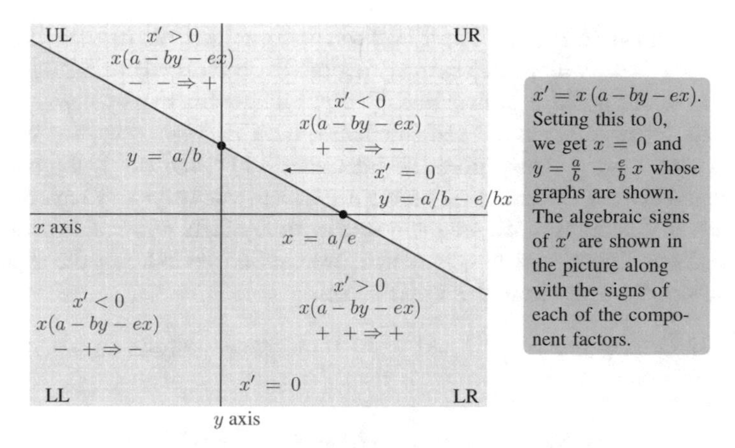

Fig. 11.1 Finding where $x' < 0$ and $x' > 0$ for the Predator–Prey self interaction model

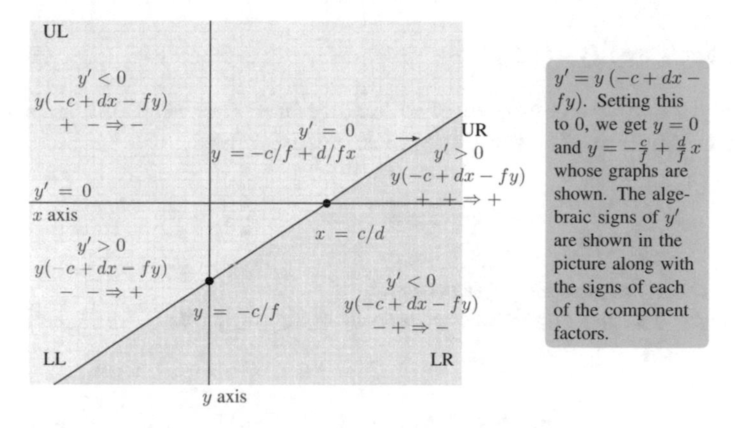

Fig. 11.2 Finding where $y' < 0$ and $y' > 0$ for the Predator–Prey self interaction model

So below the line, the factor is positive. In a similar way, the term y' has the two factors y and $-c + dx - fy$. Here the second factor is positive when

$$-c + dx - fy > 0$$
$$-\frac{c}{f} + \frac{d}{f}x > y$$

So below the line, the factor is positive. We then use this information to determine the algebraic signs in each region. In Fig. 11.1, we show these four regions (think of them as Upper Left (UL), Upper Right (UR), Lower Left (LL) and Lower Right (LR) for convenience) with the x' equation shown in each region along with the algebraic signs for each of the two factors. The y' signs are shown in Fig. 11.2.

The areas shown in Figs. 11.1 and 11.2 can be combined into one drawing. To do this, we divide the x–y plane into as many regions as needed and in each region, label x' and y' as either positive or negative. Hence, each region can be marked with an ordered pair, $(x'\pm,\ y'\pm)$. In this self-interaction case, there are three separate cases: the one where $\frac{c}{d} < \frac{a}{e}$ which gives an intersection in Quadrant 1, the one where $\frac{c}{d} = \frac{a}{e}$ which gives an intersection on the x axis and where $\frac{c}{d} > \frac{a}{e}$ which gives an intersection in Quadrant 4. We are interested in biologically reasonable solutions so if the initial conditions start in Quadrant 1, we would like to know the trajectories stay in Quadrant 1 away from the x and y axes.

Example 11.1.1 Do the $x' = 0$ and $y' = 0$ nullcline analysis separately for the model

$$x'(t) = 4\,x(t) - 5\,x(t)\,y(t) - e\,x(t)^2$$
$$y'(t) = -6\,y(t) + 2\,x(t)\,y(t) - f\,y(t)^2$$

Note we don't specify e and f.

Solution • *For x', in each region of interest, we know the term x' has the two factors x and $4 - 5y - ex$. The second factor is positive when*

$$4 - 5y - ex > 0 \Rightarrow \frac{4}{5} - \frac{e}{5}x > y$$

So below the line, the factor is positive.

- *the term y' has the two factors y and $-6 + 2x - fy$. Here the second factor is positive when*

$$-6 + 2x - fy > 0 \Rightarrow -\frac{6}{f} + \frac{2}{f}x > y$$

So below the line, the factor is positive.

- *The graphs for this solution is shown in Figs. 11.3 and 11.4.*

11.1.4 Homework

For these models, do the complete nullcline analysis for $x' = 0$ and $y' = 0$ separately with all details.

Exercise 11.1.1

$$x'(t) = 15\,x(t) - 2\,x(t)\,y(t) - 2\,x(t)^2$$
$$y'(t) = -8\,y(t) + 30\,x(t)\,y(t) - 3\,y(t)^2$$

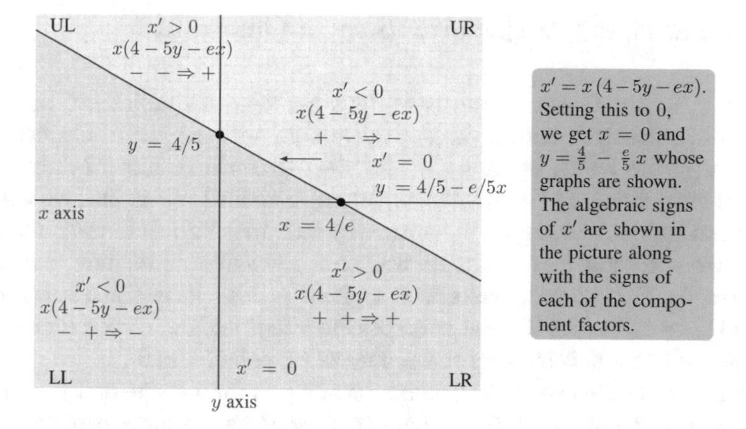

Fig. 11.3 x' nullcline for $x' = x\,(4 - 5y - ex)$

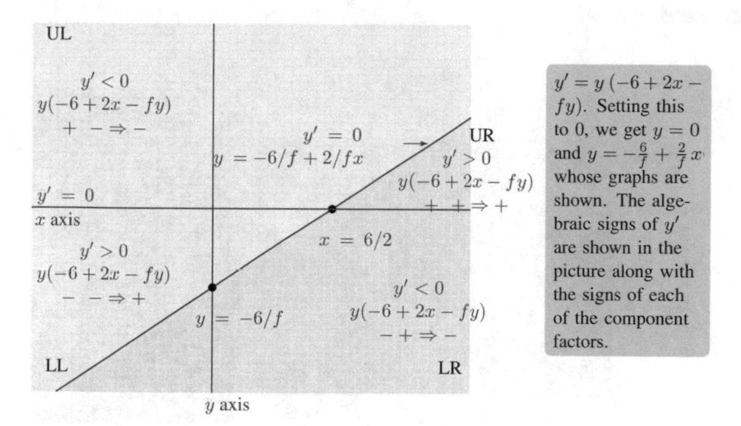

Fig. 11.4 y' nullcline for $y' = y\,(-6 + 2x - fy)$

Exercise 11.1.2

$$x'(t) = 6\,x(t) - 20\,x(t)\,y(t) - 10\,x(t)^2$$
$$y'(t) = -28\,y(t) + 30\,x(t)\,y(t) - 3\,y(t)^2$$

Exercise 11.1.3

$$x'(t) = 7\,x(t) - 6\,x(t)\,y(t) - 2\,x(t)^2$$
$$y'(t) = -10\,y(t) + 2\,x(t)\,y(t) - 3\,y(t)^2$$

Exercise 11.1.4

$$x'(t) = 8\,x(t) - 2\,x(t)\,y(t) - 2\,x(t)^2$$
$$y'(t) = -9\,y(t) + 3\,x(t)\,y(t) - 1.5\,y(t)^2$$

11.2 Quadrant 1 Trajectories Stay in Quadrant 1

To prepare for our Quadrant 1analysis, let's combine the nullclines, but only in Quadrant 1. First, let's redraw the derivative sign analysis just in Quadrant 1. In Fig. 11.5 we show the Quadrant 1 $x'+$ and $x'-$ regions in Quadrant 1 only. We will show that we only need to look at the model in Quadrant 1. To do this, we will show that the trajectories starting on the positive y axis move down towards the origin. Further, we will show trajectories starting on the positive x axis move towards the point $(a/e, 0)$. Then since trajectories can not cross this tells us that a trajectory that starts in Q1 with positive ICs can not cross the y axis and these trajectories can not cross the positive x axis but they can end up at the point $(a/e, 0)$.

In Fig. 11.6 we then show the Quadrant 1 analysis for the $y'+$ and $y'-$ regions.

We know different trajectories can not cross, so if we can show there are trajectories that stay on the x and y axes, we will know that trajectories starting in Quadrant 1 stay in Quadrant 1.

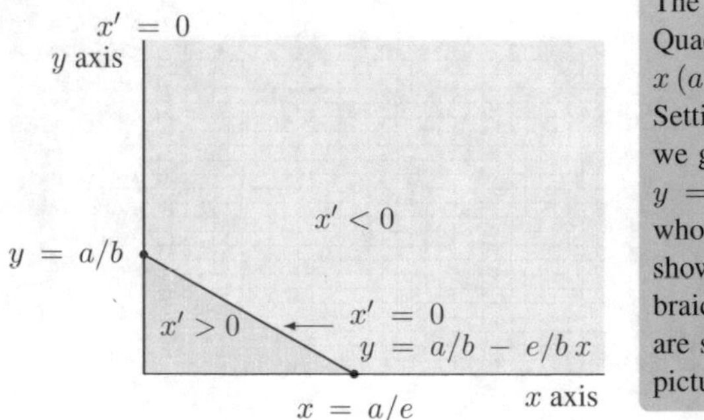

The x' equation In Quadrant 1 is $x' = x\,(a - b\,y - e\,x)$. Setting this to 0, we get $x = 0$ and $y = a/b - e/b\,x$ whose graphs are shown. The algebraic signs of x' are shown in the picture.

Fig. 11.5 The $x' < 0$ and $x' > 0$ signs in Quadrant 1 for the Predator–Prey self interaction model

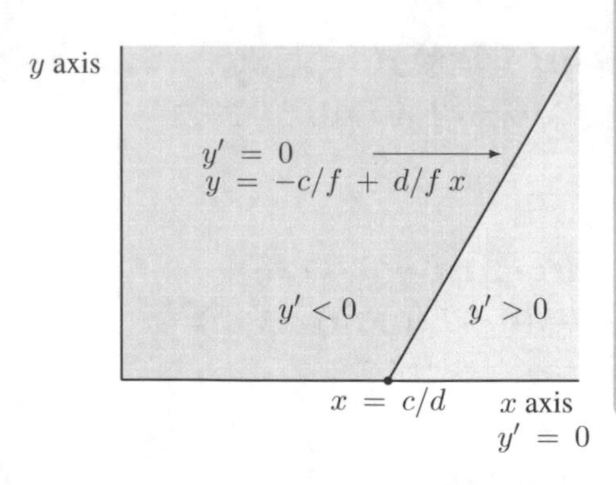

The y' equation in Quadrant 1 is $y' = y(-c + d\,x - f\,y)$. Setting this to 0, we get $y = 0$ and $y = -c/f + d/f\,x$ whose graphs are shown. The algebraic signs of y' are then shown in the picture.

Fig. 11.6 The $y' < 0$ and $y' > 0$ signs in Quadrant 1 for the Predator–Prey self interaction model

11.2.1 Trajectories Starting on the y Axis

If we start at an initial condition on the y axis, $x_0 = 0$ and $y_0 > 0$, the self-interaction model can be solved by choosing $x(t) = 0$ for all time t and solving the first order equation

$$y'(t) = y\,(-c - f\,y).$$

Let's look at the trajectories that start on the positive y axis for this model.

$$\begin{aligned} x'(t) &= 4\,x(t) - 5\,x(t)\,y(t) - e\,x(t)^2 \\ y'(t) &= -6\,y(t) + 2\,x(t)\,y(t) - f\,y(t)^2 \end{aligned}$$

- With $(x_0 = 0,\, y_0 > 0)$ then $x(t) = 0$ always and y satisfies

$$y'(t) = y\,(-6 - f\,y).$$

- Rewriting

$$\frac{y'}{y\,(6 + f\,y)} = -1.$$

- Integrating from 0 to t, we find

$$\int_0^t \frac{y'(s)}{y(s)\,(6 + f\,y(s))} = -\int_0^t ds = -t.$$

- Make the substitution $u = y(t)$

$$\int_{s=0}^{s=t} \frac{du}{u\,(c + f\,u)} = -t.$$

- This integration needs a partial fraction decomposition approach. We search for α and β so that

$$\frac{1}{u(6 + f\,u)} = \frac{\alpha}{u} + \frac{\beta}{6 + f\,u}.$$

- We want

$$\begin{aligned} \frac{1}{u\,(6 + f\,u)} &= \frac{\alpha}{u} + \frac{\beta}{6 + f\,u} \\ 1 &= \alpha\,(6 + f\,u) + \beta\,u. \end{aligned}$$

- Evaluating at $u = 0$ we get $\alpha = 1/6$ and when $u = -6/f$, $\beta = -f/6$.
- Complete the integration

$$\begin{aligned}\int_{s=0}^{s=t} \frac{1}{u(6+f\,u)}\,du &= \int_{s=0}^{s=t} \left(\frac{\alpha}{u} + \frac{\beta}{6+f\,u}\right) du \\ &= \frac{1}{6}\int_{s=0}^{s=t} \frac{1}{u}\,du - \frac{f}{6}\int_{s=0}^{s=t} \frac{1}{6+f\,u}\,du. \\ &= \frac{1}{6}\ln|u(s)|_0^t - \frac{1}{6}\ln|6+f\,u(s)|_0^t.\end{aligned}$$

- But $u = y(s)$ and here the variables are positive so absolute values are not needed.

$$\begin{aligned}\frac{1}{6}\ln|u(s)|_0^t - \frac{1}{6}\ln|6+f\,u(s)|_0^t = (1/6)\ln\left(\frac{y(t)}{y_0}\right) \\ -(1/6)\ln\left(\frac{6+fy(t)}{6+fy_0}\right)\end{aligned}$$

- Now simplify the ln.

$$\begin{aligned}&(1/6)\ln\left(\frac{y(t)}{y_0}\right) - (1/6)\ln\left(\frac{6+fy(t)}{6+fy_0}\right) \\ &= (1/6)\ln\left(\frac{y(t)}{6+fy(t)}\,\frac{6+fy_0}{y_0}\right)\end{aligned}$$

- The right hand side of the integration was $-t$, so combining

$$\ln\left(\frac{y(t)}{6+fy(t)}\,\frac{6+f\,y_0}{y_0}\right) = -6t.$$

- Exponentiate

$$\frac{y(t)}{6+fy(t)}\,\frac{6+f\,y_0}{y_0} = e^{-6t}.$$

- Solve for $y(t)$

$$\begin{aligned}y(t) &= (6+fy(t))\frac{y_0}{6+fy_0}e^{-6t} \\ \left(1 - \frac{6y_0}{6+fy_0}e^{-6t}\right)y(t) &= \frac{6y_0}{6+fy_0}e^{-6t} \\ y(t) &= \frac{\frac{6y_0}{6+fy_0}e^{-6t}}{1 - \frac{6y_0}{6+fy_0}e^{-6t}}\end{aligned}$$

- At $t \to \infty$, the numerator goes to 0 and the denominator goes to 1. So as $t \to \infty$, $y(t) \to 0$.
- So if the IC starts on the positive y axis, trajectory goes down to the origin.
- The argument is the same for other f values and other models.

We can do this argument in general using a generic c and d but you should get the idea form our example.

11.2.2 Trajectories Starting on the x Axis

Let's look at the trajectories that start on the positive x axis for the same model.

$$\begin{aligned} x'(t) &= 4\,x(t) - 5\,x(t)\,y(t) - e\,x(t)^2 \\ y'(t) &= -6\,y(t) + 2\,x(t)\,y(t) - f\,y(t)^2 \end{aligned}$$

- With $(x_0 > 0,\ y_0 = 0)$ then $y(t) = 0$ always and $x(t)$ satisfies

$$x'(t) = 4x - ex^2 = ex((4/e) - x).$$

- Hence, the x' equation is a logistics model with $L = 4/e$ and $\alpha = e$. So $x(t) \to 4/e$ as $t \to \infty$. If $x_0 > 4/e$, $x(t)$ goes down toward $4/e$ and if $x_0 < 4/e$, $x(t)$ goes up to $4/e$.
- This argument works for any e and any other model. So trajectories that start on the positive x axis move towards a/e.

We are now in a position to see what happens if we pick initial conditions in Quadrant 1.

11.2.3 Homework

Analyze the x and y positive axis trajectories as we have done in the above discussions.

Exercise 11.2.1

$$\begin{aligned} x'(t) &= 15\,x(t) - 20\,x(t)\,y(t) - 2\,x(t)^2 \\ y'(t) &= -20\,y(t) + 30\,x(t)\,y(t) - 3\,y(t)^2 \end{aligned}$$

Exercise 11.2.2

$$x'(t) = 5\,x(t) - 25\,x(t)\,y(t) - 10\,x(t)^2$$
$$y'(t) = -28\,y(t) + 30\,x(t)\,y(t) - 3\,y(t)^2$$

Exercise 11.2.3

$$x'(t) = 7\,x(t) - 7\,x(t)\,y(t) - 2\,x(t)^2$$
$$y'(t) = -10\,y(t) + 2\,x(t)\,y(t) - 3\,y(t)^2$$

Exercise 11.2.4

$$x'(t) = 8\,x(t) - 3\,x(t)\,y(t) - 3\,x(t)^2$$
$$y'(t) = -9\,y(t) + 4\,x(t)\,y(t) - 1.5\,y(t)^2$$

11.3 The Combined Nullcline Analysis in Quadrant 1

We now know that if a trajectory starts at $x_0 > 0$ and $y_0 > 0$, it can not cross the positive x or y axis. There are three cases to consider. If $\frac{c}{d} < \frac{a}{e}$, the trajectories will cross somewhere in Quadrant 1. This intersection point will play the same role as the average x and average y in the Predator–Prey model without self interaction. If $\frac{c}{d} = \frac{a}{e}$, the two nullclines intersect on the x axis at $\frac{a}{e}$. Finally, if $\frac{c}{d} > \frac{a}{e}$, the intersection occurs in Quadrant 4 which is not biologically reasonable and is not even accessible as the trajectory can not cross the x access. Let's work out the details of these possibilities.

11.3.1 The Case $\frac{c}{d} < \frac{a}{e}$

We now combine Figs. 11.5 and 11.6 to create the combined graph for the case of the intersection in Quadrant 1. We show this in Fig. 11.7. The two lines cross when

$$e\,x + b\,y = a$$
$$d\,x - f\,y = c$$

.

Solving using Cramer's rule, we find the intersection (x^*, y^*) to be

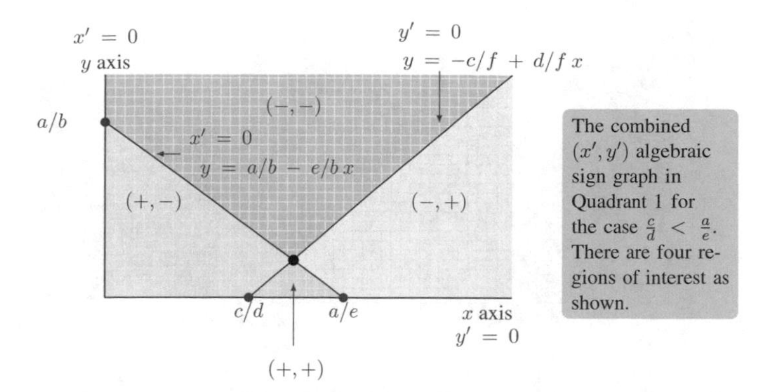

Fig. 11.7 The Quadrant 1 nullcline regions for the Predator–Prey self interaction model when $c/d < a/e$

$$x^* = \frac{det \begin{bmatrix} a & b \\ c & -f \end{bmatrix}}{det \begin{bmatrix} e & b \\ d & -f \end{bmatrix}} = \frac{a\,f + b\,c}{e\,f + b\,d}$$

$$y^* = \frac{det \begin{bmatrix} e & a \\ d & c \end{bmatrix}}{det \begin{bmatrix} e & b \\ d & -f \end{bmatrix}} = \frac{a\,d - e\,c}{e\,f + b\,d}.$$

In this case, we have $a/e > c/d$ or $a\,d - e\,c > 0$.

11.3.2 The Nullclines Touch on the X Axis

The second case is the one where the nullclines touch on the x axis. We show this situation in Fig. 11.8. This occurs when $c/d = a/e$.

11.3.3 The Nullclines Cross in Quadrant 4

The second case is the one where the nullclines do not cross in Quadrant 1. We show this situation in Fig. 11.9. This occurs when $c/d > a/e$. The two lines now cross at a negative y value, but since in this model also, trajectories that start in Quadrant 1

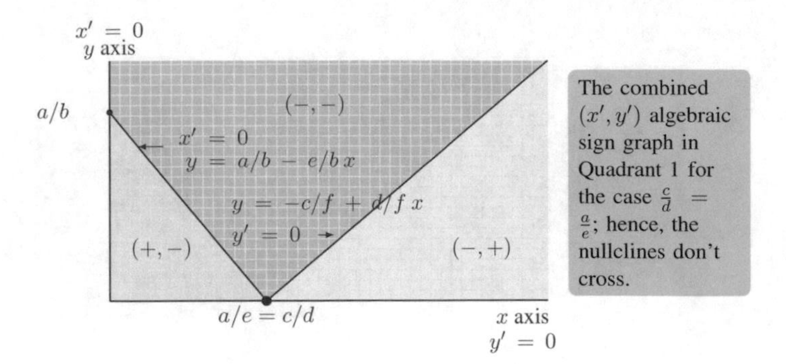

Fig. 11.8 The qualitative nullcline regions for the Predator–Prey self interaction model when $c/d = a/e$

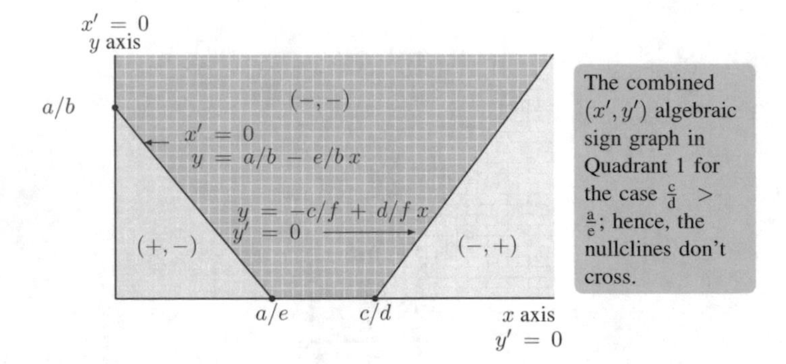

Fig. 11.9 The qualitative nullcline regions for the Predator–Prey self interaction model when $c/d > a/e$

can't cross the x or y axis, we only draw the situation in Quadrant 1. By Cramer's rule, the solution to

$$\begin{aligned} e\,x + b\,y &= a \\ d\,x - f\,y &= c \end{aligned}$$

is the pair (x^*, y^*) as shown below.

$$\begin{aligned} x^* &= \frac{a\,f + b\,c}{e\,f + b\,d} \\ y^* &= \frac{a\,d - e\,c}{e\,f + b\,d}. \end{aligned}$$

In this case, we have $a/e < c/d$ or $a\,d - e\,c < 0$ and so y^* is negative and not biologically interesting.

11.3.4 Example

Let's do the combined nullcline analysis for the model

$$x'(t) = 4\,x(t) - 5\,x(t)\,y(t) - e\,x(t)^2$$
$$y'(t) = -6\,y(t) + 2\,x(t)\,y(t) - f\,y(t)^2$$

Here, $c/d = 6/2 = 3$ and $a/e = 4/e$.

- For $c/d < a/e$, we need $3 < 4/e$ or $e < 4/3$.
- For $c/d = a/e$, we need $4 = 4/e$ or $e = 4/3$.
- For $c/d > a/e$, we need $4 > 4/e$ or $e > 4/3$.

We'll do this for various values of e so we can see all three cases.

- $e = 1$ so this is the $c/d < a/e$ case. Note the value of f not important. We show the result in Fig. 11.10.
- $e = 4/3$ so this is the $c/d = a/e$ case and is shown in Fig. 11.11.
- $e = 2$ so this is the $c/d > a/e$ case; see Fig. 11.12.

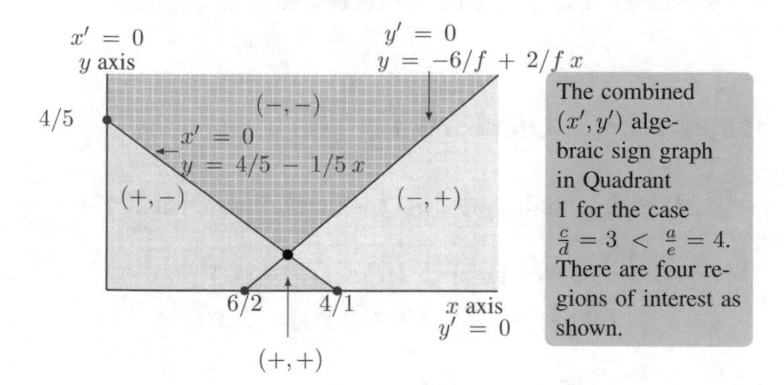

Fig. 11.10 The Quadrant 1 nullcline regions for the Predator–Prey self interaction model when $c/d < a/e$

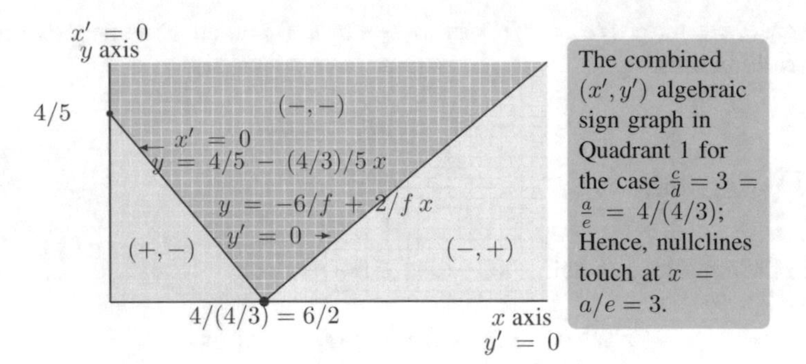

Fig. 11.11 The qualitative nullcline regions for the Predator–Prey self interaction model when $c/d = a/e$

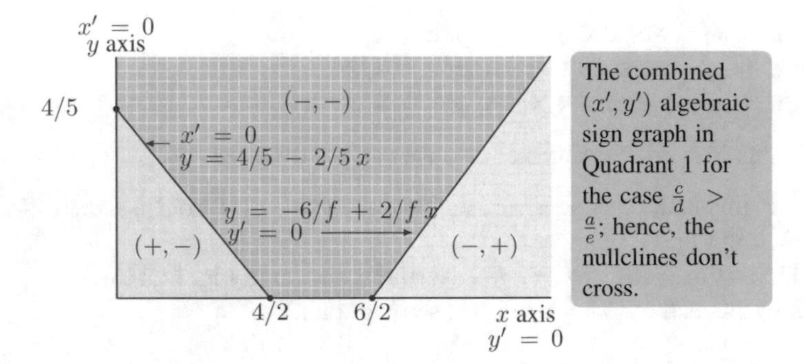

Fig. 11.12 The qualitative nullcline regions for the Predator–Prey self interaction model when $c/d > a/e$

11.4 Trajectories in Quadrant 1

We are now ready to draw trajectories in Quadrant I. For this model

$$x'(t) = 4\,x(t) - 5\,x(t)\,y(t) - e\,x(t)^2$$
$$y'(t) = -6\,y(t) + 2\,x(t)\,y(t) - f\,y(t)^2$$

we have $c/d = 6/2 = 3$ and $a/e = 4/e$.

- For $c/d < a/e$, we need $3 < 4/e$ or $e < 4/3$.
- For $c/d = a/e$, we need $4 = 4/e$ or $e = 4/3$.
- For $c/d > a/e$, we need $4 > 4/e$ or $e > 4/3$.

For the choice of $e = 1$, the nullclines cross in Quadrant I. Hence, the trajectories spiral in to the point where the nullclines cross. We find the trajectory shown in Fig. 11.13.

Next, for $e = 4/3$, the nullclines touch on the x axis and the trajectories move toward the point $(6/2, 0)$. We show this in Fig. 11.14.

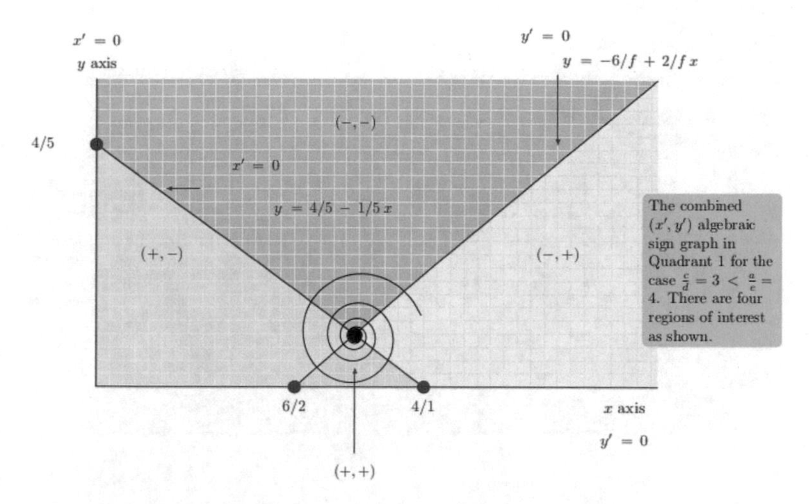

Fig. 11.13 Sample Predator–Prey model with self interaction: crossing in Q1

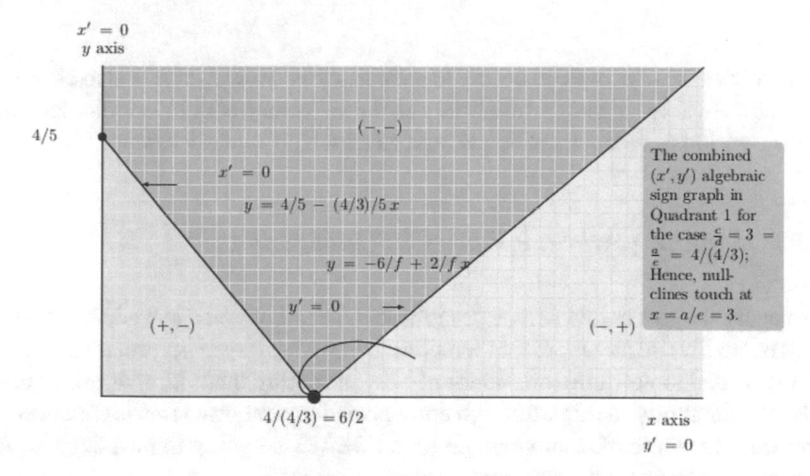

Fig. 11.14 Sample Predator–Prey model with self interaction: the nullclines cross on the x axis

Finally, for $e = 2$, the nullclines do not cross in Quadrant I and the trajectories move toward the point $(4/2, 0)$. We show the trajectories in Fig. 11.15.

11.5 Quadrant 1 Intersection Trajectories

We assume we start in Quadrant 1 in the region with (x', y') having algebraic signs $(-, -)$ or $(+, -)$. In these two regions, we know their corresponding trajectories can't cross the ones that start on the positive x or y axis. From the signs we see in Fig. 11.7, it is clear that trajectories must spiral into the point (x^*, y^*). So we don't

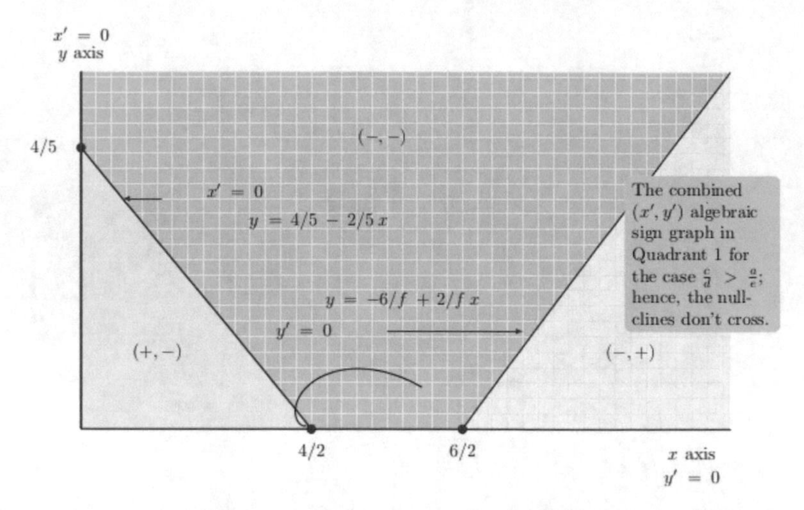

Fig. 11.15 Sample Predator–Prey model with self interaction: the nullclines cross in Quadrant 4

have to work as hard as we did before to establish this! However, it is also clear there is not a true average x and y value here as the trajectory is not periodic. However, there is a notion of an asymptotic average value which we now discuss.

11.5.1 Limiting Average Values in Quadrant 1

Now the discussions below will be complicated, but all of you can wade through it as it does not really use any more mathematics than we have seen before. It is, however, very messy and looks quite intimidating! Still, mastering these kind of things brings rewards: your ability to think through complicated logical problems is enhanced! So grab a cup of tea or coffee and let's go for a ride. We are going to introduce the idea of **limiting average** x and y **values**.

We know at any time t, the solutions $x(t)$ and $y(t)$ must be positive. Rewrite the model as follows:

$$\begin{aligned}\frac{x'}{x} + e\,x &= a - b\,y\\ \frac{y'}{y} + f\,y &= -c + d\,x.\end{aligned}$$

Now integrate from $s = 0$ to $s = t$ to obtain

$$\begin{aligned}\int_0^t \frac{x'(s)}{x(s)}\,ds + e\int_0^t x(s)\,ds &= a\,t - b\int_0^t y(s)\,ds\\ \int_0^t \frac{y'(s)}{y(s)}\,ds + f\int_0^t y(s)\,ds &= -c\,t + d\int_0^t x(s)\,ds.\end{aligned}$$

We obtain

$$\ln\left(\frac{x(t)}{x_0}\right) + e \int_0^t x(s)\, ds = a\, t - b \int_0^t y(s)\, ds$$
$$\ln\left(\frac{y(T)}{y_0}\right) + f \int_0^t y(s)\, ds = -c\, t + d \int_0^t x(s)\, ds.$$

Now the solution x and y are continuous, so the integrals

$$X(t) = \int_0^t x(s)\, ds$$
$$Y(t) = \int_0^t y(s)\, ds$$

are also continuous by the Fundamental Theorem of Calculus. Using the new variables X and Y, we can rewrite these integrations as

$$\ln\left(\frac{x(t)}{x_0}\right) = a\, t - e\, X(t) - b\, Y(t)$$
$$\ln\left(\frac{y(t)}{y_0}\right) = -c\, t + d\, X(t) - f\, Y(t).$$

Hence,

$$d\, \ln\left(\frac{x(t)}{x_0}\right) = a\, d\, t - e\, d\, X(t) - b\, d\, Y(t)$$
$$e\, \ln\left(\frac{y(t)}{y_0}\right) = -c\, e\, t + d\, e\, X(t) - e\, f\, Y(t).$$

Now add the bottom and top equation to get

$$\ln\left(\left(\frac{x(t)}{x_0}\right)^d \left(\frac{y(t)}{y_0}\right)^e\right) = \left(a\, d - c\, e\right)t - \left(e\, f + b\, d\right)Y(t).$$

Now divide through by t to get

$$\frac{1}{t}\, \ln\left(\left(\frac{x(t)}{x_0}\right)^d \left(\frac{y(t)}{y_0}\right)^e\right) = \left(a\, d - c\, e\right) - \left(e\, f + b\, d\right) \frac{1}{t}\, Y(t). \tag{11.1}$$

From Fig. 11.7, it is easy to see that no matter what (x_0, y_0) we choose in Quadrant 1, the trajectories are bounded and so there is a positive constant we will call B so that

$$\left|\ln\left(\left(\frac{x(t)}{x_0}\right)^d \left(\frac{y(t)}{y_0}\right)^e\right)\right| \leq B.$$

Thus, for all t, we have

$$\frac{1}{t}\left|\ln\left(\left(\frac{x(t)}{x_0}\right)^d\left(\frac{y(t)}{y_0}\right)^e\right)\right| \leq \frac{B}{t}.$$

Hence, if we let t grow larger and larger, B/t gets smaller and smaller, and in fact

$$\begin{aligned}\lim_{t\to\infty}\frac{1}{t}\left|\ln\left(\left(\frac{x(t)}{x_0}\right)^d\left(\frac{y(t)}{y_0}\right)^e\right)\right| &\leq \lim_{t\to\infty}\frac{B}{t}\\ &= 0.\end{aligned}$$

But the left hand side is always non-negative also, so we have

$$0 \leq \lim_{t\to\infty}\frac{1}{t}\left|\ln\left(\left(\frac{x(t)}{x_0}\right)^d\left(\frac{y(t)}{y_0}\right)^e\right)\right| \leq 0,$$

which tells us that

$$\lim_{t\to\infty}\frac{1}{t}\left|\ln\left(\left(\frac{x(t)}{x_0}\right)^d\left(\frac{y(t)}{y_0}\right)^e\right)\right| = 0.$$

Finally, the above also implies

$$\lim_{t\to\infty}\frac{1}{t}\ln\left(\left(\frac{x(t)}{x_0}\right)^d\left(\frac{y(t)}{y_0}\right)^e\right) = 0.$$

Now let t go to infinity in Eq. 11.1 to get

$$\lim_{t\to\infty}\frac{1}{t}\ln\left(\left(\frac{x(t)}{x_0}\right)^d\left(\frac{y(t)}{y_0}\right)^e\right) = \left(a\,d - c\,e\right) - \left(e\,f + b\,d\right)\lim_{t\to\infty}\frac{1}{t}Y(t).$$

The term $Y(t)/t$ is actually $(1/t)\int_0^t y(s)\,ds$ which is the average of the solution y on the interval $[0, t]$. It therefore follows that

$$0 = \lim_{t\to\infty}\frac{1}{t}\int_0^t y(s)\,ds = \frac{a\,d - c\,e}{e\,f + b\,d}.$$

But the term on the right hand side is exactly the y coordinate of the intersection of the nullclines, y^*. We conclude the **limiting average value** of the solution y is given by

$$\lim_{t\to\infty}\frac{1}{t}\int_0^t y(s)\,ds = y^*. \tag{11.2}$$

We can do a similar analysis (although there are differences in approach) to shown that the **limiting average value** of the solution x is given by

$$\lim_{t\to\infty} \frac{1}{t} \int_0^t x(s)\, ds = x^*. \tag{11.3}$$

These two results are similar to what we saw in the Predator–Prey model without self-interaction. Of course, we only had to consider the averages over the period before, whereas in the self-interaction case, we must integrate over all time. It is instructive to compare these results:

Model	Average x	Average y
No Self-Interaction	c/d	a/b
Self-Interaction	$x^* = \frac{af+bc}{ef+bd}$	$y^* = \frac{ad-ec}{ef+bd}$

11.6 Quadrant 4 Intersection Trajectories

Look back at the signs we see in Fig. 11.9. It is clear that trajectories that start to the left of c/d go up and to the left until they enter the $(-,-)$ region. The analysis we did for trajectories starting on the x axis or y axis in the crossing nullclines case are still appropriate. So we know one a trajectory is in the $(-,-)$ region, it can't hit the x axis except at a/e. Similarly, a trajectory that starts in $(+,-)$ moves right and down towards the x axis, but can't hit the x axis except at a/e. We can look at the details of the $(+,-)$ trajectories by reusing the material we figured out in the limiting averages discussion. Since this trajectory is bounded, as t grows arbitrarily large, the $x(t)$ and $y(t)$ values must approach fixed values. We will call these asymptotic x and y values x^∞ and y^∞ for convenience. The trajectory must satisfy

$$\frac{1}{t} \ln\left(\left(\frac{x(t)}{x_0}\right)^d \left(\frac{y(t)}{y_0}\right)^e\right) = \left(a\,d - c\,e\right) - \left(e\,f + b\,d\right) \frac{1}{t}\, Y(t). \tag{11.4}$$

with the big difference that the term $a\,d - c\,e$ is now negative. Exponentiate to obtain

$$\left(\frac{x(t)}{x_0}\right)^d \left(\frac{y(t)}{y_0}\right)^e = e^{(a\,d-c\,e)\,t} e^{-(e\,f+b\,d)Y(t)}. \tag{11.5}$$

Now note

- The term $e^{-(e\,f+b\,d)Y(t)}$ is bounded by 1.
- The term $e^{(a\,d-c\,e)\,t}$ goes to zero as t gets large because $a\,d - c\,e$ is negative.

Hence, as t increases to infinity, we find

$$\lim_{t\to\infty} \left(\frac{x(t)}{x_0}\right)^d \left(\frac{y(t)}{y_0}\right)^e = \left(\frac{x^\infty}{x_0}\right)^d \left(\frac{y^\infty}{y_0}\right)^e = 0.$$

Since it is easy to see that x^∞ is positive, we must have $y^\infty = 0$. We now know a trajectory starting in Region III hits the x axis as t goes to infinity. Now suppose $x^\infty < \frac{a}{e}$. Then, there is also a trajectory on the positive x axis that starts at x^∞ and moves towards $\frac{a}{e}$. Looking *backwards* from the point $(x^\infty, 0)$, we thus see two trajectories springing out of that point. The first, our Region III trajectory, and the second, our positive x axis trajectory. This is not possible. So x^∞ must be $\frac{a}{e}$.

Now is this model biologically plausible? It seems not! It doesn't seem reasonable for the predator population to shrink to 0 while the food population converges to some positive number x^∞! So adding more biological detail actually leads to a loss of biologically reasonable predictive power; food for thought!

The discussion for the middle case where the nullclines intersect on the x axis is essentially the same so we won't go over it again.

11.6.1 Homework

Draw suitable trajectories for the following Predator–Prey models with self interaction in great detail.

Exercise 11.6.1

$$\begin{aligned} x'(t) &= 15\,x(t) - 2\,x(t)\,y(t) - 2\,x(t)^2 \\ y'(t) &= -8\,y(t) + 30\,x(t)\,y(t) - 3\,y(t)^2 \end{aligned}$$

Exercise 11.6.2

$$\begin{aligned} x'(t) &= 6\,x(t) - 20\,x(t)\,y(t) - 10\,x(t)^2 \\ y'(t) &= -28\,y(t) + 30\,x(t)\,y(t) - 3\,y(t)^2 \end{aligned}$$

Exercise 11.6.3

$$\begin{aligned} x'(t) &= 7\,x(t) - 6\,x(t)\,y(t) - 2\,x(t)^2 \\ y'(t) &= -10\,y(t) + 3\,x(t)\,y(t) - 3\,y(t)^2 \end{aligned}$$

Exercise 11.6.4

$$\begin{aligned} x'(t) &= 8\,x(t) - 2\,x(t)\,y(t) - 2\,x(t)^2 \\ y'(t) &= -9\,y(t) + 3\,x(t)\,y(t) - 1.5\,y(t)^2 \end{aligned}$$

Exercise 11.6.5

$$\begin{aligned} x'(t) &= 6\,x(t) - 2\,x(t)\,y(t) - 2\,x(t)^2 \\ y'(t) &= -9\,y(t) + 5\,x(t)\,y(t) - 3\,y(t)^2 \end{aligned}$$

11.7 Summary: Working Out a Predator–Prey Self-Interaction Model in Detail

We can now summarize how you would completely solve a typical Predator–Prey self-interaction model problem from first principles. These are the steps you need to do:

1. Draw the nullclines reasonably carefully in multiple colors to make your teacher happy. Once you know what the nullclines do, you can solve these problems completely.
2. Determine if the nullclines cross as this makes a big difference in the kind of trajectories we will see. Find the place where the nullclines cross if they do.
3. Once you know what the nullclines do, you can solve these problems completely. Draw a few trajectories in each of the regions determined by the nullclines.
4. From our work in Sect. 11.5, we know the solutions to the Predator–Prey model with self-interaction having initial conditions in Quadrant 1 are always positive. You can then use this fact to derive the amazingly true statement that a solution pair, $x(t)$ and $y(t)$, satisfies

$$\ln\left(\left(\frac{x(t)}{x_0}\right)^d \left(\frac{y(t)}{y_0}\right)^e\right) = \left(a\,d - c\,e\right)t - \left(e\,f + b\,d\right) \int_0^t y(s)\,ds.$$

This derivation is the same whether the nullclines cross or not!

11.8 Self Interaction Numerically

From our theoretical investigations, we know if the ratio c/d exceeds the ratio a/e, the solutions should approach the ratio a/e as time gets large. Let's see if we get that result numerically.

Let's try this problem,

$$x'(t) = 2\,x(t) - 3\,x(t)\,y(t) - 3\,x(t)^2$$
$$y'(t) = -4\,y(t) + 5\,x(t)\,y(t) - 3\,y(t)^2$$

We can generate a full phase plane as follows

Listing 11.1: Phase Plane for $x' = 2x - 3xy - 3x^2$, $y' = -4x + 5xy - 3y^2$

```
f = @(t,x) [2*x(1)-3*x(1).*x(2)-3*x(1).^2;-4*x(2)+5*x(1).*x(2)-3*x(2)^2];
AutoPhasePlanePlot(f,.01,0,3,4,3,3,.1,4,.1,4);
```

We generate the plot as shown in Fig. 11.16. Note that here $c/d = 4/5$ and $a/e = 2/3$ so $c/d > a/e$ which tells us the nullcline intersection is in Quadrant 4. Hence, all trajectories should go toward $a/e = 2/3$ on the x-axis.

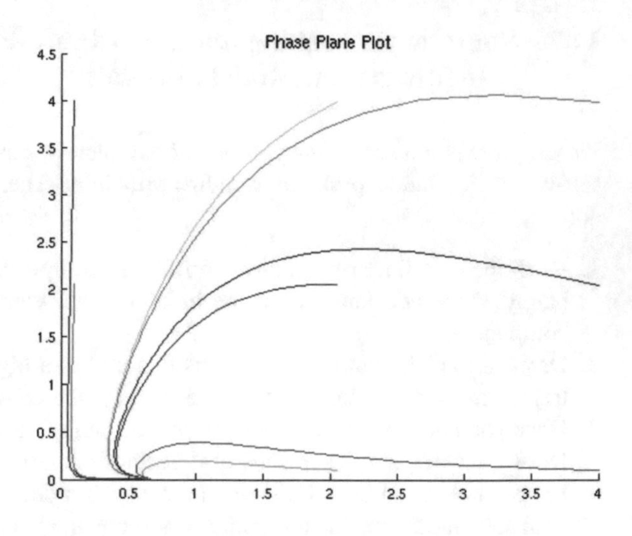

Fig. 11.16 Predator–Prey system: $x' = 2x - 3xy - 3x^2$, $y' = -4y + 5xy - 3y^2$

Now let's look at what happens when the nullclines cross. We now use the model

$$x'(t) = 2\,x(t) - 3\,x(t)\,y(t) - 1.5\,x(t)^2$$
$$y'(t) = -4\,y(t) + 5\,x(t)\,y(t) - 1.5\,y(t)^2$$

The Matlab session is now

Listing 11.2: Phase Plane for $x' = 2x - 3xy - 1.5x^2$, $y' = -4x + 5xy - 1.5y^2$

```
f = @(t,x) [2*x(1)-3*x(1).*x(2)-1.5*x(1).^2;-4*x(2)+5*x(1).*x(2)-1.5*x(2)^2];
AutoPhasePlanePlot(f,.01,0,3,4,3,3,.1,4,.1,4);
```

Since $a/e = 2/1.5$ and $c/d = 4/5$, the nullclines cross in Quadrant 1 and the trajectories should converge to

$$x^* = \frac{af + bc}{ef + bd} = \frac{2(1.5) + 3(4)}{1.5(1.5) + 3(5)} = \frac{15}{17.25} = 0.87$$
$$y^* = \frac{ad - ec}{ef + bd} = \frac{2(5) - 1.5(4)}{17.25} = \frac{4}{17.25} = 0.23$$

which is what we see in the plot shown in Fig. 11.17.

11.8.1 Homework

Generate phase plane plots for the following models.

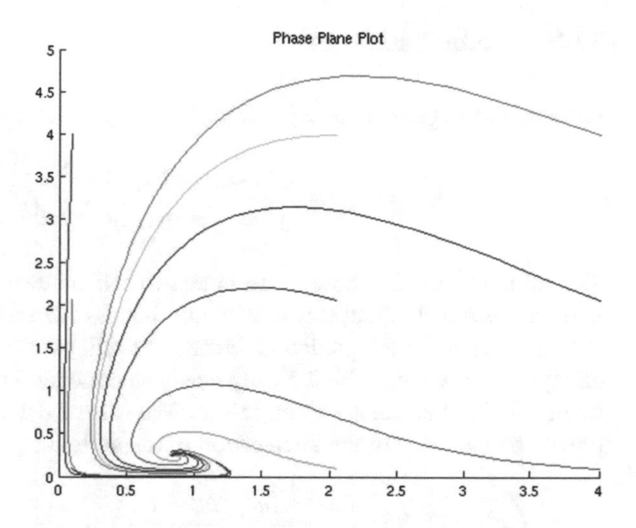

Fig. 11.17 Predator–Prey system: $x' = 2x - 3xy - 1.5x^2$, $y' = -4y + 5xy - 1.5y^2$ in this example, the nullclines cross so the trajectories moves towards a fixed point (0.23, 0.87) as shown

Exercise 11.8.1

$$x'(t) = 8\,x(t) - 4\,x(t)\,y(t) - 2\,x(t)^2$$
$$y'(t) = -10\,y(t) + 2\,x(t)\,y(t) - 1.5\,y(t)^2$$

Exercise 11.8.2

$$x'(t) = 12\,x(t) - 4\,x(t)\,y(t) - 2\,x(t)^2$$
$$y'(t) = -10\,y(t) + 2\,x(t)\,y(t) - 1.5\,y(t)^2$$

Exercise 11.8.3

$$x'(t) = 6\,x(t) - 4\,x(t)\,y(t) - 2\,x(t)^2$$
$$y'(t) = -3\,y(t) + 2\,x(t)\,y(t) - 1.5\,y(t)^2$$

Exercise 11.8.4

$$x'(t) = 6\,x(t) - 4\,x(t)\,y(t) - 2\,x(t)^2$$
$$y'(t) = -12\,y(t) + 2\,x(t)\,y(t) - 1.5\,y(t)^2$$

Exercise 11.8.5

$$x'(t) = 15\,x(t) - 4\,x(t)\,y(t) - 2\,x(t)^2$$
$$y'(t) = -10\,y(t) + 2\,x(t)\,y(t) - 1.5\,y(t)^2$$

11.9 Adding Fishing!

Let's do this in general. First, let's look at the general model

$$x' = ax - bxy - ex^2$$
$$y' = -cy + dxy - fy^2$$

The term fy^2 models how much is lost to self interaction between the predators. It seems reasonable that this loss should be less than the amount of food fish that are being eaten by the predators. Hence, we will assume in this model that $f < b$ always so that we get a biologically reasonable model. Then note adding fishing can be handled in the same way as before. The role of the average values will now be played by the value of the intersection of the nullclines. We have

$$\begin{aligned}
\left(x^*_{no}, \ y^*_{no}\right) &= \left(\frac{af + bc}{ef + bd}, \ \frac{ad - ec}{ef + bd}\right) \\
\left(x^*_{r}, \ y^*_{r}\right) &= \left(\frac{(a-r)f + b(c+r)}{ef + bd}, \ \frac{(a-r)d - e(c+r)}{ef + bd}\right) \\
&= \left(\frac{af + bc}{ef + bd}, \ \frac{ad - ec}{ef + bd}\right) + \left(\frac{-rf + br}{ef + bd}, \ \frac{-rd - er)}{ef + bd}\right) \\
&= \left(x^*_{no}, \ y^*_{no}\right) + r\left(\frac{b - f}{ef + bd}, \ -\frac{d + e}{ef + bd}\right) \\
\left(x^*_{r/2}, \ y^*_{r/2}\right) &= \left(x^*_{no}, \ y^*_{no}\right) + \frac{r}{2}\left(\frac{b - f}{ef + bd}, \ -\frac{d + e}{ef + bd}\right)
\end{aligned}$$

Now compare.

$$\begin{aligned}
x^*_{r/2} &= x^*_{no} + \frac{r}{2}\frac{b-f}{ef+bd} \\
&= x^*_{no} + r\frac{b-f}{ef+bd} - \frac{r}{2}\frac{b-f}{ef+bd} \\
&= x^*_{r} - \frac{r}{2}\frac{b-f}{ef+bd}
\end{aligned}$$

which shows us $x^*_{r/2}$ goes down with the reduction in fishing as $b > f$. Similarly,

$$\begin{aligned}
y^*_{r/2} &= y^*_{no} - \frac{r}{2}\frac{d+e}{ef+bd} \\
&= y^*_{no} - r\frac{b-f}{ef+bd} + \frac{r}{2}\frac{b-f}{ef+bd} \\
&= y^*_{r} + \frac{r}{2}\frac{b-f}{ef+bd}
\end{aligned}$$

which shows us $y^*_{r/2}$ goes up with the reduction in fishing as $b > f$. This is the same behavior we saw in the original Predator–Prey model without self-interaction.

11.10 Learned Lessons

In our study of the Predator–Prey model, we have seen the model without self-interaction was very successful at giving us insight into the fishing catch data during World War I in the Mediterranean sea. This was despite the gross nature of the model. No attempt was made to separate the food fish category into multiple classes of food fish; no attempt made to break down the predatory category into various types of predators. Yet, the modeling was ultimately successful as it provided illumination into a biological puzzle. However, the original model lacked the capacity for self-interaction and so it seemed plausible to add this feature. The self-interaction terms we use in this chapter seemed quite reasonable, but our analysis has shown it leads to completely wrong biological consequences. This tells the way we model self-interaction is wrong. The self-interaction model, in general, would be

$$
\begin{aligned}
x' &= a\,x - b\,x\,y - e\,u(x, y)\\
y' &= -c\,y + d\,x\,y - f\,v(x, y)
\end{aligned}
$$

where $u(x, y)$ and $v(x, y)$ are functions of both x and y that determine the self-interaction. To analyze this new model, we would proceed as before. We determine the nullclines

$$
\begin{aligned}
0 = x' &= a\,x - b\,x\,y - e\,u(x, y)\\
0 = y' &= -c\,y + d\,x\,y - f\,v(x, y)
\end{aligned}
$$

and begin our investigations. We will have to decide if the model generates trajectories that remain in Quadrant 1 if they start in Quadrant 1 so that they are biologically reasonable. This will require a lot of hard work!

Note, we can simply compute solutions using MatLab or some other tool. We will not know what the true solution is or even any ideas as to what general appearance it might have. You should be able to see that a balanced blend of mathematical analysis, computational study using a tool and intuition from the underlying science must be used together to solve the problems.

Chapter 12
Disease Models

We will now build a simple model of an infectious disease called the SIR model. Assume the total population we are studying is fixed at N individuals. This population is then divided into three separate pieces: we have individuals

- that are susceptible to becoming infected are called **Susceptible** and are labeled by the variable S. Hence, $S(t)$ is the number that are capable of becoming infected at time t.
- that can infect others. They are called **Infectious** and the number that are infectious at time t is given by $I(t)$.
- that have been removed from the general population. These are called **Removed** and their number at time t is labeled by $R(t)$.

We make a number of key assumptions about how these population pools interact.

- Individuals stop being *infectious* at a positive rate γ which is proportional to the number of individuals that are in the *infectious* pool. If an individual stops being infectious, this means this individual has been *removed* from the population. This could mean they have died, the infection has progressed to the point where they can no longer pass the infection on to others or they have been put into quarantine in a hospital so that further interactions with the general population is not possible. In all of these cases, these individuals are not infectious or can't cause infections and so they have been *removed* from the part of the population N which can be infected or is susceptible. Mathematically, this means we assume

$$I'_{loss} = -\gamma \, I.$$

- Susceptible individuals are those capable of catching an infection. We model the interaction of infectious and susceptible individuals in the same way we handled the interaction of food fish and predator fish in the Predator–Prey model. We assume this interaction is proportional to the product of their population sizes: i.e. SI. We assume the rate of change of **Infectious** is proportional to this interaction with positive proportionality constant r. Hence, mathematically, we assume

J.K. Peterson, *Calculus for Cognitive Scientists*,
Cognitive Science and Technology, DOI 10.1007/978-981-287-877-9_12

$$I'_{gain} = r\ S\ I.$$

We can then figure out the net rates of change of the three populations. The infectious populations gains at the rate $r\ S\ I$ and loses at the rate $\gamma\ I$. Hence, the net gain is $I'_{gain} + I'_{loss}$ or

$$I' = r\ S\ I - \gamma\ I.$$

The net change of Susceptible's is that of simple decay. Susceptibles are lost at the rate $-r\ S\ I$. Thus, we have

$$S' = -\ r\ S\ I.$$

Finally, the removed population increases at the same rate the infectious population decreases. We have

$$R' = \gamma\ I.$$

We also know that $R(t) + S(t) + I(t) = N$ for all time t because our population is constant. So only two of the three variables here are independent. We will focus on the variables I and S from now on. Our complete *Infectious Disease Model* is then

$$I' = r\ S\ I - \gamma\ I \tag{12.1}$$
$$S' = -\ r\ S\ I \tag{12.2}$$
$$I(0) = I_0 \tag{12.3}$$
$$S(0) = S_0. \tag{12.4}$$

where we can compute $R(t)$ as $N - I(t) - S(t)$.

12.1 Disease Model Nullclines

When we set I' and S' to zero, we obtain the usual nullcline equations.

$$I' = 0 = I\ (r\ S - \gamma)$$
$$S' = 0 = -\ r\ S\ I.$$

We see $I' = 0$ when $I = 0$ or when $S = \gamma/r$. This nullcline divides the I–S plane into the regions shown in Fig. 12.1.
The S' nullcline is a little simpler. S' is zero on either the S or the I axis. However, the minus sign complicates things a bit. The algebraic sign of S' in all relevant regions are shown in Fig. 12.2.

I axis
I (r S − γ)
+ −
I′ < 0
I (r S − γ)
+ +
I′ > 0
S axis
I′ = 0
I (r S − γ)
− −
I′ > 0
I (r S − γ)
− +
I′ < 0
S = γ/r
I′ = 0

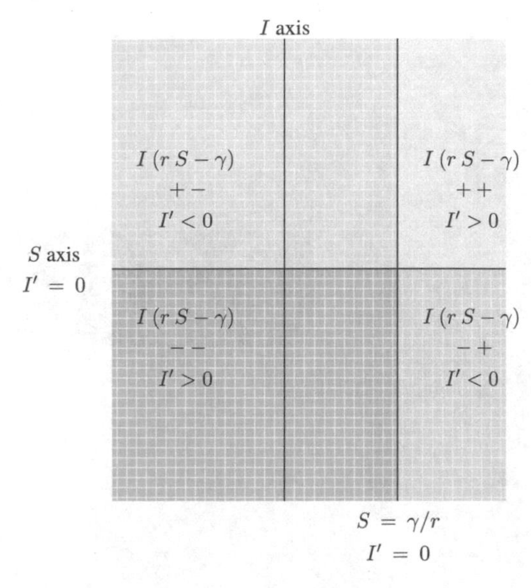

The I' equation for our system is $I' = I\,(\gamma\,S\,-\,r)$. Setting this to 0, we get $I\;=\;0$ and $S\;=\;\gamma/r$ whose graphs are shown. We show the algebraic sign of I' in all regions determined by these two lines.

Fig. 12.1 Finding where $I' < 0$ and $I' > 0$ for the disease model

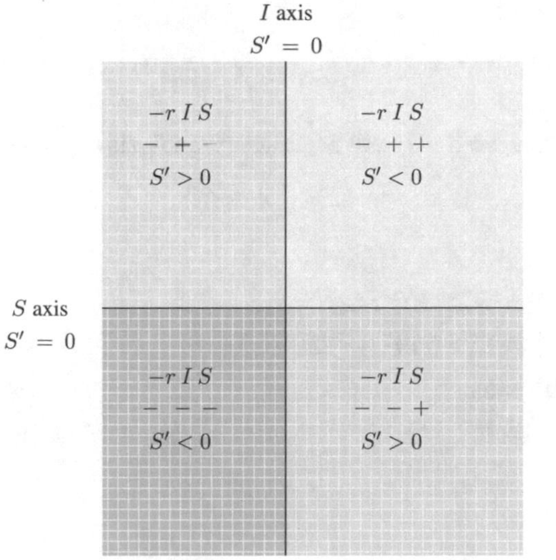

I axis
S′ = 0
−r I S
− + −
S′ > 0
−r I S
− + +
S′ < 0
S axis
S′ = 0
−r I S
− − −
S′ < 0
−r I S
− − +
S′ > 0

The S' equation for our system is $S'\;=\;-\;I\;S$. Setting this to 0, we get $S\;=\;0$ and $I\;=\;0$ whose graphs are shown. The algebraic sign of S' divides the I - S plane into the regions shown.

Fig. 12.2 Finding where $S' < 0$ and $S' > 0$ regions for the disease model

The nullcline information for $I' = 0$ and $S' = 0$ can be combined into one picture which we show in Fig. 12.3.

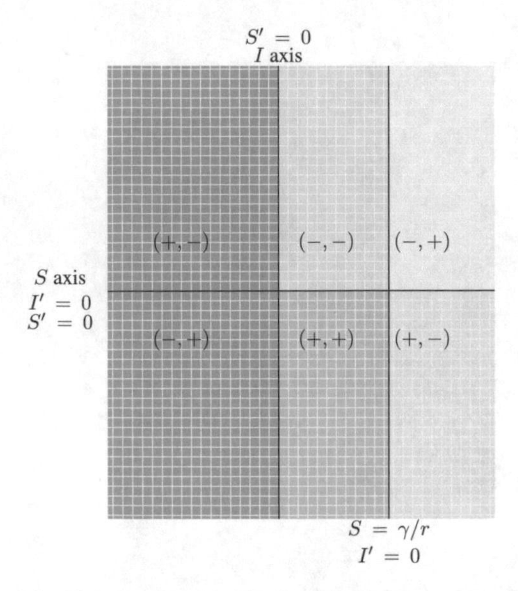

The $S' = 0$ and $I' = 0$ equations determine regions in the I - S plane with definite algebraic signs for the derivatives.

Fig. 12.3 Finding the (I', S') algebraic sign regions for the disease model

12.1.1 Homework

For the following disease models, do the I' and S' nullcline analysis separately and then assemble.

Exercise 12.1.1

$$
\begin{aligned}
S'(t) &= -15\, S(t)\, I(t) \\
I'(t) &= 15\, S(t)\, I(t) - 50\, I(t) \\
S(0) &= S_0 \\
I(0) &= I_0
\end{aligned}
$$

Exercise 12.1.2

$$
\begin{aligned}
S'(t) &= -5\, S(t)\, I(t) \\
I'(t) &= 5\, S(t)\, I(t) - 15\, I(t) \\
S(0) &= S_0 \\
I(0) &= I_0
\end{aligned}
$$

Exercise 12.1.3

$$\begin{aligned}S'(t) &= -2.4\,S(t)\,I(t)\\ I'(t) &= 2.4\,S(t)\,I(t) - 8.5\,I(t)\\ S(0) &= S_0\\ I(0) &= I_0\end{aligned}$$

Exercise 12.1.4

$$\begin{aligned}S'(t) &= -1.7\,S(t)\,I(t)\\ I'(t) &= 1.7\,S(t)\,I(t) - 1.2\,I(t)\\ S(0) &= S_0\\ I(0) &= I_0\end{aligned}$$

12.2 Only Quadrant 1 is Relevant

Consider a trajectory that starts at a point on the positive I axis. Hence, $I_0 > 0$ and $S_0 = 0$. It is easy to see that if we choose $S(t) = 0$ for all time t and I satisfying

$$\begin{aligned}I' &= -\gamma\, I\\ I(0) &= I_0\end{aligned}$$

then the pair (S, I) satisfying

$$S(t) = 0 \text{ and } I(t) = I_0\, e^{-\gamma\, t}$$

is trajectory. Since trajectories can not cross, we now know that a trajectory starting in Quadrant 1 with biologically reasonable values of $I_0 > 0$ and $S_0 > 0$ must remain on the right side of the I–S plane. Next, if we look at a trajectory which starts on the positive S axis at the point $S_0 > 0$ and $I_0 = 0$, we see immediately that the pair $S(t) = S_0$ and $I(t) = 0$ for all time t satisfies the disease model by direct calculation:

$$(S_0)' = 0 = -r\, S_0\, 0.$$

In fact, any trajectory with starting point on the positive S axis just stays there. This makes biological sense as since I_0 is 0, there is no infection and hence no disease dynamics at all. On the other hand, for the point $I_0 > 0$ and $S_0 > \gamma/r$, the algebraic signs we see in Fig. 12.3 tell us the trajectory goes to the left and upwards until it hits the line $S = \gamma/r$ and then it decays downward toward the S axis. The trajectory

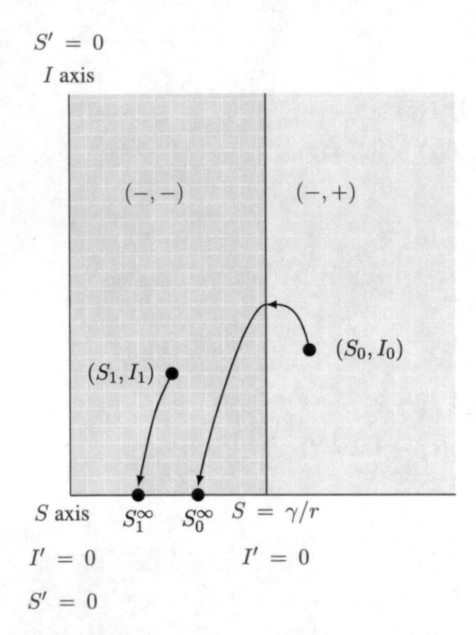

A plausible trajectory starting at the point $S_0 > \gamma/r$ and $I_0 > 0$. Another trajectory starting at $S_1 < \gamma/r$ and $I_0 > 0$ is also shown. In addition, the intersections with the S axis are labeled S_0^∞ and S_1^∞, respectively.

Fig. 12.4 The disease model in Quadrant 1

can't hit the I axis as that would cross a trajectory, so it must head downward until it hits the positive S axis. This intersection will be labeled as $(S^\infty, 0)$ and it is easy to see $S^\infty < \gamma/r$. At the point $(S^\infty, 0)$, the trajectory will stop as both the I' and S' derivatives become 0 there. Hence, we conclude we only need to look at trajectories starting in Quadrant 1 with $I_0 > 0$ as shown in Fig. 12.4.

12.2.1 Homework

For the following disease models, analyze the trajectories on the positive I and S axis and show why this means disease trajectories that start in $Q1^+$ stay there and end on the positive S axis.

Exercise 12.2.1

$$
\begin{aligned}
S'(t) &= -15\, S(t)\, I(t) \\
I'(t) &= 15\, S(t)\, I(t) - 50\, I(t) \\
S(0) &= S_0 \\
I(0) &= I_0
\end{aligned}
$$

Exercise 12.2.2

$$\begin{aligned}
S'(t) &= -5\, S(t)\, I(t) \\
I'(t) &= 5\, S(t)\, I(t) - 15\, I(t) \\
S(0) &= S_0 \\
I(0) &= I_0
\end{aligned}$$

Exercise 12.2.3

$$\begin{aligned}
S'(t) &= -2.4\, S(t)\, I(t) \\
I'(t) &= 2.4\, S(t)\, I(t) - 8.5\, I(t) \\
S(0) &= S_0 \\
I(0) &= I_0
\end{aligned}$$

Exercise 12.2.4

$$\begin{aligned}
S'(t) &= -1.7\, S(t)\, I(t) \\
I'(t) &= 1.7\, S(t)\, I(t) - 1.2\, I(t) \\
S(0) &= S_0 \\
I(0) &= I_0
\end{aligned}$$

12.3 The I Versus S Curve

We know that biologically reasonable solutions occur with initial conditions starting in Quadrant 1 and we know that our solutions satisfy $S' < 0$ always with both S and I positive until we hit the S axis. Let the time where we hit the S axis be given by t^*. Then, we can manipulate the disease model as follows. For any $t < t^*$, we can divide to obtain

$$\begin{aligned}
\frac{I'(t)}{S'(t)} &= \frac{r\, S(t)\, I(t) - \gamma\, I(t)}{-r\, S(t)\, I(t)} \\
&= -1 + \frac{\gamma}{r}\, \frac{1}{S(t)}.
\end{aligned}$$

Thus,

$$\frac{dI}{dS} = -1 + \frac{\gamma}{r}\,\frac{1}{S}$$

or integrating, we find

$$I(t) - I_0 = -\left(S(t) - S_0\right) + \frac{\gamma}{r}\,\ln\left(\frac{S(t)}{S_0}\right).$$

We can simplify this to find

$$I(t) = I_0 + S_0 - S(t) + \frac{\gamma}{r}\ln\left(\frac{S(t)}{S_0}\right).$$

Dropping the dependence on time t for convenience of notation, we see in Eq. 12.5, the functional dependence of I on S.

$$I = I_0 + S_0 - S\,\frac{\gamma}{r}\,\ln\left(\frac{S}{S_0}\right). \tag{12.5}$$

It is clear that this curve has a maximum at the critical value γ/r. This value is very important in infectious disease modeling and we call it the infectious to susceptible rate ρ. We can use ρ to introduce the idea of an **epidemic**.

Definition 12.3.1 (*A Disease Epidemic*)
For the disease model

$$\begin{aligned} I' &= r\,S\,I - \gamma\,I \\ S' &= -\,r\,S\,I \\ I(0) &= I_0 \\ S(0) &= S_0 \end{aligned}$$

the dependence of I on S is given by

$$I = I_0 + S_0 - S\,\rho\,\ln\left(\frac{S}{S_0}\right).$$

For this model, we say the infection becomes an **epidemic** if the initial value of susceptibles, S_0 exceeds the critical infectious to susceptible ratio $\rho = \frac{\gamma}{r}$ because the number of infections increases to its maximum before it begins to drop. This behavior is easy to interpret as an infection going out of control; i.e. it has entered an epidemic phase.

12.4 Homework

We are now ready to do some exercises. For the following disease models

1. Explain the meaning of the variables S and I.
2. What is the variable R and how is it related to S and I?
3. Divide the first quadrant into regions corresponding to the algebraic signs of I' and S' and explain why I and S cannot become negative. Draw a nice picture of this like Fig. 12.4.
4. What is the meaning of the constants r and γ?
5. Derive the $\frac{dI}{dS}$ equation and solve it.
6. Draw the (S, I) phase plane for this problem with a number of trajectories for the cases $S_0 > \rho$ and $S_0 < \rho$.
7. Explain why if S_0 is larger than ρ, theorists interpret this as an epidemic.
8. Explain why if S_0 is smaller than ρ, it is not interpreted as an epidemic.

Exercise 12.4.1 *For the specific model*

$$\begin{aligned} S'(t) &= -150\, S(t)\, I(t) \\ I'(t) &= 150\, S(t)\, I(t) - 50\, I(t) \\ S(0) &= S_0 \\ I(0) &= 12.7 \end{aligned}$$

1. *Is there an epidemic if S_0 is* 1.8*?*
2. *Is there an epidemic if S_0 is* 0.2*?*

Exercise 12.4.2 *For the specific model*

$$\begin{aligned} S'(t) &= -5\, S(t)\, I(t) \\ I'(t) &= 5\, S(t)\, I(t) - 25\, I(t) \\ S(0) &= S_0 \\ I(0) &= 120 \end{aligned}$$

1. *Is there an epidemic if S_0 is* 4.9*?*
2. *Is there an epidemic if S_0 is* 10.3*?*

Exercise 12.4.3 *For the specific model*

$$
\begin{aligned}
S'(t) &= -15\, S(t)\, I(t) \\
I'(t) &= 15\, S(t)\, I(t) - 35\, I(t) \\
S(0) &= S_0 \\
I(0) &= 120
\end{aligned}
$$

1. *Is there an epidemic if S_0 is* 1.9*?*
2. *Is there an epidemic if S_0 is* 4.3*?*

Exercise 12.4.4 *For the specific model*

$$
\begin{aligned}
S'(t) &= -100\, S(t)\, I(t) \\
I'(t) &= 100\, S(t)\, I(t) - 400\, I(t) \\
S(0) &= S_0 \\
I(0) &= 120
\end{aligned}
$$

1. *Is there an epidemic if S_0 is* 2.9*?*
2. *Is there an epidemic if S_0 is* 10.3*?*

Exercise 12.4.5 *For the specific model*

$$
\begin{aligned}
S'(t) &= -5\, S(t)\, I(t) \\
I'(t) &= 5\, S(t)\, I(t) - 250\, I(t) \\
S(0) &= S_0 \\
I(0) &= 120
\end{aligned}
$$

1. *Is there an epidemic if S_0 is* 40*?*
2. *Is there an epidemic if S_0 is* 70*?*

12.5 Solving the Disease Model Using Matlab

Here is a typical session to plot an SIR disease model trajectory for

$$S' = -5SI,\ I' = 5SI - 25I,\ \ S(0) = 10,\ I(0) = 5.$$

Listing 12.1: Solving $S' = -5SI$, $I' = 5SI + 25I$, $S(0) = 10$, $I(0) = 5$: Epidemic!

```
f = @(t,x) [-5*x(1).*x(2); 5*x(1).*x(2) - 25*x(2)];
T = 2.0;
h = .005;
N = ceil(T/h);
x0 = [10;5];
[ht,rk] = FixedRK(f,0,x0,h,4,N);
X = rk(1,:);
Y = rk(2,:);
xmin = min(X);
xmax = max(X);
xtop = max( abs(xmin), abs(xmax) );
ymin = min(Y);
ymax = max(Y);
ytop = max( abs(ymin), abs(ymax) );
D = max(xtop,ytop)
x = linspace(0,D,101);
GoverR = 25/5;
clf
hold on
plot([GoverR GoverR], [0 D]);
plot(X,Y,'-k');
xlabel('S axis');
ylabel('I axis');
title('Phase Plane for Disease Model S'' = -5 S I, I'' = 5 S I + 25 I, S
   (0) = 10, I(0) = 5: Epidemic!');
legend('S = 25/5','S vs I','Location','Best');
```

This generates the plot you see in Fig. 12.5.

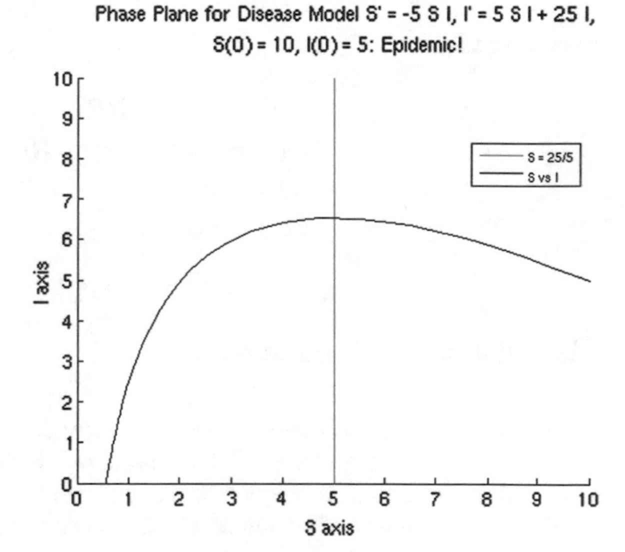

Fig. 12.5 Solution to $S' = -5SI$, $I' = 5SI + 25I$, $S(0) = 10$, $I(0) = 5$

12.5.1 Homework

For the following disease models, do the single plot corresponding to an initial condition that gives an epidemic and also draw a nice phase plane plot using **`AutoPhasePlanePlot`**.

Exercise 12.5.1

$$\begin{aligned} S'(t) &= -150\, S(t)\, I(t) \\ I'(t) &= 150\, S(t)\, I(t) - 50\, I(t) \\ S(0) &= S_0 \\ I(0) &= I_0 \end{aligned}$$

Exercise 12.5.2

$$\begin{aligned} S'(t) &= -5\, S(t)\, I(t) \\ I'(t) &= 5\, S(t)\, I(t) - 25\, I(t) \\ S(0) &= S_0 \\ I(0) &= I_0 \end{aligned}$$

Exercise 12.5.3

$$\begin{aligned} S'(t) &= -3\, S(t)\, I(t) \\ I'(t) &= 3\, S(t)\, I(t) - 4\, I(t) \\ S(0) &= S_0 \\ I(0) &= I_0 \end{aligned}$$

12.6 Estimating Parameters

During an epidemic, it is impossible to accurately determine the number of newly infected people each day or week. This is because infectious people are only recognized and removed from circulation if they seek medical attention. Indeed, we only see data on the number of people admitted to hospitals each day or week. That is, we have data on the number of newly removed people which is an estimate of R'. So to compare the results predicted by the model to data from real epidemics, we must find R' as a function of time t. Now we know

$$R' = \gamma\, I = \gamma\, (N - R - S).$$

Further, we know

$$\frac{dS}{dR} = \frac{\frac{dS}{dt}}{\frac{dR}{dt}} = \frac{-r\,S\,I}{-\gamma\,I}$$
$$= -\frac{S}{\rho}.$$

This equation we can solve to obtain

$$S(R) = S_0\,e^{-R/\rho}. \tag{12.6}$$

Hence,

$$R' = \gamma\,(N - R - S_0\,e^{-R/\rho}). \tag{12.7}$$

This differential equation is not solvable directly, so we will try some estimates.

12.6.1 Approximating the *dR*/*dt* Equation

To estimate the solution to Eq. 12.7, we would like to replace the term e which makes our integration untenable with a quadratic approximation like that of Eq. 3.5 from Sect. 3.1.3. We would approximate around the point $R = 0$, giving

$$Q\left(\frac{R}{\rho}\right) = 1 - \frac{R}{\rho} + \frac{1}{2}\left(\frac{R}{\rho}\right)^2$$

with attendant error

$$|E_Q(R, 0)| \leq \frac{1}{3}\left(\frac{R}{\rho}\right)^3.$$

We need to see if this error is not too large. Recall the I and S solution satisfies

$$I + S = I_0 + S_0 + \rho\,\ln\left(\frac{S}{S_0}\right).$$

An epidemic would start with the number of removed individuals R_0 being 0. Hence, we know initially $N = I_0 + S_0$ and so since $R = N - I - S$, we have

$$N - R = N + \rho\,\ln\left(\frac{S}{S_0}\right)$$

or

$$R = -\rho \ \ln\left(\frac{S}{S_0}\right) = \rho \ \ln\left(\frac{S_0}{S}\right).$$

We know S always decreases from its initial value of S_0, so the fraction S_0/S is larger than one; hence, the logarithm is positive. We conclude

$$-\frac{R}{\rho} = \ln\left(\frac{S}{S_0}\right) < 1.$$

Thus, the error we make in replacing $e^{R/\rho}$ by $Q(R/\rho)$ is reasonably small as

$$|E_Q(R, 0)| \leq \frac{1}{3} \ \ln\left(\frac{S_0}{S}\right)^3 \ll 1.$$

Now, let's use this approximation in Eq. 3.5 to derive an approximation to $R(t)$. The approximate differential equation to solve is

$$R' = \gamma \left(N - R - S_0 \ Q\left(\frac{R}{\rho}\right)\right) \tag{12.8}$$

$$= \gamma \left\{N - R - S_0 \left(1 - \frac{R}{\rho} + \frac{1}{2}\left(\frac{R}{\rho}\right)^2\right)\right\}. \tag{12.9}$$

This can be rewritten as follows (we will go through all the steps because it is intense!):

$$\begin{aligned} R' &= \gamma \ \frac{S_0}{2\rho^2} \left\{\left(N - S_0\right) \frac{2\rho^2}{S_0} + \left(\frac{S_0}{\rho} - 1\right) \frac{2\rho^2}{S_0} \ R - R^2\right\} \\ &= -\gamma \ \frac{S_0}{2\rho^2} \left\{R^2 - \left(\frac{S_0}{\rho} - 1\right) \frac{2\rho^2}{S_0} \ R - \left(N - S_0\right) \frac{2\rho^2}{S_0}\right\} \\ &= -\gamma \ \frac{S_0}{2\rho^2} \left\{R^2 - \left(\frac{S_0 - \rho}{S_0}\right) 2\rho R - \left(\frac{N - S_0}{S_0}\right) 2\rho^2\right\}. \end{aligned}$$

Now the next step is truly complicated. We complete the square on the quadratic. This gives

$$\begin{aligned} R' = -\gamma \ \frac{S_0}{2\rho^2} \Bigg\{ & R^2 - \left(\frac{S_0 - \rho}{S_0}\right) 2\rho R + \left(\frac{S_0 - \rho}{S_0}\right)^2 \rho^2 - \left(\frac{S_0 - \rho}{S_0}\right)^2 \rho^2 \\ & - \left(\frac{N - S_0}{S_0}\right) 2\rho^2 \Bigg\} \end{aligned}$$

$$= -\gamma \frac{S_0}{2\,\rho^2} \left\{ \left(R - \frac{S_0 - \rho}{S_0}\,\rho \right)^2 - \left(\frac{S_0 - \rho}{S_0} \right)^2 \rho^2 - \left(\frac{N - S_0}{S_0} \right) 2\,\rho^2 \right\}.$$

Egads! Now we will simplify by defining the constants α and β by

$$\alpha^2 = \left(\frac{S_0 - \rho}{S_0} \right)^2 \rho^2 + \left(\frac{N - S_0}{S_0} \right) 2\,\rho^2$$

and

$$\beta = \frac{S_0 - \rho}{S_0}\,\rho.$$

Then we can rewrite our equation is a notationally simpler form:

$$R' = -\gamma \frac{S_0}{2\,\rho^2} \left((R - \beta)^2 - \alpha^2 \right).$$

Now we can go about the business of solving the differential equation. We will use a new approach (rather than the integration by partial fraction decomposition we have already used). After separating variables, we have

$$\frac{dR}{(R - \beta)^2 - \alpha^2} = -\gamma \frac{S_0}{2\,\rho^2}\,dt.$$

Substitute $u = R - \beta$ to obtain

$$\frac{du}{u^2 - \alpha^2} = -\gamma \frac{S_0}{2\,\rho^2}\,dt.$$

We will do these integrations using what are called *hyperbolic trigonometric* functions. We make the following definitions:

Definition 12.6.1 (*Hyperbolic Functions*)
The main hyperbolic functions of the real number x are defined to be

$$\begin{aligned}
\sinh(x) &= \frac{e^x - e^{-x}}{2}, \text{ the hyperbolic sine,} \\
\cosh(x) &= \frac{e^x + e^{-x}}{2}, \text{ the hyperbolic cosine,} \\
\tanh(x) &= \frac{\sinh(x)}{\cosh(x)}, \\
&= \frac{e^x - e^{-x}}{e^x + e^{-x}}, \text{ the hyperbolic tangent,}
\end{aligned}$$

$$\begin{aligned} \text{sech}(x) &= \frac{1}{\cosh(x)}, \\ &= \frac{2}{e^x + e^{-x}}, \text{ the hyperbolic secant,} \end{aligned}$$

It is straightforward to calculate that

$$\begin{aligned} \cosh^2(x) - \sinh^2(x) &= 1, \\ 1 - \tanh^2(x) &= sech^2(x). \end{aligned}$$

Note that these definitions are similar, but different, from the ones you are used to with the standard trigonometric functions $\sin(x)$, $\cos(x)$ and so forth.

And then there are the derivatives:

Definition 12.6.2 (*Hyperbolic Function Derivatives*)
The hyperbolic functions are continuous and differentiable for all real x. We have

$$\begin{aligned} \Big(\sinh(x)\Big)' &= \cosh(x) \\ \Big(\cosh(x)\Big)' &= \sinh(x) \\ \Big(\tanh(x)\Big)' &= sech^2(x) \end{aligned}$$

Now let's go back to the differential equation we need to solve. Make the substitution $u = \alpha \tanh(z)$. Then, we have $du = \alpha \, sech^2(z) \, dz$. Making the substitution, we find

$$\frac{\alpha \, sech^2(z) \, dz}{\alpha^2(\tanh^2(z) - 1)} = -\gamma \frac{S_0}{2 \rho^2} \, dt.$$

But $\tanh^2(z) - 1 = -sech^2(z)$. Thus, the above simplifies to

$$\frac{\alpha \, sech^2(z) \, dz}{-\alpha^2 \, sech^2(z)} = \frac{-1}{\alpha} \, dz = -\gamma \frac{S_0}{2 \rho^2} \, dt.$$

The integration is now simple. We have

$$dz = \alpha \, \gamma \frac{S_0}{2 \rho^2} \, dt.$$

Integrating, we obtain

$$z(t) - z(0) = \alpha \, \gamma \frac{S_0}{2 \rho^2} \, t.$$

Just like there is an inverse tangent for trigonometric functions, there is an inverse for the hyperbolic tangent also.

Definition 12.6.3 (*Inverse Hyperbolic Function*)
It is straightforward to see that $\tanh(x)$ is always increasing and hence it has a nicely defined inverse function. We call this inverse the inverse hyperbolic tangent function and denote it by the symbol $\tanh^{-1}(x)$.

We can show using rather messy calculations the following sum and difference formulae for tanh. We will be using these in a bit.

$$\tanh(u+v) = \frac{\tanh(u)+\tanh(v)}{1+\tanh(u)\ \tanh(v)}$$
$$\tanh(u-v) = \frac{\tanh(u)-\tanh(v)}{1-\tanh(u)\ \tanh(v)}.$$

Hence, since $u = \alpha\ \tanh(z)$,

$$z = \tanh^{-1}\left(\frac{u}{\alpha}\right) = \tanh^{-1}\left(\frac{R-\beta}{\alpha}\right).$$

Our differential equation can thus be rewritten as

$$\tanh^{-1}\left(\frac{R(t)-\beta}{\alpha}\right) - \tanh^{-1}\left(\frac{R_0-\beta}{\alpha}\right) = \alpha\,\gamma\,\frac{S_0}{2\,\rho^2}\,t.$$

We have

$$\tanh^{-1}\left(\frac{R(t)-\beta}{\alpha}\right) - \tanh^{-1}\left(\frac{R_0-\beta}{\alpha}\right) = \alpha\,\gamma\,\frac{S_0}{2\,\rho^2}\,t.$$

Hence, since $R_0 = 0$ and $\tanh^{-1}$ is an odd function, we find

$$\tanh\left(\tanh^{-1}\left(\frac{R(t)-\beta}{\alpha}\right) + \tanh^{-1}\left(\frac{\beta}{\alpha}\right)\right) = \tanh\left(\alpha\,\gamma\,\frac{S_0}{2\,\rho^2}\,t\right).$$

Applying Definition 12.6.3, we have

$$\frac{\frac{R(t)-\beta}{\alpha} + \frac{\beta}{\alpha}}{1+\frac{R(t)-\beta}{\alpha}\,\frac{\beta}{\alpha}} = \tanh\left(\alpha\,\gamma\,\frac{S_0}{2\,\rho^2}\,t\right).$$

Now we want R as a function of t, so we have some algebra to suffer our way through. Grab another cup of coffee as this is going to be a rocky ride!

$$\frac{R(t)-\beta}{\alpha} + \frac{\beta}{\alpha} = \tanh\left(\alpha\,\gamma\,\frac{S_0}{2\,\rho^2}\,t\right)\left(1+\frac{R(t)-\beta}{\alpha}\,\frac{\beta}{\alpha}\right).$$

It then follows that

$$\frac{R(t)-\beta}{\alpha}\left(1-\frac{\beta}{\alpha}\,\tanh\left(\alpha\,\gamma\,\frac{S_0}{2\,\rho^2}\,t\right)\right)=-\frac{\beta}{\alpha}+\tanh\left(\alpha\,\gamma\,\frac{S_0}{2\,\rho^2}\,t\right).$$

Finally, after a division, we have

$$\frac{R(t)-\beta}{\alpha}=\frac{\tanh\left(\alpha\,\gamma\,\frac{S_0}{2\,\rho^2}\,t\right)-\frac{\beta}{\alpha}}{1-\frac{\beta}{\alpha}\,\tanh\left(\alpha\,\gamma\,\frac{S_0}{2\,\rho^2}\,t\right)}.$$

Thus, letting the number ϕ be defined by $\tanh(\phi)=\beta/\alpha$, we have

$$\begin{aligned}R(t)&=\beta+\alpha\,\frac{\tanh\left(\alpha\,\gamma\,\frac{S_0}{2\,\rho^2}\,t\right)-\tanh\phi}{1-\tanh\phi\,\tanh\left(\alpha\,\gamma\,\frac{S_0}{2\,\rho^2}\,t\right)}\\&=\beta+\alpha\,\tanh\left(\frac{\alpha\,\gamma\,S_0}{2\rho^2}\,t-\phi\right)\end{aligned}$$

using the addition formula for tanh from Definition 12.6.3. We can then find the long sought formula for R'. It is

$$R'(t)=\frac{\alpha^2\,\gamma\,S_0}{2\rho^2}sech^2\left(\frac{\alpha\,\gamma\,S_0}{2\rho^2}\,t-\phi\right)$$

12.6.2 *Using dR/dt to Estimate ρ*

Assume we have collected data for the rate of change of R with respect to time during an infectious incident. The general R' model is of the form

$$R'(t)=A\,sech^2(a\,t-b)$$

for some choice of positive constants a, b and A. We fit our R' data by choosing a, b and A carefully using some technique (these sorts of tools would be discussed in another class). We know

$$\alpha^2=\beta^2+\frac{N-S_0}{S_0}\,2\,\rho^2. \tag{12.10}$$

Our model tells us that

$$a = \frac{\alpha \gamma S_0}{2\rho^2}$$

$$A = \frac{\alpha^2 \gamma S_0}{2\rho^2} = \alpha a.$$

So we have $\alpha = A/a$. Further, since $b = \tanh^{-1}(\beta/\alpha)$, we have $\beta = \alpha \tanh(b)$. Thus, from Eq. 12.10, it follows that

$$\left(\frac{A}{a}\right)^2 = \left(\alpha \tanh(b)\right)^2 + \frac{N - S_0}{S_0} 2 \rho^2.$$

After some manipulation, we have

$$\frac{N - S_0}{S_0} 2 \rho^2 = \left(\frac{A}{a}\right)^2 \left(1 - \tanh^2(b)\right).$$

The right hand side is known from our data fit and we can assume we have an estimate of the total population N also. In addition, if we can estimate the initial susceptible value S_0, we will have an estimate for the critical value ρ from our data:

$$\rho^2 = \frac{1}{2} \frac{S_0}{N - S_0} \left(\frac{A}{a}\right)^2 \left(1 - \tanh^2(b)\right).$$

It is a lot of work to generate the approximate value of R' but the payoff is that we obtain an estimate of the γ/r ratio which determines whether we have an epidemic or not.

Chapter 13
A Cancer Model

We are going to examine a relatively simple cancer model as discussed in Novak (2006). Our model is a small fraction of what Novak shows is possible, but it should help you to continue to grow in the art of modeling. This chapter is adapted from Novak's work, so you should really go and look at his full book after you work your way through this chapter. However, we are more careful in the underlying mathematical analysis than Novak and after this chapter is completed, you should have a better understanding of why the mathematics and the science need to work together. You are not ready for more sophisticated mathematical models of cancer, but it wouldn't hurt you to look at a few to get the lay of the land, so to speak. Check out Ribba et al. (2006) and Tatabe et al. (2001) when you can. In Peterson (2015), we also discussed this cancer model, but we only looked at part of it as we had not yet learned how to handle systems of ODEs. Now we are fully prepared and we will finish our analysis. Let's review the basic facts again so that everything is easy to refer to.

We are built of individual cells that have their own reproductive machinery. These cells can sometimes revert to uncontrolled self-replication; this is, of course, a change from their normal programming. Cancer is a disease of organisms that have more than one cell. In order for organisms to have evolved multicellularity, individual cells had to establish and maintain cooperation with each other. From one point of view then, cancer is a breakdown of cellular cooperation; cells must divide when needed by the development program, but not otherwise. Complicated genetic controls on networks of cells had to evolve to make this happen. Indeed, many genes are involved to maintain integrity of the genome, to make sure cell division does not make errors and to help establish the development program that tells cells when to divide. Some of these genes monitor the cell's progress and, if necessary, induce cell death, a process called *apoptosis*. Hence, most cells listen to many signals from other cells telling them they are ok. If the signals saying this fail to arrive, the *default* program is for the cell to commit suicide; i.e. to trigger apoptosis. hence, apoptosis is a critical defense against cancer. Here is a rough timeline (adapted from Novak) that shows the important events that are relevant to our simple cancer modeling exercise.

J.K. Peterson, *Calculus for Cognitive Scientists*,
Cognitive Science and Technology, DOI 10.1007/978-981-287-877-9_13

1890 David von Hansermann noted cancer cells have *abnormal cell division events*.

1914 Theodore Boveri sees that something is wrong in the chromosomes of cancer cells: they are *aneuploid*. That is, they do not have the normal number of chromosomes.

1916 Ernst Tyzzer first applied the term *somatic mutation* to cancer.

1927 Herman Muller discovered ionizing radiation which was known to cause cancer (i.e. was *carcinogenic*) was also able to cause genetic mutations (i.e. was *mutagenic*).

1951 Herman Muller proposed cancer requires a single cell to receive multiple mutations.

1950–1959 Mathematical modeling of cancer begins. It is based on statistics.

1971 Alfred Knudson proposes the concept of a *Tumor Suppressor Gene* or **TSG**. The idea is that it takes a two point mutation to inactivate a TSG. TSG's play a central role in regulatory networks that determine the rate of cell cycling. Their inactivation modifies regulatory networks and can lead to increased cell proliferation.

1986 A *Retinoblastoma* TSG is identified which is a gene involved in a childhood eye cancer.

Since 1986, about 30 more TSG's have been found. An important TSG is **p53**. This is mutated in more than 50 % of all human cancers. This gene is at the center of a control network that monitors genetic damage such as *double stranded breaks* (**DSB**) of DNA. In a *single stranded break* (**SSB**), at some point the double stranded helix of DNA breaks apart on one strand only. In a DSB, the DNA actually separates into different pieces giving a complete gap. DSB's are often due to ionizing radiation. If a certain amount of damage is achieved, cell division is paused and the cell is given time for repair. If there is too much damage, the cell will undergo apoptosis. In many cancer cells, **p53** is inactivated. This allows these cells to divide in the presence of substantial genetic damage. In 1976, Michael Bishop and Harold Varmus introduced the idea of *oncogenes*. These are another class of genes involved in cancer. These genes increase cell proliferation if they are mutated or inappropriately expressed. Now a given gene that occupies a certain position on a chromosome (this position is called the *locus* of the gene) can have a number of alternate forms. These alternate forms are called *alleles*. The number of alleles a gene has for an individual is called that individual's *genotype* for that gene. Note, the number of alleles a gene has is therefore the number of viable DNA codings for that gene. We see then that mutations of a TSG and an oncogene increase the net reproductive rate or *somatic fitness* of a cell. Further, mutations in genetic instability genes also increase the mutation rate.

For example, mutations in mismatch repair genes lead to 50–100 fold increases in point mutation rates. These usually occur in repetitive stretches of short sequences of DNA. Such regions are called *micro satellite regions* of the genome. These regions are used as genetic markers to track inheritance in families. They are short sequences of nucleotides (i.e. **ATCG**) which are repeated over and over. Changes can occur such as increasing or decreasing the number of repeats. This type of instability is thus called a *micro satellite* or **MIN** instability. It is known that 15 % of colon cancer cells have MIN.

Another instability is *chromosomal instability* or **CIN**. This means an increase or decrease in the rate of gaining or losing whole chromosomes or large fractions of chromosomes during cell division.

Let's look at what can happen to a typical TSG. If the first allele of a TSG is inactivated by a point mutation while the second allele is inactivated by a loss of one parent's contribution to part of the cell's genome, the second allele is inactivated because it is lost in the copying process. This is an example of CIN and in this case is called *loss of heterozygosity* or **LOH**. It is known that 85 % of colon cancer cells have CIN.

13.1 Two Allele TSG Models

Let's look now at colon cancer itself. Look at Fig. 13.1. In this figure, you see a typical colon crypt. Stem cells at the bottom of the crypt differentiate and move up the walls of the crypt to the colon lining where they undergo apoptosis.
At the bottom of the crypt, a small number of *stem* cells slowly divide to produce *differentiated* cells. These differentiated cells divide a few times while migrating to the top of the crypt where they undergo apoptosis. This architecture means only a small subset of cells are at risk of acquiring mutations that become fixed in the permanent cell lineage. Many mutations that arise in the differentiated cells will be removed by apoptosis.

Colon rectal cancer is thought to arise as follows. A mutation inactivates the **Adenomatous Polyposis Coli** or **APC** TSG pathway. Ninety- five percent of colorectal cancer cells have this mutation with other mutations accounting for the other 5 %. The crypt in which the **APC** mutant cell arises becomes **dyplastic**; i.e. has abnormal growth and produces a *polyp*. Large *polyps* seem to require additional oncogene activation. Then 10–20 % of these large polyps progress to cancer.

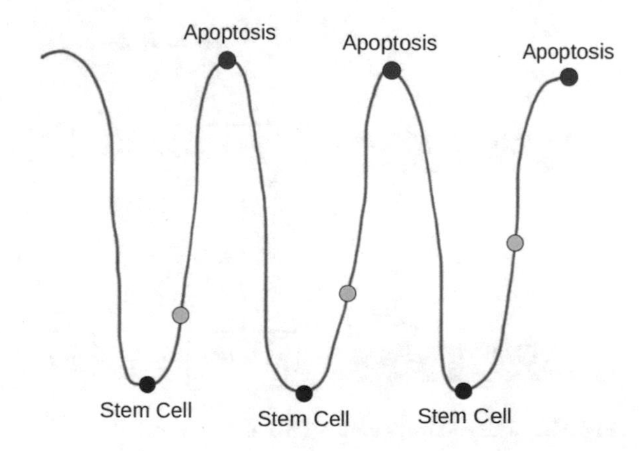

Fig. 13.1 Typical colon crypts

A general model of cancer based on TSG inactivation is as follows. The tumor starts with the inactivation of a **TSG** called **A**, in a small compartment of cells. A good example is the inactivation of the **APC** gene in a colonic crypt, but it could be another gene. Initially, all cells have two active alleles of the **TSG**. We will denote this by $\boldsymbol{A}^{+/+}$ where the superscript "$+/+$" indicates both alleles are active. One of the alleles becomes inactivated at mutation rate u_1 to generate a cell type denoted by $\boldsymbol{A}^{+/-}$. The superscript $+/-$ tells us one allele is inactivated. The second allele becomes inactivated at rate $\hat{u}_2$ to become the cell type $\boldsymbol{A}^{-/-}$. In addition, $\boldsymbol{A}^{+/+}$ cells can also receive mutations that trigger **CIN**. This happens at the rate u_c resulting in the cell type $\boldsymbol{A}^{+/+\ CIN}$. This kind of a cell can inactivate the first allele of the **TSG** with normal mutation rate u_1 to produce a cell with one inactivated allele (i.e. a $+/-$) which started from a *CIN* state. We denote these cells as $\boldsymbol{A}^{+/-\ CIN}$. We can also get a cell of type $\boldsymbol{A}^{+/-\ CIN}$ when a cell of type $\boldsymbol{A}^{+/-}$ receives a mutation which triggers **CIN**. We will assume this happens at the same rate u_c as before. The $\boldsymbol{A}^{+/-\ CIN}$ cell then rapidly undergoes **LOH** at rate $\hat{u}_3$ to produce cells of type $\boldsymbol{A}^{-/-\ CIN}$. Finally, $\boldsymbol{A}^{-/-}$ cells can experience **CIN** at rate u_c to generate $\boldsymbol{A}^{-/-\ CIN}$ cells. We show this information in Fig. 13.2.
Let N be the population size of the compartment. For colonic crypts, the typical value of N is 1000–4000. The first allele is inactivated by a point mutation. The rate at which this occurs is modeled by the rate u_1 as shown in Fig. 13.2. We make the following assumptions:

Assumption 13.1.1 (*Mutation rates for u_1 and u_c are population independent*)
The mutations governed by the rates u_1 and u_c are **neutral**. This means that these rates do not depend on the size of the population N.

Assumption 13.1.2 (*Mutation rates $\hat{u}_2$ and $\hat{u}_3$ give selective advantage*)
The events governed by $\hat{u}_2$ and $\hat{u}_3$ give what is called **selective advantage**. This means that the size of the population size does matter.

Using these assumptions, we will model $\hat{u}_2$ and $\hat{u}_3$ like this:

$$\hat{u}_2 = N\, u_2$$

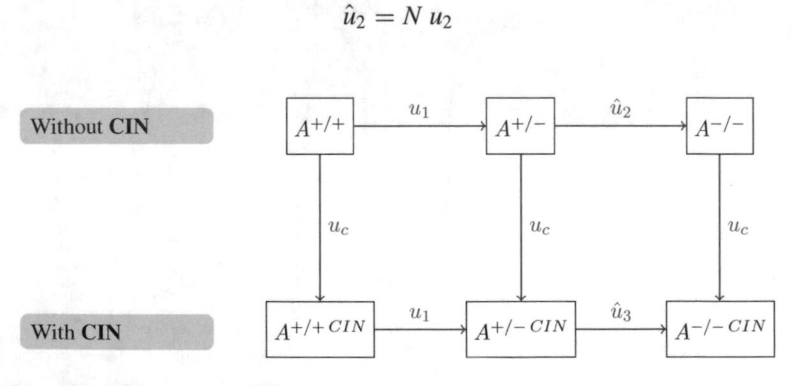

Fig. 13.2 The pathways for the **TSG** allele losses

and

$$\hat{u}_3 = N\, u_3.$$

where u_2 and u_3 are neutral rates. We can thus redraw our figure as Fig. 13.3. The mathematical model is then setup as follows. Let

$X_0(t)$ is the probability a cell in cell type $A^{+/+}$ at time t.
$X_1(t)$ is the probability a cell in cell type $A^{+/-}$ at time t.
$X_2(t)$ is the probability a cell in cell type $A^{-/-}$ at time t.
$Y_0(t)$ is the probability a cell in cell type $A^{+/+\,CIN}$ at time t.
$Y_1(t)$ is the probability a cell in cell type $A^{+/-\,CIN}$ at time t.
$Y_2(t)$ is the probability a cell in cell type $A^{-/-\,CIN}$ at time t.

Looking at Fig. 13.3, we can generate rate equations. First, let's rewrite Fig. 13.3 using our variables as Fig. 13.4.

To generate the equations we need, note each box has arrows coming into it and arrows coming out of it. The **arrows in** are **growth** terms for the net change of the variable in the box and the **arrows out** are the **decay or loss** terms. We model **growth** as **exponential growth** and **loss** as **exponential decay**. So X_0 only has arrows going out which tells us it only has **loss** terms. So we would say $\left(X_0'\right)_{loss} = -u_1 X_0 - u_c X_0$ which implies $X_0' = -(u_1 + u_c)X_0$. Further, X_1 has arrows going in and out which tells us it has **growth** and **loss** terms. So we would say $\left(X_1'\right)_{loss} = -Nu_2 X_1 - u_c X_1$ and $\left(X_1'\right)_{growth} = u_1 X_0$ which implies $X_1' = u_1 X_0 - (Nu_2 + u_c)X_1$. We can continue

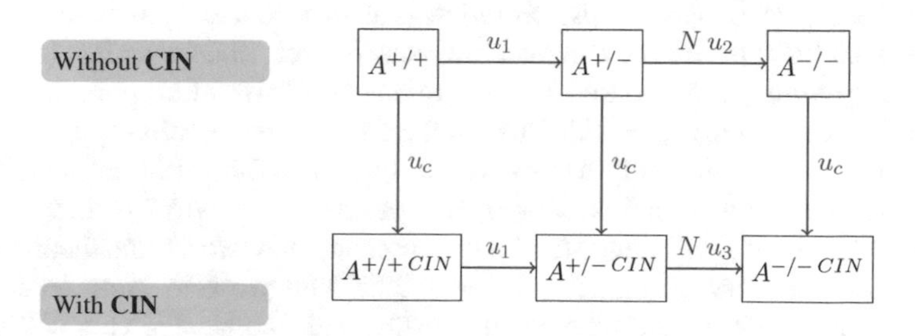

Fig. 13.3 The pathways for the **TSG** allele losses rewritten using selective advantage

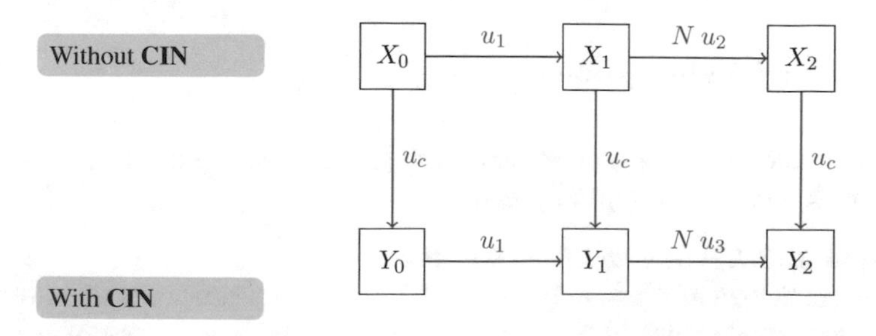

Fig. 13.4 The pathways for the **TSG** allele losses rewritten using mathematical variables

in this way to find all the model equations. We can then see the *Cancer Model* rate equations are

$$X_0' = -(u_1 + u_c)\, X_0 \tag{13.1}$$
$$X_1' = u_1\, X_0 - (u_c + N\, u_2)\, X_1 \tag{13.2}$$
$$X_2' = N\, u_2\, X_1 - u_c\, X_2 \tag{13.3}$$
$$Y_0' = u_c\, X_0 - u_1\, Y_0 \tag{13.4}$$
$$Y_1' = u_c\, X_1 + u_1\, Y_0 - N\, u_3\, Y_1 \tag{13.5}$$
$$Y_2' = N\, u_3\, Y_1 + u_c\, X_2 \tag{13.6}$$

Initially, at time 0, all the cells are in the state X_0, so we have

$$X_0(0) = 1, \;\; X_1(0) = 0, \;\; X_2(0) = 0 \tag{13.7}$$
$$Y_0(0) = 0, \;\; Y_1(0) = 0, \;\; Y_2(0) = 0. \tag{13.8}$$

The problem we want to address is this one:

Question *Under what circumstances is the CIN pathway to cancer the dominant one?*

In order to answer this, we need to analyze the trajectories of this model. Note, if we were interested in the asymptotic behavior of this model as t goes to infinity, then it is clear everything ends up with value 0. However, our interest is over the typical lifetime of a human being and thus, we never reach the asymptotic state. Thus, our analysis in the sections that follow is always concerned with the values of our six variables at the end of a human life span. In Sect. 13.6 we do this for the cell populations without CIN and then we use those results to develop approximations in Sect. 13.7 to cell populations with CIN. Along the way, we develop approximations to the true solutions that help us answer our question. But, make no mistake, the analysis is a tricky and requires a fair amount of intellectual effort. Still, we think at the end of it all, you will understand and appreciate how we use mathematics and science together to try to come to grips with a difficult problem. Now, let's look at the results first, before we do the detailed mathematical analysis. This will let you see the larger picture.

13.2 Model Assumptions

Since our interest in these variables is over the typical lifetime of a human being, we need to pick a maximum typical lifetime.

Assumption 13.2.1 (*Average human lifetime*)
The average human life span is 100 years. We also assume that cells divide once per day and so a good choice of time unit is *days*. The final time for our model will be denoted by T and hence

$$T = 3.65 \times 10^4 \text{ days.}$$

Next, recall our colonic crypt, N is from 1000 to 4000 cells. For estimation purposes, we often think of N as the upper value, $N = 4 \times 10^3$.

Assumption 13.2.2 (*Loss of allele rates for neutral mutations*)
We assume

$$\begin{aligned} u_1 &\approx 10^{-7} \\ u_2 &\approx 10^{-7}. \end{aligned}$$

We will assume the rate $N\,u_3$ is quite rapid and so it is close to 1. We will set u_3 as follows:

Assumption 13.2.3 (*Losing the second allele due to CIN is close to probability one*)
We assume

$$N\,u_3 \approx 1 - r.$$

for small positive values of r.

Hence, once a cell reaches the Y_1 state, it will rapidly transition to the end state Y_2 if r is sufficiently small.

We are not yet sure how to set the magnitude of u_c, but it certainly is at least u_1. For convenience, we will assume

Assumption 13.2.4 (u_c *is proportional to* u_1)
We assume

$$u_c = R\,u_1.$$

where R is a number at least 1. For example, if $u_c = 10^{-5}$, this would mean $R = 100$.

13.3 Preliminary Analysis

The mathematical model can be written in matrix vector form as usual. This gives

$$\begin{bmatrix} X_0' \\ X_1' \\ X_2' \\ Y_0' \\ Y_1' \\ Y_2' \end{bmatrix} = \begin{bmatrix} -(u_1 + u_c) & 0 & 0 & 0 & 0 & 0 \\ u_1 & -(u_c + Nu_2) & 0 & 0 & 0 & 0 \\ 0 & Nu_2 & -u_c & 0 & 0 & 0 \\ u_c & 0 & 0 & -u_1 & 0 & 0 \\ 0 & u_c & 0 & u_1 & -Nu_3 & 0 \\ 0 & 0 & u_c & 0 & Nu_3 & 0 \end{bmatrix} \begin{bmatrix} X_0 \\ X_1 \\ X_2 \\ Y_0 \\ Y_1 \\ Y_2 \end{bmatrix}$$

$$\begin{bmatrix} X_0(0) \\ X_1(0) \\ X_2(0) \\ Y_0(0) \\ Y_1(0) \\ Y_2(0) \end{bmatrix} = \begin{bmatrix} 1 \\ 0 \\ 0 \\ 0 \\ 0 \\ 0 \end{bmatrix}$$

We can find eigenvalues and associated eigenvectors for this system as we have done for 2×2 models in the past. However, it is clear, we can solve the X_0, X_1 system directly and then use that solution to find X_2. Here are the details.

13.4 Solving the Top Pathway Exactly

We will now solve the top pathway model exactly using the tools we have developed in this course.

13.4.1 The X_0–X_1 Subsystem

The X_0–X_1 subsystem is a familiar 2×2 system.

$$\begin{bmatrix} X_0' \\ X_1' \end{bmatrix} = \begin{bmatrix} -(u_1 + u_c) & 0 \\ u_1 & -(u_c + Nu_2) \end{bmatrix} \begin{bmatrix} X_0 \\ X_1 \end{bmatrix}$$

$$\begin{bmatrix} X_0(0) \\ X_1(0) \end{bmatrix} = \begin{bmatrix} 1 \\ 0 \end{bmatrix}$$

The characteristic equation is

$$\Big(r + (u_1 + u_c)\Big)\Big(r + (u_c + Nu_2)\Big) = 0.$$

Hence, the two eigenvalues are $r_1 = -(u_1+u_c)$ and $r_2 = -(u_c+Nu_2)$. The associated eigenvectors are straightforward to find:

$$\boldsymbol{E}_1 = \begin{bmatrix} 1 \\ \frac{u_1}{Nu_2 - u_1} \end{bmatrix} \text{ and } \boldsymbol{E}_2 = \begin{bmatrix} 0 \\ 1 \end{bmatrix}$$

Applying the initial conditions, we find

$$\begin{bmatrix} X_0(t) \\ X_1(t) \end{bmatrix} = \begin{bmatrix} 1 \\ \frac{u_1}{Nu_2 - u_1} \end{bmatrix} e^{-(u_1+u_c)t} - \frac{u_1}{Nu_2 - u_1} \begin{bmatrix} 0 \\ 1 \end{bmatrix} e^{-(u_c+Nu_2)t}$$

From this, we see

$$X_0(t) = e^{-(u_1+u_c)t} \tag{13.9}$$

$$X_1(t) = \frac{u_1}{Nu_2 - u_1}\left(e^{-(u_1+u_c)t} - e^{-(u_c+Nu_2)t}\right) \tag{13.10}$$

Note, we don't have to use all the machinery of eigenvalues and eigenvectors here. It is clear that solving X_0 is a simple integration.

$$X_0' = -(u_1 + u_c)X_0, \;\; X_0(0) = 1 \Rightarrow X_0(t) = e^{-(u_1+u_c)t}.$$

The next solutions all use the integrating factor method. The next four variables all satisfy models of the form. $u'(t) = -au(t) + f(t)$; $u(t) = 0$ where f is the external data. Using the Integrating factor approach, we find

$$\begin{aligned} e^{at}\left(u'(t) + au(t)\right) &= e^{at}f(t) \\ \left(e^{at}u(t)\right)' &= e^{at}f(t) \\ e^{at}u(t) &= \int_0^t e^{as}f(s)ds \end{aligned}$$

Hence,

$$u(t) = e^{-at}\int_0^t e^{as}f(s)ds.$$

So, we can find X_1 directly. We solve $X_1' = u_1\, X_0 - (u_c + N\, u_2)\, X_1$ with $X_1(0) = 0$. Using the Integrating factor formula with $f(t) = u_1X_0(t)$ and $a = (u_c + Nu_2)$, we have $X_1' = -(u_c + Nu_2)X_1 + u_1X_0$ implies

$$\begin{aligned} X_1(t) &= e^{-(u_c+Nu_2)t}\int_0^t e^{(u_c+Nu_2)s}\, u_1X_0(s)\, ds. \\ &= e^{-(u_c+Nu_2)t}\int_0^t e^{(u_c+Nu_2)s}\, u_1e^{-(u_1+u_c)s}\, ds \\ &= u_1\, e^{-(u_c+Nu_2)t}\int_0^t e^{(Nu_2-u_1)s}\, ds \\ &= \frac{u_1}{Nu_2-u_1}\, e^{-(u_c+Nu_2)t}\left(e^{(Nu_2-u_1)s}\right)\Big|_0^t \\ &= \frac{u_1}{Nu_2-u_1}\, e^{-(u_c+Nu_2)t}\left(e^{(Nu_2-u_1)t} - 1\right) \end{aligned}$$

$$= \frac{u_1}{Nu_2 - u_1}\left(e^{-(u_c+u_1)t} - e^{(u_c+Nu_2)t}\right)$$

Using the X_1 solution, we have to solve $X_2' = N\, u_2\, X_1 - u_c\, X_2$. This leads to

$$X_2(t) = \frac{Nu_1u_2}{Nu_2 - u_1}\left(\frac{1}{u_1}\left(e^{-u_ct} - e^{-(u_1+u_c)t}\right)\right.$$
$$\left. - \frac{1}{Nu_2}\left(e^{-u_ct} - e^{-(u_c+Nu_2)t}\right)\right)$$

The same integrating factor technique with $Y_0' = -u_1 Y_0 + u_c X_0$ with $Y(0) = 0$ using $a = u_1$ and $f(t) = u_c X_0(t)$ leads to

$$Y_0(t) = e^{-u_1t} \int_0^t e^{u_1s}\, u_c X_0(s)\, ds.$$

However, the solutions for the models Y_1 and Y_2 are very messy and so we will try to see what is happening approximately.

13.5 Approximation Ideas

To see how to approximate these solution, let's recall ideas from Sect. 3.1 and apply them to the approximation of the difference of two exponentials. Let's look at the function $f(t) = e^{-rt} - e^{-(r+a)t}$ for positive r and a. To approximate this difference, we expand each exponential function into the second order approximation plus the error as usual.

$$e^{-rt} = 1 - rt + r^2\frac{t^2}{2} - r^3 e^{-rc_1}\frac{t^3}{6}$$
$$e^{-(r+a)t} = 1 - (r+a)t + (r+a)^2\frac{t^2}{2} - (r+a)^3 e^{-(r+a)c_2}\frac{t^3}{6}$$

for some c_1 and c_2 between 0 and t. Subtracting, we have

$$e^{-rt} - e^{-(r+a)t} = \left(1 - rt + r^2\frac{t^2}{2} - r^3 e^{-rc_1}\frac{t^3}{6}\right)$$
$$- \left(1 - (r+a)t + (r+a)^2\frac{t^2}{2}(r+a)^3 e^{-(r+a)c_2}\frac{t^3}{6}\right)$$
$$= at - (a^2 + 2ar)\frac{t^2}{2} + \left(-r^3 e^{-rc_1} + (r+a)^3 e^{-(r+a)c_2}\right)\frac{t^3}{6}$$

We conclude

$$e^{-rt} - e^{-(r+a)t} = at - (a^2 + 2ar)\frac{t^2}{2} + \left(-r^3 e^{-rc_1} + (r+a)^3 e^{-(r+a)c_2}\right)\frac{t^3}{6} \quad (13.11)$$

We can also approximate the function $g(t) = e^{-(r+a)t} - e^{-(r+b)t}$ for positive r, a and b. Using a first order tangent line type approximation, we have

$$\begin{aligned} e^{-(r+a)t} &= 1 - (r+a)t + (r+a)^2 e^{-(r+a)c_1}\frac{t^2}{2} \\ e^{-(r+b)t} &= 1 - (r+b)t + (r+b)^2 e^{-(r+a)c_2}\frac{t^2}{2} \end{aligned}$$

for some c_1 and c_2 between 0 and t. Subtracting, we find

$$\begin{aligned} e^{-(r+a)t} - e^{-(r+b)t} &= \left(1 - (r+a)t + (r+a)^2 e^{-(r+a)c_1}\frac{t^2}{2}\right) \\ &\quad - \left(1 - (r+b)t + (r+b)^2 e^{-(r+b)c_2}\frac{t^2}{2}\right) \\ &= (-a+b)t + \left((r+a)^2 e^{-(r+a)c_1} - (r+b)^2 e^{-(r+b)c_2}\right)\frac{t^2}{2} \end{aligned}$$

We conclude

$$e^{-(r+a)t} - e^{-(r+b)t} = (-a+b)t + \left((r+a)^2 e^{-(r+a)c_1} - (r+b)^2 e^{-(r+b)c_2}\right)\frac{t^2}{2} \quad (13.12)$$

13.5.1 Example

Example 13.5.1 Approximate $e^{-1.0t} - e^{-1.1t}$ using Eq. 13.11.

Solution *We know*

$$e^{-rt} - e^{-(r+a)t} = at - (a^2 + 2ar)\frac{t^2}{2} + \left(-r^3 e^{-rc_1} + (r+a)^3 e^{-(r+a)c_2}\right)\frac{t^3}{6}$$

and here $r = 1.0$ and $a = 0.1$. So we have

$$e^{-1.0t} - e^{-1.1t} \approx 0.1t - (0.01 + 0.2)\frac{t^2}{2} = 0.1t - 0.105t^2$$

and the error is on the order of t^3 which we write as $O(t^3)$ where O stands for order.

Example 13.5.2 Approximate $e^{-2.0t} - e^{-2.1t}$ using Eq. 13.11.

Solution *Again, we use our equation*

$$e^{-rt} - e^{-(r+a)t} = at - (a^2 + 2ar)\frac{t^2}{2} + \left(-r^3 e^{-rc_1} + (r+a)^3 e^{-(r+a)c_2}\right)\frac{t^3}{6}$$

and here $r = 2.0$ *and* $a = 0.1$. *So we have*

$$e^{-2.0t} - e^{-2.1t} \approx 0.1t - (0.01 + 2(0.1)(2))\frac{t^2}{2} = 0.1t - 0.21t^2$$

and the error is $O(t^3)$.

Example 13.5.3 Approximate $e^{-2.1t} - e^{-2.2t}$ using Eq. 13.12.

Solution *Now, we use our equation*

$$e^{-(r+a)t} - e^{-(r+b)t} = (-a+b)t + \left((r+a)^2 e^{-(r+a)c_1} - (r+b)^2 e^{-(r+b)c_2}\right)\frac{t^2}{2}$$

and here $r = 2.0$, $a = 0.1$ *and* $b = 0.2$. *So we have*

$$e^{-2.1t} - e^{-2.2t} \approx (0.2 - 0.1)t = 0.1t$$

and the error is $O(t^2)$.

Example 13.5.4 Approximate $\left(e^{-2.1t} - e^{-2.2t}\right) - \left(e^{-1.1t} - e^{-1.2t}\right)$ using Eq. 13.12.

Solution *We have*

$$\left(e^{-2.1t} - e^{-2.2t}\right) - \left(e^{-1.1t} - e^{-1.2t}\right) \approx 0.1t - 0.1t = 0$$

plus $O(t^2)$ *which is not very useful. Of course, if the numbers had been a little different, we would have not gotten* 0. *If we instead approximate using Eq. 13.11 we find*

$$\begin{aligned}\left(e^{-2.1t} - e^{-2.2t}\right) - \left(e^{-1.1t} - e^{-1.2t}\right) &\approx \left(0.1t - (0.01 + 2(0.1)(2.1)(t^2/2)\right) \\ &\quad - \left(0.1t - (0.01 + 2(0.1)(1.1)(t^2/2)\right) \\ &= -0.215t^2 + 0.230t^2 = 0.015t^2\end{aligned}$$

plus $O(t^3)$ *which is better. Note if the numbers are just right lots of stuff cancels!*

13.5.2 Homework

Exercise 13.5.1 *Approximate* $e^{-3.0t} - e^{-3.2t}$ *using Eq. 13.11.*

Exercise 13.5.2 *Approximate* $e^{-0.8t} - e^{-0.85t}$ *using Eq. 13.11.*

Exercise 13.5.3 *Approximate* $e^{-1.3t} - e^{-1.5t}$ *using Eq. 13.11.*

Exercise 13.5.4 *Approximate* $e^{-1.1t} - e^{-1.3t}$ *using Eq. 13.12.*

Exercise 13.5.5 *Approximate* $e^{-0.07t} - e^{-0.09t}$ *using Eq. 13.12.*

Exercise 13.5.6 *Approximate* $\left(e^{-1.1t} - e^{-1.3t}\right) - \left(e^{-0.7t} - e^{-0.8t}\right)$ *using Eqs. 13.11 and 13.12.*

Exercise 13.5.7 *Approximate* $\left(e^{-2.2t} - e^{-2.4t}\right) - \left(e^{-1.8t} - e^{-1.9t}\right)$ *using Eqs. 13.11 and 13.12.*

Exercise 13.5.8 *Approximate* $\left(e^{-0.03t} - e^{-0.035t}\right) - \left(e^{-0.02t} - e^{-0.025t}\right)$ *using Eqs. 13.11 and 13.12.*

13.6 Approximation of the Top Pathway

We will want to solve the Y_0–Y_2 models in addition to solving the equations for the top pathway. To do this, we can take advantage of some key approximations.

13.6.1 Approximating X_0

We have

$$X_0(t) = 1 - (u_1 + u_c)t + +(u_1 + u_c)^2 e^{-(u_c+u_1)c_1} \frac{t^2}{2}$$

for some c_1 between 0 and t. Hence, $X_0(t) \approx 1 - (u_1 + u_c)t$ with error $E_0(t)$

$$\begin{aligned} E_0(t) &= (u_1 + u_c)^2 e^{-(u_c+u_1)c_1} \frac{t^2}{2} \\ &\leq (u_1 + u_c)^2 \frac{T^2}{2}. \end{aligned}$$

Thus, the maximum error made in approximating $X_0(t)$ by $1-(u_1+u_c)t$ is E_0 given by

$$E_0 = (u_1+u_c)^2\frac{T^2}{2} = \left(10^{-7}(1+R)\right)^2 6.67\times 10^8 = 6.67(1+R)^2\,10^{-6}.$$

We want this estimate for X_0 be reasonable; i.e. first, give a positive number over the human life time range and second, the discrepancy between the true $X_0(t)$ and this approximation is small. Hence, we will assume we want the maximum error in X_0 to be 0.05, we which implies

$$E_0 \leq (1+R)^2\,6.67\times 10^{-6} < 0.01.$$

This implies that

$$(1+R)^2 < 15.0\times 10^2$$

or

$$R < 38.7$$

Since $u_c = R\,u_1$, we se

$$u_c < 3.87\ \times 10^{-6}.$$

to have a good X_0 estimate.

13.6.2 Approximating X_1

Recall

$$X_1(t) = \frac{u_1}{Nu_2-u_1}\left(e^{-(u_c+u_1)t} - e^{-(u_c+Nu_2)t}\right)$$

Since $X_1(t)$ is written as the difference of two exponentials, we can use a first order approximation as discussed in Eq. 13.12, to find

$$\begin{aligned}X_1(t) &= \frac{u_1}{Nu_2-u_1}\left((Nu_2-u_1)t + \left((u_c+u_1)^2e^{-(u_c+u_1)c_1}\right.\right.\\&\quad\left.\left.-(u_c+Nu_2)^2e^{-(u_c+Nu_2)c_2}\right)\frac{t^2}{2}\right)\\&= u_1t + \frac{u_1}{Nu_2-u_1}\left((u_c+u_1)^2e^{-(u_c+u_1)c_1} - (u_c+Nu_2)^2e^{-(u_c+Nu_2)c_2}\right)\frac{t^2}{2}.\end{aligned}$$

Hence, approximating with $X_1(t) \approx u_1 t$, gives a maximum error of

$$E_1 = \max_{0 \le t \le T} \frac{u_1}{Nu_2 - u_1}\left(\left((u_c + u_1)^2 e^{-(u_c+u_1)c_1} - (u_c + Nu_2)^2 e^{-(u_c+Nu_2)c_2}\right)\frac{t^2}{2}\right)$$
$$\le \frac{u_1}{Nu_2 - u_1}\left((u_c + u_1)^2 + (u_c + Nu_2)^2\right)\frac{T^2}{2}.$$

We have already found the $R < 39$ and for N at most 4000 with $u_1 = u_2 = 10^{-7}$, we see $Nu_2 - u_1 \approx Nu_2$. Further, $u_c = Ru_1 \le 10^{-5}$ and $Nu_2 \le 4 \times 10^{-4}$ so that in the second term, Nu_2 is dominant. We therefore have

$$E_1 \le \frac{u_1}{Nu_2}\left((1+R)^2(u_1)^2 + (Nu_2)^2\right)\frac{T^2}{2}.$$

But the term u_1^2 is negligible in the second term, so we have

$$E_1 \approx \frac{u_1}{Nu_2}(Nu_2)^2\frac{T^2}{2}$$
$$= Nu_1 u_2 \frac{T^2}{2} = 4000 \times 10^{-14} \times 6.67 \times 10^8 = 0.027.$$

13.6.3 Approximating X_2

Now here comes the messy part. Apply the second order difference of exponentials approximation from Eq. 13.11 above to our X_2 solution. To make our notation somewhat more manageable, we will define the error term $E(r, a, t)$ by

$$E(r, a, t) = \left(-r^3 e^{-rc_1} + (r+a)^3 e^{-(r+a)c_2}\right)\frac{t^3}{6} \le 2(r+a)^3\frac{t^3}{6}$$

Note the maximum error over human life time is thus $E(r, a)$ which is

$$E(r, a) = 2(r+a)^3\frac{T^3}{6}. \tag{13.13}$$

Now let's try to find an approximation for $X_2(t)$. We have

$$X_2(t) = \frac{Nu_1u_2}{Nu_2 - u_1}\left(\frac{1}{u_1}\left(e^{-u_c t} - e^{-(u_1+u_c)t}\right) - \frac{1}{Nu_2}\left(e^{-u_c t} - e^{-(u_c+Nu_2)t}\right)\right)$$
$$= \frac{Nu_1u_2}{Nu_2 - u_1}\left(\frac{1}{u_1}\left(u_1 t - (u_1^2 + 2u_1u_c)\frac{t^2}{2} + E(u_c, u_1, t)\right)\right)$$

$$- \frac{Nu_1u_2}{Nu_2 - u_1}\left(\frac{1}{Nu_2}\left(Nu_2t - ((Nu_2)^2 + 2Nu_2u_c)\frac{t^2}{2} + E(u_c, Nu_2, t)\right)\right)$$

Now do the divisions above to obtain

$$\begin{aligned} X_2(t) &= \frac{Nu_1u_2}{Nu_2 - u_1}\left(\left(t - (u_1 + 2u_c)\frac{t^2}{2} + \frac{E(u_c, u_1, t)}{u_1}\right)\right) \\ &- \frac{Nu_1u_2}{Nu_2 - u_1}\left(\left(t - (Nu_2 + 2u_c)\frac{t^2}{2} + \frac{E(u_c, Nu_2, t)}{Nu_2}\right)\right) \end{aligned}$$

We can then simplify a bit to get

$$\begin{aligned} X_2(t) &= \frac{Nu_1u_2}{Nu_2 - u_1}\left(\left(t - (u_1 + 2u_c)\frac{t^2}{2} + \frac{E(u_c, u_1, t)}{u_1} - t \right.\right. \\ &\left.\left. + (Nu_2 + 2u_c)\frac{t^2}{2} - \frac{E(u_c, Nu_2, t)}{Nu_2}\right)\right) \\ &= \frac{Nu_1u_2}{Nu_2 - u_1}\left(\left((Nu_2 - u_1)\frac{t^2}{2} + \frac{E(u_c, u_1, t)}{u_1} - \frac{E(u_c, Nu_2, t)}{Nu_2}\right)\right) \\ &= Nu_1u_2\frac{t^2}{2} + \frac{Nu_2}{Nu_2 - u_1}E(u_c, u_1, t) - \frac{u_1}{Nu_2 - u_1}E(u_c, Nu_2, t). \end{aligned}$$

Hence, we see $X_2(t) \approx Nu_1u_2\frac{t^2}{2}$ with maximum error E_2 over human life time T given by

$$\begin{aligned} E_2 &= \max_{0 \le t \le T}\left(\frac{Nu_2}{Nu_2 - u_1}E(u_c, u_1, t) - \frac{u_1}{Nu_2 - u_1}E(u_c, Nu_2, t).\right) \\ &\le \frac{Nu_2}{Nu_2 - u_1}E(u_c, u_1) + \frac{u_1}{Nu_2 - u_1}E(u_c, Nu_2) \\ &= \left(\frac{Nu_2}{Nu_2 - u_1}\, 2(u_1 + u_c)^3 + \frac{u_1}{Nu_2 - u_1} 2(u_c + Nu_2)^3\right)\frac{T^3}{6} \end{aligned}$$

From our model assumptions, an upper bound on N is 4000. Since $u_1 = u_2 = 10^{-7}$, we see the term $Nu_2 - u_1 \approx Nu_2 = 4 \times 10^{-4}$. Also, $u_1 + u_c$ is dominated by u_c. Thus,

$$\begin{aligned} E_2 &\approx \left(2u_c^3 + \frac{u_1}{Nu_2 - u_1}(Nu_2)^3\right)\frac{T^3}{6} \\ &\approx \left(2u_c^3 + \frac{u_1}{Nu_2}(Nu_2)^3\right)\frac{T^3}{6} \\ &\approx \left(2u_c^3 + u_1(Nu_2)^2\right)\frac{T^3}{6} \end{aligned}$$

Table 13.1 The Non CIN Pathway Approximations with error estimates

Approximation	Maximum error
$X_0(t) \approx 1 - (u_1 + u_c)\, t$	$(u_1 + u_c)^2 \, \frac{T^2}{2}$
$X_1(t) \approx u_1\, t$	$\frac{u_1}{Nu_2 - u_1} \left((u_c + u_1)^2 + (u_c + Nu_2)^2 \right) \frac{T^2}{2}$
$X_2(t) \approx N\, u_1\, u_2\, \frac{t^2}{2}$	$2u_c^3 \frac{T^3}{6}$

But the term $u_1 (Nu_2)^2$ is very small ($\approx 1.6e - 14$) and can also be neglected. So, we have

$$E_2 \approx 2u_c^3 \frac{T^3}{6}$$

Thus, if $R \leq 39$, we have $R^3 = 5.9 \times 10^4$ and

$$\begin{aligned} E_2 &\approx 2 \times (5.8 \times 10^{-17}) \times 8.1 \times 10^{12} \\ &= 94 \times 10^{-5} = 0.00094. \end{aligned}$$

We summarize our approximation results for the top pathway in Table 13.1.

13.7 Approximating the CIN Pathway Solutions

We now know that

$$X_0(t) \approx 1 - (u_1 + u_c)\, t \tag{13.14}$$
$$X_1(t) \approx u_1\, t \tag{13.15}$$
$$X_2(t) \approx N\, u_1\, u_2\, \frac{t^2}{2}. \tag{13.16}$$

We can use these results to approximate the CIN cell populations Y_0, Y_1 and Y_2. Let's recall the relevant equations.

$$Y_0' = u_c\, X_0 - u_1\, Y_0 \tag{13.17}$$
$$Y_1' = u_c\, X_1 + u_1\, Y_0 - N\, u_3\, Y_1 \tag{13.18}$$
$$Y_2' = N\, u_3\, Y_1 + u_c\, X_2. \tag{13.19}$$

Now, replace X_0, X_1 and X_2 in the CIN cell population dynamics equations to give us the approximate equations to solve.

$$Y_0' = u_c\,(1 - (u_1 + u_c)\,t) - u_1\,Y_0 \tag{13.20}$$

$$Y_1' = u_c\,(u_1\,t) + u_1\,Y_0 - N\,u_3\,Y_1 \tag{13.21}$$

$$Y_2' = N\,u_3\,Y_1 + u_c\left((N\,u_1\,u_2\,\frac{t^2}{2}\right). \tag{13.22}$$

13.7.1 The Y_0 Estimate

First, if we replace X_0 by its approximation, we have $Y_0' = u_c\,(1 - (u_1 + u_c)\,t + E(t,0)) - u_1\,Y_0$ where $E(t,0)$ is the error we make at time t. The approximate model is then $Y_0' = u_c\,(1 - (u_1 + u_c)\,t) - u_1\,Y_0$. The difference of these two models give us a way of understanding how much error we make in replacing the full solution X_0 by its approximation. Let ϵ be this difference. We find

$$\epsilon'(t) = u_c E(t,0),\ \epsilon(0) = 0$$

is our model of the difference. We know $E(t,0) = (u_1 + u_c)^2 e^{-(u_1+u_c)\beta}\frac{t^2}{2}$ here, so the difference model is bounded by the envelope models:

$$\epsilon'(t) = \pm\, u_c(u_1 + u_c)^2 \frac{t^2}{2},\ \epsilon(0) = 0$$

The maximum error we can have due to this difference over human life time is then found by integrating. We have

$$\epsilon(T) = u_c(u_1 + u_c)^2 \frac{T^3}{6},$$

which is about 0.0005. Hence, this contribution to the error is a bit small. Next, let's think about the other approximation error. After rearranging, the Y_0 approximate dynamics are

$$Y_0' + u_1\,Y_0 = u_c - u_c\,(u_1 + u_c)\,t.$$

We solve this equation using the integrating factor method, with factor $e^{u_1 t}$. This yields

$$\left(Y_0(t)\,e^{u_1 t}\right)' = u_c\,e^{u_1 t} - u_c\,(u_1 + u_c)\,t\,e^{u_1 t}.$$

Integrating and using the fact that $Y_0(0) = 0$, we have

$$Y_0(t)\,e^{u_1 t} = u_c \int_0^t e^{u_1 s}\,ds - u_c \int_0^t (u_1 + u_c)\,s\,e^{u_1 s}\,ds$$

$$= \frac{u_c}{u_1}\left(e^{u_1 t} - 1\right) - u_c\,\frac{u_1 + u_c}{u_1}\,s\,e^{u_1 s}\Big|_0^t + u_c\int_0^t \frac{u_1 + u_c}{u_1}\,e^{u_1 s}\,ds$$

$$= \frac{u_c}{u_1}\left(e^{u_1 t} - 1\right) - u_c\,\frac{u_1 + u_c}{u_1}\,t\,e^{u_1 t} + u_c\,\frac{u_1 + u_c}{u_1^2}\left(e^{u_1 t} - 1\right).$$

Now group the $e^{u_1 t}$ and $t\,e^{u_1 t}$ terms together. We have

$$Y_0(t)\,e^{u_1 t} = \frac{2\,u_1\,u_c + u_c^2}{u_1^2}\,e^{u_1 t} - \frac{u_1\,u_c + u_c^2}{u_1}\,t\,e^{u_1 t} - \frac{2\,u_1\,u_c + u_c^2}{u_1^2}.$$

Now multiply through by $e^{-u_1 t}$ to find

$$Y_0(t) = \frac{2\,u_1\,u_c + u_c^2}{u_1^2} - \frac{u_1\,u_c + u_c^2}{u_1}\,t - \frac{2\,u_1\,u_c + u_c^2}{u_1^2}\,e^{-u_1 t}$$

To see what is going on here, we split this into two pieces as follows:

$$Y_0(t) = \frac{u_c}{u_1}\left(1 - e^{-u_1 t}\right) + \frac{u_1\,u_c + u_c^2}{u_1^2}\left(1 - u_t\,t - e^{-u_1 t}\right). \tag{13.23}$$

The critical thing to recall now is that

$$e^{-u_1 t} \approx 1 - u_1\,t$$

with error $E(t, 0)$ given by

$$E(t, 0) = u_1^2\,e^{-u_1\,\beta}\,\frac{t^2}{2}$$

where β is some number between 0 and t. Then, as before, the largest possible error over a human lifetime uses maximum time $T = 3.65 \times 10^4$ giving

$$|E(t, 0)| \leq u_1^2\,\frac{T^2}{2}.$$

Hence, we can rewrite Eq. 13.23 as

$$Y_0(t) = \frac{u_c}{u_1}\left(u_1\,t - E(t, 0)\right) + \frac{u_1\,u_c + u_c^2}{u_1^2}\left(-\,E(t, 0)\right) \tag{13.24}$$

$$= u_c\,t - \frac{2\,u_1\,u_c + u_c^2}{u_1^2}\,E(t, 0). \tag{13.25}$$

We conclude that $Y_0(t) \approx u_c\, t$ with maximum error given by

$$|Y_0(t) - u_c\, t| \leq \frac{2\, u_1\, u_c + u_c^2}{u_1^2}\, u_1^2\, \frac{T^2}{2} = (2\, u_1\, u_c + u_c^2)\, \frac{T^2}{2}. \qquad (13.26)$$

For our chosen value of $R \leq 39$, we see the magnitude of the Y_0 error is given by

$$E = (2 + R)\, R\, (6.67 \times 10^{-6}).$$

For $R < 39$, we have

$$E < 1.067 \times 10^{-2} = 0.01067.$$

If we add the error due to replacing the true dynamics by the approximation, we find the total error is about $0.01067 + 0.0005 = 0.0105$.

13.7.2 The Y_1 Estimate

We can do the error due to the replacement of the true solutions by their approximations here too, but the story is similar; the error is small. We will focus on how to approximate Y_1 using the approximate dynamics. The Y_1 dynamics are

$$Y_1' = u_c\, X_1 + u_1\, Y_0 - N\, u_3\, Y_1.$$

Over the effective lifetime of a human being, we can use our approximations for X_1 and Y_0 to obtain the dynamics that are relevant. This yields

$$Y_1' = u_c\, (u_1\, t) + u_1\, (u_c\, t) - N\, u_3\, Y_1 = 2\, u_1\, u_c\, t - N\, u_3\, Y_1.$$

This can be integrated using the integrating factor $e^{N\, u_3\, t}$. Hence,

$$\begin{aligned} Y_1(t)\, e^{N\, u_3\, t} &= 2\, u_1\, u_c \int_0^t s\, e^{N\, u_3\, s}\, ds \\ &= 2\, u_1\, u_c \left(\frac{1}{N\, u_3} t\, e^{N\, u_3\, t} - \frac{1}{(N\, u_3)^2}\, (e^{N\, u_3\, t} - 1) \right) \\ &= \frac{2\, u_1\, u_c}{(N\, u_3)^2} \left((N\, u_3\, t - 1)\, e^{N\, u_3\, t} + 1 \right). \end{aligned}$$

Multiplying through by the integrating factor, we have

$$Y_1(t) = \frac{2\, u_1\, u_c}{(N\, u_3)^2} \left((N\, u_3\, t - 1) + e^{-N\, u_3\, t} \right).$$

Now group the terms as follows:

$$\begin{aligned} Y_1(t) &= \frac{2\,u_1\,u_c}{(N\,u_3)^2}\left(N\,u_3\,t\, + (e^{-N\,u_3\,t}\, - 1)\right) \\ &= \frac{2\,u_1\,u_c}{N\,u_3}\,t + \frac{2\,u_1\,u_c}{(N\,u_3)^2}\,(e^{-N\,u_3\,t}\, - 1). \end{aligned}$$

We know that we can approximate $e^{-N\,u_3\,t}$ as usual as

$$e^{-N\,u_3\,t} = 1 + \,E(t, 0).$$

The error term E is then given by

$$E(t, 0) = -N\,u_3\,e^{-N\,u_3\,\beta}\,t$$

for some β in $[0, t]$. As usual, this error is largest when $\beta = 0$ and t is the lifetime of our model. Thus

$$|E(t, 0)| \le N\,u_3\,T.$$

Thus,

$$\begin{aligned} \left|Y_1(t) - \frac{2\,u_1\,u_c}{N\,u_3}\,t\right| &\le \frac{2\,u_1\,u_c}{(N\,u_3)^2}\,(N\,u_3)\,T \\ &= \frac{2\,u_1\,u_c}{N\,u_3}\,T \end{aligned}$$

However, we know $Nu_3 \approx 1$ and so

$$\begin{aligned} |Y_1(t) - \frac{2\,u_1\,u_c}{N\,u_3}\,t| &\le 2u_1u_cT \\ &\le 2\,(10^{-7})\,R\,(10^{-7})\,T, \end{aligned}$$

using our value for u_1 and our model for u_c. We already know to have reasonable error in X_0 we must have $R < 39$. Since our average human lifetime is $T = 36{,}500$ days, we see

$$\begin{aligned} |Y_1(t) - \frac{2\,u_1\,u_c}{N\,u_3}\,t| &\le 2\,(10^{-7})\,39\,(10^{-7})\,3.65\,(10^4) \\ &\le 2.85 \times 10^{-8}. \end{aligned}$$

Thus, the error in the Y_1 estimate is quite small.

13.7.3 The Y_2 Estimate

All that remains is to estimate Y_2. The dynamics are

$$Y_2' = N\,u_3\,Y_1 + u_c\left(N\,u_1\,u_2\,\frac{t^2}{2}\right).$$

Using our approximations, we obtain

$$\begin{aligned} Y_2' &= N\,u_3\left(\frac{2\,u_1\,u_c}{N\,u_3}\,t\right) + u_c\left(N\,u_1\,u_2\,\frac{t^2}{2}\right) \\ &= 2\,u_1\,u_c\,t + \frac{N\,u_1\,u_2\,u_c}{2}\,t^2. \end{aligned}$$

The integration (for a change!) is easy. We find

$$Y_2(t) = u_1\,u_c\,t^2 + \frac{N\,u_1\,u_2\,u_c}{6}\,t^3.$$

Note that

$$\left|Y_2(t) - u_1\,u_c\,t^2\right| = \frac{N\,u_1\,u_2\,u_c}{6}\,t^3.$$

The error term is largest at the effective lifetime of a human being. Thus, we can say (using our assumptions on the sizes of u_1, u_2 and u_c)

$$\left|Y_2(t) - u_1\,u_c\,t^2\right| \leq (N\,u_1\,u_2\,u_c)\,\frac{T^3}{6}.$$

The magnitude estimate for Y_2 can now be calculated. For $R < 39$, we have

$$N\,R\,8.11 \times 10^{-9} < 316.3\,N \times 10^{-9}.$$

For $N \leq 4000$, this becomes

$$N\,R\,8.11 \times 10^{-9} < 1.265 \times 10^{-3}.$$

We summarize our approximation results for the top pathway in Table 13.2.

Table 13.2 The CIN Model Approximations with error estimates

Approximation	Maximum error
$Y_0(t) \approx u_c\, t$	$(2\, u_1\, u_c + u_c^2)\, \frac{T^2}{2}$
$Y_1(t) \approx \frac{2\, u_1\, u_c}{N\, u_3}\, t$	$\frac{2\, u_1\, u_c}{N\, u_3}\, T$
$Y_2(t) \approx u_1\, u_c\, t^2$	$N\, u_1\, u_2\, u_c\, \frac{T^3}{6}$

13.8 Error Estimates

In Sect. 13.4, we solved for X_0, X_1 and X_2 exactly and then we developed their approximations in Sect. 13.6. The approximations to X_0, X_1 and X_2 then let us develop approximations to the CIN solutions. We have found that for reasonable errors in $X_0(t)$, we need $R < 38.7$. We were then able to calculate the error bounds for all the variables. For convenience, since u_1 and u_2 are equal, let's set them to be the common value u. Then, by assumption $u_c = R\, u$ for some $R \geq 1$. Our error estimates can then be rewritten as seen in Table 13.3. This uses our assumptions that $u_1 = u_2 = 10^{-7}$ and $u_c = Ru_1 \approx 4 \times 10^{-6}$.

The error magnitude estimates are summarized in Table 13.4.

Table 13.3 The Non CIN and CIN Model Approximations with error estimates using $u_1 = u_2 = u$ and $u_c = R\, u$

Approximation	Maximum error
$X_0(t) \approx 1\, - (u_1 + u_c)\, t$	0.01
$X_1(t) \approx u_1\, t$	0.027
$X_2(t) \approx N\, u_1\, u_2\, \frac{t^2}{2}$	0.0009
$Y_0(t) \approx u_c\, t$	0.0016
$Y_1(t) \approx \frac{2\, u_1\, u_c}{N\, u_3}\, t$	2.9×10^{-8}
$Y_2(t) \approx u_1\, u_c\, t^2$	0.0013

Table 13.4 The Non CIN and CIN Model Approximations Dependence on population size N and the CIN rate for $R \approx 39$ with $u_1 = 10^{-7}$ and $u_c = R\, u_1$

Approximation	Maximum error
$X_0(t) \approx 1\, - (u_1 + u_c)\, t$	$(1 + R)^2\, 6.67 \times 10^{-6} < 0.01$
$X_1(t) \approx u_1\, t$	$\frac{u_1^2}{N} \left((1 + R)^2 + (R + N)^2 \right) \frac{T^2}{2} < 0.027$
$X_2(t) \approx N\, u_1\, u_2\, \frac{t^2}{2}$	$(3\, N\, R + \frac{N}{N-1}\, (N^2 + 1)\, e^{-RuT})\, 8.11 \times 10^{-9} < 0.0009$
$Y_0(t) \approx u_c\, t$	$(2 + R)\, R\, 6.67 \times 10^{-6} < 0.0016$
$Y_1(t) \approx \frac{2\, u_1\, u_c}{N\, u_3}\, t$	$2Ru_1^2 T < 2.9 \times 10^{-8}$
$Y_2(t) \approx u_1\, u_c\, t^2$	$RN\, u_1^3\, \frac{T^3}{6} = RN\, 8.11 \times 10^{-9} < 0.0013$

13.9 When Is the CIN Pathway Dominant?

We think of N as about 4000 for a colon cancer model, but the loss of two allele model is equally valid for different population sizes. Hence, we can think of N as a variable also. In the estimates we generated in Sects. 13.6 and 13.7, we used the value 4000 which generated even larger error magnitudes than what we would have gotten for smaller N. To see if the CIN pathway dominates, we can look at the ratio of the Y_2 output to the X_2 output. The ratio of Y_2 to X_2 tells us how likely the loss of both alleles is due to CIN or without CIN. We have, for $R < 39$, that

$$\frac{Y_2(T)}{X_2(T)} = \frac{u_1 u_c T^2 + E(T)}{(1/2)N u_1 u_2 T^2 + F(T)}$$

where $E(T)$ and $F(T)$ are the errors associated with our approximations for X_2 and Y_2. We assume $u_1 = u_2$ and so we can rewrite this as

$$\frac{Y_2(T)}{X_2(T)} = \frac{(2R/N) + (2/N u_1 u_2 T^2)E(T)}{1 + (2/N u_1 u_2 T^2)F(T)}$$

For $N = 4000$ and $u + 1 = u_2 = 10^{-7}$ and $T = 36,500$ days, we find $2/N u_1 u_2 T^2 = 37.5$ and hence we have

$$\frac{Y_2(T)}{X_2(T)} = \frac{(2R/N) + 37.5\, E(T)}{1 + 37.5\, F(T)}$$

Now $E(T) \approx 0.0009$ and $F(T) \approx 0.0013$ and so

$$\frac{Y_2(T)}{X_2(T)} \approx \frac{(2R/N) + 0.0388}{1.0488}$$

We can estimate how close this is to the ratio $2R/N$ and find Now $E(T) \approx 0.00009$ and $F(T) \approx 0.0013$ and so

$$\left| \frac{(2R/N) + 0.04}{1.05} - \frac{2R}{N} \right| \approx \left| \frac{0.04 - 0.05(2R/N)}{1.05} \right|$$

Here $R < 39$ and N is at least 1000, so $(2R/N) \approx 0.08$. Hence, the numerator is about $|0.04 - 0.0004| \approx 0.04$. Thus, we see the error we make in using $(2R/N)$ as an estimate for $Y_2(T)/X_2(T)$ is about 0.04 which is fairly small. Hence, we can be reasonably confident that the critical ratio $(2R)/N$ is the same as the ratio $Y_2(T)/X_2(T)$ as the error over human life time is small. Our analysis only works for $R < 39$ though so we should be careful in applying it. Hence, we can say

$$\frac{Y_2(T)}{X_2(T} \approx \frac{2\,R}{N}.$$

Table 13.5 The CIN decay rates, u_c required for CIN dominance. with $u_1 = u_2 = 10^{-7}$ and $u_c = R\,u_1$ for $R \geq 1$

N	Permissible u_c For CIN dominance	R value
100	$>5.0 \times 10^{-6}$	50
170	$>8.5 \times 10^{-6}$	85
200	$>1.0 \times 10^{-5}$	100
500	$>2.5 \times 10^{-5}$	250
800	$>4.0 \times 10^{-5}$	400
1000	$>5.0 \times 10^{-5}$	500
2000	$>10.0 \times 10^{-5}$	1000
4000	$>20.0 \times 10^{-5}$	2000

The third column shows the R value needed for a good CIN dominance

Hence, the pathway to Y_2 is the most important if $2\,R > N$. This implies the CIN pathway is dominant if

$$R > \frac{N}{2}. \tag{13.27}$$

For the fixed value of $u_1 = 10^{-7}$, we calculate in Table 13.5 possible u_c values for various choices of N. We have CIN dominance if $R > \frac{N}{2}$ and the approximations are valid if $R < 39$.

Our estimates have been based on a cell population size N of 1000–4000. We also know for a good $X_0(t)$ estimate we want $R < 39$. Hence, as long as $u_c < 3.9 \times 10^{-6}$, we have a good $X_0(t)$ approximation and our other estimates are valid. From Table 13.5, it is therefore clear we will not have the CIN pathway dominant. We would have to drop the population size to 70 or so to find CIN dominance. Now our model development was based on the loss of alleles in a TSG with two possible alleles, but the mathematical model is equally valid in another setting other than the colon cancer one. If the population size is smaller than what we see in our colon cancer model, it is therefore possible for the CIN pathway to dominate! **However, in our cancer model, it is clear that the non CIN pathway dominates.**

Note we can't even do this ratio analysis unless we are confident that our approximations to Y_2 and X_2 are reasonable. The only way we know if they are is to do a careful analysis using the approximation tools developed in Sect. 3.1. **The moral is that we need to be very careful about how we use estimated solutions when we try to do science!** However, with caveats, it is clear that our simple model gives an interesting inequality, $2u_c > N u_2$, which helps us understand when the CIN pathway dominates the formation of cancer.

13.9.1 Homework

We are assuming $u_1 = u_2 = 10^{-7}$ and $u_c = Ru_1$ where R is a parameter which needs to be less than 39 or so in order for our approximations to be ok. We have the human life time is $T = 36{,}500$ days. and finally, we have N, the total number of cells in the population which is 1000–4000. Now let's do some calculations using different T, N, u_1 and R values. For the choices below,

- Find the approximate values of all variables at the given T.
- Interpret the number at T as the number of cells in the that state after T days. This means take your values for say $X_2(T)$ and multiple by N to get the number of cells.
- Determine if the top or bottom pathway to cancer is dominant Recall, if our approximations are good, the equation $2R/N > 1$ implies the top pathway is dominant.

As you answer these questions, note that we can easily use the equation above for R and N values for which our approximations are probably not good. Hence, it is always important to know the guideline we use to answer our question can be used! So for example, most of the work in the book suggests that $R \leq 39$ or so, so when we use these equations for $R = 110$ etc. we can't be sure our approximations are accurate to let us do that!

Exercise 13.9.1 *$R = 32$, $N = 1500$, $u_1 = u_2 = 6.3 \times 10^{-8}$ and $T = 100$ years but express in days.*

Exercise 13.9.2 *$R = 70$, $N = 4000$, $u_1 = u_2 = 8.3 \times 10^{-8}$ and $T = 100$ years but express in days.*

Exercise 13.9.3 *$R = 110$, $N = 500$, $u_1 = u_2 = 1 \times 10^{-7}$ and $T = 105$ years but express in days.*

Exercise 13.9.4 *$R = 20$, $N = 180$, $u_1 = u_2 = 1 \times 10^{-7}$ and $T = 95$ years but express in days.*

13.10 A Little Matlab

Let's look at how we might solve a cancer model using Matlab. Our first attempt might be something like this. First, we set up the dynamics as

Listing 13.1: The cancer model dynamics

```
f = @(t,x) [-(u1+uc)*x(1);...
             u1*x(1)-(uc+N*u2)*x(2);...
             N*u2*x(2)-uc*x(3);...
             uc*x(1)-u1*x(4);...
5            uc*x(2)+u1*x(4)-N*u3*x(5);...
             N*u3*x(5)+uc*x(3)];
```

But to make this work, we also must initialize all of our parameters before we try to use the function. We have annotated the code lightly as most of it is pretty common place to us now. We set the final time, as usual, to be human lifetime of 100 years or 36,500 days. Our mutation rate $u_1 = 1.0 \times 10^{-7}$ is in units of base pairs/day and this requires many steps. If we set the step size to 0.5 as we do here, 73,000 steps are required and even on a good laptop it will take a long time. Let's do the model for a value of $R = 80$ which is much higher than we can handle with our estimates!

Listing 13.2: A First Cancer Model Attempt

```
% Set the parameters
u1 = 1.0e-7;
u2 = u1;
N = 1000;
R = 80;
uc = R*u1;
r = 1.0e-4;
u3 = (1-r)/N;
% set the dynamics
f = @(t,x) [-(u1+uc)*x(1);...
                      u1*x(1)-(uc+N*u2)*x(2);...
                      N*u2*x(2)-uc*x(3);...
                      uc*x(1)-u1*x(4);...
                      uc*x(2)+u1*x(4)-N*u3*x(5);...
                      N*u3*x(5)+uc*x(3)];
% set the final time and stepsize
T = 36500;
h = .5;
% find the RK 4 approximations
M = ceil(T/h);
xinit = [1;0;0;0;0;0];
[htime,rk,frk] = FixedRK(f,0,xinit,h,4,M);
% save the approximations in variables named
% like the text
X0 = rk(1,:);
X1 = rk(2,:);
X2 = rk(3,:);
Y0 = rk(4,:);
Y1 = rk(5,:);
Y2 = rk(6,:);
% plot Number of A-- and A--CIN cells
plot(htime,N*X2,htime,N*Y2);
legend('A--','A--CIN','location','west');
xlabel('Time days');
ylabel('A--, A--CIN cells');
title('N = 1000, R = 80; Top always dominant');
% find the last entry in each variable
[rows,cols] = size(rk);
% find Y2 at final time and X2 at final time
Y2(cols)
X2(cols)
% Find Y2(T)/X2(T)
Y2(cols)/X2(cols)
```

The plot of the number of cells in the A^{--} and A^{--CIN} state is seen in Fig. 13.5.

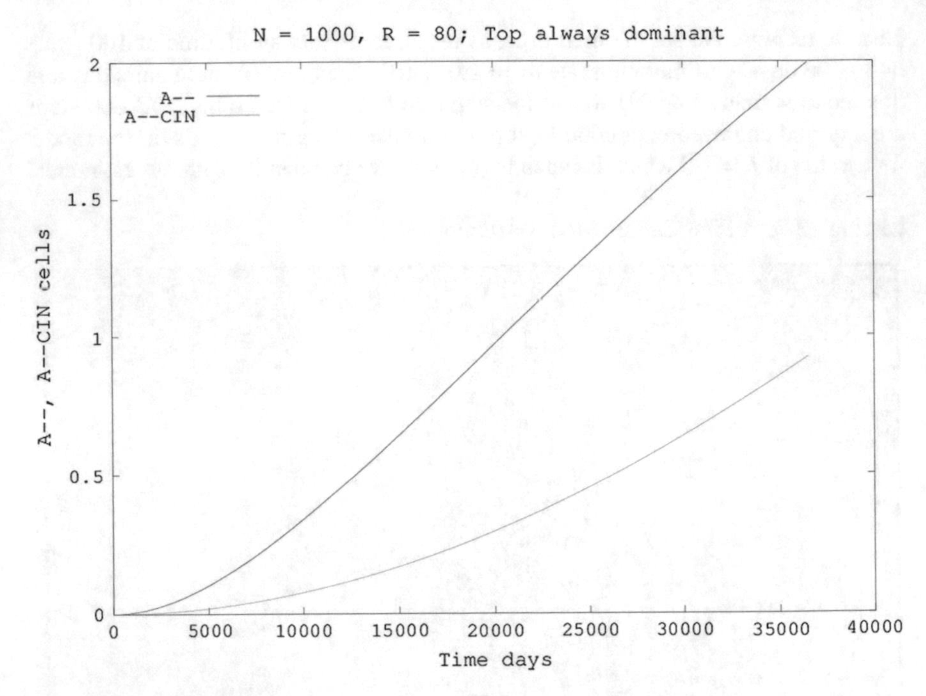

Fig. 13.5 The number of cells in A^{--} and A^{--CIN} state versus time

If we could use our theory, it would tell us that since $2R/N = 160/1000 = 0.16$, the top pathway to cancer is dominant. Our numerical results give us

Listing 13.3: Our First Results

```
cols =
   73000
Y2(cols) =
   9.2256e-04
X2(cols) =
   0.0020
Y2(cols)/X2(cols) =
   0.4624
```

so the numerical results, while giving a $Y2(T)/X2(T)$ different from $2R/N$ still predict the top pathway to cancer is dominant. So it seems our general rule that $2R/N < 1$ which we derived using complicated approximation machinery is working well. We can change our time units to years to cut down on the number of steps we need to take. To do this, we convert base pairs/day to base pairs/year by multiplying u_1 by 365. The new code is then

Listing 13.4: Switching to time units to years: step size is one half year

```
u1 = 1.0e-7*365;
u2 = u1;
N = 1000;
R = 80;
uc = R*u1;
r = 1.0e-4;
u3 = (1-r)/N;
f = @(t,x) [-(u1+uc)*x(1);...
             u1*x(1)-(uc+N*u2)*x(2);...
             N*u2*x(2)-uc*x(3);...
             uc*x(1)-u1*x(4);...
             uc*x(2)+u1*x(4)-N*u3*x(5);...
             N*u3*x(5)+uc*x(3)];
T = 100;
h = .5;
M = ceil(T/h);
[htime,rk,frk] = FixedRK(f,0,[1;0;0;0;0;0],h,4,M);
X0 = rk(1,:);
X1 = rk(2,:);
X2 = rk(3,:);
Y0 = rk(4,:);
Y1 = rk(5,:);
Y2 = rk(6,:);
[rows,cols] = size(rk)
N*Y2(cols)
N*X2(cols)
% Find Y2(T)/X2(T)
Y2(cols)/X2(cols)
2*R/N
Y2(cols)
   9.0283e-04
X2(cols)
   0.0020
Y2(cols)/X2(cols)
   0.4548
```

We get essentially the same results using a time unit of years rather than days! The example above uses a time step of $h = 0.5$ which is a half year step. We can do equally well using a step size of $h = 1$ which is a year step. We can do it again using $h = 1$ (i.e. the step is one year now)

Listing 13.5: Switching the step size to one year

```
h = 1; M = ceil(T/h); %ans =  100
[htime,rk,frk] = FixedRK(f,0,[1;0;0;0;0;0],h,4,M);
X0 = rk(1,:); X1 = rk(2,:); X2 = rk(3,:);
Y0 = rk(4,:); Y1 = rk(5,:); Y2 = rk(6,:);
[rows,cols] = size(rk);
Y2(cols)/X2(cols)
ans =   0.45291
2*R/N
ans =   0.16000
Y2(cols)
ans =   8.9435e-04
N*Y2(cols)
ans =   0.89435
X2(cols)
ans =   0.0019747
N*X2(cols)
ans =   1.9747
```

We can do similar things for different choices of N, R and h.

13.10.1 Homework

We again assume $u_1 = u_2 = 10^{-7}$ and $u_c = Ru_1$ where R is a parameter. We have the human life time is $T = 36{,}500$ days. and finally, we have N, the total number of cells in the population which is 1000–4000. Now let's do some calculations using different N and R values. For the choices below,

- Solve the problem in MatLab using time measured in days. This is the hardest case as it is slow. Calculate $2R/N$ and $Y2(T)/X2(T)$ in your MatLab session and determine which pathway to cancer is dominant.
- Do the same but now measure time in years and use $h = 1$. The calculations are much faster now and the results are similar.

As you answer these questions, note the rule $2R/N$ seems valid for a much larger value of R than our approximations suggest. But remember, our estimates give worst case errors, so it is not surprising if the rule performs well even if R exceeds 39.

Exercise 13.10.1 $R = 100$, $N = 1500$.

Exercise 13.10.2 $R = 200$, $N = 4000$.

Exercise 13.10.3 $R = 250$, $N = 400$.

Exercise 13.10.4 $R = 40$, $N = 60$.

13.11 Insight Is Difficult to Achieve

We have had to work hard to develop some insight into the relative dominance of the CIN pathway in this two allele cancer model. It has been important to solve our model for arbitrary parameters u_1, u_2, u_3, u_c and N. This need to do everything in terms of parameters treated as variables complicated much of our analysis. Could we do it another way? Well, we could perform a parametric study. We could solve the model for say 10 choices of each of the 5 parameters using MatLab for the values $X_2(T)$ and $Y_2(T)$ which, of course, depend on the values of u_1, u_2, u_3, u_c and N used in the computation. Hence, we should label this final time values as $X_2(T, u_1, u_2, u_3, u_c, N)$ and $Y_2(T, u_1, u_2, u_3, u_c, N)$ to denote this dependence. After 10^5 separate MatLab computations, we would then have a data set consisting of 10^5 values of the variables $X_2(T, u_1, u_2, u_3, u_c, N)$ and $Y_2(T, u_1, u_2, u_3, u_c, N)$. We could then try to do statistical modeling to see if we could tease out the CIN dominance relation $2\,u_c > N\,u_2$. But, choosing only 10 values for each parameter might be too coarse for our purposes. If we choose 100 values for each parameter, we would have to do 10^{10} computations to develop an immense table of 10^{10} entries of $X_2(T)$ and $Y_2(T)$ values! You should be able to see that the theoretical approach we have taken here, while hard to work through, has some benefits!

With all the we have said so far, look at the following exposition of our cancer model; you might read the following in a research report. Over the typical lifetime of a human being, the variables in our model have functional dependencies on time (denoted by the symbol $\sim$) given as follows:

$$
\begin{aligned}
X_0(t) &\sim 1 \; - (u_1 + u_c)\, t \\
X_1(t) &\sim u_1\, t \\
X_2(t) &\sim N\, u_1\, u_2\, \frac{t^2}{2} \\
Y_0(t) &\sim u_c\, t \\
Y_1(t) &\sim \frac{2\, u_1\, u_c}{N\, u_3}\, t \\
Y_2(t) &\sim u_1\, u_c\, t^2.
\end{aligned}
$$

It is then noted that the ratio Y_2 to X_2 gives interesting information about the dominance of the CIN pathway. CIN dominance requires the ratio exceeds one giving us the fundamental inequality

$$\frac{2\, u_c}{N\, u_2} > 1.$$

Note there is no mention in this derivation about the approximation error magnitudes that must be maintained for the ratio tool to be valid! So because no details are presented, perhaps we should be wary of accepting this model! if we used it without being sure it was valid to make decisions, we could be quite wrong.

References

M. Novak, *Evolutionary Dynamics: Exploring the Equations of Life* (Belknap Press, Cambridge, 2006)

J. Peterson, *Calculus for Cognitive Scientists: Derivatives, Integration and Modeling*, Springer Series on Cognitive Science and Technology (Springer Science+Business Media Singapore Pte Ltd, Singapore, 2015 In press)

B. Ribba, T. Colin, S. Schnell, A multiscale mathematical model of cancer, and its use in analyzing irradiation therapies. Theor. Biol. Med. Model. **3**, 1–19 (2006)

Y. Tatabe, S. Tavare, D. Shibata, Investigating stem cells in human colon using methylation patterns. Proc. Natl. Acad. Sci. **98**, 10839–10844 (2001)

Part V
Nonlinear Systems Again

Chapter 14
Nonlinear Differential Equations

We are now ready to solve nonlinear systems of nonlinear differential equations using our new tools. Our new tools will include

1. The use of linearization of nonlinear ordinary differential equation systems to gain insight into their long term behavior. This requires the use of partial derivatives.
2. More extended qualitative graphical methods.

Recall, this means we are using the ideas of approximating functions of two variables by tangent planes which results in tangent plane error. This error is written in terms of Hessian like terms. The difference now is that x' has its dynamics f and y' has its dynamics g and so we have to combine tangent plane approximations to both f and g into one Hessian like term to estimate the error. So the discussions below are similar to the ones in the past, but a bit different.

14.1 Linear Approximations to Nonlinear Models

Consider the nonlinear system

$$\begin{aligned} x' &= f(x, y) \\ y' &= g(x, y) \\ x(0) &= x_0 \\ y(0) &= y_0 \end{aligned}$$

where both f and g possess continuous partial derivatives up to the second order. For any fixed Δx and Δy, define the function h by

$$h(t) = \begin{bmatrix} f(x_0 + t\Delta x, y_0 + t\Delta y) \\ g(x_0 + t\Delta x, y_0 + t\Delta y) \end{bmatrix}$$

J.K. Peterson, *Calculus for Cognitive Scientists*,
Cognitive Science and Technology, DOI 10.1007/978-981-287-877-9_14

Then we can expand h in a second order approximation to find

$$h(t) = h(0) + h'(0)t + h''(c)\frac{t^2}{2}.$$

Using the chain rule, we find

$$h'(t) = \begin{bmatrix} f_x(x_0 + t\Delta x, y_0 + t\Delta y)t\Delta x + f_y(x_0 + t\Delta x, y_0 + t\Delta y)t\Delta y \\ g_x(x_0 + t\Delta x, y_0 + t\Delta y)t\Delta x + g_y(x_0 + t\Delta x, y_0 + t\Delta y)t\Delta y \end{bmatrix}$$

and in terms of the Hessian's of f and g

$$h''(t) = \begin{bmatrix} \begin{bmatrix} t\Delta x \; t\Delta y \end{bmatrix}^T \begin{bmatrix} f_{xx}(x_0 + t\Delta x, y_0 + t\Delta y) & f_{yx}(x_0 + t\Delta x, y_0 + t\Delta y) \\ f_{xy}(x_0 + t\Delta x, y_0 + t\Delta y) & f_{yy}(x_0 + t\Delta x, y_0 + t\Delta y) \end{bmatrix} \begin{bmatrix} t\Delta x \\ t\Delta y \end{bmatrix} \\ \\ \begin{bmatrix} t\Delta x \; t\Delta y \end{bmatrix}^T \begin{bmatrix} g_{xx}(x_0 + t\Delta x, y_0 + t\Delta y) & g_{yx}(x_0 + t\Delta x, y_0 + t\Delta y) \\ g_{xy}(x_0 + t\Delta x, y_0 + t\Delta y) & g_{yy}(x_0 + t\Delta x, y_0 + t\Delta y) \end{bmatrix} \begin{bmatrix} t\Delta x \\ t\Delta y \end{bmatrix} \end{bmatrix}$$

or

$$h''(t) = \begin{bmatrix} \begin{bmatrix} t\Delta x \; t\Delta y \end{bmatrix}^T \boldsymbol{H}_f(x_0 + t\Delta x, y_0 + t\Delta y) \begin{bmatrix} t\Delta x \\ t\Delta y \end{bmatrix} \\ \\ \begin{bmatrix} t\Delta x \; t\Delta y \end{bmatrix}^T \boldsymbol{H}_g(x_0 + t\Delta x, y_0 + t\Delta y) \begin{bmatrix} t\Delta x \\ t\Delta y \end{bmatrix} \end{bmatrix}$$

Thus, since the approximation is known to be

$$h(1) = h(0) + h'(0)(1 - 0) + h''(c)\frac{1}{2}$$

for some c between 0 and 1, we have

$$\begin{bmatrix} f(x_0 + \Delta x, y_0 + \Delta y) \\ \\ g(x_0 + \Delta x, y_0 + \Delta y) \end{bmatrix}$$
$$= \begin{bmatrix} f(x_0, y_0) + f_x(x_0, y_0)\Delta x + f_y(x_0, y_0)\Delta y + \frac{1}{2} \begin{bmatrix} \Delta x \\ \Delta y \end{bmatrix}^T \boldsymbol{H}_f(x^*, y^*) \begin{bmatrix} \Delta x \\ \Delta y \end{bmatrix} \\ \\ g(x_0, y_0) + g_x(x_0, y_0)\Delta x + g_y(x_0, y_0)\Delta y + \frac{1}{2} \begin{bmatrix} \Delta x \\ \Delta y \end{bmatrix}^T \boldsymbol{H}_g(x^*, y^*) \begin{bmatrix} \Delta x \\ \Delta y \end{bmatrix} \end{bmatrix}$$

where $x^* = x_0 + c\Delta x$ and $y^* = y_0 + c\Delta y$. Hence, we can rewrite our nonlinear model as

$$x' = f(x_0, y_0) + f_x(x_0, y_0)\Delta x + f_y(x_0, y_0)\Delta y + \boldsymbol{E}_f(x^*, y^*)$$
$$y' = f(x_0, y_0) + g_x(x_0, y_0)\Delta x + g_y(x_0, y_0)\Delta y + \boldsymbol{E}_g(x^*, y^*)$$

where the error terms $\boldsymbol{E}_f(x^*, y^*)$ and $\boldsymbol{E}_g(x^*, y^*)$ are the usual Hessian based terms

$$\boldsymbol{E}_f(x^*, y^*) = \frac{1}{2}\begin{bmatrix}\Delta x\\ \Delta y\end{bmatrix}^T \boldsymbol{H}_f(x^*, y^*)\begin{bmatrix}\Delta x\\ \Delta y\end{bmatrix}$$
$$\boldsymbol{E}_g(x^*, y^*) = \frac{1}{2}\begin{bmatrix}\Delta x\\ \Delta y\end{bmatrix}^T \boldsymbol{H}_g(x^*, y^*)\begin{bmatrix}\Delta x\\ \Delta y\end{bmatrix}$$

With sufficient knowledge of the Hessian terms, we can have some understanding of how much error we make but, of course, it is difficult in interesting problems to make this very exact. Roughly speaking though, for our functions with continuous second order partial derivatives, in any closed circle of radius R around the base point (x_0, y_0), there is a constant B_R so that

$$\boldsymbol{E}_f(x^*, y^*) \leq \frac{1}{2}B_R(|\Delta x| + |\Delta y|)^2 \text{ and } \boldsymbol{E}_f(x^*, y^*) \leq \frac{1}{2}B_R(|\Delta x| + |\Delta y|)^2.$$

It is clear that the simplest approximation arises at those special points (x_0, y_0) where both $f(x_0, y_0) = 0$ **and** $g(x_0, y_0) = 0$. The points are called **equilibrium points** of the model. At such points, we can a linear approximation to the true dynamics of this form

$$x' = f_x(x_0, y_0)\Delta x + f_y(x_0, y_0)\Delta y$$
$$y' = g_x(x_0, y_0)\Delta x + g_y(x_0, y_0)\Delta y$$

and this linearization of the model is going to give us trajectories that are *close* to the true trajectories when are deviations Δx and Δy are small *enough*! We can rewrite our linearizations again using $x - x_0 = \Delta x$ and $y - y_0 = \Delta y$ as follows

$$\begin{bmatrix}x'\\ y'\end{bmatrix} = \begin{bmatrix}f_x(x_0, y_0) & f_y(x_0, y_0)\\ g_x(x_0, y_0) & g_y(x_0, y_0)\end{bmatrix}\begin{bmatrix}x - x_0\\ y - y_0\end{bmatrix}.$$

The special matrix of first order partials of f and g is called the Jacobian of our model and is denoted by $\boldsymbol{J}(x, y)$. Hence, in general

$$\boldsymbol{J}(x, y) = \begin{bmatrix}f_x(x, y) & f_y(x, y)\\ g_x(x, y) & g_y(x, y)\end{bmatrix}$$

so that the linearization of the model at the equilibrium point (x_0, y_0) has the form

$$\begin{bmatrix} x' \\ y' \end{bmatrix} = \boldsymbol{J}(x_0, y_0) \begin{bmatrix} x - x_0 \\ y - y_0 \end{bmatrix}$$

Next, if we make the change of variables $u = x - x_0$ and $v = y - y_0$, we find at each equilibrium point, there is a standard linear model (which we know how to study) of the form

$$\begin{bmatrix} u' \\ v' \end{bmatrix} = \boldsymbol{J}(x_0, y_0) \begin{bmatrix} u \\ v \end{bmatrix}$$

as $\boldsymbol{J}(x_0, y_0)$ is simply as 2×2 real matrix which will have real distinct, repeated or complex conjugate pair eigenvalues which we know how to deal with after our discussions in Chap. 8. We are now ready to study interesting nonlinear models using the tools of linearization.

14.1.1 A First Nonlinear Model

Let's consider this model

$$\begin{aligned} x' &= (1 - x)x - \frac{2xy}{1 + x} \\ y' &= \left(1 - \frac{y}{1 + x}\right) y \end{aligned}$$

which has multiple equilibrium points. Clearly, we need to stay away from the line $x = -1$ for initial conditions as there the dynamics themselves are not defined! However, we can analyze at other places. We define the nonlinear function f and g then by

$$\begin{aligned} f(x, y) &= (1 - x)x - \frac{2xy}{1 + x} \\ g(x, y) &= \left(1 - \frac{y}{1 + x}\right) y \end{aligned}$$

We find that $f(x, y) = 0$ when either $x = 0$ or $2y = 1 - x^2$. When $x = 0$, we find $g(0, y) = (1 - y)\, y$ which is 0 when $y = 0$ or $y = 1$. So we have equilibrium points at (0, 0 and (0, 1). Next, when $2y = 1 - x^2$, we find g becomes

$$g\left(x, \frac{1 - x^2}{2}\right) = \frac{1}{4}\,(1 + x)(1 + x)(1 - x).$$

We need to discard $x = -1$ as the dynamics are not defined there. So the last equilibrium point is at $(1, 0)$. We can then use MatLab to find the Jacobians and the associated eigenvalues and eigenvectors are each equilibrium point. We encode the Jacobian

$$J(x, y) = \begin{bmatrix} 1 - 2x - 2\frac{y}{(1+x)^2} & -2\frac{x}{(1+x)} \\ \frac{y^2}{(1+x)^2)} & 1 - 2\frac{y}{1+x} \end{bmatrix}$$

in the file **Jacobian.m** as

Listing 14.1: Jacobian

```
function A = Jacobian(x,y)
A = [1-2*x - 2*y/(1 + x*x), -2*x/(1+x);...
     y*y/((1+x)*(1+x)), 1-2*y/(1+x)];
```

We can them find the local linear systems at each equilibrium.

14.1.1.1 Local Analysis

We will use MatLab to help with our analysis of these equilibrium points. Here are the MatLab sessions for all the equilibrium points. The MatLab command **eig** is used to find the eigenvalues and eigenvectors of a matrix.

Equilibrium Point (0,0) For the first equilibrium point $(0, 0)$, we find the Jacobian at $(0, 0)$ and the associated eigenvalues and eigenvectors with the following MatLab commands.

Listing 14.2: Jacobian at equilibrium point (0; 0)

```
J0 = Jacobian(0,0)
J0 =
        1     0
        0     1
[V0,D0] = eig(J0)
V0 =
        1     0
        0     1
D0 =
        1     0
        0     1
```

Hence, there is a repeated eigenvalue, $r = 1$ but there are two different eigenvectors:

$$E_1 = \begin{bmatrix} 1 \\ 0 \end{bmatrix}, \ E_2 = \begin{bmatrix} 0 \\ 1 \end{bmatrix}$$

with general local solution near (0, 0) of the form

$$\begin{bmatrix} x(t) \\ y(t) \end{bmatrix} = a \begin{bmatrix} 1 \\ 0 \end{bmatrix} e^{t} + b \begin{bmatrix} 0 \\ 1 \end{bmatrix} e^{t}$$

where a and b are arbitrary. Hence, trajectories move away from the origin locally. Recall, the local linear system is

$$\begin{bmatrix} x'(t) \\ y'(t) \end{bmatrix} = \begin{bmatrix} 1 & 0 \\ 0 & 1 \end{bmatrix} \begin{bmatrix} x(t) \\ y(t) \end{bmatrix}$$

which is the same as the *local variable* system using the change of variables $u = x$ and $v = y$.

$$\begin{bmatrix} u'(t) \\ v'(t) \end{bmatrix} = \begin{bmatrix} 1 & 0 \\ 0 & 1 \end{bmatrix} \begin{bmatrix} u(t) \\ v(t) \end{bmatrix}$$

Equilibrium Point (1,0) For the second equilibrium point $(1, 0)$, we find the Jacobian at $(1, 0)$ and the associated eigenvalues and eigenvectors in a similar way.

Listing 14.3: Jacobian at equilibrium point (1; 0)

```
J1 = Jacobian(1,0)
J1 =
    -1    -1
     0     1
[V1,D1] = eig(J1)
V1 =
    1.0000    -0.4472
         0     0.8944
D1 =
    -1     0
     0     1
```

Now there is a two different eigenvalues, $r_1 = -1$ and $r_2 = 1$ with associated eigenvectors

$$E_1 = \begin{bmatrix} 1 \\ 0 \end{bmatrix}, \quad E_2 = \begin{bmatrix} -0.4472 \\ 0.8944 \end{bmatrix}$$

with general local solution near (1, 0) of the form

$$\begin{bmatrix} x(t) \\ y(t) \end{bmatrix} = a \begin{bmatrix} 1 \\ 0 \end{bmatrix} e^{-t} + b \begin{bmatrix} -0.4472 \\ 0.8944 \end{bmatrix} e^{t}$$

where a and b are arbitrary. Hence, trajectories move away from $(1, 0)$ locally for all trajectories except those that start on E_1. Recall, the local linear system is

$$\begin{bmatrix} x'(t) \\ y'(t) \end{bmatrix} = \begin{bmatrix} -1 & -1 \\ 0 & 1 \end{bmatrix} \begin{bmatrix} x(t) - 1 \\ y(t) \end{bmatrix}$$

which, using the change of variables $u = x - 1$ and $v = y$, is the *local variable* system.

$$\begin{bmatrix} u'(t) \\ v'(t) \end{bmatrix} = \begin{bmatrix} -1 & -1 \\ 0 & 1 \end{bmatrix} \begin{bmatrix} u(t) \\ v(t) \end{bmatrix}$$

Equilibrium Point (0,1) For the third equilibrium point $(0, 1)$, we again find the Jacobian at $(0, 1)$ and the associated eigenvalues and eigenvectors in a similar way.

Listing 14.4: Jacobian at equilibrium point (0; 1)

```
J2 = Jacobian(0,1)
J2 =
    -1      0
     1     -1
[V2,D2] = eig(J2)
V2 =
         0    0.0000
    1.0000   -1.0000
D2 =
    -1      0
     0     -1
```

Now there is again a repeated eigenvalue, $r_1 = -1$. If you look at the $D2$ matrix, you see both the columns are the same. In this case, MatLab does not give us useful information. We can use the first column as our eigenvector E_1, but we still must find the other vector F.

Recall, the local linear system is

$$\begin{bmatrix} x'(t) \\ y'(t) \end{bmatrix} = \begin{bmatrix} -1 & 0 \\ 1 & -1 \end{bmatrix} \begin{bmatrix} x(t) \\ y(t) - 1 \end{bmatrix}$$

which, using the change of variables $u = x$ and $v = y - 1$, is the *local variable* system

$$\begin{bmatrix} u'(t) \\ v'(t) \end{bmatrix} = \begin{bmatrix} -1 & 0 \\ 1 & -1 \end{bmatrix} \begin{bmatrix} u(t) \\ v(t) \end{bmatrix}$$

Recall, the general solution to a model with a repeated eigenvalue with only one eigenvector is given by

$$\begin{bmatrix} x(t) \\ y(t) \end{bmatrix} = a\, E_1\, e^{-t} + b\Big(F\, e^{-t} + E_1\, t\, e^{-t} \Big)$$

where F solves $(-I - A)F = -E_1$. Here, that gives

$$F = \begin{bmatrix} 1 \\ 0 \end{bmatrix}$$

and so the local solution near (0, 1) has the form

$$\begin{bmatrix} x(t) \\ y(t) \end{bmatrix} = a \begin{bmatrix} 0 \\ 1 \end{bmatrix} e^{-t} + b\left(\begin{bmatrix} 1 \\ 0 \end{bmatrix} e^{-t} + \begin{bmatrix} 0 \\ 1 \end{bmatrix} t\, e^{-t} \right)$$

where a and b are arbitrary. Hence, trajectories move toward from (0, 1) locally.

14.1.1.2 Generating a Phase Plane Portrait

Piecing together the global behavior from the local trajectories is difficult, so it is helpful to write scripts in MatLab to help us. We can use the **AutoPhasePlanePlot()** function from before, but this time we use different dynamics. The dynamics are stored in the file **autonomousfunc.m** which encodes the right hand side of the model

$$\begin{aligned} x' &= (1 - x)x - \frac{2xy}{1 + x} \\ y' &= \left(1 - \frac{y}{1 + x}\right) y \end{aligned}$$

in the usual way.

Listing 14.5: Autonomousfunc

```
function f = autonomousfunc(t,x)
f = [(1-x(1)).*x(1) -2.*x(1).*x(2)./(1+x(1));...
     x(2) - x(2).*x(2)./(1+x(1))];
```

Then, to generate a nice phase plane portrait, we try a variety of $[xmin, xmax] \times [ymin, ymax]$ initial condition boxes until it *looks* right! Here, we avoid any initial conditions that have negative values as for those the trajectories go off to infinity and the plots are not manageable. We show the plot in Fig. 14.1.

Listing 14.6: Sample Phase Plane Plot

```
AutoPhasePlanePlot('autonomousfunc',0.1,0.0,3.9,4,8,8,0.0,2.5,0.0,2.5);
```

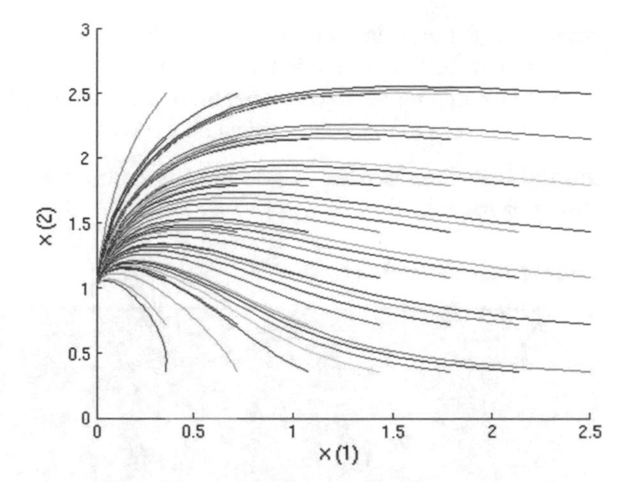

Fig. 14.1 Nonlinear phase plot for multiple initial conditions!

You should play with this function. You'll see it involves a lot of trial and error. Any box $[xmin, xmax] \times [ymin, ymax]$ which includes trajectories whose x or y values increase exponentially causes the overall plot's x and y ranges to be skewed toward those large numbers. This causes a huge loss in resolution on all other trajectories! So it takes time and a bit of skill to generate a nice collection of phase plane plots!

14.1.2 Finding Equilibrium Points Numerically

We can find the equilibrium points using the root finding methods called **bisection** and **Newton**'s method.

14.1.2.1 The Bisection Method

We need a simple function to find the root of a *nice* function f of the real variable x using what is called bisection. The method is actually quite simple. We know that if f is a continuous function on the finite interval $[a, b]$ then f must have a zero inside the interval $[a, b]$ if f has a different algebraic sign at the endpoints a and b. This means the product $f(a)\ f(b)$ is not zero. So we assume we can find an interval $[a, b]$ on which this change in sign satisfies $f(a)\ f(b) \leq 0$ (which we can do by switching to $-f$ if we have to!) and then if we divide the interval $[a, b]$ into two equal pieces $[a, m]$ and $[m, b]$, $f(m)$ can't have the same sign as both $f(a)$ and $f(b)$ because of the assumed sign difference. So at least one of the two halves has a sign change.
Note that if $f(a)$ and $f(b)$ was zero then we still have $f(a)\ f(b) \leq 0$ and either a or b could be our chosen root and either half interval works fine. If only one of the

endpoint function values is zero, then the bisection of $[a, b]$ into the two halves still finds the one half interval that has the root.

So our prototyping Matlab code should use tests like $f(x)\, f(y) \leq 0$ rather than $f(x)\, f(y) < 0$ to make sure we catch the root.

Simple Bisection MatLab Code Here is a simple Matlab function to perform the Bisection routine.

Listing 14.7: Bisection Code

```
function root = Bisection(fname,a,b,delta)
%
% fname is a string that gives the name of a
%       continuous function f(x)
% a,b   this is the interval [a,b] on
%       which f is defined and for which
%       we assume that the product
%       f(a)*f(b) <= 0
% delta this is a non negative real number
%
% root  this is the midpoint of the interval
%       [alpha,beta] having the property that
%       f(alpha)*f(beta) <= 0 and
%       |beta-alpha| <= delta + eps * max(|alpha|,|beta|)
%       and eps is machine zero
%
 disp(' ')
 disp('       k  |        a(k)        |        b(k)        |        b(k) - a(k) ')
 k = 1;
 disp(sprintf(' %6d  | %12.7f  | %12.7f  | %12.7f',k,a,b,b-a));
 fa = feval(fname,a);
 fb = feval(fname,b);
 while abs(a-b) > delta + eps*max(abs(a),abs(b))
   mid = (a+b)/2;
   fmid = feval(fname,mid);
   if fa*fmid <= 0
     % there is a root in [a,mid]
     b = mid;
     fb = fmid;
   else
     % there is a root in [mid,b]
     a = mid;
     fa = fmid;
   end
   k = k+1;
   disp(sprintf(' %6d  | %12.7f  | %12.7f  | %12.7f',k,a,b,b-a));
 end
 root = (a+b)/2;
end
```

We should look at some of these lines more closely. First, to use this routine, we need to write a function definition for the function we want to apply bisection to. We will do this in a file called **`func.m`** (Inspired Name, eh?) An example would be the one we wrote for the function

$$f(x) = \tan\left(\frac{x}{4}\right) - 1;$$

which is coded in Matlab by

Listing 14.8: Function Definition In MatLab

```
function y = func(x)
%
% x   real input
% y   real output
%
  y = tan(x/4)-1;
end
```

So to apply bisection to this function on the interval [2, 4] with a stopping tolerance of say 10^{-4}, in Matlab, we would type the command

Listing 14.9: Applying bisection

```
root = Bisection('func',2,4,10^-4)}
```

Note that the name of our supplied function, the uninspired choice *func*, is passed in as the first argument in single quotes as it is a string. Also, in the Bisection routine, we have added the code to print out what is happening at each iteration of the while loop. Matlab handles prints to the screen a little funny, so do set up a table of printed values we use this syntax:

Listing 14.10: Bisection step by step code

```
% this prints a blank line and then a table heading.
% note disp prints a string only
disp(' ')
disp('        k  |          a(k)          |          b(k)          |          b(k) - a(k) ')
% now to print the k, a, b and b-a, we must first
% put their values into a string using the c like function
% sprintf and then use disp to disply that string.
% so we do this
% disp( sprintf(' output specifications here ',variables here))
% so inside the while loop we use
disp(sprintf(' %6d  | %12.7f  | %12.7f  | %12.7f',k,a,b,b-a));
```

Running the Code As mentioned above, we will test this code on the function

$$f(x) = \tan\left(\frac{x}{4}\right) - 1;$$

on the interval [2, 4] with a stopping tolerance of $\delta = 10^{-6}$. Our function has been written as the Matlab function **func** supplied in the file **func.m**. The Matlab run time looks like this:

Listing 14.11: Sample Bisection run step by step

```
root = Bisection('func',2,4,10^-6)

      k  |      a(k)      |      b(k)      | b(k) - a(k)
      1  |   2.0000000    |   4.0000000    |   2.0000000
      2  |   3.0000000    |   4.0000000    |   1.0000000
      3  |   3.0000000    |   3.5000000    |   0.5000000
      4  |   3.0000000    |   3.2500000    |   0.2500000
      5  |   3.1250000    |   3.2500000    |   0.1250000
      6  |   3.1250000    |   3.1875000    |   0.0625000
      7  |   3.1250000    |   3.1562500    |   0.0312500
      8  |   3.1406250    |   3.1562500    |   0.0156250
      9  |   3.1406250    |   3.1484375    |   0.0078125
     10  |   3.1406250    |   3.1445312    |   0.0039062
     11  |   3.1406250    |   3.1425781    |   0.0019531
     12  |   3.1406250    |   3.1416016    |   0.0009766
     13  |   3.1411133    |   3.1416016    |   0.0004883
     14  |   3.1413574    |   3.1416016    |   0.0002441
     15  |   3.1414795    |   3.1416016    |   0.0001221
     16  |   3.1415405    |   3.1416016    |   0.0000610
     17  |   3.1415710    |   3.1416016    |   0.0000305
     18  |   3.1415863    |   3.1416016    |   0.0000153
     19  |   3.1415863    |   3.1415939    |   0.0000076
     20  |   3.1415901    |   3.1415939    |   0.0000038
     21  |   3.1415920    |   3.1415939    |   0.0000019
     22  |   3.1415920    |   3.1415930    |   0.0000010
root =
    3.1416
```

Homework Well, you have to practice this stuff to see what is going on. So here are two problems to sink your teeth into!

Exercise 14.1.1 Use bisection to find the first five positive solutions of the equation $x = \tan(x)$. You can see where this is roughly by graphing $\tan(x)$ and x simultaneously. Do this for tolerances $\{10^{-1}, 10^{-2}, 10^{-3}, 10^{-4}, 10^{-5}, 10^{-6}, 10^{-7}\}$. For each root, choose a reasonable bracketing interval $[a, b]$, explain why you chose it, provide a table of the number of iterations to achieve the accuracy and a graph of this number versus accuracy.

Exercise 14.1.2 Use the Bisection Method to find the largest real root of the function $f(x) = x^6 - x - 1$. Do this for tolerances $\{10^{-1}, 10^{-2}, 10^{-3}, 10^{-4}, 10^{-5}, 10^{-6}, 10^{-7}\}$. Choose a reasonable bracketing interval $[a, b]$, explain why you chose it, provide a table of the number of iterations to achieve the accuracy and a graph of this number versus accuracy.

14.1.2.2 Newton's Method

Newton's method is based on the tangent line and rapidly converges to a zero of the function f if the original guess is *reasonable*. Of course, that is the problem. A bad initial guess is a great way to generate random numbers! So usually, we find a good interval where the root might reside by first using bisection. The following code uses a simple test to see which we should do in our zero finding routine.

Listing 14.12: Should We Do A Newton Step?

```
function ok = StepIsIn(x,fx,fpx,a,b)
%
% x       current approximate root
% fx      function f value at approximate root
% fpx     derivative f' value at approximate root
% a,b     root is in this interval
%
% ok      1 if the Newton Step x - fx/fpx is in [a,b]
%         0 if not
%
if fpx > 0
  ok = ((a-x)*fpx <= -fx) & (-fx <= (b-x)*fpx);
elseif fpx < 0
  ok = ((a-x)*fpx >= -fx) & (-fx >= (b-x)*fpx);
else
  ok = 0;
end

end
```

Then, once the bisection steps have given us an interval where the root might be, we switch to Newton's method. This takes the current guess, say x_1, and finds the tangent line, $T(x)$, to the function at that point. This gives

$$T(x) = f(x_0) + f'(x_0)(x - x_0).$$

If we find the value of x where the tangent line crosses the x axis, this becomes our next guess x_1. We find

$$x_1 = x_0 - \frac{f(x_0)}{f'(x_0)}.$$

In essence this is Newton's Method which can be rephrased for the scalar function case as

$$x_{n+1} = x_N - \frac{f(x_n)}{f'(x_n)}.$$

and it is clear the method fails if $f'(x_n) = 0$ or is close to 0 at any iteration. MatLab code to implement this method is given next.

A Global Newton Method The code for a global Newton method is pretty straightforward. Here is the listing.

Listing 14.13: Global Newton Method

```
function [x,fx,nEvals,aF,bF] = ...
         GlobalNewton(fName,fpName,a,b,tolx,tolf,nEvalsMax)
%
% fName          a string that is the name of the function f(x)
% fpName         a string that is the name of the functions derivative
%                f'(x)
% a,b            we look for the root in the interval [a,b]
% tolx           tolerance on the size of the interval
% tolf           tolerance of f(current approximation to root)
% nEvalsMax      Maximum Number of derivative Evaluations
%
% x              Approximate zero of f
% fx             The value of f at the approximate zero
% nEvals         The Number of Derivative Evaluations Needed
% aF, bF         the final interval the approximate root lies in,
%                [aF,bF]
%
% Termination    Interval [a,b] has size < tolx
%                |f(approximate root)| < tolf
%                Have exceeded nEvalsMax derivative Evaluations
%
fa = feval(fName,a);
fb = feval(fName,b);
x = a;
fx = feval(fName,x);
fpx = feval(fpName,x);

nEvals = 1;
k = 1;
disp(' ')
disp(' Step      |      k  |      a(k)      |      x(k)      |      b(k) ')
disp(sprintf('Start       | %6d  | %12.7f  | %12.7f  | %12.7f',k,a,x,b));
while  (abs(a-b)>tolx) && (abs(fx)> tolf) && (nEvals<nEvalsMax) || (nEvals==1)
  %[a,b] brackets a root and x=a or x=b
  check = StepIsIn(x,fx,fpx,a,b);
  if check
    %Take Newton Step
    x = x - fx/fpx;
  else
    %Take a Bisection Step:
    x = (a+b)/2;
  end
  fx = feval(fName,x);
  fpx = feval(fpName,x);
  nEvals = nEvals+1;
  if fa*fx<=0
    %there is a root in [a,x]. Use right endpoint.
    b = x;
    fb = fx;
  else
    %there is a root in [x,b]. Bring in left endpoint.
    a = x;
    fa = fx;
  end
  k = k+1;
  if(check)
    disp(sprintf('Newton     | %6d  | %12.7f  | %12.7f  | %12.7f',k,a,x,b));
  else
    disp(sprintf('Bisection | %6d  | %12.7f  | %12.7f  | %12.7f',k,a,x,b));
  end
end
aF = a;
bF = b;

end
```

A Run Time Example We will apply our global newton method root finding code to a simple example: find a root for $f(x) = \sin(x)$ in the interval $[\frac{-7\pi}{2}, 15\pi + 0.1]$.

We code the function and its derivative in two simple Matlab files; **f1.m** and **f1p.m**. These are

Listing 14.14: Global Newton Function

```
function y = f1(x)
y = sin(x);
end
```

and

Listing 14.15: Global Newton Function Derivative

```
function y = f1p(x)
y = cos(x);
end
```

To run this code on this example, we would then type a phrase like the one below:

Listing 14.16: Global Newton Sample

```
[x,fx,nEvals,aLast,bLast] = GlobalNewton('f1','f1p',-7*pi/2,...
                                    15*pi+.1,10^-6,10^-8,200)
```

Here is the runtime output:

Listing 14.17: Global Newton runtime results

```
[x,fx,nEvals,aLast,bLast] = GlobalNewton('f1','f1p',...
                              -7*pi/2,15*pi+.1,...
                              10^-6,10^-8,200)
 Step      |   k |   a(k)       |   x(k)       |   b(k)
Start      |   1 | -10.9955743  | -10.9955743  | 47.2238898
Bisection  |   2 | -10.9955743  | 18.1141578   | 18.1141578
Bisection  |   3 | -10.9955743  |  3.5592917   |  3.5592917
Newton     |   4 |   3.1154761  |  3.1154761   |  3.5592917
Newton     |   5 |   3.1154761  |  3.1415986   |  3.1415986
Newton     |   6 |   3.1415927  |  3.1415927   |  3.1415986
x =
    3.1416
fx =
   1.2246e-16
nEvals =
     6
aLast =
    3.1416
bLast =
    3.1416
```

Homework

Exercise 14.1.3 Use the Global Newton Method to find the first five positive solutions of the equation $x = \tan(x)$. You can see where this is roughly by graphing $\tan(x)$ and x simultaneously. Do this for tolerances $\{10^{-1}, 10^{-2}, 10^{-3}, 10^{-4}, 10^{-5}, 10^{-6}, 10^{-7}\}$. For each root, choose a reasonable bracketing interval $[a, b]$, explain why you chose it, provide a table of the number of iterations to achieve the accuracy and a graph of this number versus accuracy.

Exercise 14.1.4 Use the Global Newton Method to find the largest real root of the function $f(x) = x^6 - x - 1$. Do this for tolerances $\{10^{-1}, 10^{-2}, 10^{-3}, 10^{-4}, 10^{-5}, 10^{-6}, 10^{-7}\}$. Choose a reasonable bracketing interval $[a, b]$, explain why you chose it, provide a table of the number of iterations to achieve the accuracy and a graph of this number versus accuracy.

14.1.2.3 Adding Finite Difference Approximations to the Derivative

We can also choose to replace the derivative function for f with a finite difference approximation. We will use

$$f'(x) \approx \frac{f(x_c + \delta_c) - f(x_c)}{\delta_c}$$

to approximate the value of the derivative at the point x_c. As we have discussed earlier, some care is required to pick a size for δ_c so that round-off errors do not destroy the accuracy of our finite difference approximation to f'.
The simple Matlab code to implement this is given below:

Listing 14.18: Finite difference approximation to the derivative

```
fval = feval(fname,x);
fpval = (feval(fname,x+delta) - fval)/delta;
```

We can also use a secant approximation as follows:

$$f'(x) \approx \frac{f(x_c) - f(x_-)}{x_c - x_-}$$

where x_- is the previous iterate from our routine. The Matlab fragment we need is then:

Listing 14.19: Secant approximation to the derivative

```
fpc = (fc - f_)/(xc - x_);
```

14.1.2.4 A Finite Difference Global Newton Method

We add the finite difference routines into our Global Newton's Method as follows:

Listing 14.20: Finite Difference Global Newton Method

```
function [x,fx,nEvals,aF,bF] = ...
         GlobalNewtonFD(fName,a,b,tolx,tolf,nEvalsMax)
%
% fName        a string that is the name of the function f(x)
% a,b          we look for the root in the interval [a,b]
% tolx         tolerance on the size of the interval
% tolf         tolerance of f(current approximation to root)
% nEvalsMax    Maximum Number of derivative Evaluations
%
% x            Approximate zero of f
% fx           The value of f at the approximate zero
% nEvals       The Number of Derivative Evaluations Needed
% aF, bF       the final interval the approximate root lies in,
%              [aF,bF]
%
% Termination  Interval [a,b] has size < tolx
%              |f(approximate root)| < tolf
%              Have exceeded nEvalsMax derivative Evaluations
%
fa = feval(fName,a);
fb = feval(fName,b);
x = a;
fx = feval(fName,x);
delta = sqrt(eps)* abs(x);
fpval = feval(fName,x+delta);
fpx = (fpval-fx)/delta;

nEvals = 1;
k = 1;
%disp(' ')
%disp(' Step      |      k  |      a(k)      |      x(k)      |      b(k) ')
%disp(sprintf('Start     | %6d | %12.7f | %12.7f | %12.7f',k,a,x,b));
while  (abs(a-b)>tolx) && (abs(fx)> tolf) && (nEvals<nEvalsMax) || (nEvals==1)
  %[a,b] brackets a root and x=a or x=b
  check = StepIsIn(x,fx,fpx,a,b);
  if check
    %Take Newton Step
    x = x - fx/fpx;
  else
    %Take a Bisection Step:
    x = (a+b)/2;
  end
  fx = feval(fName,x);
  fpval = feval(fName,x+delta);
  fpx = (fpval-fx)/delta;
  nEvals = nEvals+1;
  if fa*fx<=0
    %there is a root in [a,x].  Use right endpoint.
    b = x;
    fb = fx;
  else
    %there is a root in [x,b].  Bring in left endpoint.
    a = x;
    fa = fx;
  end
  k = k+1;
  %if(check)
  %  disp(sprintf('Newton     | %6d | %12.7f | %12.7f | %12.7f',k,a,x,b));
  %else
  %  disp(sprintf('Bisection  | %6d | %12.7f | %12.7f | %12.7f',k,a,x,b));
  %end
end
aF = a;
bF = b;

end
```

Note, we use for our finite difference stepsize $\sqrt{\epsilon_{machine}}|x|$.

A Run Time Example We will apply our finite difference global newton method root finding code to the same simple example: find a root for $f(x) = \sin(x)$ in the interval $[\frac{-7\pi}{2}, 15\pi + 0.1]$. We only need the code for the function now which is as usual in the file **f1.m**.
To run this code on this example, we would then type a phrase like the one below:

Listing 14.21: Sample GlobalNewton Finite Difference solution

```
[x,fx,nEvals,aLast,bLast] = GlobalNewtonFD('f1',-7*pi/2,15*pi+.1,...
                                   10^-6,10^-8,200)
```

Here is the runtime output:

Listing 14.22: Global Newton Finite Difference runtime results

```
  [x,fx,nEvals,aLast,bLast] = GlobalNewtonFD('f1',-7*pi/2,15*pi+.1,...
                                   10^-6,10^-8,200)
3  Step       |  k  |     a(k)      |     x(k)      |     b(k)
   Start      |  1  |  -10.9955743  |  -10.9955743  |  47.2238898
   Bisection  |  2  |  -10.9955743  |   18.1141578  |  18.1141578
   Bisection  |  3  |  -10.9955743  |    3.5592917  |   3.5592917
   Newton     |  4  |    3.1154761  |    3.1154761  |   3.5592917
8  Newton     |  5  |    3.1154761  |    3.1415986  |   3.1415986
   Newton     |  6  |    3.1154761  |    3.1415927  |   3.1415927
   x =
       3.1416
   fx =
13   -4.3184e-15
   nEvals =
        6
   aLast =
       3.1155
18 bLast =
       3.1416
```

Homework

Exercise 14.1.5 Use the Finite Difference Global Newton Method to find the second positive solution of the equation $x = \tan(x)$. Do this for tolerances 10^{-8}. This time alter the GlobalNewtonFD code to allow the finite difference step size **delta** to be a parameter and do a parametric study on the effects of **delta**. Note that the code now uses the reasonable choice of $\sqrt{\epsilon_{machine}}|x|$ but you need to use the additional δ choices $\{10^{-4}, 10^{-6}, 10^{-8}, 10^{-10}\}$. This will give you five δ choices. Provide a table and a graph of δ versus accuracy of the root approximation.

Exercise 14.1.6 Use the Finite Difference Global Newton Method to find the largest real root of the function $f(x) = x^6 - x - 1$. Do this for tolerances 10^{-8}. Again use altered GlobalNewtonFD code with the finite difference step size **delta** as a parameter and do a parametric study on the effects of **delta**. Note that the code now uses the reasonable choice of $\sqrt{\epsilon_{machine}}|x|$ but you need to use the additional δ

choices $\{10^{-4}, 10^{-6}, 10^{-8}, 10^{-10}\}$. This will give you five δ choices. Provide a table and a graph of δ versus accuracy of the root approximation.

Exercise 14.1.7 Do the same thing for the problems above, but replace the Finite Difference Global Newton Code with a Secant Global Newton Code. This will only require a few lines of code to change really, so don't freak out!

14.1.3 A Second Nonlinear Model

Now let's look at this model

$$\begin{aligned} x' &= 0.5(-h(x) + y) \\ y' &= 0.2(-x - 1.5y + 1.2) \end{aligned}$$

for

$$h(x) = 17.76x - 103.79x^2 + 229.62x^3 - 226.31x^4 + 83.72x^5,$$

This is model of how an electrical component called a diode behaves called the **trigger model** and the details are not really important as we are just investigating how to use our code and theoretical ideas. The equilibrium points are the solutions to the simultaneous equations

$$\begin{aligned} y &= 17.76x - 103.79x^2 + 229.62x^3 - 226.31x^4 + 83.72x^5 \\ y &= -\frac{2}{3}x + \frac{4}{5}. \end{aligned}$$

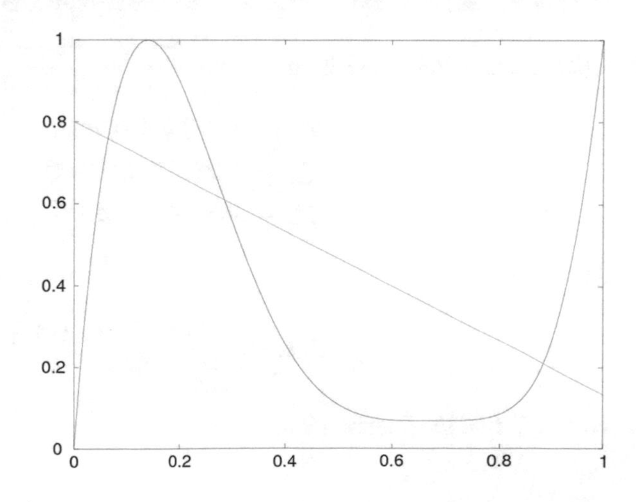

Fig. 14.2 Equilibrium points graphically for the trigger model

We can see these solutions graphically by plotting the two curves simultaneously and using the cursor to locate the roots and read off the (x, y) values from the plot. This is not quite accurate so a better way is to find the roots numerically. The plot which shows the equilibrium points graphically is shown in Fig. 14.2.
We do this with the following MatLab/Octave session.

Listing 14.23: Finding equilibrium points numerically

```
g = @(x) (12 - 10*x)/15;
h = @(x) 17.76*x -103.79*x.^2...
  + 229.62*x.^3 - 226.31*x.^4 + 83.72*x.^5;
X = linspace(0,1,101);
Y1 = h(X);
Y2 = g(X);
plot(X,Y1,X,Y2);
q = @(x) h(x) - g(x);
[x1,fx1,Nevals1,aF1,bF1] = ...
GlobalNewtonFD(q,0,.1,10^-6,10^-8,10);
[x2,fx2,Nevals2,aF2,bF2] = ...
GlobalNewtonFD(q,.2,.4,10^-6,10^-8,10);
[x3,fx3,Nevals3,aF3,bF3] = ...
GlobalNewtonFD(q,.4,.9,10^-6,10^-8,10);
y1 = h(x1);
y2 = h(x2);
y3 = h(x3);
EP = {[x1,y1],[x2,y2],[x3,y3]};
[x3,fx3,Nevals3,aF3,bF3] = ...
GlobalNewtonFD(q,.4,.9,10^-6,10^-8,10);
y3 = h(x3);
EP = {[x1,y1],[x2,y2],[x3,y3]};
EP
EP =

   [1,1] =
      0.062695    0.758672
   [1,2] =
      0.28537    0.60975
   [1,3] =
      0.88443    0.21038
}
```

There are now three equilibrium points

$$\begin{aligned} Q_1 &= (0.062695, 0.758672) \\ Q_2 &= (0.28537, 0.60975) \\ Q_3 &= (0.88443, 0.21038) \end{aligned}$$

The Jacobian is

$$J(x, y) = \begin{bmatrix} -0.5h'(x) & 0.5 \\ -0.2 & -0.3 \end{bmatrix}$$

Now let's find the linearizations.

14.1.3.1 Equilibrium Point $Q1$

For the equilibrium point $Q1$, we have

Listing 14.24: Jacobian at $Q1$

```
J(x1,y1)
ans =

   -3.61840    0.50000
   -0.20000   -0.30000
[V1,D1] = eig(J(x1,y1))
V1 =

   -0.998155   -0.150341
   -0.060715   -0.988634

D1 =

Diagonal Matrix

   -3.58798          0
          0   -0.33041
```

Hence, there are two real distinct eigenvalues, $r_1 = 3.58798$ and $r_2 = -0.33041$. The eigenvectors are

$$E_1 = \begin{bmatrix} -0.998155 \\ -0.060715 \end{bmatrix} \quad \text{and} \quad E_2 = \begin{bmatrix} -0.150341 \\ -0.988634 \end{bmatrix}$$

The dominant eigenvector is E_1 and it is easy to plot the resulting trajectories. We modify our function **`linearsystem`** to the new function **`linearsystemep`** so that we can plot the trajectories centered at the equilibrium point $Q1$.

Listing 14.25: Linearsystemep

```
function f = linearsystemep(p,t,x)
a = p(1);
b = p(2);
c = p(3);
d = p(4);
ex = p(5);
ey = p(6);
u = x(1) - ex;
v = x(2) - ey;
f = [a*u+b*v;c*u+d*v];
```

We set up the plot as follows. We define the coefficient matrix **`A1`** of our linearization and set up the parameter **`p1`** by listing all the entries of **`A1`** followed by the coordinates of $Q1$. We then generate the automatic phase plane plot.

Listing 14.26: Phase plane plot for $Q1$ linearization

```
A1 = J(x1,y1);
p1 = [A1(1,1),A1(1,2),A1(2,1),A1(2,2),x1,y1]
p1 =

  -3.618397    0.500000   -0.200000   -0.300000    0.062695    0.758672

AutoPhasePlanePlotLinearSystemRKF5('x axis','y axis',...
         'Q1 Plot',...
         1.0e-6,1.0e-6,.01,.2,...
         'linearsystemep',p1,.01,0,.4,12,12,0,1,0,1);
```

This generates the plot shown in Fig. 14.3.

14.1.3.2 Equilibrium Point $Q2$

For equilibrium point $Q2$, we find the linearization just like we did for $Q1$. First, we find the Jacobian at this point.

Listing 14.27: Jacobian at $Q2$

```
J(x2,y2)
ans =

1.82012     0.50000
  -0.20000   -0.30000
[V2,D2] = eig(J(x2,y2))
V2 =
   0.995373   -0.234595
  -0.096085    0.972093
D2 =
   1.77185          0
         0   -0.25173
```

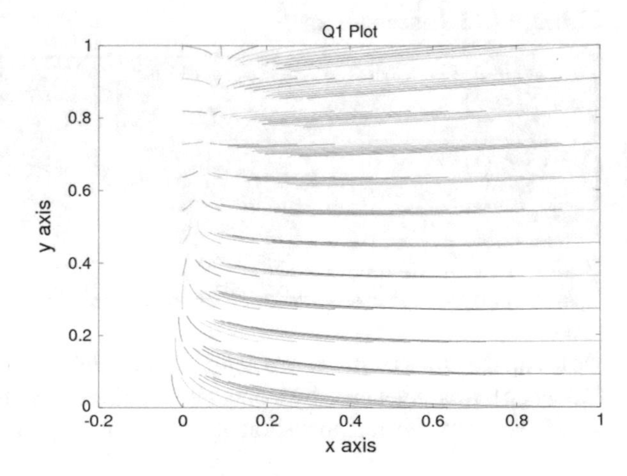

Fig. 14.3 Trigger model linearized about Q1

Hence, there are two real distinct eigenvalues, $r_1 = 1.77185$ and $r_2 = -0.25173$. The eigenvectors are

$$\boldsymbol{E}_1 = \begin{bmatrix} 0.995373 \\ -0.096085 \end{bmatrix} \text{ and } \boldsymbol{E}_2 = \begin{bmatrix} -0.234595 \\ 0.972093 \end{bmatrix}$$

The dominant eigenvector is again $\boldsymbol{E}_1$. We set the coefficient matrix **A2** of our linearization and the parameter **p2** of our model and generate the automatic phase plane plot.

Listing 14.28: Phase plane for $Q2$ linearization

```
A2 = J(x2,y2);
p2 = [A2(1,1),A2(1,2),A2(2,1),A2(2,2),x2,y2]
p2 =

   1.82012    0.50000   -0.20000   -0.30000    0.28537    0.60975
AutoPhasePlanePlotLinearSystemRKF5('x axis',...
        'y axis','Q1 Plot',...
        1.0e-6,1.0e-6,.01,.2,...
        'linearsystemep',p2,.01,0,.4,12,12,-0.5,1,0,1);
```

This generates the plot shown in Fig. 14.4.

14.1.3.3 Equilibrium Point $Q3$

Finally, we analyze the model near the point $Q3$. The Jacobian is now

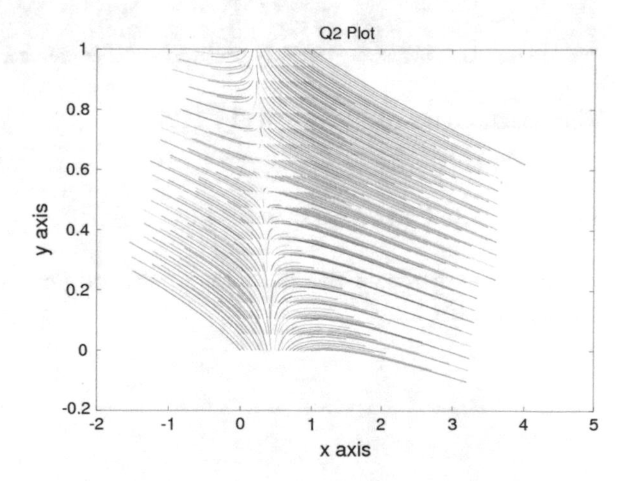

Fig. 14.4 Trigger model linearized about Q2

Listing 14.29: Jacobian at $Q3$

```
J(x3,y3)
ans =
   -1.43702    0.50000
   -0.20000   -0.30000
[V3,D3] = eig(J(x3,y3))
V3 =
   -0.98204   -0.43297
   -0.18868   -0.90141
D3 =
   -1.34096          0
          0   -0.39607
```

Hence, there are two real distinct eigenvalues, $r_1 = -1.34096$ and $r_2 = -0.39607$. The eigenvectors are

$$\boldsymbol{E}_1 = \begin{bmatrix} -0.98204 \\ -0.18868 \end{bmatrix} \quad \text{and} \quad \boldsymbol{E}_2 = \begin{bmatrix} -0.43297 \\ -0.90141 \end{bmatrix}$$

The dominant eigenvector is now $\boldsymbol{E}_2$. We set the coefficient matrix **A3** of our linearization and the parameter **p3** of our model and generate the automatic phase plane plot.

Listing 14.30: Phase plane for $Q3$ linearization

```
A3 = J(x3,y3);
p3 = [A3(1,1),A3(1,2),A3(2,1),A3(2,2),x3,y3]
p3 =
   -1.43702    0.50000   -0.20000   -0.30000    0.88443    0.21038
AutoPhasePlanePlotLinearSystemRKF5('x axis',...
         'y axis','Q1 Plot',...
         1.0e-6,1.0e-6,.01,.2,'linearsystemep',...
         p3,.01,0,.4,12,12,0,1,0,1);
```

This generates the plot shown in Fig. 14.5.

14.1.3.4 The Full Phase Plane

We can also generate the full plot of the original system using the function **triggermodel**

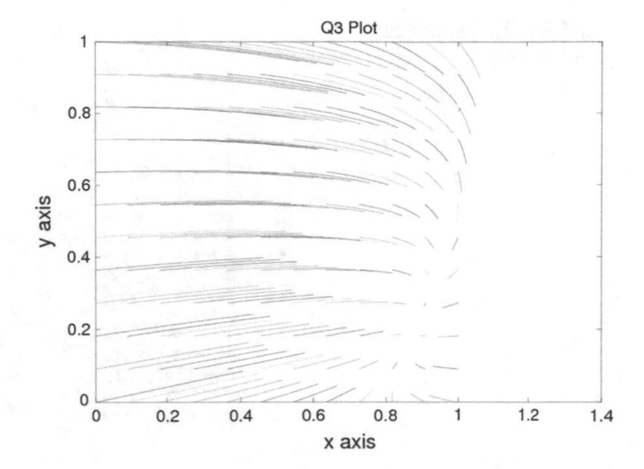

Fig. 14.5 Trigger model linearized about Q3

Listing 14.31: Trigger model

```
function y = triggermodel(t,x)
%
u = x(1);
v = x(2);
y = [-0.5*(17.76*u -103.79*u.^2 + 229.62*u.^3 - 226.31*u.^4 + 83.72*u.^5 -
    v);...
     0.2*(-u - 1.5*v + 1.2)];
```

We then plot the trajectories using **AutoPhasePlanePlotRKF5NoP**. This gives us Fig. 14.6. You can see in this plot how trajectories do not stay near $Q2$; instead, they move to $Q1$ or to $Q3$. Hence, you can a *trigger* a move from Q_1 to Q_3 or vice versa by choosing the right Initial condition! So this gives us a way to implement computer memory. Choose say Q_1 to represent a binary "1" and Q_3 to represent a binary "0" or to implement a model of how emotional states can switch quickly.

Listing 14.32: Phase plane plot for trigger model

```
AutoPhasePlanePlotRKF5NoP('x axis',...
        'y axis','Trigger Model Phase Plane',...
        1.0e-6,1.0e-6,.01,.05,...
        'triggermodel',.01,0,2,20,20,-0.2,1,0,1);
```

14.1.4 The Predator–Prey Model

The next nonlinear system is the familiar Predator–Prey model. Consider the following example

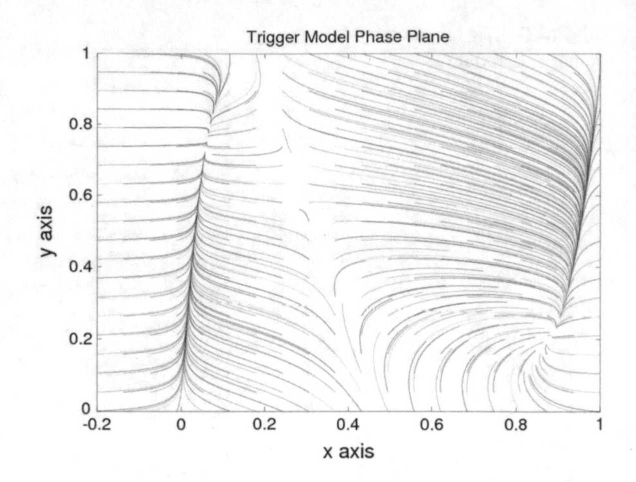

Fig. 14.6 Trigger model phase plane plot

$$x'(t) = 3\,x(t) - 4\,x(t)\,y(t)$$
$$y'(t) = -5\,y(t) + 7\,x(t)\,y(t)$$

The equilibrium points (x, y) solve the equations

$$x\,(3 - 4y) = 0$$
$$y\,(-5 + 7x) = 0$$

which has the familiar solutions $Q_1 = (0, 0)$ and $Q_2 = (\frac{5}{7}, \frac{3}{4})$. The Jacobian here is

$$\boldsymbol{J}(x, y) = \begin{bmatrix} 3 - 4y & -4x \\ 7y & -5 + 7x \end{bmatrix}$$

14.1.4.1 Equilibrium Point $Q1$

For the equilibrium point $Q1$, we have

Listing 14.33: Jacobian at $Q1$

```
A1 = J(0,0)
A1 =
   3  -0
   0  -5
[V1,D1] = eig(A1)
V1 =
   0   1
   1   0
D1 =
  -5   0
   0   3
```

Hence, there are two real distinct eigenvalues, $r_1 = -5$ and $r_2 = 3$. The eigenvectors are

$$E_1 = \begin{bmatrix} 0 \\ 1 \end{bmatrix} \quad \text{and} \quad E_2 = \begin{bmatrix} 1 \\ 0 \end{bmatrix}$$

The dominant eigenvector is thus E_2 and it is easy to plot the resulting trajectories. Using the function **linearsystemep** we set up the plot. Using the coefficient matrix **A1** of our linearization and the coordinates of Q_1, we set the parameter **p1** by listing all the entries of **A1** followed by the coordinates of $Q1$. We then generate the automatic phase plane plot.

Listing 14.34: Phase plane plot for $Q1$ linearization

```
A1 = J(x1,y1);
p1 = [A1(1,1),A1(1,2),A1(2,1),A1(2,2),x1,y1]
p1 =

   3  -0   0  -5   0   0

AutoPhasePlanePlotLinearSystemRKF5('x axis',...
        'y axis','Q1 Plot',...
        1.0e-6,1.0e-6,.01,.2,...
        'linearsystemep',p1,.01,0,.4,12,12,-1,1,-1,1);
```

This generates the plot shown in Fig. 14.7.
Note the linearization shows a set of trajectories that moves out from the origin along the x axis depending on the quadrant the initial condition comes from. We also know that no trajectory of the Predator–Prey model that starts in Quadrant One can cross the x and y axis, so the lesson that the linearization is an approximation to the true trajectories really comes home here.

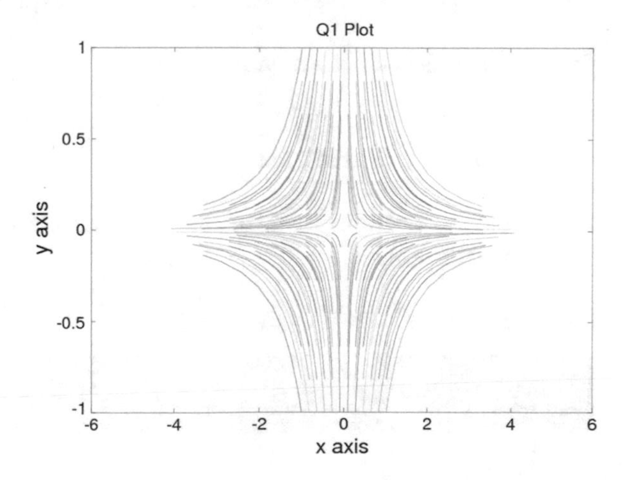

Fig. 14.7 Predator–Prey model linearized about Q1

14.1.4.2 Equilibrium Point $Q2$

For equilibrium point $Q2$, we find the Jacobian at this point.

Listing 14.35: Jacobian at $Q2$

```
A2 = J(x2,y2)
A2 =
   0.00000   -2.85714
   5.25000    0.00000
[V2,D2] = eig(A2)
V2 =
   0.00000 + 0.59365i   0.00000 - 0.59365i
   0.80472 + 0.00000i   0.80472 - 0.00000i
D2 =
   0.0000 + 3.8730i              0
                 0   0.0000 - 3.8730i
```

Hence, there is now a complex conjugate eigenvalue pair which has zero real part. Thus, these trajectories will be circles about Q_2. The eigenvalues are $r_1 = 3.8730i$ and $r_2 = -3.8730i$. The eigenvectors are not really needed for our phase plane. We set the coefficient matrix **A2** of our linearization and the parameter **p2** of our model and generate the automatic phase plane plot.

Listing 14.36: Phase plane for $Q2$ linearization

```
A2 = J(x2,y2);
p2 = [A2(1,1),A2(1,2),A2(2,1),A2(2,2),x2,y2]
p2 =

   0.00000   -2.85714   5.25000   0.00000   0.71429   0.75000
AutoPhasePlanePlotLinearSystemRKF5('x axis',...
         'y axis','Q2 Plot',...
         1.0e-6,1.0e-6,.01,.2,...
         'linearsystemep',p2,.01,0,2.4,12,12,-0.5,.1,0,.1);
```

This generates the plot shown in Fig. 14.8.

14.1.4.3 The Full Phase Plane

The full plot of the original system uses the standard function **PredPrey** modified for the model at hand.

Listing 14.37: Predator–Prey dynamics

```
function f = PredPrey(t,x)
f = [3*x(1)-4*x(1)*x(2);-5*x(2)+7*x(1)*x(2)];
```

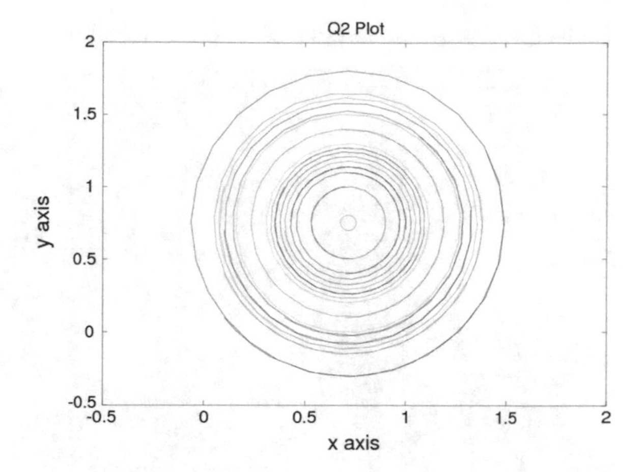

Fig. 14.8 Predator–Prey model linearized about Q2

We then plot the trajectories using **`AutoPhasePlanePlotRKF5NoPMultiple`**. This gives us Fig. 14.9. We had to write a new function to generate the plot as the equilibrium point Q_1 generates trajectories with large x and y values which complicate the calculations. So the new function manually builds four separate plots, one for each quadrant, and stays carefully away from the trajectories on the x and y axis. It is pretty hard to automate this kind of plot! In each quadrant, the number of trajectories and the size of the box from which the initial conditions come must be chosen inside the code. Also, the length of time the calculations run in a given quadrant must be handpicked. Hence, there is a lot of tweaking here!

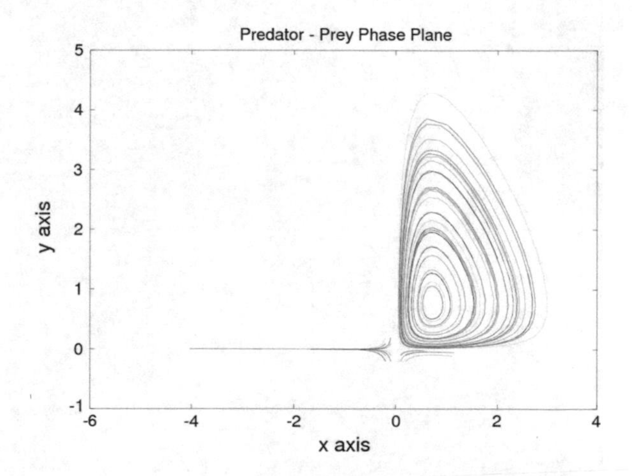

Fig. 14.9 Predator–Prey model phase plane plot

Listing 14.38: AutoPhasePlanePlotRKF5NoPMultiple

```
function AutoPhasePlanePlotRKF5NoPMultiple(Hlabel,Vlabel,Tlabel,errortol,steptol,
    minstep,maxstep,...
                                     fname,hinit,Q1,Q2,Q3,Q4)
  % fname is the name of the model dynamics
  % errortol is tolerance in RKF5 algorithm
5 % steptol is another tolerance in RKF5 algorithm
  % minstep is the minimum stepsize allowed
  % maxstep is the maximum stepsize allowed
  % hinit is the initial step size
  % Q1 is 1 if we want to generate the plot in quadrant 1
10 % Q2 is 1 if we want to generate the plot in quadrant 2
  % Q3 is 1 if we want to generate the plot in quadrant 3
  % Q4 is 1 if we want to generate the plot in quadrant 4
  % tend is the final time
  %
15 %hold plot and cycle line colors
  newplot;
  hold all;
  if (Q3 == 1)
  endLL = 2;
20 uLL = linspace(-.2,-.1,endLL);
  vLL = linspace(-.2,-.1,endLL);
  u = uLL;
  v = vLL;
  for i=1:endLL
25   for j=1:endLL
      y0 = [u(i);v(j)];
      [tvals,yvals,fvals,hvals] = RKF5NoP(errortol,steptol,minstep,maxstep,...
                                          fname,hinit,0,0.5,y0);
      plot(yvals(1,:),yvals(2,:));
30   end
  end
  end
  if (Q2 == 1)
  endUL = 2;
35 uUL = linspace(-.2,-.1,endUL);
  vUL = linspace(.1,.2,endUL);
  u = uUL;
  v = vUL;
  for i=1:endUL
40   for j=1:endUL
      y0 = [u(i);v(j)];
      [tvals,yvals,fvals,hvals] = RKF5NoP(errortol,steptol,minstep,maxstep,...
                                          fname,hinit,0,0.6,y0);
      plot(yvals(1,:),yvals(2,:));
45   end
  end
  end
  if (Q4 == 1)
  endLR = 2;
50 uLR = linspace(.1,.2,endLR);
  vLR = linspace(-.2,-.1,endLR);
  u = uLR
  v = vLR
  for i=1:endLR
55   for j=1:endLR
      y0 = [u(i);v(j)];
      [tvals,yvals,fvals,hvals] = RKF5NoP(errortol,steptol,minstep,maxstep,...
                                          fname,hinit,0,0.5,y0);
      plot(yvals(1,:),yvals(2,:));
60   end
  end
  end
  if (Q1 == 1)
  endUR = 4;
65 uUR = linspace(.1,2.0,endUR);
  vUR = linspace(.1,2.0,endUR);
  u = uUR;
  v = vUR;
  for i=1:endUR
70   for j=1:endUR
      y0 = [u(i);v(j)];
      [tvals,yvals,fvals,hvals] = RKF5NoP(errortol,steptol,minstep,maxstep,...
                                          fname,hinit,0,2.4,y0);
      plot(yvals(1,:),yvals(2,:));
75   end
  end
  end
  xlabel(Hlabel);
  ylabel(Vlabel);
80 title(Tlabel);
  hold off;
```

Listing 14.39: A sample phase plane plot

```
AutoPhasePlanePlotRKF5NoPMultiple('x axis' ,...
        'y axis','Predator - Prey Phase Plane' ,...
        1.0e-6,1.0e-6,.01,.05,'PredPrey',.01,1,1,1,1);
```

14.1.5 The Predator–Prey Model with Self Interaction

The next nonlinear system is the Predator–Prey model with self interaction. Consider the following example

$$x'(t) = 3\,x(t) - 4\,x(t)\,y(t) - e\,x(t)^2$$
$$y'(t) = -5\,y(t) + 7\,x(t)\,y(t) - f\,y(t)^2$$

where e and f are positive numbers. From our earlier discussions, we know that there are essentially two interesting cases here. The x' and y' nullclines cross in the first quadrant leading to spiral in trajectories towards that common point or they cross in the fourth quadrant which leads to all trajectories converging to a point on the positive x axis. We can now look at this model using our new tools. The equilibrium points (x, y) solve the equations

$$x\,(3 - 4y - ex) = 0$$
$$y\,(-5 + 7x - fy) = 0$$

Three of the equilibrium points are then $Q_1 = (0, 0)$, $Q_2 = (0, -5/f)$ and $Q_3 = (3/e, 0)$. These equilibrium points coincide with trajectories that we do not see if we choose positive initial conditions in Quadrant I.

14.1.5.1 The Nullclines Intersect in Quadrant I

This situation occurs when $3/e > 5/7$ so as an example, let's choose the value $e = 1$. The value of f is not very important, so let's choose $f = 1$ also just to make it easy. Then, the intersection occurs when

$$x + 4y = 3$$
$$7x - y = 5$$

We can find this point of intersection many ways. The old fashioned way is by elimination. We find $x = 23/29$ and $y = 16/29$. Let $Q_4 = (23/29, 16/29)$. The Jacobian in general is

$$J(x, y) = \begin{bmatrix} 3 - 4y - 2ex & -4x \\ 7y & -5 + 7x - 2fy \end{bmatrix}.$$

Thus, for $e = 1$ and $f = 1$, we have

$$J(x, y) = \begin{bmatrix} 3 - 4y - 2x & -4x \\ 7y & -5 + 7x - 2y \end{bmatrix}.$$

We will only look at the local linearizations for equilibrium points Q_4 and then Q_3. The other two are similar to what we have done before.

Equilibrium Point $Q4$ For the equilibrium point $Q4$, we can find the Jacobian at Q_4 and the corresponding eigenvalues and eigenvectors. As expected, the eigenvalues are complex with negative real part implying the local linearization gives spiral in trajectories.

Listing 14.40: Jacobian at $Q4$

```
J = @(x,y) [3 - 4*y-2*x,-4*x;7*y,-5 + 7*x - 2*y];
J(23/29,16/29)
ans =
   -0.79310   -3.17241
    3.86207   -0.55172
A4 = J(23/29,16/29);
[V4,D4] = eig(A4)
V4 =
   -0.02315 + 0.67115i   -0.02315 - 0.67115i
    0.74096 + 0.00000i    0.74096 - 0.00000i
D4 =
Diagonal Matrix
   -0.6724 + 3.4982i                    0
                   0   -0.6724 - 3.4982i
```

Hence, there is a complex eigenvalue pair, $r = -0.6724 \pm 3.4982i$. Using **linear systemep** we generate the plot using the coefficient matrix **A4** of our linearization and the coordinates of Q_4. We set the parameter **p4** by listing all the entries of **A4** followed by the coordinates of $Q4$. We then generate the automatic phase plane plot.

Listing 14.41: Phase plane for $Q4$ linearization

```
A4 = J(x1,y1);
p4 = [A4(1,1),A4(1,2),A4(2,1),A4(2,2),23/29,16/29];
AutoPhasePlanePlotLinearSystemRKF5('x axis',...
          'y axis','Q4 Plot',...
          1.0e-6,1.0e-6,.01,.2,...
          'linearsystemep',p4,.01,0,.4,12,12,-1,1,-1,1);
```

This generates the plot shown in Fig. 14.10.

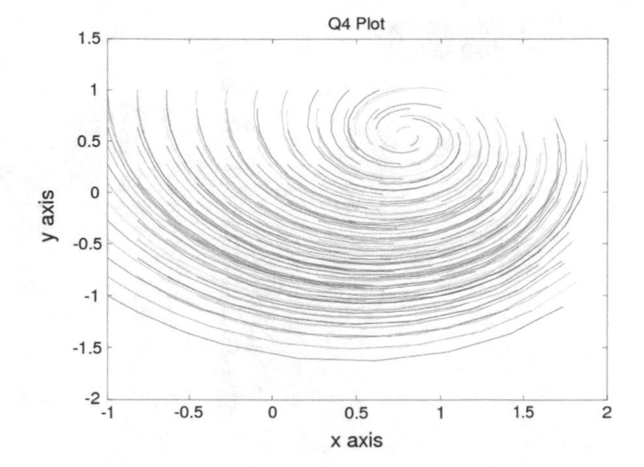

Fig. 14.10 Predator–Prey model with self interaction with nullcline intersection Q4 in Quadrant I, linearized about Q4

Equilibrium Point $Q3$ For equilibrium point $Q3 = (3/e, 0) = (3, 0)$ here, we find the Jacobian, eigenvalues and eigenvectors at this point.

Listing 14.42: Jacobian at $Q3$

```
A3 = J(3,0);
[V3,D3] = eig(A3)
V3 =
    1.00000   -0.53399
    0.00000    0.84549
D3 =
Diagonal Matrix
   -3    0
    0   16
```

The eigenvalues are both real, $r_1 = -3$ and $r_2 = 16$. Hence, the dominant eigenvector is $\boldsymbol{E_2}$ where

$$\boldsymbol{E_2} = \begin{bmatrix} -0.53399 \\ 0.84549 \end{bmatrix}$$

So we will see all the trajectories moving parallel or towards the $\boldsymbol{E_2}$ line. We set the coefficient matrix **A3** of our linearization and the parameter **p3** of our model and generate the automatic phase plane plot. It took a bit of experimentation to generate this plot so that it looked reasonably good. The eigenvalue of 16 causes very fast growth!

MatLab! A Predator–Prey Model with Self Interaction! the nullclines intersect in Quadrant 1: session code for phase plane portrait for equilibrium point Q3

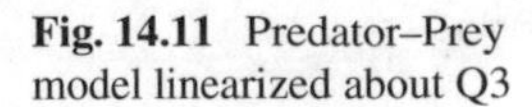

Fig. 14.11 Predator–Prey model linearized about Q3

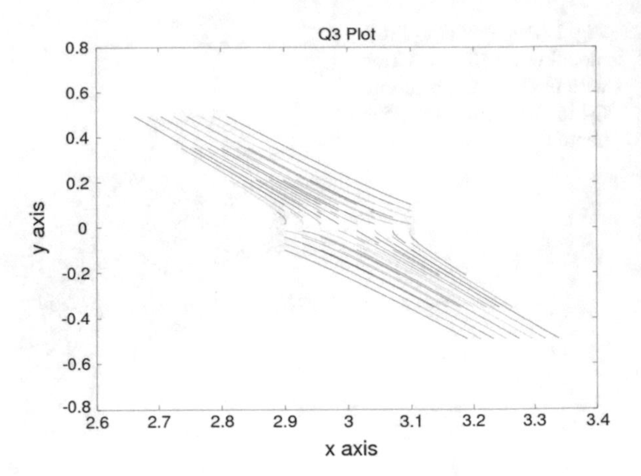

Listing 14.43: Phase plane for $Q3$ linearization

```
p3 = [A3(1,1),A3(1,2),A3(2,1),A3(2,2),3,0];
AutoPhasePlanePlotLinearSystemRKF5('x axis','y axis','Q3 Plot',...
        1.0e-6,1.0e-6,.001,.01,'linearsystemep',...
        p3,.001,0,.1,8,8,2.9,3.1,-.1,.1);
```

This generates the plot shown in Fig. 14.11.

The Full Phase Plane The full plot of the original system uses the standard function **`PredPrey`** modified for the model at hand.

MatLab! A Predator–Prey Model with Self Interaction! the nullclines intersect in Quadrant 1: session code for full phase plane portrait: dynamics

Listing 14.44: PredPreySelf

```
function f = PredPreySelf(t,x)
f = [3*x(1)-4*x(1)*x(2)-x(1)*x(1);-5*x(2)+7*x(1)*x(2)-x(2)*x(2)];
```

To generate the full phase plane in all quadrants is difficult due to the growth rates in all areas of the plane other than quadrant I.

Listing 14.45: Actual Phase plane

```
AutoPhasePlanePlotRKF5NoPMultiple('x axis','y axis',...
'Predator - Prey Self Interaction Phase Plan Plot',...
1.0e-6,1.0e-6,.01,.1,...
'PredPreySelf',.01,1,1,1,1);
```

This generates the plot shown in Fig. 14.12.

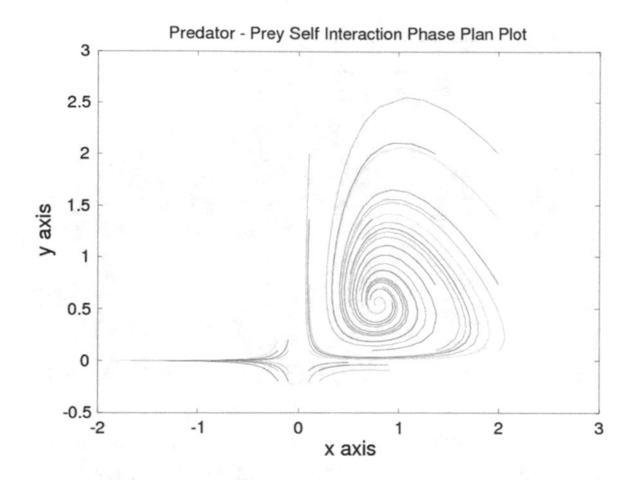

Fig. 14.12 Predator–Prey self interaction model with intersection in quadrant I

14.1.5.2 The Nullclines Intersect in Quadrant IV

This situation occurs when $3/e < 5/7$ so as an example, let's choose the value $e = 5$. The value of f is not very important, so let's choose $f = 5$ also just to make it easy. Then, the intersection occurs when

$$\begin{aligned} 5x + 4y &= 3 \\ 7x - 5y &= 5 \end{aligned}$$

We find $x = 35/53$ and $y = -4/53$. Let $Q_4 = (35/53, -4/53)$. The Jacobian for $e = 5$ and $f = 5$ is

$$\boldsymbol{J}(x, y) = \begin{bmatrix} 3 - 4y - 10x & -4x \\ 7y & -5 + 7x - 10y \end{bmatrix}.$$

The other equilibrium points are $Q_1 = (0, 0)$, $Q_2 = (0, -5/f)$ and $Q_3 = (3/e, 0)$. Equilibrium points coincide with trajectories that we do not see if we choose positive initial conditions in Quadrant I. However, equilibrium point Q_3 is different now. All the trajectories that start in the positive first quadrant will converge to Q_3. Again, we will only look at the local linearizations for equilibrium points Q_3 and Q_4.

Equilibrium Point $Q4$ For the equilibrium point $Q4$, we can find the Jacobian at Q_4 and the corresponding eigenvalues and. eigenvectors

Listing 14.46: Jacobian at $Q4$

```
J = @(x,y) [3-4*y-10*x,-4*y;7*y,-5+7*x-10*y];
J(35/53,-4/53)
ans =
   -3.30189    0.30189
   -0.52830    0.37736
A4 = J(35/53,-4/53);
[V4,D4] = eig(A4)
V4 =
   -0.989605   -0.082757
   -0.143812   -0.996570
D4 =
   -3.25802          0
          0    0.33349
```

Hence, there are two real roots that are distinct and the dominant eigenvector is $\boldsymbol{E_2}$. Using the coefficient matrix **A4** of our linearization and the coordinates of Q_4, we set the parameter **p4** and generate the automatic phase plane plot.

Listing 14.47: Phase plane plot for $Q4$ linearization

```
A4 = J(x1,y1);
p4 = [A4(1,1),A4(1,2),A4(2,1),A4(2,2),35/53,-4/53];
AutoPhasePlanePlotLinearSystemRKF5('x axis',...
'y axis','Q4 Plot',...
1.0e-6,1.0e-6,.01,.2,'linearsystemep',...
p4,.01,0,.4,12,12,-1,1,-1,1);
```

This generates the plot shown in Fig. 14.13.

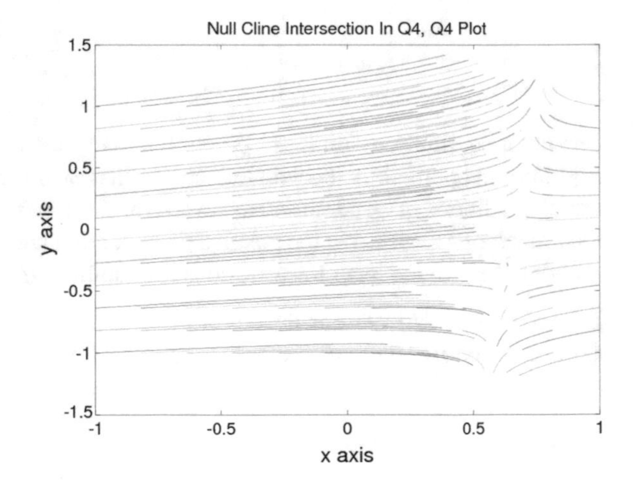

Fig. 14.13 Predator–Prey model with self interaction with nullcline intersection Q4 in Quadrant IV, linearized about Q4

Equilibrium Point $Q3$ For equilibrium point $Q3 = (3/e, 0) = (3/5, 0)$ here, we find the Jacobian, eigenvalues and eigenvectors at this point.

Listing 14.48: Jacobian at $Q3$

```
A3 = J(.6,0);
[V3,D3] = eig(A3)
V3 =
   1   0
   0   1
D3 =
   -3.00000          0
          0   -0.80
```

The eigenvalues are now both real and negative, $r_1 = -3$ and $r_2 = -0.8$. The dominant eigenvector is again $\boldsymbol{E}_2$. So we will see all the trajectories moving parallel or towards the $\boldsymbol{E}_2$ line.

Listing 14.49: Phase plane for $Q3$ linearization

```
p3 = [A3(1,1),A3(1,2),A3(2,1),A3(2,2),.6,0];
AutoPhasePlanePlotLinearSystemRKF5('x axis','y axis','Q3 Plot',...
,1.0e-6,1.0e-6,.001,.01,'linearsystemep',p3,...
.001,0,.1,8,8,2.9,3.1,-.1,.1);
```

This generates the plot shown in Fig. 14.14.

The Full Phase Plane The full plot of the original system uses the standard function `PredPrey` modified for the model at hand.

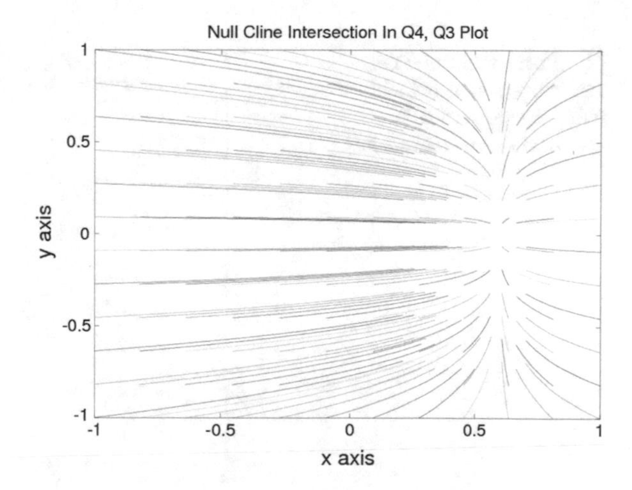

Fig. 14.14 Predator–Prey model with self interaction with nullcline intersection in Quadrant IV, linearized about Q3

Listing 14.50: PredPreySelf

```
function f = PredPreySelf(t,x)
f = [3*x(1)-4*x(1)*x(2)-5*x(1)*x(1);-5*x(2)+7*x(1)*x(2)-5*x(2)*x(2)];
```

To generate the full phase plane in all quadrants is as always difficult and we have to play with the settings in the function **AutoPhasePlanePlotRKF5NoPMultiple**.

Listing 14.51: The actual phase plane

```
AutoPhasePlanePlotRKF5NoPMultiple('x axis','y axis',...
'Predator - Prey Self Interaction Model with null
cline intersection in Quadrant IV Phase Plan Plot',...
1.0e-6,1.0e-6,.01,.1,...
'PredPreySelf',.005,1,1,1,1);
```

This generates the plot shown in Fig. 14.15.

14.1.6 Problems

For each of the following systems, do this:

1. Graph the nullclines $x' = 0$ and $y' = 0$ and show on the $x - y$ plane the regions where x' and y' take on their various algebraic signs.
2. Find the equilibrium points.

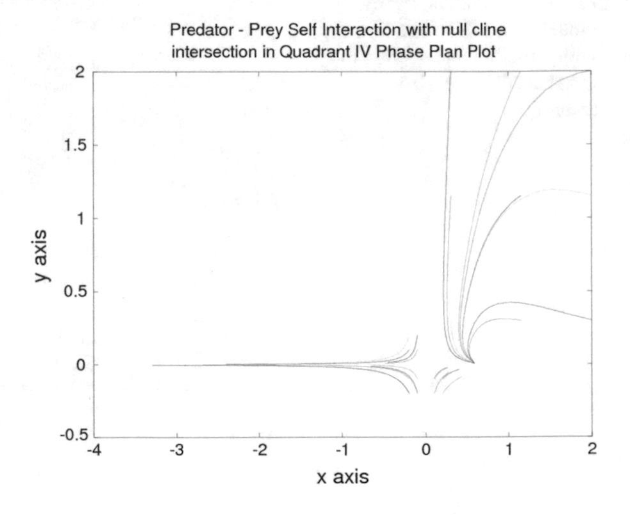

Fig. 14.15 Predator–Prey self interaction model with nullcline intersection in Quadrant IV phase plane plot

3. At each equilibrium point, find the Jacobian of the system and analyze the linearized system we have discussed in class. This means:
   - find eigenvalues and eigenvectors if the system has real eigenvalues. You don't need the eigenvectors if the eigenvalues are complex.
   - sketch a graph of the linearized solutions near the equilibrium point.
4. Using 1 and 3 to combine all this information into full graph of the system.

Exercise 14.1.8

$$x' = y$$
$$y' = -x + \frac{x^3}{6} - y$$

Exercise 14.1.9

$$x' = -x + y$$
$$y' = 0.1x - 2y - x^2 - 0.1x^3$$

Exercise 14.1.10

$$x' = y$$
$$y' = -x + y\,(1 - 3x^2 - 2y^2)$$

Chapter 15
An Insulin Model

We are now going to discuss a very nice model of diabetes detection or equivalently, a model of insulin regulation, which was presented in the classic text on applied mathematical modeling by Braun, **Differential Equations and Their Applications** (Braun 1978).

In diabetes there is too much sugar in the blood and the urine. This is a metabolic disease and if a person has it, they are not able to use up all the sugars, starches and various carbohydrates because they don't have enough **insulin**. Diabetes can be diagnosed by a **glucose tolerance test** (GTT). If you are given this test, you do an overnight fast and then you are given a large dose of sugar in a form that appears in the bloodstream. This sugar is called **glucose**. Measurements are made over about five hours or so of the concentration of glucose in the blood. These measurements are then used in the diagnosis of diabetes. It has always been difficult to interpret these results as a means of diagnosing whether a person has diabetes or not. Hence, different physicians interpreting the same data can come up with a different diagnosis, which is a pretty unacceptable state of affairs!

In this chapter, we are going to discuss a criterion developed in the 1960s by doctors at the Mayo Clinic and the University of Minnesota that was fairly reliable. It showcases a lot of our modeling in this course and will give you another example of how we use our tools. We start with a simple model of the blood glucose regulatory system.

Glucose plays an important role in vertebrate metabolism because it is a source of energy. For each person, there is an optimal blood glucose concentration and large deviations from this leads to severe problems including death. Blood glucose levels are autoregulated via standard forward and backward interactions like we see in many biological systems. An example is the signal that is used to activate the creation of a protein which we discussed earlier. The signaling molecules are typically either bound to another molecule in the cell or are free. The equilibrium concentration of free signal is due to the fact that the rate at which signaling molecules bind equals the rate at which they split apart from their binding substrate. When an external message comes into the cell called a trigger, it induces a change in this careful balance which temporarily upgrades or degrades the equilibrium signal concentration. This then

J.K. Peterson, *Calculus for Cognitive Scientists*,
Cognitive Science and Technology, DOI 10.1007/978-981-287-877-9_15

influence the protein concentration rate. Blood glucose concentrations work like this too, although the details differ. The blood glucose concentration is influenced by a variety of signaling molecules just like the protein creation rates can be. Here are some of them. The hormone that **decreases** blood glucose concentration is **insulin**. **Insulin** is a hormone secreted by the β cells of the pancreas. After we eat carbohydrates, our gastrointestinal tract sends a signal to the pancreas to secrete insulin. Also, the glucose in our blood directly stimulates the β cells to secrete insulin. We think insulin helps cells pull in the glucose needed for metabolic activity by attaching itself to membrane walls that are normally impenetrable. This attachment increases the ability of glucose to pass through to the inside of the cell where it can be used as *fuel*. So, if there is not enough insulin, cells don't have enough energy for their needs. The other hormones we will focus on all tend to **change** blood glucose concentrations also.

- **Glucagon** is a hormone secreted by the α cells of the pancreas. Excess glucose is stored in the liver in the form of **Glycogen**. There is the usual equilibrium amount of storage caused by the rate of glycogen formation being equal to the rate of the reverse reaction that moves glycogen back to glucose. Hence the glycogen serves as a reservoir for glucose and when the body needs glucose, the rate balance is tipped towards conversion back to glucose to release needed glucose to the cells. The hormone **glucagon** increases the rate of the reaction that converts glycogen back to glucose and so serves an important regulatory function. **Hypoglycemia** (low blood sugar) and **fasting** tend to increase the secretion of the hormone glucagon. On the other hand, if the blood glucose levels increase, this tends to suppress glucagon secretion; i.e. we have another back and forth regulatory tool.
- **Epinephrine** also called **adrenalin** is a hormone secreted by the adrenal medulla. It is part of an emergency mechanism to quickly increase the blood glucose concentration in times of extremely low blood sugar levels. Hence, epinephrine also increases the rate at which glycogen converts to glucose. It also directly inhibits how much glucose is able to be pulled into muscle tissue because muscles use a lot of energy and this energy is needed elsewhere more urgently. It also acts on the pancreas directly to inhibit insulin production which keeps glucose in the blood. There is also another way to increase glucose by converting lactate into glucose in the liver. Epinephrine increases this rate also so the liver can pump this extra glucose back into the blood stream.
- **Glucocorticoids** are hormones like **cortisol** which are secreted by the adrenal cortex which influence how carbohydrates are metabolized which is turn increase glucose if the the metabolic rate goes up.
- **Thyroxin** is a hormone secreted by the thyroid gland and it helps the liver form glucose from sources which are not carbohydrates such as glycerol, lactate and amino acids. So another way to up glucose!
- **Somatotropin** is called the growth hormone and it is secreted by the anterior pituitary gland. This hormone directly affect blood glucose levels (i.e. an increase in Somatotropin increases blood glucose levels and vice versa) but it also inhibits the effect of insulin on muscle and fat cell's permeability which diminishes insulin's

ability to help those cells pull glucose out of the blood stream. These actions can therefore increase blood glucose levels.

Now net hormone concentration is the sum of insulin plus the others. Let $\boldsymbol{H}$ denote this net hormone concentration. At normal conditions, call this concentration $\boldsymbol{H_0}$. There have been studies performed that show that under close to normal conditions, the interaction of the one hormone **insulin** with blood glucose completely dominates the net hormonal activity. That is normal blood sugar levels primarily depend on insulin-glucose interactions.

So if insulin increases from normal levels, it increases net hormonal concentration to $\boldsymbol{H_0} + \Delta \boldsymbol{H}$ and decreases glucose blood concentration. On the other hand, if other hormones such as cortisol increased from base levels, this will make blood glucose levels go up. Since insulin dominates all activity at normal conditions, we can think of this increase in cortisol as a decrease in insulin with a resulting drop in blood glucose levels. A decrease in insulin from normal levels corresponds to a drop in net hormone concentration to $\boldsymbol{H_0} - \Delta \boldsymbol{H}$. Now let $\boldsymbol{G}$ denote blood glucose level. Hence, in our model an increase in $\boldsymbol{H}$ means a drop in $\boldsymbol{G}$ and a decrease in $\boldsymbol{H}$ means an increase in $\boldsymbol{G}$! Note our lumping of all the hormone activity into a single net activity is very much like how we modeled food fish and predator fish in the predator–prey model.

The idea of our model for diagnosing diabetes from the GTT is to find a simple dynamical model of this complicated blood glucose regulatory system in which the values of two parameters would give a nice criterion for distinguishing normal individuals from those with mild diabetes or those who are pre diabetic. Here is what we will do. We describe the model as

$$\begin{aligned} \boldsymbol{G}'(t) &= F_1(\boldsymbol{G}, \boldsymbol{H}) + \boldsymbol{J}(t) \\ \boldsymbol{H}'(t) &= F_2(\boldsymbol{G}, \boldsymbol{H}) \end{aligned}$$

where the function $\boldsymbol{J}$ is the external rate at which blood glucose concentration is being increased. There are two nonlinear interaction functions F_1 and F_2 because we know $\boldsymbol{G}$ and $\boldsymbol{H}$ have complicated interactions.

Let's assume $\boldsymbol{G}$ and $\boldsymbol{H}$ have achieved optimal values $\boldsymbol{G_0}$ and $\boldsymbol{H_0}$ by the time the fasting patient has arrived at the hospital. Hence, we don't expect to have any contribution to $\boldsymbol{G}'(0)$ and $\boldsymbol{H}'(0)$; i.e. $F_1(\boldsymbol{G_0}, \boldsymbol{H_0}) = 0$ and $F_2(\boldsymbol{G_0}, \boldsymbol{H_0}) = 0$.

We are interested in the deviation of $\boldsymbol{G}$ and $\boldsymbol{H}$ from their optimal values $\boldsymbol{G_0}$ and $\boldsymbol{H_0}$, so let $\boldsymbol{g} = \boldsymbol{G} - \boldsymbol{G_0}$ and $\boldsymbol{h} = \boldsymbol{H} - \boldsymbol{H_0}$. We can then write $\boldsymbol{G} = \boldsymbol{G_0} + \boldsymbol{g}$ and $\boldsymbol{H} = \boldsymbol{H_0} + \boldsymbol{h}$. The model can then be rewritten as

$$\begin{aligned} (\boldsymbol{G_0} + \boldsymbol{g})'(t) &= F_1(\boldsymbol{G_0} + \boldsymbol{g}, \boldsymbol{H_0} + \boldsymbol{h}) + \boldsymbol{J}(t) \\ (\boldsymbol{H_0} + \boldsymbol{h})'(t) &= F_2(\boldsymbol{G_0} + \boldsymbol{g}, \boldsymbol{H_0} + \boldsymbol{h}) \end{aligned}$$

or

$$
\begin{aligned}
\boldsymbol{g}'(t) &= F_1(\boldsymbol{G_0}+\boldsymbol{g}, \boldsymbol{H_0}+\boldsymbol{h}) + \boldsymbol{J}(t) \\
\boldsymbol{h}'(t) &= F_2(\boldsymbol{G_0}+\boldsymbol{g}, \boldsymbol{H_0}+\boldsymbol{h})
\end{aligned}
$$

We know the tangent plane to a function $F(x, y)$ at the point (x_0, y_0) is given by

$$
T(x, y) = F(x_0, y_0) + F_x(x_0, y_0)(x - x_0) + F_y(x_0, y_0)(y - y_0) + E_F
$$

where the error is E_F. We use this idea on our functions F_1 and F_2 at the optimal values $\boldsymbol{G_0}$ and $\boldsymbol{H_0}$. We have

$$
\begin{aligned}
F_1(\boldsymbol{G_0}+\boldsymbol{g}, \boldsymbol{H_0}+\boldsymbol{h}) &= F_1(\boldsymbol{G_0}, \boldsymbol{H_0}) + \frac{\partial F_1}{\partial \boldsymbol{g}}(\boldsymbol{G_0}, \boldsymbol{H_0})\,\boldsymbol{g} + \frac{\partial F_1}{\partial \boldsymbol{h}}(\boldsymbol{G_0}, \boldsymbol{H_0})\,\boldsymbol{h} + E_{F_1} \\
F_2(\boldsymbol{G_0}+\boldsymbol{g}, \boldsymbol{H_0}+\boldsymbol{h}) &= F_2(\boldsymbol{G_0}, \boldsymbol{H_0}) + \frac{\partial F_2}{\partial \boldsymbol{g}}(\boldsymbol{G_0}, \boldsymbol{H_0})\,\boldsymbol{g} + \frac{\partial F_2}{\partial \boldsymbol{h}}(\boldsymbol{G_0}, \boldsymbol{H_0})\,\boldsymbol{h} + E_{F_2}
\end{aligned}
$$

but the terms $F_1(\boldsymbol{G_0}, \boldsymbol{H_0}) = 0$ and $F_1(\boldsymbol{G_0}, \boldsymbol{H_0}) = 0$, so we can simplify to

$$
\begin{aligned}
F_1(\boldsymbol{G_0}+\boldsymbol{g}, \boldsymbol{H_0}+\boldsymbol{h}) &= \frac{\partial F_1}{\partial \boldsymbol{g}}(\boldsymbol{G_0}, \boldsymbol{H_0})\,\boldsymbol{g} + \frac{\partial F_1}{\partial \boldsymbol{h}}(\boldsymbol{G_0}, \boldsymbol{H_0})\,\boldsymbol{h} + E_{F_1} \\
F_2(\boldsymbol{G_0}+\boldsymbol{g}, \boldsymbol{H_0}+\boldsymbol{h}) &= \frac{\partial F_2}{\partial \boldsymbol{g}}(\boldsymbol{G_0}, \boldsymbol{H_0})\,\boldsymbol{g} + \frac{\partial F_2}{\partial \boldsymbol{h}}(\boldsymbol{G_0}, \boldsymbol{H_0})\,\boldsymbol{h} + E_{F_2}
\end{aligned}
$$

It seems reasonable to assume that since we are so close to ordinary operating conditions, the errors E_{F_1} and E_{F_2} will be negligible. Thus our model approximation is

$$
\begin{aligned}
\boldsymbol{g}'(t) &= \frac{\partial F_1}{\partial \boldsymbol{g}}(\boldsymbol{G_0}, \boldsymbol{H_0})\,\boldsymbol{g} + \frac{\partial F_1}{\partial \boldsymbol{h}}(\boldsymbol{G_0}, \boldsymbol{H_0})\,\boldsymbol{h} + \boldsymbol{J}(t) \\
\boldsymbol{h}'(t) &= \frac{\partial F_2}{\partial \boldsymbol{g}}(\boldsymbol{G_0}, \boldsymbol{H_0})\,\boldsymbol{g} + \frac{\partial F_2}{\partial \boldsymbol{h}}(\boldsymbol{G_0}, \boldsymbol{H_0})\,\boldsymbol{h}
\end{aligned}
$$

We can reason out the algebraic signs of the four partial derivatives to be

$$
\begin{aligned}
\frac{\partial F_1}{\partial \boldsymbol{g}}(\boldsymbol{G_0}, \boldsymbol{H_0}) &= - \\
\frac{\partial F_1}{\partial \boldsymbol{h}}(\boldsymbol{G_0}, \boldsymbol{H_0}) &= - \\
\frac{\partial F_2}{\partial \boldsymbol{g}}(\boldsymbol{G_0}, \boldsymbol{H_0}) &= + \\
\frac{\partial F_2}{\partial \boldsymbol{h}}(\boldsymbol{G_0}, \boldsymbol{H_0}) &= -
\end{aligned}
$$

The arguments for these algebraic signs come from our understanding of the physiological processes that are going on here. Let's look at a small positive deviation$\boldsymbol{g}$ from the optimal value $\boldsymbol{G_0}$ while letting the net hormone concentration be fixed at $\boldsymbol{H_0}$. At this point, we are not adding an external input, so here $\boldsymbol{J}(t) = 0$. Then our model approximation is

$$\boldsymbol{g}'(t) = \frac{\partial F_1}{\partial \boldsymbol{g}}(\boldsymbol{G_0}, \boldsymbol{H_0})\, \boldsymbol{g}$$

At a state where we have an increase in blood sugar levels over optimal, i.e. $\boldsymbol{g} > 0$, the other hormones such as cortisol and glucagon will try to regulate the blood sugar level down by increasing their concentrations and for example storing more sugar into glycogen. Hence, the term $\frac{\partial F_1}{\partial \boldsymbol{g}}(\boldsymbol{G_0}, \boldsymbol{H_0})$ should be negative as here $\boldsymbol{g}'$ is negative as $\boldsymbol{g}$ should be decreasing. So we model this as $\frac{\partial F_1}{\partial \boldsymbol{g}}(\boldsymbol{G_0}, \boldsymbol{H_0}) = -m_1$ for some positive number m_1. Now consider a positive change in $\boldsymbol{h}$ from the optimal level while keeping at the optimal level $\boldsymbol{G_0}$. Then the model is

$$\boldsymbol{g}'(t) = \frac{\partial F_1}{\partial \boldsymbol{h}}(\boldsymbol{G_0}, \boldsymbol{H_0})\, \boldsymbol{h}$$

and since $\boldsymbol{h} > 0$, this means the net hormone concentration is up which we interpret as insulin above normal. This means blood sugar levels go down which implies $\boldsymbol{g}'$ is negative again. Thus, $\frac{\partial F_1}{\partial \boldsymbol{h}}(\boldsymbol{G_0}, \boldsymbol{H_0})$ must be negative which means we model it as $\frac{\partial F_1}{\partial \boldsymbol{h}}(\boldsymbol{G_0}, \boldsymbol{H_0}) = -m_2$ for some positive m_2.

Now look at the $\boldsymbol{h}'$ model in these two cases. If we have a small positive deviation $\boldsymbol{g}$ from the optimal value $\boldsymbol{G_0}$ while letting the net hormone concentration be fixed at $\boldsymbol{H_0}$, we have

$$\boldsymbol{h}'(t) = \frac{\partial F_2}{\partial \boldsymbol{g}}(\boldsymbol{G_0}, \boldsymbol{H_0})\, \boldsymbol{g}.$$

Again, since $\boldsymbol{g}$ is positive, this means we are above normal blood sugar levels which implies mechanisms are activated to bring the level down. Hence $\boldsymbol{h}' > 0$ as we have increasing net hormone levels. Thus, we must have $\frac{\partial F_2}{\partial \boldsymbol{g}}(\boldsymbol{G_0}, \boldsymbol{H_0}) = m_3$ for some positive m_3. Finally, if we have a positive deviation $\boldsymbol{h}$ from optimal while blood sugar levels are optimal, the model is

$$\boldsymbol{h}'(t) = \frac{\partial F_2}{\partial \boldsymbol{h}}(\boldsymbol{G_0}, \boldsymbol{H_0})\, \boldsymbol{h}.$$

Since $\boldsymbol{h}$ is positive, we have the concentrations of the hormones that pull glucose out of the blood stream are above optimal. This means that too much sugar is being removed as so the regulatory mechanisms will act to stop this action implying $\boldsymbol{h}' < 0$. This tells us $\frac{\partial F_2}{\partial \boldsymbol{g}}(\boldsymbol{G_0}, \boldsymbol{H_0}) = -m_4$ for some positive constant m_4. Hence, the four partial derivatives at the optimal points can be defined by four positive numbers m_1, m_2, m_3 and m_4 as follows:

$$\frac{\partial F_1}{\partial \boldsymbol{g}}(\boldsymbol{G_0}, \boldsymbol{H_0}) = -m_1$$
$$\frac{\partial F_1}{\partial \boldsymbol{h}}(\boldsymbol{G_0}, \boldsymbol{H_0}) = -m_2$$
$$\frac{\partial F_2}{\partial \boldsymbol{g}}(\boldsymbol{G_0}, \boldsymbol{H_0}) = +m_3$$
$$\frac{\partial F_2}{\partial \boldsymbol{h}}(\boldsymbol{G_0}, \boldsymbol{H_0}) = -m_4$$

Our model dynamics are thus approximated by

$$\boldsymbol{g}'(t) = -m_1\, \boldsymbol{g} - m_2\, \boldsymbol{h} + \boldsymbol{J}(t)$$
$$\boldsymbol{h}'(t) = m_3\, \boldsymbol{g} - m_4\, \boldsymbol{h}$$

This implies

$$\boldsymbol{g}''(t) = -m_1\, \boldsymbol{g}' - m_2\, \boldsymbol{h}' + \boldsymbol{J}'(t)$$

Now plug in the formula for $\boldsymbol{h}'$ to get

$$\begin{aligned} \boldsymbol{g}''(t) &= -m_1\, \boldsymbol{g}' - m_2\, (m_3\, \boldsymbol{g} - m_4\, \boldsymbol{h}) + \boldsymbol{J}'(t) \\ &= -m_1\, \boldsymbol{g}' - m_2 m_3 \boldsymbol{g} + m_2 m_4 \boldsymbol{h} + \boldsymbol{J}'(t). \end{aligned}$$

But we can use the $\boldsymbol{g}'$ equation to solve for $\boldsymbol{h}$. This gives

$$m_2\, \boldsymbol{h} = -\boldsymbol{g}'(t) - m_1\, \boldsymbol{g} + \boldsymbol{J}(t)$$

which leads to

$$\begin{aligned} \boldsymbol{g}''(t) &= -m_1\, \boldsymbol{g}' - m_2 m_3 \boldsymbol{g} + m_4 \left(-\boldsymbol{g}'(t) - m_1\, \boldsymbol{g} + \boldsymbol{J}(t)\right) + \boldsymbol{J}'(t) \\ &= -(m_1 + m_4)\, \boldsymbol{g}' - (m_1 m_4 + m_2 m_3)\boldsymbol{g} + m_4 \boldsymbol{J}(t) + \boldsymbol{J}'(t). \end{aligned}$$

So our final model is

$$\boldsymbol{g}''(t) + (m_1 + m_4)\, \boldsymbol{g}' + (m_1 m_4 + m_2 m_3)\boldsymbol{g} = m_4 \boldsymbol{J}(t) + \boldsymbol{J}'(t).$$

Let $\alpha = (m_1 + m_4)/2$ and $\omega^2 = m_1 m_4 + m_2 m_3$ and we can rewrite as

$$\boldsymbol{g}''(t) + 2\alpha\, \boldsymbol{g}' + \omega^2 \boldsymbol{g} = S(t).$$

where $S(t) = m_4 \boldsymbol{J}(t) + \boldsymbol{J}'(t)$. Now the right hand side here is zero except for the very short time interval when the glucose load is being ingested. Hence, we can simply search for the solution to the homogeneous model

$$\boldsymbol{g}''(t) + 2\alpha\, \boldsymbol{g}' + \omega^2 \boldsymbol{g} = 0.$$

The roots of the characteristic equation here are

$$r = \frac{-2\alpha \pm \sqrt{4\alpha^2 - 4\omega^2}}{2} = -\alpha \pm \sqrt{\alpha^2 - \omega^2}.$$

The most interesting case is if we have complex roots. In that case, $\alpha^2 - \omega^2 < 0$. Let $\Omega^2 = |\alpha^2 - \omega^2|$. Then, the general phase shifted solution has the form $\boldsymbol{g} = Re^{-\alpha t} \cos(\Omega t - \delta)$ which implies

$$\boldsymbol{G} = \boldsymbol{G_0} + Re^{-\alpha t} \cos(\Omega t - \delta).$$

Hence, our model has five unknowns to find: $\boldsymbol{G_0}$, R, α, Ω and δ, The easiest way to do this is to measure $\boldsymbol{G_0}$, the patient's initial blood glucose concentration, when the patient arrives. Then measure the blood glucose concentration N more times giving the data pairs $(t_1, \boldsymbol{G_1})$, $(t_2, \boldsymbol{G_2})$ and so on out to $(t_N, \boldsymbol{G_N})$. Then form the least squares error function

$$E = \sum_{i=1}^{N} \left(\boldsymbol{G_i} - \boldsymbol{G_0} - Re^{-\alpha t_i} \cos(\Omega t_i - \delta)\right)^2$$

and find the five parameter values that make this error a minimum. This can be done using MatLab using some tools that outside of the scope of our text. Numerous experiments have been done with this model and if we let $T_0 = 2\pi/\Omega$, it has been found that if $T_0 < 4$ h, the patient is normal and if T_0 is much larger than that, the patient has mild diabetes.

15.1 Fitting the Data

Here is some typical glucose versus time data.

Time	Glucose Level
0	95
1	180
2	155
3	140

We will now try to find the parameter values which minimize the nonlinear least squares problem we have here. This appears to be a simple problem, but you will see all numerical optimization problems are actually fairly difficult. Our problem is to find the free parameters $\boldsymbol{G_o}$, R, α, Ω and δ which minimize

$$E(\boldsymbol{G_0}, R, \alpha, \Omega, \delta) = \sum_{i=1}^{N} \left(\boldsymbol{G_i} - \boldsymbol{G_0} - Re^{-\alpha t_i} \cos(\Omega t_i - \delta)\right)^2$$

For convenience, let $\boldsymbol{X} = [\boldsymbol{G_0}, R, \alpha, \Omega, \delta]'$ and $f_i(\boldsymbol{X}) = \boldsymbol{G_i} - \boldsymbol{G_0} - Re^{-\alpha t_i} \cos(\Omega t_i - \delta)$; then we can rewrite the error function as $E(\boldsymbol{X}) = \sum_{i=1}^{N} f_i^2(\boldsymbol{X})$. Gradient descent requires the gradient of this error function. This is just a messy calculation;

$$\frac{\partial E}{\partial \boldsymbol{G_o}} = 2 \sum_{i=1}^{N} f_i(\boldsymbol{X}) \frac{\partial f_i}{\partial \boldsymbol{G_o}}$$
$$\frac{\partial E}{\partial R} = 2 \sum_{i=1}^{N} f_i(\boldsymbol{X}) \frac{\partial f_i}{\partial R}$$
$$\frac{\partial E}{\partial \alpha} = 2 \sum_{i=1}^{N} f_i(\boldsymbol{X}) \frac{\partial f_i}{\partial \alpha}$$
$$\frac{\partial E}{\partial \Omega} = 2 \sum_{i=1}^{N} f_i(\boldsymbol{X}) \frac{\partial f_i}{\partial \Omega}$$
$$\frac{\partial E}{\partial \delta} = 2 \sum_{i=1}^{N} f_i(\boldsymbol{X}) \frac{\partial f_i}{\partial \delta}$$

where the f_i partials are given by

$$\frac{\partial f_i}{\partial \boldsymbol{G_o}} = -1$$
$$\frac{\partial f_i}{\partial R} = -e^{-\alpha t_i} \cos(\Omega t_i - \delta)$$
$$\frac{\partial f_i}{\partial \alpha} = t_i \, Re^{-\alpha t_i} \cos(\Omega t_i - \delta),$$
$$\frac{\partial f_i}{\partial \Omega} = t_i \, Re^{-\alpha t_i} \sin(\Omega t_i - \delta)$$
$$\frac{\partial f_i}{\partial \delta} = -Re^{-\alpha t_i} \sin(\Omega t_i - \delta)$$

and so

$$\frac{\partial E}{\partial \boldsymbol{G_o}} = -2 \sum_{i=1}^{N} f_i(\boldsymbol{X})$$

$$\frac{\partial E}{\partial R} = -2 \sum_{i=1}^{N} f_i(\boldsymbol{X})\, e^{-\alpha t_i} \cos(\Omega t_i - \delta)$$

$$\frac{\partial E}{\partial \alpha} = 2 \sum_{i=1}^{N} f_i(\boldsymbol{X})\, t_i\, Re^{-\alpha t_i} \cos(\Omega t_i - \delta),$$

$$\frac{\partial E}{\partial \Omega} = 2 \sum_{i=1}^{N} f_i(\boldsymbol{X})\, t_i\, Re^{-\alpha t_i} \sin(\Omega t_i - \delta)$$

$$\frac{\partial E}{\partial \delta} = -2 \sum_{i=1}^{N} f_i(\boldsymbol{X})\, Re^{-\alpha t_i} \sin(\Omega t_i - \delta)$$

Now suppose we are at the point $\boldsymbol{X_0}$ and we want to know how much of the descent vector D to use. Note, if we use the amount ξ of the descent vector at $\boldsymbol{X_0}$, we compute the new error value $E(\boldsymbol{X_0} - \xi D(\boldsymbol{X_0}))$. Let $g(\xi) = E(\boldsymbol{X_0} - \xi D(\boldsymbol{X_0}))$. We see $g(0) = E(\boldsymbol{X_0})$ and given a first choice of $\xi = \lambda$, we have $g(\lambda) = E(\boldsymbol{X_0} - \lambda D(\boldsymbol{X_0}))$. Next, let $\boldsymbol{Y} = \boldsymbol{X_0} - \xi\, D(\boldsymbol{X_0})$. Then, using the chain rule, we can calculate the derivative of g at 0. First, we have

$$g'(\xi) = -< \nabla(E), D(\boldsymbol{X_0}) >$$

and using the normalized gradient of E as the descent vector, we find

$$g'(0 = -||\nabla(E)||.$$

Now let's approximate g using a quadratic model. Since we are trying for a minimum, in general we try to take a step in the direction of the negative gradient which makes the error function go down. Then, we have $g(0) = E(\boldsymbol{X_0})$ is less than $g(\lambda) = E(\boldsymbol{X_0} - \lambda D(\boldsymbol{X_0}))$ and the directional derivative gives $g'(0) = -||\nabla(E)|| < 0$. Hence, if we approximate g by a simple quadratic model, $g(\xi) = A + B\xi + C\xi^2$, this model will have a unique minimizer and we can use the value of ξ where the minimum occurs as our next choice of descent step. This technique is called a **Line Search Method** and it is quite useful. To summarize, we fit our g model and find

$$g(0) = E(\boldsymbol{X_0})$$
$$g'(0) = -||\nabla(E)||$$
$$g(\lambda) = A + B\lambda + C\lambda^2 \Longrightarrow C = \frac{E(\boldsymbol{X_0} - \lambda D(\boldsymbol{X_0})) - E(\boldsymbol{X_0}) + ||\nabla(E)||\lambda}{\lambda^2}$$

The minimum of this quadratic occurs at $\lambda^* = \frac{B}{2C}$ and this will give us our next descent direction $\boldsymbol{X_0} - \lambda^* D(\boldsymbol{X_0})$.

15.1.1 Gradient Descent Implementation

Let's get started on how to find the optimal parameters numerically. Along the way, we will show you how hard this is. We start with a minimal implementation. We have already discussed some root finding codes in Chap. 14 so we have seen code kind of similar. But this will be a little different and it is good to have you see a bit about it. What is complicated here is that we have lots of functions that depend on the data we are trying to fit. So the number of functions depends on the size of the data set which makes it harder to set up.

Listing 15.1: Nonlinear LS For Diabetes Model: Version One

```
function [Error,G0,R,alpha,Omega,delta,normgrad,update] = DiabetesGradOne(Initial,
    lambda,maxiter,data)
%
% Initial guess for G0, R, alpha, Omega, delta
% data = collection of (time, glucose) pairs
% maxiter = maximum number of iterations to use
% lambda = how much of the descent vector to use
%
% setup least squares error function
N = length(data);
E = 0.0;
Time = data(:,1);
G = data(:,2);
% Initial = [initial G, initial R, initial alpha, initial Omega, initial delta]
g = Initial(1); r = Initial(2); a = Initial(3); o = Initial(4); d = Initial(5);
f = @(t,equilG,r,a,o,d) equilG + r*exp(-a^2*t).*(cos(o*t-d));
E = DiabetesError(f,g,r,a,o,d,Time,G)
for i = 1:maxiter
   % calculate error
   E = DiabetesError(f,g,r,a,o,d,Time,G);
   if (i == 1)
     Error(1) = E;
   end
   Error = [Error;E];
   % find grad of E
   [gradE, normgrad] = ErrorGradient(Time,G,g,r,a,o,d);
   % find descent direction D
   if(normgrad>1)
      Descent = gradE/normgrad;
   else
      Descent = gradE;
   end
   Del = [g;r;a;o;d]-lambda*Descent;
   g = Del(1); r = Del(2); a = Del(3); o = Del(4); d = Del(5);
end
G0 = g; R = r; alpha = a; Omega = o; delta = d;
E = DiabetesError(f,g,r,a,o,d,Time,G)
update = [G0;R;alpha;Omega;delta];
end
```

Note inside this function, we call another function to calculate the gradient of the norm. This is given below and implements the formulae we presented earlier for these partial derivatives.

Listing 15.2: The Error Gradient Function

```
function [gradE, normgrad] = ErrorGradient(Time,G,g,r,a,o,d)
%
% Time is the time data
% G is the glucose data
% g = current equilG
% r = current R
% a = current alpha
% o = current Omega
% d = current delta
%
% Calculate Error gradient
ferror = @(t,G,g,r,a,o,d) G - g - r*exp(-a^2*t).*(cos(o*t-d));
gerror = @(t,r,a,o,d) r*exp(-a^2*t).*(cos(o*t-d));
herror = @(t,r,a,o,d) r*exp(-a^2*t).*(sin(o*t-d));
N = length(Time);
  sum = 0;
  for k=1:N
    sum = sum + ferror(Time(k),G(k),g,r,a,o,d);
  end
  pEpequilg = -2*sum;
  sum = 0;
  for k=1:N
    sum = sum + ferror(Time(k),G(k),g,r,a,o,d)*gerror(Time(k),r,a,o,d)/r;
  end
  pEpR = -2*sum;
  sum = 0;
  for k=1:N
    sum = sum + ferror(Time(k),G(k),g,r,a,o,d)*Time(k)*2*a*Time(k)*gerror(Time(k
       ),r,a,o,d);
  end
  pEpa = 2*sum;
  sum = 0;
  for k=1:N
    sum = sum + ferror(Time(k),G(k),g,r,a,o,d)*Time(k)*herror(Time(k),r,a,o,d);
  end
  pEpo = 2*sum;
  sum = 0;
  for k=1:N
    sum = sum + ferror(Time(k),G(k),g,r,a,o,d)*herror(Time(k),r,a,o,d);
  end
  pEpd = -2*sum;
  gradE = [pEpequilg;pEpR;pEpa;pEpo;pEpd];
  normgrad = norm(gradE);
end
```

We also need code for the error calculations which is given here.

Listing 15.3: Diabetes Error Calculation

```
function E = DiabetesError(f,g,r,a,o,d,Time,G)
%
% T = Time
% G = Glocuose values
% f = nonlinear insulin model
% g,a,r,o,d = parameters in diabetes nonlinear model
% N = size of data
N = length(G);
E = 0.0;
% calculate error function
for i = 1:N
  E = E +( G(i) - f(Time(i),g,r,a,o,d) )^2;
end
```

Let's look at some run time results using this code.

Listing 15.4: Run time results for gradient descent on the original data

```
Data = [0,95;1,180;2,155;3,140];
Time = Data(:,1);
G = Data(:,2);
f = @(t,equilG,r,a,o,d) equilG + r*exp(-a^2*t).*(cos(o*t-d));
time = linspace(0,3,41);
RInitial = 53.64;
GOInitial = 95 + RInitial;
AInitial = sqrt(log(17/5));
OInitial = pi;
dInitial = -pi;
Initial = [GOInitial;RInitial;AInitial;OInitial;dInitial];
[Error,G0,R,alpha,Omega,delta,normgrad,update] = DiabetesGrad(Initial,5.0e
    -4,20000,Data,0);
InitialE =  463.94
E =  376.40
octave:16> [Error,G0,R,alpha,Omega,delta,normgrad,update] = DiabetesGrad(
    Initial,5.0e-4,40000,Data,0);
InitialE =  463.94
E =  377.77
octave:145> [Error,G0,R,alpha,Omega,delta,normgrad,update] = DiabetesGrad(
    Initial,5.0e-4,100000,Data,0);
InitialE =  463.94
E =  377.77
```

After 100,000 iterations we still do not have a good fit. Note we start with a small constant $\lambda = 5.0e - 4$ here. Try it yourself. If you let this value be larger, the optimization spins out of control. Also, we have not said how we chose our initial values. We actually looked at the data on a sheet of paper and did some rough calculations to try for some decent values. We will leave that to you to figure out. If the initial values are poorly chosen, gradient descent optimization is a great way to generate really bad values! So be warned. You will have to exercise due diligence to find a sweet starting spot.

We can see how we did by looking the resulting curve fit in Fig. 15.1.

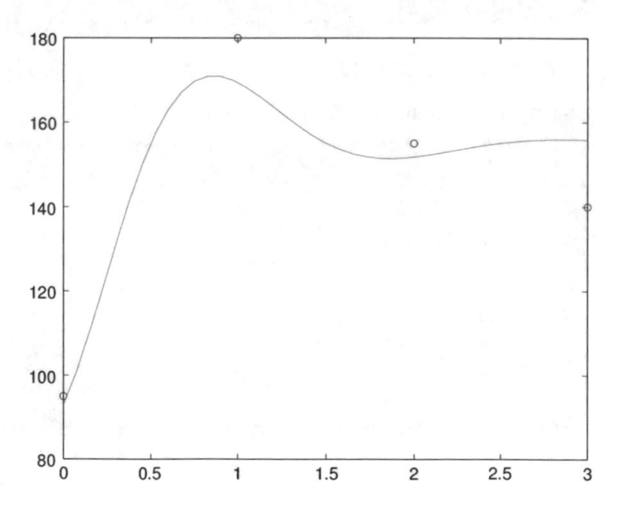

Fig. 15.1 The diabetes model *curve* fit: no line search

15.1.2 Adding Line Search

Now let's add line search and see if it gets better. We will also try scaling the data so all the variables in question are roughly the same size. For us, a good choice is to scale the $\boldsymbol{G_0}$ and the R value by 50, although we could try other choices. We have already discussed line search for our problem, but here it is again in a quick nutshell. If we are minimizing a function of M variables, say $f(\boldsymbol{X})$, then if we are at the point $\boldsymbol{X_0}$, we can look at the *slice* of this function if we move out from the base point $\boldsymbol{X_0}$ is the direction of the negative gradient, $\nabla(f(\boldsymbol{X_0}) = \nabla(f^0)$. Define a function of the single variable ξ as $g(\xi) = f(\boldsymbol{X_0} + \xi\, \nabla(f^0))$. Then, we can try to approximate g as a quadratic, $g(\xi) \approx= A + B\xi + C\xi^2$. Of course, the actual function might not be approximated nicely by such a quadratic, but it is worth a shot! Once we fit the parameters A, B and C, we see this quadratic model is minimized at $\lambda^* = -\frac{B}{2C}$. The code now adds the line search code which is contained in the block

Listing 15.5: Line Search Code

```
function [lambdastar,DelOptimal] = LineSearch(Del,EStart,EFullStep,Base,Descent,
    normgrad,lambda)
%
% Del is the update due to the step lambda
% Estart is E at the base value
% EFullStep is E at the given lambda value
% Base is the current parameter vector
% Descent is the current descent vector
% normgrad of grad of E at the base value
% lambda is the current step
%
% we have enough information to do a line search here
  A = EStart;
  BCheck = -normgrad;
  C = ( EFullStep- A - BCheck*lambda )/(lambda^2);
  lambdastar = -BCheck/(2*C);
  if(C<0 || lambdastar < 0)
    % we are going to a maximum on the line search; reject
    DelOptimal = Del;
  else
    % we have a minimum on the line search
    DelOptimal= Base-lambdastar*Descent;
  end
end
```

Listing 15.6: Nonlinear LS For Diabetes Model

```
function [Error,G0,R,alpha,Omega,delta,normgrad,update] = DiabetesGrad(Initial,
    lambda,maxiter,data,dolinesearch)
%
% Initial guess for G0, R, alpha, Omega, delta
% data = collection of (time, glucose) pairs
% tol = stop tolerance
% maxiter = maximum number of iterations to use
% lambda = how much of the descent vector to use
%
% setup least squares error function
N = length(data);
E = 0.0;
Time = data(:,1);
G = data(:,2);
% Initial = [initial G, initial R, initial alpha, initial Omega, initial delta]
g = Initial(1);
r = Initial(2);
a = Initial(3);
o = Initial(4);
d = Initial(5);
f = @(t,equilG,r,a,o,d) equilG + r*exp(-a^2*t).*(cos(o*t-d));
InitialE = DiabetesError(f,g,r,a,o,d,Time,G)
for i = 1:maxiter
   % calculate error
   E = DiabetesError(f,g,r,a,o,d,Time,G);
   if (i == 1)
     Error(1) = E;
   end
   Error = [Error;E];
   % Calculate Error gradient
   [gradE, normgrad] = ErrorGradient(Time,G,g,r,a,o,d);
   % find descent direction
   if(normgrad>1)
      Descent = gradE/normgrad;
   else
      Descent = gradE;
   end
   Base = [g;r;a;o;d];
   Del = Base-lambda*Descent;
   newg = Del(1); newr = Del(2); newa = Del(3); newo = Del(4); newd = Del(5);
   EStart = E;
   EFullStep = DiabetesError(f,newg,newr,newa,newo,newd,Time,G);
   if(dolinesearch==1)
      [lambdastar,DelOptimal] = LineSearch(Del,EStart,EFullStep,Base,Descent,
           normgrad,lambda);
      g = DelOptimal(1); r = DelOptimal(2); a = DelOptimal(3); o = DelOptimal(4);
           d = DelOptimal(5);
      EOptimalStep = DiabetesError(f,g,r,a,o,d,Time,G);
   else
      g = Del(1); r = Del(2); a = Del(3); o = Del(4); d = Del(5);
   end
end
G0 = g; R = r; alpha = a; Omega = o; delta = d;
E = DiabetesError(f,g,r,a,o,d,Time,G)
update = [G0;R;alpha;Omega;delta];
end
```

We can see how this is working by letting some of the temporary calculations print. Here are two iterations of line search printing out the A, B and C and the relevant energy values. Our initial values don't matter much here as we are just checking out the line search algorithm.

Listing 15.7: Some Details of the Line Search

```
Data = [0,95;1,180;2,155;3,140];
Time = Data(:,1);
G = Data(:,2);
f = @(t,equilG,r,a,o,d) equilG + r*exp(-a^2*t).*(cos(o*t-d));
time = linspace(0,3,41);
RInitial = 85*17/21;
GOInitial = 95 + RInitial;
AInitial = sqrt(log(17/4));
OInitial = pi;
dInitial = -pi;
Initial = [GOInitial;RInitial;AInitial;OInitial;dInitial];
[Error,G0,R,alpha,Omega,delta,normgrad,update] = DiabetesGrad(Initial,5.0e
    -4,2,Data,1);
InitialE =  635.38
EFullStep =  635.31
A =  635.38
BCheck = -595.35
C =     9.1002e+05
lambdastar =      3.2711e-04
EFullStep =  635.27
A =  635.33
BCheck = -592.34
C =     9.0725e+05
lambdastar =      3.2645e-04
E =  635.29
```

Now let's remove those prints and let it run for awhile. We are using the original data here to try to find the fit.

Listing 15.8: The Full Run with Line Search

```
octave:90> [Error,G0,R,alpha,Omega,delta,normgrad,update] = DiabetesGrad(
    Initial,5.0e-4,20000,Data,1);
InitialE =  635.38
E =  389.07
octave:93> [Error,G0,R,alpha,Omega,delta,normgrad,update] = DiabetesGrad(
    Initial,5.0e-4,30000,Data,1);
InitialE =  635.38
E =  0.067171
```

We have success! The line search got the job done in 30,000 iterations while the attempt using just gradient descent without line search failed. but remember, we do additional processing at each step. We show the resulting curve fit in Fig. 15.2.

The qualitative look of this fit is a bit different. We leave it you to think about how we are supposed to choose which fit is better; i.e. which fit is a better one to use for the biological reality we are trying to model? This is a really hard question. Finally, the optimal values of the parameters are

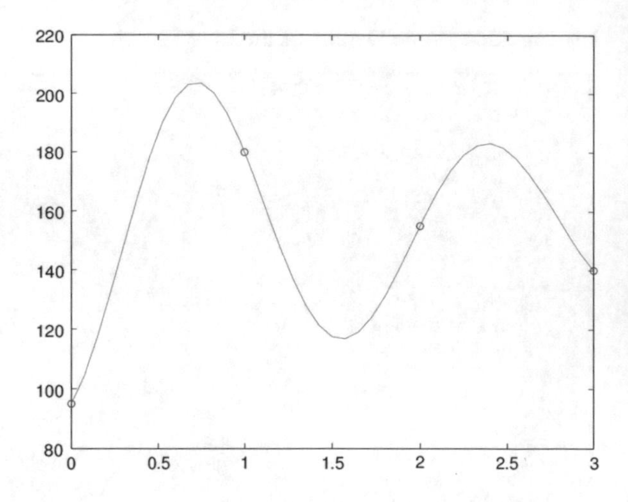

Fig. 15.2 The diabetes model *curve* fit with line search on unscaled data

Listing 15.9: Optimal Parameter Values

```
octave:14> update
update =
   154.37240
    62.73116
     0.57076
     3.77081
    -3.47707
```

The critical value of $2\pi/\Omega = 1.6663$ here which is less than 4 so this patient is normal!

Also, note these sorts of optimizations are very frustrating, If we use the scaled version of the first `Initial =[GOInitial;.RInitial;AInitial; OInitial;.dInitial];` we make no progress even though we run for 60, 000 iterations and these iterations a bit more expensive because we use line search. So let's perturb the starting point a bit and see what happens.

Listing 15.10: Runtime Results

```
% use scaled data
Data = [0,95/50;1,180/50;2,155/50;3,140/50];
Time = Data(:,1);
G = Data(:,2);
RInitial = 53.64/50;
GOInitial = 95/50 + RInitial;
AInitial = sqrt(log(17/5));
OInitial = pi;
dInitial = -pi;
% perturb the start point a bit
Initial = [GOInitial;.7*RInitial;AInitial;1.1*OInitial;.9*dInitial];
[Error,G0,R,alpha,Omega,delta,normgrad,update] = DiabetesGrad(Initial,5.0e
    -4,40000,Data,1);
InitialE = 0.36485
E =      1.5064e-06
```

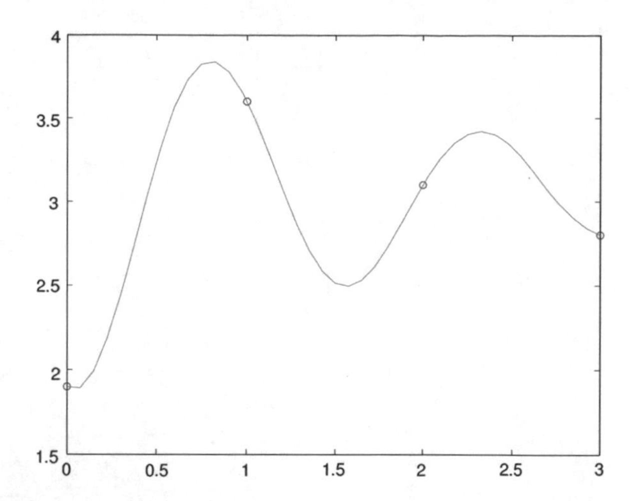

Fig. 15.3 The diabetes model *curve* fit with line search on scaled data

We have success! We see the fit to the data in Fig. 15.3.

The moral here is that it is quite difficult to automate our investigations. This is why truly professional optimization code is so complicated. In our problem here, we have a hard time finding a good starting point and we even find that scaling the data—which seems like a good idea—is not as helpful as we thought it would be. Breathe a deep sigh and accept this as our lot!

Reference

M. Braun, *Differential Equations and Their Applications* (Springer, Berlin, 1978)

Part VI
Series Solutions to PDE Models

Chapter 16
Series Solutions

Let's look at another tool we can use to solve models. Sometimes our models involve partial derivatives instead of normal derivatives; we call such models partial differential equation models or **PDE** models. They are pretty important and you should have a beginning understanding of them. Let's get started with a common tool called the Separation of Variables Method.

16.1 The Separation of Variables Method

A common PDE model is the general cable model which is given below is fairly abstract form.

$$\begin{aligned}
\beta^2 \frac{\partial^2 \Phi}{\partial x^2} - \Phi - \alpha \frac{\partial \Phi}{\partial t} &= 0, \quad \text{for } 0 \le x \le L, \quad t \ge 0, \\
\frac{\partial \Phi}{\partial x}(0, t) &= 0, \\
\frac{\partial \Phi}{\partial x}(L, t) &= 0, \\
\Phi(x, 0) &= f(x).
\end{aligned}$$

for positive constants α and β. The domain is the usual half infinite $[0, L] \times [0, \infty)$ where the spatial part of the domain corresponds to the length of the dendritic cable in an excitable nerve cell. We won't worry too much about the details of where this model comes from as we will discuss that in another volume. The boundary conditions $u_x(0, t) = 0$ and $u_x(L, t) = 0$ are called *Neumann Boundary conditions*. The conditions $u(0, t) = 0$ and $u(L, t) = 0$ are known as *Dirichlet Boundary conditions*. The solution to a model such as this is a function $\Phi(x, t)$ which is sufficiently smooth to have partial derivatives with respect to the needed variables continuous for all the orders required. For these problems, the highest order we need

J.K. Peterson, *Calculus for Cognitive Scientists*,
Cognitive Science and Technology, DOI 10.1007/978-981-287-877-9_16

is the second order partials. One way to find the solution is to assume we can separate the variables so that we can write $\Phi(x, t) = u(x)w(t)$. If we make this separation assumption, we will find solutions that must be written as what are called infinite series and to solve the boundary conditions, we will have to be able to express boundary functions as series expansions. Hence, we will have to introduce some new ideas in order to understand these things. Let's motivate what we need to do by applying the separation of variables technique to the cable equation. This will shows the ideas we need to use in a specific example. Then we will step back and go over the new mathematical ideas of series and then return to the cable model and finish finding the solution.

We assume a solution of the form $\Phi(x, t) = u(x)\, w(t)$ and compute the needed partials. This leads to a the new equation

$$\beta^2 \frac{d^2u}{dx^2}\, w(t) - u(x)w(t) - \alpha u(x)\frac{dw}{dt} = 0.$$

Rewriting, we find for all x and t, we must have

$$w(t)\left(\beta^2 \frac{d^2u}{dx^2} - u(x)\right) = \alpha u(x)\frac{dw}{dt}.$$

This tells us

$$\frac{\beta^2 \frac{d^2u}{dx^2} - u(x)}{u(x)} = \frac{\alpha \frac{dw}{dt}}{w(t)}, \quad 0 \leq x \leq L, \quad t > 0.$$

The only way this can be true is if both the left and right hand side are equal to a constant that is usually called the **separation constant** Θ. This leads to the decoupled Eqs. 16.1 and 16.2.

$$\alpha \frac{dw}{dt} = \Theta\, w(t), \quad t > 0, \tag{16.1}$$

$$\beta^2 \frac{d^2u}{dx^2} = (1 + \Theta)\, u(x), \quad 0 \leq x \leq L, \tag{16.2}$$

We also have boundary conditions. Our assumption leads to the following boundary conditions in x:

$$\frac{du}{dx}(0)\, w(t) = 0, \quad t > 0,$$

$$\frac{du}{dx}(L)\, w(t) = 0, \quad t > 0.$$

Since these equations must hold for all t, this forces

$$\frac{du}{dx}(0) = 0, \tag{16.3}$$

$$\frac{du}{dx}(L) = 0. \tag{16.4}$$

Equations 16.1–16.4 give us the boundary value problem in $u(x)$ we need to solve. Then, we can find w.

16.1.1 Determining the Separation Constant

The model is then

$$\begin{aligned} u'' - \frac{1+\Theta}{\beta^2} u &= 0 \\ \frac{du}{dx}(0) &= 0, \\ \frac{du}{dx}(L) &= 0. \end{aligned}$$

We are looking for nonzero solutions, so any choice of separation constant Θ that leads to a zero solution will be rejected.

16.1.1.1 Case I: $1 + \Theta = \omega^2$, $\omega \neq 0$

The model to solve is

$$\begin{aligned} u'' - \frac{\omega^2}{\beta^2} u &= 0 \\ u'(0) &= 0, \\ u'(L) &= 0. \end{aligned}$$

with characteristic equation $r^2 - \frac{\omega^2}{\beta^2} = 0$ with the real roots $\pm \frac{\omega}{\sqrt{D}}$. The general solution of this second order model is given by

$$u(x) = A \cosh\left(\frac{\omega}{\beta} x\right) + B \sinh\left(\frac{\omega}{\beta} x\right)$$

which tells us

$$u'(x) = A\,\frac{\omega}{\beta}\,\sinh\left(\frac{\omega}{\beta}x\right) + B\,\frac{\omega}{\beta}\,\cosh\left(\frac{\omega}{\beta}x\right)$$

Next, apply the boundary conditions, $u'(0) = 0$ and $u'(L) = 0$. Hence,

$$\begin{aligned} u'(0) &= 0 = B \\ u'(L) &= 0 = A\,\sinh\left(L\frac{\omega}{\beta}\right) \end{aligned}$$

Hence, $B = 0$ and $A\,\sinh\left(L\frac{\omega}{\beta}\right) = 0$. Since sinh is never zero when ω is not zero, we see $A = 0$ also. Hence, the only u solution is the trivial one and we can reject this case.

16.1.1.2 Case II: $1 + \Theta = 0$

The model to solve is now

$$\begin{aligned} u'' &= 0 \\ u'(0) &= 0, \\ u'(L) &= 0. \end{aligned}$$

with characteristic equation $r^2 = 0$ with the double root $r = 0$. Hence, the general solution is now

$$u(x) = A + B\,x$$

Applying the boundary conditions, $u(0) = 0$ and $u(L) = 0$. Hence, since $u'(x) = B$, we have

$$\begin{aligned} u'(0) &= 0 = B \\ u'(L) &= 0 = BL \end{aligned}$$

Hence, $B = 0$ but the value of A can't be determined. Hence, any arbitrary constant which is not zero is a valid non zero solution. Choosing $B = 1$, let $u_0(x) = 1$ be our chosen nonzero solution for this case.

We now need to solve for w in this case. Since $\Theta = -1$, the model to solve is

$$\begin{aligned} \frac{dw}{dt} &= -\frac{1}{\alpha}w(t), \quad 0 < t, \\ w(L) &= 0. \end{aligned}$$

The general solution is $w(t) = Ce^{-\frac{1}{\alpha}t}$ for any value of C. Choose $C = 1$ and we set

$$w_0(y) = e^{-\frac{1}{\alpha}t}.$$

Hence, the product $\phi_0(x, t) = u_0(x)\, w_0(t)$ solves the boundary conditions. That is

$$\phi_0(x, t) = e^{-\frac{1}{\alpha}t}.$$

is a solution.

16.1.1.3 Case III: $1 + \Theta = -\omega^2, \omega \neq 0$

$$\begin{aligned} u'' + \frac{\omega^2}{\beta^2} u &= 0 \\ u'(0) &= 0, \\ u'(L) &= 0. \end{aligned}$$

The general solution is given by

$$u(x) = A\, \cos\left(\frac{\omega}{\beta}x\right) + B\, \sin\left(\frac{\omega}{\beta}x\right)$$

and hence

$$u'(x) = -A\frac{\omega}{\beta}\sin\left(\frac{\omega}{\beta}x\right) + B\frac{\omega}{\beta}\cos\left(\frac{\omega}{\beta}x\right)$$

Next, apply the boundary conditions to find

$$\begin{aligned} u'(0) &= 0 = B \\ u'(L) &= 0 = A\sin\left(L\frac{\omega}{\beta}\right) \end{aligned}$$

Hence, $B = 0$ and $A\sin\left(L\frac{\omega}{\beta}\right) = 0$. Thus, we can determine a unique value of A only if $\sin\left(L\frac{\omega}{\beta}\right) \neq 0$. If $\omega \neq \frac{n\pi\beta}{L}$, we can solve for B and find $B = 0$, but otherwise, B can't be determined. So the only solutions are the trivial or zero solutions unless $\omega L = n\pi\beta$. Letting $\omega_n = \frac{n\pi\beta}{L}$, we find a non zero solution for each nonzero value of B of the form

$$u_n(x) = B\cos\left(\frac{\omega_n}{\beta}x\right) = B\cos\left(\frac{n\pi}{L}x\right).$$

For convenience, let's choose all the constants $B = 1$. Then we have an infinite family of nonzero solutions $u_n(x) = \cos\left(\frac{n\pi}{L}x\right)$ and an infinite family of separation constants $\Theta_n = -1 - \omega_n^2 = -1 - \frac{n^2\pi^2 D}{L^2}$.

We can then solve the w equation. We must solve

$$\frac{dw}{dt} = -\frac{(1+\omega_n^2)}{\alpha}\, w(t), \quad t \geq 0.$$

The general solution is

$$w(t) = B_n\, e^{-\frac{1+\omega_n^2}{\alpha}t} = B_n\, e^{-\frac{1+n^2\pi^2\beta^2}{\alpha L^2}t}$$

Choosing the constants $B_n = 1$, we obtain the w_n functions

$$w_n(t) = e^{-\frac{n^2\pi^2\beta^2}{\alpha L^2}t}$$

Hence, any product

$$\phi_n(x,t) = u_n(x)\, w_n(t)$$

will solve the model with the x boundary conditions and any finite sum of the form, for arbitrary constants A_n

$$\begin{aligned}\Psi_N(x,t) &= \sum_{n=1}^{N} A_n \phi_n(x,t) = \sum_{n=1}^{N} A_n u_n(x)\, w_n(t)\\ &= \sum_{n=1}^{N} A_n \cos\left(\frac{n\pi}{L}x\right) e^{-\frac{1+n^2\pi^2\beta^2}{\alpha L^2}t}\end{aligned}$$

Adding in the $1 + \Theta = 0$ case, we find the most general finite term solution has the form

$$\begin{aligned}\Phi_N(x,t) &= A_0\phi_0(x,t) + \sum_{n=1}^{N} A_n\phi_n(x,t) = A_0 u_0(x) w_0(t) + \sum_{n=1}^{N} A_n u_n(x)\, w_n(t)\\ &= A_0 e^{-\frac{1}{\alpha}t} + \sum_{n=1}^{N} A_n \cos\left(\frac{n\pi}{L}x\right) e^{-\frac{1+n^2\pi^2\beta^2}{\alpha L^2}t}.\end{aligned}$$

Now these finite term solutions do solve the boundary conditions $\frac{\partial\Phi}{\partial x}(0,t) = 0$ and $\frac{\partial\Phi}{\partial x}(L,t) = 0$, but how do we solve the remaining condition $\Phi(x,0) = f(x)$? To do this, we note since we can assemble the finite term solutions for any value of N, no matter how large, it is clear we should let $N \to \infty$ and express the solution as

$$\Phi(x,t) = A_0\phi_0(x,t) + \sum_{n=1}^{\infty} A_n\phi_n(x,t) = A_0u_0(x)w_0(t) + \sum_{n=1}^{\infty} A_n u_n(x)\, w_n(t)$$

$$= A_0 e^{-\frac{1}{\alpha}t} + \sum_{n=1}^{\infty} A_n \cos\left(\frac{n\pi}{L}x\right) e^{-\frac{1+n^2\pi^2\beta^2}{\alpha L^2}t}.$$

This is the form that will let us solve the remaining boundary condition. We need to step back now and talk more about this idea of a **series** solution to our model.

16.2 Infinite Series

Let's look at sequences of functions made up of building blocks of the form $u_n(x) = \cos\left(\frac{n\pi}{L}\right)$ or $v_n(x) = \sin\left(\frac{n\pi}{L}\right)$ for various values of the integer n. The number L is a fixed value here. We can combine these functions into finite sums: let $U_N(x)$ and $V_N(x)$ be defined as follows:

$$U_N(x) = \sum_{n=1}^{N} a_n \sin\left(\frac{n\pi}{L}x\right).$$

and

$$V_N(x) = b_0 + \sum_{n=1}^{N} b_n \cos\left(\frac{n\pi}{L}x\right).$$

If we fixed the value of x to be say, x_0, the collection of numbers

$$\{U_1(x_0),\ U_2(x_0),\ U_3(x_0),\ \ldots,\ U_n(x_0),\ \ldots\}$$

and

$$\{V_0(x_0),\ V_1(x_0),\ V_2(x_0),\ V_3(x_0),\ \ldots,\ V_n(x_0),\ \ldots\}$$

are the partial sums formed from the sequences of cosine and sine numbers. However, the underlying sequences can be negative, so these are not sequences of non negative terms like we previously discussed. These sequences of partial sums may or may not have a finite supremum value. Nevertheless, we still represent the supremum using the same notation: i.e. the supremum of $\left(U_i(x_0)\right)_{i=1}^{\infty}$ and the supremum of $\left(V_i(x_0)\right)_{i=0}^{\infty}$ can be written as $\sum_{n=1}^{\infty} a_n \sin\left(\frac{n\pi}{L}x\right)$ and $b_0 + \sum_{n=1}^{\infty} b_n \cos\left(\frac{n\pi}{L}x\right)$

Let's consider the finite sequence

$$\{U_1(x_0),\ U_2(x_0),\ U_3(x_0),\ \ldots,\ U_n(x_0)\}.$$

This sequence of real numbers converges to a possibly different number for each x_0; hence, let's call this possible limit $S(x_0)$. Now the limit may not exist, of course. We will write $\lim_{n\to} U_n(x_0) = S(x_0)$ when the limit exists. If the limit does not exist for some value of x_0, we will understand that the value $S(x_0)$ is not defined in some way. Note, from our discussion above, this could mean the limiting value flips between a finite set of possibilities, the limit approaches ∞ or the limit approaches $-\infty$. In any case, the value $S(x_0)$ is not defined as a finite value. We would say this precisely as follows: given any positive tolerance ϵ, there is a positive integer N so that

$$n > N \Longrightarrow \left|\sum_{i=1}^{n} a_i \sin\left(\frac{i\pi}{L}\right) - S(x_0)\right| < \epsilon.$$

We use the notation of the previous section and write this as

$$\lim_{n->\infty} \sum_{i=1}^{n} a_i \sin\left(\frac{i\pi}{L}\right) = S(x_0).$$

with limiting value, $S(x_0)$ written as

$$S(x_0) = \sum_{i=1}^{\infty} a_i \sin\left(\frac{i\pi}{L}\right),$$

As before, this symbol is called an *infinite series* and we see we get a potentially different series at each point x_0. The error term $S(x_0) - U_n(x_0)$ is then written as

$$S(x_0) - \sum_{i=1}^{n} a_i \sin\left(\frac{i\pi}{L}x_0\right) = \sum_{i=n+1}^{\infty} a_i \sin\left(\frac{i\pi}{L}x_0\right),$$

which you must remember is just a short hand for this error.

Now that we have an infinite series notation defined, we note the term $U_n(x_0)$, which is the sum of n terms, is also called the nth partial sum of the series $\sum_{i=1}^{\infty} a_i \sin\left(\frac{i\pi}{L}x_0\right)$. Note we can define the convergence at a point x_0 for the partial sums of the cos functions in a similar manner.

16.3 Independant Objects

Let's go back and think about vectors in $\Re^2$. As you know, we think of these as arrows with a tail fixed at the origin of the two dimensional coordinate system we call the x–y plane. They also have a length or magnitude and this arrow makes an angle with the positive x axis. Suppose we look at two such vectors, $\boldsymbol{E}$ and $\boldsymbol{F}$. Each vector has an x and a y component so that we can write

$$\boldsymbol{E} = \begin{bmatrix} a \\ b \end{bmatrix}, \ \boldsymbol{F} = \begin{bmatrix} c \\ d \end{bmatrix}$$

The cosine of the angle between them is proportional to the inner product $< \boldsymbol{E}, \boldsymbol{F} >= ac + bd$. If this angle is 0 or π, the two vectors lie along the same line. In any case, the angle associated with $\boldsymbol{E}$ is $\tan^{-1}(\frac{b}{a})$ and for $\boldsymbol{F}$, $\tan^{-1}(\frac{d}{c})$. Hence, if the two vectors lie on the same line, $\boldsymbol{E}$ must be a multiple of $\boldsymbol{F}$. This means there is a number β so that

$$\boldsymbol{E} = \beta\, \boldsymbol{F}.$$

We can rewrite this as

$$\begin{bmatrix} a \\ b \end{bmatrix} = \beta \begin{bmatrix} c \\ d \end{bmatrix}$$

Now let the number 1 in front of $\boldsymbol{E}$ be called $-\alpha$. Then the fact that $\boldsymbol{E}$ and $\boldsymbol{F}$ lie on the same line implies there are 2 constants α and β, both not zero, so that

$$\alpha\, \boldsymbol{E} + \beta\, \boldsymbol{F} = 0.$$

Note we could argue this way for vectors in $\Re^3$ and even in $\Re^n$. Of course, our ability to think of these things in terms of lying on the same line and so forth needs to be extended to situations we can no longer draw, but the idea is essentially the same. Instead of thinking of our two vectors as lying on the same line or not, we can *rethink* what is happening here and try to identify what is happening in a more abstract way. If our two vectors lie on the same line, they are not *independent* things in the sense one is a multiple of the other. As we saw above, this implies there was a linear equation connecting the two vectors which had to add up to 0. Hence, we might say the vectors were *not linearly independent* or simply, they are *linearly dependent*. Phrased this way, we are on to a way of stating this idea which can be used in many more situations. We state this as a definition.

Definition 16.3.1 (*Two Linearly Independent Objects*)
Let $\boldsymbol{E}$ and $\boldsymbol{F}$ be two mathematical objects for which addition and scalar multiplication is defined. We say $\boldsymbol{E}$ and $\boldsymbol{F}$ are **linearly dependent** if we can find non zero constants α and β so that

$$\alpha\,\boldsymbol{E} + \beta\,\boldsymbol{F} = 0.$$

Otherwise, we say they are **linearly independent**.

We can then easily extend this idea to any finite collection of such objects as follows.

Definition 16.3.2 (*Finitely many Linearly Independent Objects*)
Let $\{\boldsymbol{E}_i \,:\, 1 \leq i \leq N\}$ be N mathematical objects for which addition and scalar multiplication is defined. We say $\boldsymbol{E}$ and $\boldsymbol{F}$ are **linearly dependent** if we can find non zero constants α_1 to α_N, not all 0, so that

$$\alpha_1\,\boldsymbol{E}_1 + \cdots + \alpha_N\,\boldsymbol{E}_N = 0.$$

Note we have changed the way we define the constants a bit. When there are more than two objects involved, we can't say, in general, that *all* of the constants must be non zero.

16.3.1 Independent Functions

Now let's apply these ideas to functions f and g defined on some interval I. By this we mean either

- I is all of $\Re$, i.e. $a = -\infty$ and $b = \infty$,
- I is half-infinite. This means $a = -\infty$ and b is finite with I of the form $(-\infty, b)$ or $(-\infty, b]$. Similarly, I could have the form (a, ∞) or $[a, \infty$,
- I is an interval of the form (a, b), $[a, b)$, $(a, b]$ or $[a, b]$ for finite $a < b$.

We would say f and g are linearly independent on the interval I if the equation

$$\alpha_1\, f(t) + \alpha_2\, g(t) = 0, \quad \text{for all } t \in I.$$

implies α_1 and α_2 must both be zero. Here is an example. The functions $\sin(t)$ and $\cos(t)$ are linearly independent on $\Re$ because

$$\alpha_1\, \cos(t) + \alpha_2\, \sin(t) = 0, \quad \text{for all } t,$$

also implies the above equation holds for the derivative of both sides giving

$$-\alpha_1\, \sin(t) + \alpha_2\, \cos(t) = 0, \quad \text{for all } t,$$

This can be written as the system

$$\begin{bmatrix} \cos(t) & \sin(t) \\ -\sin(t) & \cos(t) \end{bmatrix} \begin{bmatrix} \alpha_1 \\ \alpha_2 \end{bmatrix} = \begin{bmatrix} 0 \\ 0 \end{bmatrix}$$

for all t. The determinant of the matrix here is $\cos^2(t) + \sin^2(t) = 1$ and so picking any t we like, we find the unique solution is $\alpha_1 = \alpha_2 = 0$. Hence, these two functions are linearly independent on $\Re$. In fact, they are linearly independent on any interval I.

This leads to another important idea. Suppose f and g are linearly independent differentiable functions on an interval I. Then, we know the system

$$\begin{bmatrix} f(t) & g(t) \\ f'(t) & g'(t) \end{bmatrix} \begin{bmatrix} \alpha_1 \\ \alpha_2 \end{bmatrix} = \begin{bmatrix} 0 \\ 0 \end{bmatrix}$$

only has the unique solution $\alpha_1 = \alpha_2 = 0$ for all t in I. This tells us

$$\det\left(\begin{bmatrix} f(t) & g(t) \\ f'(t) & g'(t) \end{bmatrix}\right) \neq 0$$

for all t in I. This determinant comes up a lot and it is called the **Wronskian** of the two functions f and g and it is denoted by the symbol $W(f, g)$. Hence, we have the implication: if f and g are linearly independent differentiable functions, then $W(f, g) \neq 0$ for all t in I. What about the converse? If the Wronskian is never zero on I, then the system

$$\begin{bmatrix} f(t) & g(t) \\ f'(t) & g'(t) \end{bmatrix} \begin{bmatrix} \alpha_1 \\ \alpha_2 \end{bmatrix} = \begin{bmatrix} 0 \\ 0 \end{bmatrix}$$

must have the unique solution $\alpha_1 = \alpha_2 = 0$ at each t in I also. So the converse is true: if the Wronskian is not zero on I, then the differentiable functions f and g are linearly independent on I. We can state this formally as a theorem.

Theorem 16.3.1 (Two Functions are Linearly Independent if and only if their Wronskian is not zero) *If f and g are differentiable functions on I, the* **Wronskian** *of f and g is defined to be*

$$W(f, g) = \det\left(\begin{bmatrix} f(t) & g(t) \\ f'(t) & g'(t) \end{bmatrix}\right).$$

where $W(f, g)$ is the symbol for the Wronskian of f and g. Sometimes, this is just written as W, if the context is clear. Then f and g are linearly independent on I if and only if $W(f, g)$ is non zero on I.

Proof See the discussions above. ■

If f, g and h are twice differentiable on I, the Wronskian uses a third row of second derivatives and the statement that these three functions are linearly independent on I if and only if their Wronskian is non zero on I is proved essentially the same way. The appropriate theorem is

Theorem 16.3.2 (Three Functions are Linearly Independent if and only if their Wronskian is not zero) *If f, g and h are twice differentiable functions on I, the* **Wronskian** *of f, g and h is defined to be*

$$W(f, g, h) = \det\left(\begin{bmatrix} f(t) & g(t) & h(t) \\ f'(t) & g'(t) & h'(t) \\ f''(t) & g''(t) & h''(t) \end{bmatrix}\right).$$

where $W(f, g, h)$ is the symbol for the Wronskian of f and g. Then f, g and h are linearly independent on I if and only if $W(f, g, h)$ is non zero on I.

Proof The arguments are similar, although messier. ■

For example, to show the three functions $f(t) = t$, $g(t) = \sin(t)$ and $h(t) = e^{2t}$ are linearly independent on $\Re$, we could form their Wronskian

$$W(f, g, h) = \det\left(\begin{bmatrix} t & \sin(t) & e^{2t} \\ 1 & \cos(t) & 2e^{2t} \\ 0 & -\sin(t) & 4e^{2t} \end{bmatrix}\right) = t\begin{bmatrix} \cos(t) & 2e^{2t} \\ -\sin(t) & 4e^{2t} \end{bmatrix} - \begin{bmatrix} \sin(t) & e^{2t} \\ -\sin(t) & 4e^{2t} \end{bmatrix}$$

$$= t\left(e^{2t}(4\cos(t) + 2\sin(t)\right) - \left(e^{2t}(4\sin(t) + \sin(t)\right)$$

$$= e^{2t}\left(4t\cos(t) + 2t\sin(t) - 5\sin(t)\right).$$

Since, e^{2t} is never zero, the question becomes is

$$4t\cos(t) + 2t\sin(t) - 5\sin(t)$$

zero for all t? If so, that would mean the functions $t\sin(t)$, $t\cos(t)$ and $\sin(t)$ are linearly dependent. We could then form another Wronskian for these functions which would be rather messy. To see these three new functions are linearly independent, it is easier to just pick *three* points t from $\Re$ and solve the resulting linearly dependence equations. Since $t = 0$ does not give any information, let's try $t = -\pi$, $t = \frac{\pi}{4}$ and $t = \frac{\pi}{2}$. This gives the system

$$\begin{bmatrix} -4\pi & 0 & 0 \\ \pi & 2\frac{\pi}{4}\frac{\sqrt{2}}{2} & -5\frac{\sqrt{2}}{2} \\ 0 & 2\frac{\pi}{2} & -5 \end{bmatrix}\begin{bmatrix} \alpha_1 \\ \alpha_2 \\ \alpha_3 \end{bmatrix} = \begin{bmatrix} 0 \\ 0 \\ 0 \end{bmatrix}$$

in the unknowns α_1, α_2 and α_3. We see immediately $\alpha - 1 = 0$ and the remaining two by two system has determinant $\frac{\sqrt{2}}{2}(-10\frac{\pi}{4}) + 10\frac{\pi}{2} \neq 0$. Hence, $\alpha_2 = \alpha_3 = 0$ too. This shows $t\sin(t)$, $t\cos(t)$ and $\sin(t)$ are linearly independent and show the line $4t\cos(t) + 2t\sin(t) - 5\sin(t)$ is not zero for all t. Hence, the functions $f(t) = t$, $g(t) = \sin(t)$ and $h(t) = e^{2t}$ are linearly independent. As you can see, these calculations become messy quickly. Usually, the Wronskian approach for more than two functions is too hard and we use the *"pick three suitable points t_i, from I"* approach and solve the resulting linear system. If we can show the solution is always 0, then the functions are linearly independent.

16.3.2 Homework

Exercise 16.3.1 *Prove e^t and e^{-t} are linearly independent on $\Re$.*

Exercise 16.3.2 *Prove e^t and e^{2t} are linearly independent on $\Re$.*

Exercise 16.3.3 *Prove $f(t) = 1$ and $g(t) = t^2$ are linearly independent on $\Re$.*

Exercise 16.3.4 *Prove e^t, e^{2t} and e^{3t} are linearly independent on $\Re$. Use the pick three points approach here.*

Exercise 16.3.5 *Prove $\sin(t)$, $\sin(2t)$ and $\sin(3t)$ are linearly independent on $\Re$. Use the pick three points approach here.*

Exercise 16.3.6 *Prove 1, t and t^2 are linearly independent on $\Re$. Use the pick three points approach here.*

16.4 Vector Spaces and Basis

We can make the ideas we have been talking about more formal. If we have a set of objects $\boldsymbol{u}$ with a way to add them to create new objects in the set and a way to *scale* them to make new objects, this is formally called a **Vector Space** with the set denoted by $\mathscr{V}$. For our purposes, we scale such objects with either real or complex numbers. If the scalars are real numbers, we say $\mathscr{V}$ is a vector space over the reals; otherwise, it is a vector space over the complex field.

Definition 16.4.1 (*Vector Space*) Let $\mathscr{V}$ be a set of objects $\boldsymbol{u}$ with an additive operation $\oplus$ and a scaling method $\odot$. Formally, this means

1. Given any $\boldsymbol{u}$ and $\boldsymbol{v}$, the operation of adding them together is written $\boldsymbol{u} \oplus \boldsymbol{v}$ and results in the creation of a new object w in the vector space. This operation is *commutative* which means the order of the operation is not important; so $\boldsymbol{u} \oplus \boldsymbol{v}$ and $\boldsymbol{v} \oplus \boldsymbol{u}$ give the same result. Also, this operation is associative as we can group any two objects together first, perform this addition $\oplus$ and then do the others and the order of the grouping does not matter.

2. Given any $\boldsymbol{u}$ and any number c (either real or complex, depending on the type of vector space we have), the operation $c \odot \boldsymbol{u}$ creates a new object. We call such numbers *scalars*.
3. The scaling and additive operations are nicely compatible in the sense that order and grouping is not important. These are called the *distributive* laws for scaling and addition. They are

$$\begin{aligned} c \odot (\boldsymbol{u} \oplus \boldsymbol{v}) &= (c \odot \boldsymbol{u}) \oplus (c \odot \boldsymbol{v}) \\ (c+d) \odot \boldsymbol{u} &= (c \odot \boldsymbol{u}) \oplus (d \odot \boldsymbol{u}). \end{aligned}$$

4. There is a special object called $\boldsymbol{o}$ which functions as a *zero* so we always have $\boldsymbol{o} \oplus \boldsymbol{u} = \boldsymbol{u} \oplus \boldsymbol{o} = \boldsymbol{u}$.
5. There are *additive inverses* which means to each $\boldsymbol{u}$ there is a unique object $\boldsymbol{u}^\dagger$ so that $\boldsymbol{u} \oplus \boldsymbol{u}^\dagger = \boldsymbol{o}$.

Comment 16.4.1 *These laws imply*

$$(0+0) \odot \boldsymbol{u} = (0 \odot \boldsymbol{u}) \oplus (0 \odot \boldsymbol{u})$$

which tells us $0 \odot \boldsymbol{u} = 0$. *A little further thought then tells us that since*

$$\begin{aligned} \mathbf{0} &= (1-1) \odot \boldsymbol{u} \\ &= (1 \odot \boldsymbol{u}) \oplus (-1 \odot \boldsymbol{u}) \end{aligned}$$

we have the additive inverse $\boldsymbol{u}^\dagger = -1 \odot \boldsymbol{u}$.

Comment 16.4.2 *We usually say this much simpler. The set of objects* $\mathscr{V}$ *is a vector space over its scalar field if there are two operations which we denote by* $\mathbf{u} + \mathbf{v}$ *and* $c\mathbf{u}$ *which generate new objects in the vector space for any* $\mathbf{u}$, $\mathbf{v}$ *and scalar* c. *We then just add that these operations satisfy the usual commutative, associative and distributive laws and there are unique additive inverses.*

Comment 16.4.3 *The objects are often called* vectors *and sometimes we denote them by* $\mathbf{u}$ *although this notation is often too cumbersome.*

Comment 16.4.4 *To give examples of vector spaces, it is usually enough to specify how the additive and scaling operations are done.*

- *Vectors in* $\Re^2$, $\Re^3$ *and so forth are added and scaled by components.*
- *Matrices of the same size are added and scaled by components.*
- *A set of functions of similar characteristics uses as its additive operator, pointwise addition. The new function* $(f \oplus g)$ *is defined pointwise by* $(f \oplus g)(t) = f(t) + g(t)$. *Similarly, the new function* $c \odot f$ *is defined by* $c \odot f$ *is the function whose value at* t *is* $(cf)(t) = cf(t)$. *Classic examples are*

1. $C[a,b]$ *is the set of all functions whose domain is* $[a,b]$ *that are continuous on the domain.*

2. *$C^1[a, b]$ is the set of all functions whose domain is $[a, b]$ that are continuously differentiable on the domain.*
3. *$R[a, b]$ is the set of all functions whose domain is $[a, b]$ that are Riemann integrable on the domain.*

There are many more, of course.

Vector spaces have two other important ideas associated with them. We have already talked about linearly independent objects. Clearly, the kinds of objects we were focusing on were from some vector space $\mathscr{V}$. The first idea is that of the span of a set.

Definition 16.4.2 (*The Span Of A Set Of Vectors*) Given a finite set of vectors in a vector space $\mathscr{V}$, $\mathscr{W} = \{\boldsymbol{u}_1, \ldots, \boldsymbol{u}_N\}$ for some positive integer N, the span of $\mathscr{W}$ is the collection of all new vectors of the form $\sum_{i=1}^{N} c_i \boldsymbol{u}_i$ for any choices of scalars $c_1, \ldots, c_N$. It is easy to see $\mathscr{W}$ is a vector space itself and since it is a subset of $\mathscr{V}$, we call it a *vector subspace*. The span of the set $\mathscr{W}$ is denoted by $\boldsymbol{Sp}\boldsymbol{W}$. If the set of vectors $\mathscr{W}$ is not finite, the definition is similar but we say the span of $\mathscr{W}$ is the set of all vectors which can be written as $\sum_{i=1}^{N} c_i \boldsymbol{u}_i$ for some finite set of vectors $\boldsymbol{u}_1, \ldots \boldsymbol{u}_N$ from $\mathscr{W}$.

Then there is the notion of a *basis* for a vector space. First, we need to extend the idea of linear independence to sets that are not necessarily finite.

Definition 16.4.3 (*Linear Independence For Non Finite Sets*) Given a set of vectors in a vector space $\mathscr{V}$, $\mathscr{W}$, we say $\mathscr{W}$ is a linearly independent subset if every finite set of vectors from $\mathscr{W}$ is linearly independent in the usual manner.

Definition 16.4.4 (*A Basis For A Vector Space*) Given a set of vectors in a vector space $\mathscr{V}$, $\mathscr{W}$, we say $\mathscr{W}$ is a *basis* for $\mathscr{V}$ if the span of $\mathscr{W}$ is all of $\mathscr{V}$ and if the vectors in $\mathscr{W}$ are linearly independent. Hence, a basis is a linearly independent spanning set for $\mathscr{V}$. The number of vectors in $\mathscr{W}$ is called the *dimension* of $\mathscr{V}$. If $\mathscr{W}$ is not finite is size, then we say $\mathscr{V}$ is an *infinite dimensional vector space.*

Comment 16.4.5 *In a vector space like $\Re^n$, the maximum size of a set of linearly independent vectors is n, the dimension of the vector space.*

Comment 16.4.6 *Let's look at the vector space $C[0, 1]$, the set of all continuous functions on $[0, 1]$. Let $\mathscr{W}$ be the set of all powers of t, $\{1, t, t^2, t^3, \ldots\}$. We can use the derivative technique to show this set is linearly independent even though it is infinite in size. Take any finite subset from $\mathscr{W}$. Label the resulting powers as $\{n_1, n_2, \ldots, n_p\}$. Write down the linear dependence equation*

$$c_1 t^{n_1} + c_2 t^{n_2} + \cdots + c_p t^{n_p} = 0.$$

Take n_p derivatives to find $c_p = 0$ and then backtrack *to find the other constants are zero also. Hence $C[0, 1]$ is an infinite dimensional vector space. It is also clear that $\mathscr{W}$ does not span $C[0, 1]$ as if this was true, every continuous function on $[0, 1]$ would be a polynomial of some finite degree. This is not true as $\sin(t)$, e^{-2t} and many others are not finite degree polynomials.*

16.4.1 Inner Products in Function Spaces

Now there is an important result that we use a lot in applied work. If we have an object $\boldsymbol{u}$ in a Vector Space $\mathscr{V}$, we often want to find to *approximate* $\boldsymbol{u}$ using an element from a given subspace $\mathscr{W}$ of the vector space. To do this, we need to add another property to the vector space. This is the notion of an *inner product*. We already know what an inner product is in a simple vector space like $\Re^n$. Many vector spaces can have an inner product structure added easily. For example, in $C[a, b]$, since each object is continuous, each object is Riemann integrable. Hence, given two functions f and g from $C[a, b]$, the real number given by $\int_a^b f(s)g(s)ds$ is well-defined. It satisfies all the usual properties that the inner product for finite dimensional vectors in $\Re^n$ does also. These properties are so common we will codify them into a definition for what an inner product for a vector space $\mathscr{V}$ should behave like.

Definition 16.4.5 (*Real Inner Product*) Let $\mathscr{V}$ be a vector space with the reals as the scalar field. Then a mapping ω which assigns a pair of objects to a real number is called an inner product on $\mathscr{V}$ if

1. $\omega(\boldsymbol{u}, \boldsymbol{v}) = \omega(\boldsymbol{v}, \boldsymbol{u})$; that is, the order is not important for any two objects.
2. $\omega(c \odot \boldsymbol{u}, \boldsymbol{v}) = c\omega(\boldsymbol{u}, \boldsymbol{v})$; that is, scalars in the *first slot* can be pulled out.
3. $\omega(\boldsymbol{u} \oplus \boldsymbol{w}, \boldsymbol{v}) = \omega(\boldsymbol{u}, \boldsymbol{v}) + \omega(\boldsymbol{w}, \boldsymbol{v})$, for any three objects.
4. $\omega(\boldsymbol{u}, \boldsymbol{u}) \geq 0$ and $\omega(\boldsymbol{u}, \boldsymbol{u}) = 0$ if and only if $u = 0$.

These properties imply that $\omega(\boldsymbol{u}, c \odot \boldsymbol{v}) = c\omega(\boldsymbol{u}, \boldsymbol{v})$ as well. A vector space $\mathscr{V}$ with an inner product is called an inner product space.

Comment 16.4.7 *The inner product is usually denoted with the symbol* $<, >$ *instead of* $\omega(,)$. *We will use this notation from now on.*

Comment 16.4.8 *When we have an inner product, we can* measure *the* size *or* magnitude *of an object, as follows. We define the analogue of the euclidean norm of an object* $\boldsymbol{u}$ *using the usual* $|| \; ||$ *symbol as*

$$||\boldsymbol{u}|| = \sqrt{< \boldsymbol{u}, \boldsymbol{u} >}.$$

This is called the norm induced by the inner product *of the object. In* $C[a, b]$, *with the inner product* $<f, g> = \int_a^b f(s)g(s)ds$, *the* norm *of a function* f *is thus* $||f|| = \sqrt{\int_a^b f^2(s)ds}$. *This is called the* L_2 *norm of* f.

It is possible to prove the Cauchy–Schwartz inequality in this more general setting also.

Theorem 16.4.1 (General Cauchy–Schwartz Inequality)
If $\mathscr{V}$ *is an inner product space with inner product* $<, >$ *and induced norm* $|| \; ||$, *then*

$$| < \boldsymbol{u}, \boldsymbol{v} > | \leq ||\boldsymbol{u}|| \; ||\boldsymbol{v}||$$

with equality occurring if and only if $\boldsymbol{u}$ *and* $\boldsymbol{v}$ *are linearly dependent.*

Proof The proof is different than the one you would see in a Calculus text for $\Re^2$, of course, and is covered in a typical course on beginning linear analysis. ■

Comment 16.4.9 *We can use the Cauchy–Schwartz inequality to define a notion of angle between objects exactly like we would do in $\Re^2$. We define the angle θ between $\boldsymbol{u}$ and $\boldsymbol{v}$ via its cosine as usual.*

$$\cos(\theta) = \frac{< \boldsymbol{u}, \boldsymbol{v} >}{||\boldsymbol{u}||\ ||\boldsymbol{v}||}.$$

Hence, objects can be perpendicular or orthogonal even if we can not interpret them as vectors in $\Re^2$. We see two objects are orthogonal if their inner product is 0.

Comment 16.4.10 *If $\mathscr{W}$ is a finite dimensional subspace, a basis for $\mathscr{W}$ is said to be an orthonormal basis if each object in the basis has L_2 norm 1 and all of the objects are mutually orthogonal. This means $< \boldsymbol{u}_i, \boldsymbol{u}_j >$ is 1 if $i = j$ and 0 otherwise. We typically let the Kronecker delta symbol δ_{ij} be defined by $\delta_{ij} = 1$ if $i = j$ and 0 otherwise so that we can say this more succinctly as $< \boldsymbol{u}_i, \boldsymbol{u}_j >= \delta_{ij}$.*

16.5 Fourier Series

A general trigonometric series $S(x)$ has the following form

$$S(x) = b_0 + \sum_{i=1}^{\infty} \left(a_i \sin\left(\frac{i\pi}{L}x\right) + b_i \cos\left(\frac{i\pi}{L}x\right)\right)$$

for any numbers a_n and b_n. Of course, there is no guarantee that this series will converge at any x! If we start with a function f which is continuous on the interval $[0, L]$, we can define the trigonometric series associated with f as follows

$$\begin{aligned} S(x) &= \frac{1}{L} < f, \mathbf{1} > \\ &+ \sum_{i=1}^{\infty} \left(\frac{2}{L}\left\langle f(x), \sin\left(\frac{i\pi}{L}x\right)\right\rangle \sin\left(\frac{i\pi}{L}x\right) + \frac{2}{L}\left\langle f(x), \cos\left(\frac{i\pi}{L}x\right)\right\rangle \cos\left(\frac{i\pi}{L}x\right)\right). \end{aligned}$$

where the symbol $<, >$ is the inner product in the set of functions $C[0, L]$ defined by $< u, v >= \int_0^L u(s)v(s)ds$. The coefficients in the Fourier series for f are called the *Fourier coefficients* of f. Since these coefficients are based on inner products with scaled sin and cos functions, we call these the *normalized* Fourier coefficients. Let's be clear about this and a bit more specific. The nth *Fourier* sin *coefficient*, $n \geq 1$, of f is as follows:

$$a_n(f) = \frac{2}{L} \int_0^L f(x) \sin\left(\frac{i\pi}{L}x\right) dx$$

The nth *Fourier* cos *coefficient*, $n \geq 0$, of f are defined similarly:

$$b_0(f) = \frac{1}{L}\int_0^L f(x)\,dx$$
$$b_n(f) = \frac{2}{L}\int_0^L f(x)\sin\left(\frac{i\pi}{L}x\right)dx, \quad n \geq 1.$$

16.5.1 The Cable Model Infinite Series Solution

We now know series of the form

$$A_0 e^{-\frac{1}{\alpha}t} + \sum_{n=1}^{N} A_n \cos\left(\frac{n\pi}{L}x\right) e^{-\frac{1+n^2\pi^2\beta^2}{\alpha L^2}t}.$$

are like Fourier series although in terms of two variables. We can show these series converge pointwise for x in $[0, L]$ and all t. We can also show that we can take the partial derivative of this series solutions term by term (see the discussions in Peterson (2015) for details) to obtain

$$\sum_{n=1}^{N} -A_n \frac{n\pi}{L}\sin\left(\frac{n\pi}{L}x\right) e^{-\frac{1+n^2\pi^2\beta^2}{\alpha L^2}t}.$$

This series evaluated at $x = 0$ and $x = L$ gives 0 and hence the Neumann conditions are satisfied. Hence, the solution $\Phi(x, t)$ given by

$$\Phi(x,t) = A_0 e^{-\frac{1}{\alpha}t} + \sum_{n=1}^{N} A_n \cos\left(\frac{n\pi}{L}x\right) e^{-\frac{1+n^2\pi^2\beta^2}{\alpha L^2}t}.$$

for the arbitrary sequence of constants (A_n) is a well-behaved solution on our domain. The remaining boundary condition is

$$\Phi(x, 0) = f(x), \quad \text{for } 0 \leq x \leq L$$

and

$$\Phi(x, 0) = A_0 + \sum_{n=1}^{\infty} A_n \cos\left(\frac{n\pi}{L}x\right).$$

Rewriting in terms of the series solution, for $0 \leq x \leq L$, we find

$$A_0 + \sum_{n=1}^{\infty} A_n \cos\left(\frac{n\pi}{L}x\right) = f(x)$$

The Fourier series for f is given by

$$f(x) = B_0 + \sum_{n=1}^{\infty} B_n \cos\left(\frac{n\pi}{L}x\right)$$

with

$$B_0 = \frac{1}{L}\int_0^L f(x),$$
$$B_n = \frac{2}{L}\int_0^L f(x)\cos\left(\frac{n\pi}{L}x\right)dx.$$

Then, setting these series equal, we find that the solution is given by $A_n = B_n$ for all $n \geq 0$. The full details of all of this is outside the scope of our work here, but this will give you a taste of how these powerful tools can help us solve PDE models.

Reference

J. Peterson. *Calculus for Cognitive Scientists: Partial Differential Equation Models*, Springer Series on Cognitive Science and Technology (Springer Science+Business Media Singapore Pte Ltd., Singapore, 2015 in press)

Part VII
Summing It All Up

Chapter 17
Final Thoughts

How can we train people to think in an interdisciplinary manner using these sorts of tools? It is our belief we must foster a new mindset within practicing scientists. We can think of this as the development of a new integrative discipline we could call **How It All Fits Together Science**. Of course, this name is not formal enough to serve as a new disciplinary title, but keeping to its spirit, we could use the name **Integrative Science**. Now the name *Integrative Biology* has occurred here and there in the literature over the last few decades and it is not clear to us it has the right tone. So, we will begin calling our new integrative point of view **Building Scientific Bridges** or ***BSB***.

17.1 Fostering Interdisciplinary Appreciation

We would like the typical scientist to have a deeper appreciation of the use of the triad of biology, mathematics and computer science (**BMC**) in this attempt to build bridges between the many disparate areas of biology, cognitive sciences and the other sciences. The sharp disciplinary walls that have been built in academia hurt everyone's chances at developing an active and questing mind that is able to both be suspicious of the status—quo and also have the tools to challenge it effectively. Indeed, we have longed believed that all research requires a rebellious mind. If you revere the expert opinion of others too much, you will always be afraid to forge a new path for yourself. So respect and disrespect are both part of the toolkit of our budding scientists.

There is currently an attempt to create a new **Astrobiology** program at the University of Washington which is very relevant to our discussion. Astrobiology is an excellent example of how science and a little mathematics can give insight into issues such as the creation of life and its abundance in the universe. Read its primer edited by Linda Billings and others, (Billings 2006) for an introduction to this field. Careful arguments using chemistry, planetary science and many other fields inform what is probable. It is a useful introduction not only to artificial life issues, but also to the

J.K. Peterson, *Calculus for Cognitive Scientists*,
Cognitive Science and Technology, DOI 10.1007/978-981-287-877-9_17

process by which we marshal ideas form disparate fields to form interesting models. Indeed, if you look at the new textbook, **Planets and Life: The Emerging Science of Astrobiology**, (Sullivan and Baross 2007) you will be inspired at the wealth of knowledge such a field must integrate. This integration is held back by students who are not trained in both **BMC** and ***BSB***. We paraphrase some of the important points made by the graduate students enrolled in this new program about interdisciplinary training and research. Consider what the graduate students in this new program have said Sullivan and Baross (2007, p. 548):

> …some of the ignorance exposed by astrobiological questions reveals not the boundaries of scientific knowledge, but instead the boundaries of individual disciplines. Furthermore, collaboration by itself does not address this ignorance, but instead compounds it by encouraging scientists to rely on each other's authority. Thus, anachronistic disciplinary borders are reinforced rather than overrun. In contrast **astrobiology** can motivate challenges to disciplinary isolation and the appeals to authority that such isolation fosters.

Indeed, studying problems that require points of view from many places is of great importance to our society and from Sullivan and Baross (2007, p. 548) we hear that

> many different disciplines should now be applied to a class of questions perceived as broadly unified and that such an amalgamation justifies a new discipline (or even meta discipline) such as **astrobiology**.

Now simply replace the key word **astrobiology** by **biomathematics** or ***BSB*** or cognitive science and reread the sentence again. Aren't we trying to do just this when we design any textual material such as this book and its earlier companion (Peterson 2015) to integrate separate disciplines? Since we believe we can bring additional illumination to problems we wish to solve in cognitive sciences by adding new ideas from mathematics and computer science to the mix, we are asking explicitly for such an amalgamation and interdisciplinary blending. We would like to create a small cadre of like minded people who believe as the nascent astrobiology graduate students do (Sullivan and Baross 2007, p. 549) that

> Dissatisfaction with disciplinary approaches to fundamentally interdisciplinary questions also led many of us to major in more than one field. This is not to deny the importance of reductionist approaches or the advances stimulated by them. Rather, as undergraduates we wanted to integrate the results of reductionist science. Such synthesis is often poorly accommodated by disciplines that have evolved, especially in academia, to become insular, autonomous departments. Despite the importance of synthesis for many basic scientific questions, it is rarely attempted in research…

We believe, as the astrobiology students (Sullivan and Baross 2007, p. 550) do that

> [***BSB*** enriched by **BMC**] can change this by challenging the ignorance fostered by disciplinary structure while pursuing the creative ignorance underlying genuine inquiry. Because of its integrative questions, interdisciplinary nature,…, [***BSB*** enriched by **BMC**] emerges as an ideal vehicle for scientific education at the graduate, undergraduate and even high school levels. [It] permits treatment of traditionally disciplinary subjects as well as areas where those subjects converge (and, sometimes, fall apart!) At the same time, [it] is well suited to reveal the creative ignorance at scientific frontiers that drives discovery.

where in the original quote, we have replaced *astrobiology* by the phrase ***BSB*** enriched by **BMC** and enclosed our changes in brackets. The philosophy of this course has been to encourage thinking that is suitable to the ***BSB*** enriched by **BMC**. As you have seen, we have used the following point of view:

- All mathematical concepts are tied to real biological and scientific need. Hence, after preliminary calculus concepts have been discussed, there is a careful development of the exponential and natural logarithm functions. This is done in such a way that all properties are derived so there are no mysteries, no references to "this is how it is" and we don't go through the details because the mathematical derivation is beyond you. We insist that learning to think with the language of mathematics is an important skill. Once the exponential function is understood, it is immediately tied to the simplest type of real biological model: exponential growth and decay.
- Mathematics is subordinate to the science in the sense that the texts build the mathematical knowledge needed to study interesting nonlinear biological and cognitive models. We emphasize that to add more interesting science always requires more difficult mathematics and concomitant intellectual resources.
- Standard mathematical ideas from linear algebra such as eigenvalues and eigenvectors can be used to study systems of linear differential equations using both numerical tools (MatLab based currently) and graphical tools. We always stress how the mathematics and the biology interact.
- Nonlinear models are important to study scientific things. We begin with the logistics equation and progress to the Predator–Prey model and a standard SIR disease model. We emphasize how we must abstract out of biological and scientific complexity the variables necessary to build the model and how we can be wrong. We show how the original Predator–Prey model, despite being a gross approximation of the biological reality of fish populations in the sea, gives great explanatory insight. Then, we equally stress that adding self-interaction to the model leads to erroneous biological predictions. The SIR disease model is also clearly explained as a gross simplification of the immunological reality which nevertheless illuminates the data we see.
- Difficult questions can be hard to formulate quantitatively but if we persevere, we can get great insights. We illustrate this point of view with a model of cancer which requires 6 variables but which is linear. We show how much knowledge is both used and thrown away when we make the abstractions necessary to create the model. At this point, the students also know that models can easily have more variables in them than we can graph comfortably. Also, numerical answers delivered via MatLab are just not very useful. We need to derive information about the functional relationships between the parameters of the model and that requires a nice blend of mathematical theory and biology.

The use of **BMC** is therefore fostered in this text and its earlier companion by introducing the students to concepts from the traditional Math-Two-Engineering (just a little as needed), Linear Algebra (a junior level mathematics course) and Differential

Equations (a sophomore level mathematics course) as well as the rudiments of using a language such as MatLab for computational insight. We deliberately try to use scientific examples from biology and behavior in many places to foster the ***BSB*** point of view.

We believe just as the astrobiology students do (Sullivan and Baross 2007, p. 550) that

> The ignorance motivating scientific inquiry will never wholly be separated from the ignorance hindering it. The disciplinary organization of scientific education encourages scientists to be experts on specialized subjects and silent on everything else. By creating scientists dependent of each other's specializations, this approach is self-reinforcing. A discipline of [***BSB*** enriched by **BMC**] should attempt something more ambitious: it should instead encourage scientists to master for themselves what formerly they deferred to their peers.

Indeed, there is more that can be said (Sullivan and Baross 2007, p. 552)

> What [***BSB*** enriched by **BMC**] can mean as a science and discipline is yet to be decided, for it must face the two-fold challenge of cross-disciplinary ignorance that disciplinary education itself enforces. First, ignorance cannot be skirted by deferral to experts, or by other implicit invocations of the disciplinary mold that [***BSB*** enriched by **BMC**] should instead critique. Second, ignorance must actually be *recognized*. This is not trivial: how do you know what you do not know? Is it possible to understand a general principle without also understanding the assumptions and caveats underlying it? Knowledge superficially "understood" is self-affirming. For example, the meaning of the molecular tree of life may appear unproblematic to an astronomer who has learned that the branch lengths represent evolutionary distance, but will the astronomer even know to consider the hidden assumptions about rate constancy by which the tree is derived? Similarly, images from the surface of Mars showing evidence of running water are prevalent in the media, yet how often will a biologist be exposed to alternative explanations for these geologic forms, or to the significant evidence to the contrary? [There is a need] for a way to discriminate between science and ... uncritically accepted results of science.

A first attempt at developing a first year curriculum for the graduate program in astrobiology led to an integrative course in which specialists from various disciplines germane to the study of astrobiology gave lectures in their own areas of expertise and then left as another expert took over. This was disheartening to the students in the program. They said that (Sullivan and Baross 2007, p. 553)

> As a group, we realized that we could not speak the language of the many disciplines in astrobiology and that we lacked the basic information to consider their claims critically. Instead, this attempt at an integrative approach provided only a superficial introduction to the major contributions of each discipline to astrobiology. How can critical science be built on a superficial foundation? Major gaps in our backgrounds still needed to be addressed. In addition, we realized it was necessary to direct ourselves toward a more specific goal. What types of scientific information did we most need? What levels of mastery should we aspire to? At the same time, catalyzed by our regular interactions in the class, we students realized that we learned the most (and enjoyed ourselves the most) in each other's interdisciplinary company. While each of us had major gaps in our basic knowledge, as a group we could begin to fill many of them.

Now paraphrase as follows:

> Students can not speak the language of the many disciplines ***BSB*** enriched by **BMC** requires and they lack the basic information to consider their claims critically. An attempt at an integrative approach that provided only a superficial introduction to the major contributions of each discipline can not lead to the ability to do critical science. Still, major gaps in their backgrounds need to be addressed. What types of scientific information are most needed? What levels of mastery should they aspire to? It is clear the students learned the most (and enjoyed themselves the most) in each other's interdisciplinary company. While each has major gaps in basic knowledge, as a group they can begin to fill many of them.

Given the above, it is clear we must introduce and tie disparate threads of material together with care. In many things, we favor theoretical approaches that attempt to put an overarching theory of everything together into a discipline. Since biology and cognitive science is so complicated, we will always have to do this. You can see examples of this way of thinking explored in many areas of current research. For example, it is very hard to understand how organs form in the development process. A very good book on these development issues is first the one by Davies (2005). Davis talks about the complexity that emerges from local interactions. This is, of course, difficult to define precisely, yet, it is a useful principle that helps us understand the formation of a kidney and other large scale organs. Read this book and then read the one by Schmidt-Rhaesa (2007) and you will see that the second book makes more sense given the ideas you have mastered from Davis.

Finally, it is good to always keep in mind (Sullivan and Baross 2007, p. 552)

> Too often, the scientific community and academia facilely redress ignorance by appealing to the testimony of experts. This does not resolve ignorance. It fosters it.

This closing comment sums it up nicely. We want to give the students and ourselves an atmosphere that fosters solutions and our challenge is to find a way to do this.

So how do we become a good user of the three fold way of mathematics, science and computer tools? We believe it is primarily a question of deep respect for the balance between these disciplines. The basic idea is that once we abstract from biology or some other science how certain quantities interact, we begin to phrase these interactions in terms of mathematics. It is very important to never forget that once the mathematical choices have been made, the analysis of the mathematics alone will lead you to conclusions which may or may not be biologically relevant. You must always be willing to give up a mathematical model or a computer science model if it does not lead to useful insights into the original science.

We can quote from Randall (2005, pp. 70–71). She works in Particle Physics which is the experimental arm that gives us data to see if string theory or loop gravity is indeed a useful model of physical reality. It is not necessary to know physics here to get the point. Notice what she says:

> The term "model" might evoke a small scale battleship or castle you built in your childhood. Or you might think of simulations on a computer that are meant to reproduce known dynamics—how a population grows, for example, or how water moves in the ocean. Modeling in particle physics is not the same as either of these definitions. Particle physics models are guesses at alternate physical theories that might underlie the standard model... Different

> assumptions and physical concepts distinguish theories, as do the distance or energy scales at which a theory's principles apply. Models are a way at getting at the heart of such distinguishing features. They let you explore a theory's potential implications. If you think of a theory as general instructions for making a cake, a model would be a precise recipe. The theory would say to add sugar, a model would specify whether to add half a cup or two cups.

Now substitute *Biological* for *Particle Physics* and so forth and you can get a feel for what a model is trying to do. Of course, biological models are much more complicated than physics ones!

The primary message of this course is thus to teach you to think deeply and carefully. The willingness to attack hard problems with multiple tools is what we need in young scientists. We hope this course teaches you a bit more about that.

We believe that we learn all the myriad things we need to build reasonable models over a lifetime of effort. Each model we design which pulls in material from disparate areas of learning enhances our ability to develop the kinds of models that give insight. As Pierrehumbert (2010, p. xi) says about climate modeling

> When it comes to understanding the whys and wherefores of climate, there is an infinite amount one needs to know, but life affords only a finite time in which to learn it…It is a lifelong process. [We] attempt to provide the student with a sturdy scaffolding upon which a deeper understanding may be built later.
>
> The climate system [and other biological systems we may study] is made up of building blocks which in themselves are based on elementary…principles, but which have surprising and profound collective behavior when allowed to interact on the [large] scale. In this sense, the "climate game" [the biological modeling game] is rather like the game of Go, where interesting structure emerges from the interaction of simple rules on a big playing field, rather than complexity in the rules themselves.

References

L. Billings, Education paper: the astrobiology primer: an outline of general knowledge-version 1, 2006. Astrobiology **6**(5), 735–813 (2006)

J. Davies, *Mechanisms of Morphogenesis* (Academic Press Elsevier, Massachusetts, 2005)

J. Peterson, *Calculus for Cognitive Scientists: Derivatives, Integration and Modeling*, Springer Series on Cognitive Science and Technology (Springer Science+Business Media Singapore Pte Ltd, Singapore, 2015 in press)

R. Pierrehumbert, *Principles of Planetary Climate* (Cambridge University Press, Cambridge, 2010)

L. Randall, *Warped Passages: Unraveling the Mysteries of the Universe's Hidden Dimensions* (Harper Collins, New York, 2005)

A. Schmidt-Rhaesa, *The Evolution of Organ Systems* (Oxford University Press, Oxford, 2007)

W. Sullivan, J. Baross (eds.), *Planets and Life: The Emerging Science of Astrobiology* (Cambridge University Press, Cambridge, 2007)

Part VIII
Advise to the Beginner

Chapter 18
Background Reading

To learn more about how the combination of mathematics, computational tools and science have been used with profit, we have some favorite books that have tried to do this. Reading these books have helped us design biologically inspired algorithms and models and has inspired how we wrote this text! A selection of these books would include treatment of regulatory systems in genomics such as Davidson (2001, 2006). An interesting book on how the architecture of genomes may have developed which gives lots of food for thought for model abstraction is found in Lynch (2007). Since the study of how to build cognitive systems is our big goal, the encyclopediac treatment of how nervous systems evolved given in J. Kass's four volumes (Kaas and Bullock 2007a, b; Kaas and Krubitzer 2007; Kaas and Preuss 2007) is very useful. To build a cognitive system that would be deployable as an autonomous device in a robotic system requires us to think carefully about how to build a computer system using computer architectural choices and hardware that has some plasticity. Clues as to how to do this can be found by looking at how simple nervous systems came to be. After all, the complexity of the system we can build will of necessity be fair less than that of a real neural system that controls a body; hence, any advice we can get from existing biological systems is welcome! With that said, as humans we process a lot of environmental inputs and the thalamus plays a vital role in this. If you read Murray Sherman's discussion of the thalamus in Sherman and Guillery (2006), you will see right away how you could begin to model some of the discussed modules in the thalamus which will be useful in later work in the next two volumes.

Let's step back now and talk about our ultimate goal which is to study cognitive systems. You now have two self study texts under your belt (ahem, Peterson 2015a, b), and you have been exposed to a fair amount of coding in MatLab. You have seen a programming language is just another tool we can use to gain insight into answering very difficult questions. The purpose of this series of four books (the next ones are Peterson 2015c, d) are to prime you to always think about attacking complicated and difficult questions using a diverse toolkit. You have been trained enough now that you can look at all that you read and have a part of you begin the model building process for the area you are learning about in your reading.

J.K. Peterson, *Calculus for Cognitive Scientists*,
Cognitive Science and Technology, DOI 10.1007/978-981-287-877-9_18

We believe firmly that the more you are exposed to ideas from multiple disciplines, the better you will be able to find new ways to solve our most challenging problems. Using a multidisciplinary toolkit does not always go over big with your peers, so you must not let that hold you back. A great glimpse into how hard it is to get ideas published (even though they are great) is to look at how this process went for others. We studied Hamilton's models of altruism in the previous book, so it might surprise you to know he had a hard time getting them noticed. His biography (Segerstrale 2013) is fascinating reading as is the collection of stories about biologists who thought differently (Harman and Dietrich 2013). Also, although everyone knows who Francis Crick was, reading his biography (Olby 2009) gives a lot of insight into the process of thinking creatively outside of the established paradigms. Crick believed, as we do, that theoretical models are essential to developing a true understanding of a problem.

The process of piecing together a story or model of a biological process indirectly from many disparate types of data and experiments is very hard. You can learn much about this type of work by reading carefully the stories of people who are doing this. Figuring out what dinosaurs were actually like even though they are not alive now requires that we use a lot of indirect information. The book on trace fossils (i.e. footprints and coprolites—fossilized poop, etc.) Martin (2014) shows you how much can be learned from those sources of data. In another reconstruction process, we know enough now about how hominids construct their bodies; i.e. how thick the fat is on top of muscle at various places on the face and so forth, to make a good educated guess at the building a good model of extinct hominids such as Australopithecus. This process is detailed in Gurchie (2013). Gurchie is an artist and it is fascinating to read how he uses the knowledge scientists can give him and his own creative insight to create the stunning models you can see at the Smithsonian now in the Hall of Humanity. We also know much more about how to analyze DNA sample found in very old fossils and the science behind how we can build a reasonable genome for a Neanderthal as detailed in Pääbo (2014) is now well understood. Read Martin's Gurchie's, Pääbo's discussions and apply what you have been learning about mathematics, computation and science to what they are saying. You should be seeing it all come together much better now.

Since our ultimate aim is to build cognitive models with our tools, it is time for you to read some about the brains that have been studied over the years. Kaas's books mentioned about nice, but almost too intense. Study some of these others and apply the new intellectual lens we have been working to develop.

- Stiles (2008) has written a great book on brain development which can be read in conjunction with another evolutionary approach by Streidter (2005). Read them both and ask yourself how you could design and implement even a minimal model that can handle such modular systems interacting in complex ways. Learning the right language to do that is the province of what we do in Peterson (2015c, d), so think of this reading as wetting your appetite! Another lower level and a bit more layman in tone is the nice text by Allman (2000) which is good to read as a overview of what the others are saying. The last one you should tackle is Allen

(2009) book about the evolution of the brain. Compare and contrast what all these books are saying: learn to think critically!
- On a slightly different note, Taylor (2012) tells you about the various indirect ways we use to measure what is going on in a functioning neural system. At the bottom of all these tools is a lot of mathematics, computation and science which you are much better prepared to read about now.
- It is very hard to understand how to model altruism as we have found out in this book. It is also very hard to model quantitatively sexual difference. As you are reading about the brain in the books we have been mentioning, you can dip into Jordan-Young (2010) to see just how complicated the science behind sexual difference is and how difficult it is to get good data. Further, there are many mysteries concerning various aspects of human nature which are detailed in Barash (2012). Ask yourself, how would you build models of such things? Another really good read here on sex is by Ridley (1993) which is on sex and how human nature evolved. It is a bit old now, but still very interesting.
- Once you have read a bit about the brain in general and from an evolutionary viewpoint, you are ready to learn more about how our behavior is modulated by neurotransmitters and drugs. S. Snyder has a nice layman discussion of those ideas in Snyder (1996).

To understand cognition in some ways to try to understand how biological processes evolved to handle information flow. Hence, reading about evolution in general—going beyond what we touched on in our discussions of altruism—is a good thing. An important part of our developmental chain is controlled by stem cells and a nice relatively non technical introduction to those ideas is in Fox (2007) which is very interesting also because it includes a lot of discussion on how stem cell manipulation is a technology we have to come to grips with. A general treatment of human development can be found in Davies (2014) which you should read as well. Also, the ways in which organ development is orchestrated by regulatory genes helps you again see the larger picture in which many modules of computation interact. You should check out A. Schmidt-Raisa's book on how organ systems evolved as well as J. Davies's treatment of how an organism determines its shape in Davies (2005).

The intricate ways that genes work together is hard to model and we must make many abstractions and approximations to make progress. G. Wagner has done a terrific job of explaining some of the key ideas in evolutionary innovation in Wagner (2014) and J. Archibald shows how symbiosis is a key element in evolution in Archibald (2014).

Processing books like these will help hone your skills and acquaint you with lots of material outside of your usual domain. This is a good thing! So keep reading and we hope you join us in the next volumes!

All of these readings have helped us to see the *big picture*!

References

J. Allen, *The Lives of the Brain: Human Evolution and the Organ of Mind* (The Belknap Press of Harvard University Press, Cambridge, 2009)
J. Allman, *Evolving Brains* (Scientific American Library, New York, 2000)
J. Archibald, *One PlusOne Equals One* (Oxford University Press, Oxford, 2014)
D. Barash, *Homo Mysterious: Evolutionary Puzzles of Human Nature* (Oxford University Press, Oxford, 2012)
E. Davidson, *Genomic Regulatory Systems: Development and Evolution* (Academic Press, San Diego, 2001)
E. Davidson, *The Regulatory Genome: Gene Regulatory Networks in Development and Evolution* (Academic Press Elsevier, Burlington, 2006)
J. Davies, *Mechanisms of Morphogenesis* (Academic Press Elsevier, Boston, 2005)
J. Davies, *Life Unfolding: How the Human Body Creates Itself* (Oxford University Press, Oxford, 2014)
C. Fox, *Cell of Cells: The Global Race To Capture and Control the Stem Cell* (W. H. Norton and Company, New York, 2007)
J. Gurchie, *Shaping Humanity: How Science, Art, and Imagination Help Us Understand Our Origins* (Yale University Press, New Haven, 2013)
O. Harman, M. Dietrich, *Outsider Scientists: Routes to Innovation in Biology* (University of Chicago Press, Chicago, 2013)
R. Jordan-Young, *Brainstorm: The Flaws in the Science of Sex Differences* (Harvard University Press, Brainstorm: The Flaws in the Science of Sex Differences, 2010)
J. Kaas, T. Bullock, (eds). *Evolution of Nervous Systems: A Comprehensive Reference Editor J. Kaas (Volume 1: Theories, Development, Invertebrates).* (Academic Press Elsevier, Amsterdam, 2007a)
J. Kaas, T. Bullock, (eds). *Evolution of Nervous Systems: A Comprehensive Reference Editor J. Kaas (Volume 2: Non-Mammalian Vertebrates).* (Academic Press Elsevier, Amsterdam, 2007b)
J. Kaas, L. Krubitzer, (eds). *Evolution of Nervous Systems: A Comprehensive Reference Editor J. Kaas (Volume 3: Mammals).* (Academic Press Elsevier, Amsterdam, 2007)
J. Kaas, T. Preuss, (eds). *Evolution of Nervous Systems: A Comprehensive Reference Editor J. Kaas (Volume 4: Primates).* (Academic Press Elsevier, Amsterdam, 2007)
M. Lynch, *The Origins of Genome Architecture* (Sinauer Associates, Inc., Sunderland, 2007)
A. Martin, *Dinosaurs Without Bones: Dinosaur Lives Revealed By Their Trace Fossils* (Pegasus Books, New York, 2014)
S. Murray Sherman, R. Guillery, *Exploring The Thalamus and Its Role in Cortical Function* (The MIT Press, Cambridge, 2006)
R. Olby, *Francis Crick: Hunter of Life's Secrets* (ColdSpring Harbor Laboratory Press, New York, 2009)
S. Pääbo, *Neanderthal Man: In Search of Lost Genomes* (Basic Books, New York, 2014)
J. Peterson, *Calculus for Cognitive Scientists: Derivatives, Integration and Modeling*, Springer Series on Cognitive Science and Technology (Springer Science+Business Media Singapore Pte Ltd, Singapore, 2015a in press)
J. Peterson, *Calculus for Cognitive Scientists: Higher Order Models and Their Analysis*, Springer Series on Cognitive Science and Technology (Springer Science+Business Media Singapore Pte Ltd., Singapore, 2015b in press)
J. Peterson, *Calculus for Cognitive Scientists: Partial Differential Equation Models*, Springer Series on Cognitive Science and Technology (Springer Science+Business Media Singapore Pte Ltd., Singapore, 2015c in press)
J. Peterson, *BioInformation Processing: A Primer On Computational Cognitive Science*, Springer Series on Cognitive Science and Technology (Springer Science+Business Media Singapore Pte Ltd., Singapore, 2015d in press)

M. Ridley, *The Red Queen: Sex and the Evolution of Human Nature* (Harper Perennial, New York, 1993)
U. Segerstrale, *Nature's Oracle: The Life and Work of W.D. Hamilton* (Oxford University Press, Oxford, 2013)
S. Snyder, *Drugs and the Brain* (Scientific American Library, New York, 1996)
J. Stiles, *The Fundamentals of Brain Development* (Harvard University Press, Cambridge, 2008)
G. Streidter, *Principles of Brain Evolution* (Sinauer Associates, Inc., Sunderland, 2005)
K. Taylor, *The Brain Supremacy: Notes from the frontiers of neuroscience* (Oxford University Press, Oxford, 2012)
G. Wagner, *Homology, Genes and Evolutionary Innovation* (Princeton University Press, Oxford, 2014)

Glossary

B

Biological Modeling This is the study of biological systems using a combination of mathematical, scientific and computational approaches, p. 4.

C

Cancer Model A general model of cancer based on TSG inactivation is as follows. The tumor starts with the inactivation of a **TSG** called **A**, in a small compartment of cells. A good example is the inactivation of the **APC** gene in a colonic crypt, but it could be another gene. Initially, all cells have two active alleles of the **TSG**. We will denote this by $\boldsymbol{A}^{+/+}$ where the superscript "$+/+$" indicates both alleles are active. One of the alleles becomes inactivated at mutation rate u_1 to generate a cell type denoted by $\boldsymbol{A}^{+/-}$. The superscript $+/-$ tells us one allele is inactivated. The second allele becomes inactivated at rate $\hat{u}_2$ to become the cell type $\boldsymbol{A}^{-/-}$. In addition, $\boldsymbol{A}^{+/+}$ cells can also receive mutations that trigger **CIN**. This happens at the rate u_c resulting in the cell type $\boldsymbol{A}^{+/+\,CIN}$. This kind of a cell can inactivate the first allele of the **TSG** with normal mutation rate u_1 to produce a cell with one inactivated allele (i.e. a $+/-$) which started from a *CIN* state. We denote these cells as $\boldsymbol{A}^{+/-\,CIN}$. We can also get a cell of type $\boldsymbol{A}^{+/+\,CIN}$ when a cell of type $\boldsymbol{A}^{+/-}$ receives a mutation which triggers **CIN**. We will assume this happens at the same rate u_c as before. The $\boldsymbol{A}^{+/-\,CIN}$ cell then rapidly undergoes **LOH** at rate $\hat{u}_3$ to produce cells of type $\boldsymbol{A}^{-/-\,CIN}$. Finally, $\boldsymbol{A}^{-/-}$ cells can experience **CIN** at rate u_c to generate $\boldsymbol{A}^{-/-\,CIN}$ cells. The first allele is inactivated by a point mutation. The rate at which this occurs is modeled by the rate u_1. We assume the mutations governed by the rates u_1 and u_c are **neutral**. This means that these rates do not depend on the size of the population N. The events governed by $\hat{u}_2$ and $\hat{u}_3$ give

J.K. Peterson, *Calculus for Cognitive Scientists*,
Cognitive Science and Technology, DOI 10.1007/978-981-287-877-9

what is called **selective advantage**. This means that the size of the population size does matter. Using these assumptions, we therefore model $\hat{u}_2$ and $\hat{u}_3$ as

$$\hat{u}_2 = N\, u_2$$

and

$$\hat{u}_3 = N\, u_3.$$

where u_2 and u_3 are neutral rates. The mathematical model is then setup as follows. Let

$X_0(t)$ is the probability a cell in cell type $A^{+/+}$ at time t.
$X_1(t)$ is the probability a cell in cell type $A^{+/-}$ at time t.
$X_2(t)$ is the probability a cell in cell type $A^{-/-}$ at time t.
$Y_0(t)$ is the probability a cell in cell type $A^{+/+CIN}$ at time t.
$Y_1(t)$ is the probability a cell in cell type $A^{+/-CIN}$ at time t.
$Y_2(t)$ is the probability a cell in cell type $A^{-/-CIN}$ at time t.

We can then derive rate equations to be

$$\begin{aligned}
X_0' &= -(u_1 + u_c)\, X_0 \\
X_1' &= u_1\, X_0 - (u_c + N\, u_2)\, X_1 \\
X_2' &= N\, u_2\, X_1 - u_c\, X_2 \\
Y_0' &= u_c\, X_0 - u_1\, Y_0 \\
Y_1' &= u_c\, X_1 + u_1\, Y_0 - N\, u_3\, Y_1 \\
Y_2' &= N\, u_3\, Y_1 + u_c\, X_2
\end{aligned}$$

We are interested in analyzing this model over a typical human life span of 100 years, p. 406.

Cauchy Fundamental Theorem of Calculus Let G be any antiderivative of the Riemann integrable function f on the interval $[a, b]$. Then $G(b) - G(a) = \int_a^b f(t)\, dt$, p. 129.

Characteristic Equation of a linear second order differential equation For a linear second order differential equation,

$$a\, u''(t) + b\, u'(t) + c\, u(t) = 0$$

assume that e^{rt} is a solution and try to find what values of r might work. We see for $u(t) = e^{rt}$, we find

$$\begin{aligned} 0 &= a\,u''(t) + b\,u'(t) \;+\; c\,u(t) \\ &= a\,r^2\,e^{rt} + b\,r\,e^{rt} + c\,e^{rt} \\ &= \left(a\,r^2 + b\,r + c\right) e^{rt}. \end{aligned}$$

Since e^{rt} can never be 0, we must have

$$0 = a\,r^2 + b\,r + c.$$

The roots of the quadratic equation above are the only values of r that will work as the solution e^{rt}. We call this quadratic equation the *characteristic equation* for this differential equation, p. 152.

Characteristic Equation of the linear system For a linear system

$$\begin{bmatrix} x'(t) \\ y'(t) \end{bmatrix} = A \begin{bmatrix} x(t) \\ y(t) \end{bmatrix}$$
$$\begin{bmatrix} x(0) \\ y(0) \end{bmatrix} = \begin{bmatrix} x_0 \\ y_0 \end{bmatrix}.$$

assume the solution is $\boldsymbol{V}\,e^{rt}$ for some nonzero vector $\boldsymbol{V}$. This implies that r and $\boldsymbol{V}$ must satisfy

$$\left(r\,I - A\right)\boldsymbol{V}\,e^{rt} = \begin{bmatrix} 0 \\ 0 \end{bmatrix}.$$

Since e^{rt} is never zero, we must have

$$\left(r\,I - A\right)\boldsymbol{V} = \begin{bmatrix} 0 \\ 0 \end{bmatrix}.$$

The only way we can get nonzero vectors $\boldsymbol{V}$ as solutions is to choose the values of r so that

$$\det\left(r\,I - A\right) = 0.$$

The second order polynomial we obtain is called the *characteristic equation* associated with this linear system. Its roots are called the *eigenvalues* of the system and any nonzero vector $\boldsymbol{V}$ associated with an eigenvalue r is an *eigenvector* for r, p. 178.

Complex number This is a number which has the form of $a + b\,\boldsymbol{i}$ where a and b are arbitrary real numbers and the letter i represents a very abstract concept: a number whose square $i^2 = -1$! We usually draw a complex number in a standard

x–y Cartesian plane with the y axis labeled as iy instead of the usual y. Then the number $5 + 4\,\boldsymbol{i}$ would be graphed just like the two dimensional coordinate $(4, 5)$, p. 141.

Continuity A function f is continuous at a point p if for all positive tolerances ϵ, there is a positive δ so that $\mid f(t) - f(p) \mid < \epsilon$ if t is in the domain of f and $\mid t - p \mid < \delta$. You should note continuity is something that is only defined at a point and so functions in general can have very few points of continuity. Another way of defining the continuity of f at the point p is to say the $\lim_{t \to p} f(t)$ exists and equals $f(p)$, p. 129.

D

Differentiability A function f is differentiable at a point p if there is a number L so that for all positive tolerances ϵ, there is a positive δ so that

$$\mid \frac{f(t) - f(p)}{t - p} - L \mid < \epsilon \quad \text{if } t \text{ is in the domain of } f \text{ and } \mid t - p \mid < \delta$$

You should note differentiability is something that is only defined at a point and so functions in general can have very few points of differentiability. Another way of defining the differentiability of f at the point p is to say the $\lim_{t \to p} \frac{f(t)-f(p)}{t-p}$ exists. At each point p where this limit exists, we can define a new function called the derivative of f at p. This is usually denoted by $f'(p)$ or $\frac{df}{dt}(p)$, p. 129.

E

Exponential growth Some biological systems can be modeled using the idea of exponential growth. This means the variable of interest, x, has growth proportional to its rate of change. Mathematically, this means $x' \propto r\,x$ for some proportionality constant r, p. 149.

F

Fundamental Theorem of Calculus Let f be Riemann Integrable on $[a, b]$. Then the function F defined on $[a, b]$ by $F(x) = \int_a^x f(t)\,dt$ satisfies

1. F is continuous on all of $[a, b]$
2. F is differentiable at each point x in $[a, b]$ where f is continuous and $F'(x) = f(x)$, p. 129.

H

Half life The amount of time it takes a substance x to lose half its original value under exponential decay. It is denoted by $t_{1/2}$ and can also be expressed as $t_{1/2} = \ln(2)/r$ where r is the decay rate in the differential equation $x'(t) = -r\,x(t)$, p. 149.

I

Infectious Disease Model Assume the total population we are studying is fixed at N individuals. This population is then divided into three separate pieces: we have individuals

- that are susceptible to becoming infected are called **Susceptible** and are labeled by the variable S. Hence, $S(t)$ is the number that are capable of becoming infected at time t.
- that can infect others. They are called **Infectious** and the number that are infectious at time t is given by $I(t)$.
- that have been removed from the general population. These are called **Removed** and their number at time t is labeled by $R(t)$.

We make a number of key assumptions about how these population pools interact.

- Individuals stop being *infectious* at a positive rate γ which is proportional to the number of individuals that are in the *infectious* pool. If an individual stops being infectious, this means this individual has been *removed* from the population. This could mean they have died, the infection has progressed to the point where they can no longer pass the infection on to others or they have been put into quarantine in a hospital so that further interactions with the general population is not possible. In all of these cases, these individuals are not infectious or can't cause infections and so they have been *removed* from the part of the population N which can be infected or is susceptible. Mathematically, this means we assume

$$I' = -\gamma\,I.$$

- Susceptible individuals are those capable of catching an infection. We model the interaction of infectious and susceptible individuals in the same way we handled the interaction of food fish and predator fish in the Predator–Prey model. We assume this interaction is proportional to the product of their population sizes: i.e. SI. We assume the rate of change of **Susceptible's** is proportional to this interaction with positive proportionality constant r. Hence, mathematically, we assume

$$S' = r\,S\,I.$$

We can then figure out the net rates of change of the three populations. The infectious populations gains at the rate $r\ S\ I$ and loses at the rate $\gamma\ I$. Hence

$$I' = r\ S\ I - \gamma\ I.$$

The net change of Susceptible's is that of simple decay. Susceptibles are lost at the rate $r\ S\ I$. Thus, we have

$$S' = -\ r\ S\ I.$$

Finally, the removed population increases at the same rate the infectious population decreases. We have

$$R' = \gamma\ I.$$

We also know that $R(t) + S(t) + I(t) = N$ for all time t because our population is constant. So only two of the three variables here are independent. We typically focus on the variables I and S for that reason. Our complete *Infectious Disease Model* is then

$$\begin{aligned} I' &= r\ S\ I - \gamma\ I \\ S' &= -r\ S\ I \\ I(0) &= I_0 \\ S(0) &= S_0. \end{aligned}$$

where we can compute $R(t)$ as $N - I(t) - S(t)$, p. 382.

Insulin regulation Glucose plays an important role in vertebrate metabolism because it is a source of energy. For each person, there is an optimal blood glucose concentration and large deviations from this leads to severe problems including death. Blood glucose levels are autoregulated via standard forward and backward interactions like we see in many biological systems. The blood glucose concentration is influenced by a variety of signaling molecules just like the protein creation rates can be. Here are some of them. The hormone that **decreases** blood glucose concentration is **insulin**. **Insulin** is a hormone secreted by the β cells of the pancreas. After we eat carbohydrates, our gastrointestinal tract sends a signal to the pancreas to secrete insulin. Also, the glucose in our blood directly stimulates the β cells to secrete insulin. We think insulin helps cells pull in the glucose needed for metabolic activity by attaching itself to membrane walls that are normally impenetrable. This attachment increases the ability of glucose to pass through to the inside of the cell where it can be used as *fuel*. So, if there is not enough insulin, cells don't have enough energy for their needs. There are other hormones that tend to **change** blood glucose concentrations also, such as glucagon, epinephrine, glucocorticoids, thyroxin and somatotropin. Net hormone concentration is the sum of insulin plus the others and we let $\boldsymbol{H}$ denote the net hormone concentration. At normal conditions, call this concentration $\boldsymbol{H_0}$. Under close to normal conditions, the interaction of the one hormone **insulin** with blood glucose completely dominates the net hormonal activity; so normal blood sugar levels primarily depend on insulin–glucose interactions. Hence, if insulin increases from normal levels, it increases net hormonal concentration to $\boldsymbol{H_0} + \boldsymbol{\Delta H}$ and decreases glucose blood concentration. On the other hand, if other hormones such as cortisol increased from base levels, this will make blood glucose levels go up. Since insulin dominates all activity at normal conditions, we can think of this increase in cortisol as a decrease in insulin with a resulting drop in blood glucose levels. A decrease in insulin from normal levels corresponds to a drop in net hormone concentration to $\boldsymbol{H_0} - \boldsymbol{\Delta H}$. Now let $\boldsymbol{G}$ denote blood glucose level. Hence, in our model an increase in $\boldsymbol{H}$ means a drop in $\boldsymbol{G}$ and a decrease in $\boldsymbol{H}$ means an increase in $\boldsymbol{G}$! Note our lumping of all the hormone activity into a single net activity is very much like how we modeled food fish and predator fish in the predator–prey model. We describe the model as

$$\begin{aligned} \boldsymbol{G}'(t) &= F_1(\boldsymbol{G}, \boldsymbol{H}) + \boldsymbol{J}(t) \\ \boldsymbol{H}'(t) &= F_2(\boldsymbol{G}, \boldsymbol{H}) \end{aligned}$$

where the function $\boldsymbol{J}$ is the external rate at which blood glucose concentration is being increased in a glucose tolerance test. There are two nonlinear interaction functions F_1 and F_2 because we know $\boldsymbol{G}$ and $\boldsymbol{H}$ have complicated interactions. Let's assume $\boldsymbol{G}$ and $\boldsymbol{H}$ have achieved optimal values $\boldsymbol{G_0}$ and $\boldsymbol{H_0}$ by the time the fasting patient has arrived at the hospital. Hence, we don't expect to have any contribution to $\boldsymbol{G}'(0)$ and $\boldsymbol{H}'(0)$; i.e. $F_1(\boldsymbol{G_0}, \boldsymbol{H_0}) = 0$ and $F_2(\boldsymbol{G_0}, \boldsymbol{H_0}) = 0$. We are interested in the deviation of $\boldsymbol{G}$ and $\boldsymbol{H}$ from their optimal values $\boldsymbol{G_0}$ and $\boldsymbol{H_0}$, so let $\boldsymbol{g} = \boldsymbol{G} - \boldsymbol{G_0}$ and $\boldsymbol{h} = \boldsymbol{H} - \boldsymbol{H_0}$. We can then write $\boldsymbol{G} = \boldsymbol{G_0} + \boldsymbol{g}$ and $\boldsymbol{H} = \boldsymbol{H_0} + \boldsymbol{h}$. The model can then be rewritten as

$$\begin{aligned} \boldsymbol{g}'(t) &= F_1(\boldsymbol{G_0} + \boldsymbol{g}, \boldsymbol{H_0} + \boldsymbol{h}) + \boldsymbol{J}(t) \\ \boldsymbol{h}'(t) &= F_2(\boldsymbol{G_0} + \boldsymbol{g}, \boldsymbol{H_0} + \boldsymbol{h}) \end{aligned}$$

and we can then approximate these dynamics using tangent plane approximations to F_1 and F_2 giving

$$\begin{aligned} \boldsymbol{g}'(t) &\approx \frac{\partial F_1}{\partial \boldsymbol{g}}(\boldsymbol{G_0}, \boldsymbol{H_0})\, \boldsymbol{g} + \frac{\partial F_1}{\partial \boldsymbol{h}}(\boldsymbol{G_0}, \boldsymbol{H_0})\, \boldsymbol{h} + \boldsymbol{J}(t) \\ \boldsymbol{h}'(t) &\approx \frac{\partial F_2}{\partial \boldsymbol{g}}(\boldsymbol{G_0}, \boldsymbol{H_0})\, \boldsymbol{g} + \frac{\partial F_2}{\partial \boldsymbol{h}}(\boldsymbol{G_0}, \boldsymbol{H_0})\, \boldsymbol{h} \end{aligned}$$

It is this linearized system of equations we can analyze to give some insight into how to interpret the results of a glucose tolerance test, p. 475.

L

Linear Second Order Differential Equations These have the general form

$$a\,u''(t) + b\,u'(t) + c\,u(t) = 0$$
$$u(0) = u_0, \quad u'(0) = u_1$$

where we assume a is not zero, p. 149.

Linear Systems of differential equations These are systems of differential equations of the form

$$\begin{aligned} x'(t) &= a\,x(t) + b\,y(t) \\ y'(t) &= c\,x(t) + d\,y(t) \\ x(0) &= x_0 \\ y(0) &= y_0 \end{aligned}$$

which can be written in matrix–vector form as

$$\begin{bmatrix} x'(t) \\ y'(t) \end{bmatrix} = \begin{bmatrix} a & b \\ c & d \end{bmatrix} \begin{bmatrix} x(t) \\ y(t) \end{bmatrix}$$
$$\begin{bmatrix} x(0) \\ y(0) \end{bmatrix} = \begin{bmatrix} x_0 \\ y_0 \end{bmatrix}.$$

We typically call the coefficient matrix A so that the system is, p. 171.

$$\begin{bmatrix} x'(t) \\ y'(t) \end{bmatrix} = A \begin{bmatrix} x(t) \\ y(t) \end{bmatrix}$$
$$\begin{bmatrix} x(0) \\ y(0) \end{bmatrix} = \begin{bmatrix} x_0 \\ y_0 \end{bmatrix},$$

P

Predator–Prey This is a classical model developed to model the behavior of the interacting population of a food source and its predator. Let the variable $x(t)$ denote the population of food and $y(t)$, the population of predators at time t. This is, of course, a *coarse* model. For example, food fish could be divided into categories like halibut, mackerel with a separate variable for each and the predators could be divided into different classes like sharks, squids and so forth. Hence, instead of dozens of variables for both the food and predator population, everything is lumped together. We then make the following assumptions:

1. The food population grows exponentially. Letting x'_g denote the growth rate of the food, we have

$$x'_g = a\,x$$

for some positive constant a.

2. The number of contacts per unit time between predators and prey is proportional to the product of their populations. We assume the food are eaten by the predators at a rate proportional to this contact rate. Letting the decay rate of the food be denoted by x'_d, we see

$$x'_d = -b\,x\,y$$

for some positive constant b.

Thus, the net rate of change of food is $x' = x'_g + x'_d$ giving

$$x' = a\,x - b\,x\,y.$$

for some positive constants a and b. We then make assumptions about the predators as well.

1. Predators naturally die following an exponential decay; letting this decay rate be given by y'_d, we have

$$y'_d = -c\,y$$

for some positive constant c.

2. We assume the predators grow proportional to how much they eat. In turn, how much they eat is assumed to be proportional to the rate of contact between food and predator fish. We model the contact rate just like before and let y'_g be the growth rate of the predators. We find

$$y'_g = d\,x\,y$$

for some positive constant d.

Thus, the net rate of change of predators is $y' = y'_g + y'_d$ giving

$$y' = -c\,y + d\,x\,y.$$

for some positive constants c and d. The full model is thus, p. 292.

$$
\begin{aligned}
x' &= \ a\, x - b\, x\, y \\
y' &= -c\, y + d\, x\, y \\
x(0) &= x_0 \\
y(0) &= y_0,
\end{aligned}
$$

Predator–Prey Self Interaction The original Predator–Prey model does not include self-interaction terms. These are terms that model how the food and predator populations interaction with themselves. We can model these effects by assuming their magnitude is proportional to the interaction. Mathematically, we assume these are both *decay* terms giving us

$$
\begin{aligned}
x'_{self} &= -e\, x\, x \\
y'_{self} &= -f\, y\, y.
\end{aligned}
$$

for positive constants e and f. We are thus led to the new self-interaction model given below, p. 355:

$$
\begin{aligned}
x'(t) &= a\, x(t) \ - \ b\, x(t)\, y(t) \ - \ e\, x(t)^2 \\
y'(t) &= -c\, y(t) \ + \ d\, x(t)\, y(t) \ - \ f\, y(t)^2,
\end{aligned}
$$

Predator–Prey with fishing If we add fishing at the rate r to the Predator–Prey model, we obtain, p. 338:

$$
\begin{aligned}
x'(t) &= a\, x(t) \ - \ b\, x(t)\, y(t) \ - \ r\, x(t) \\
y'(t) &= -c\, y(t) \ + \ d\, x(t)\, y(t) \ - \ r\, y(t),
\end{aligned}
$$

Primitives The primitive of a function f is any function F which is differentiable and satisfies $F'(t) = f(t)$ at all points in the domain of f, p. 129.

Protein Modeling The gene ***Y*** is a string of nucleotides (**A**, **C**, **T** and **G**) with a special starting string in front of it called the **promoter**. The nucleotides in the gene ***Y*** are *read* three at a time to create the amino acids which form the protein ***Y**** corresponding to the gene. The process is this: a special **RNA** polymerase, **RNAp,** which is a complex of several proteins, binds to the promoter region. Once **RNAp** binds to the promoter, messenger **RNA**, **mRNA**, is synthesized that corresponds to the specific nucleotide triplets in the gene ***Y***. The process of forming this **mRNA** is called **transcription**. Once the **mRNA** is formed, the protein ***Y**** is then made. The protein creation process is typically regulated. A single **regulator** works like this. An activator called ***X*** is a protein which increases the rate of **mRNA** creation when ti binds to the promoter. The activator ***X*** switches between and active and inactive version due to a signal S_X. We let the active form be denoted by ***X****. If ***X**** binds in front of the promoter, **mRNA** creation increases implying an increase in the creation of the protein ***Y**** also. Once the signal S_X appears, ***X*** rapidly transitions to its state ***X****, binds with the front of the promoter and protein ***Y****

begins to accumulate. We let β denote the rate of protein accumulation which is **constant** once the signal S_X begins. However, proteins also degrade due to two processes:

- proteins are destroyed by other proteins in the cell. Call this rate of destruction α_{des}.
- the concentration of protein in the cell goes down because the cell grows and therefore its volume increases. Protein is usually measured as a concentration and the concentration goes down as the volume goes up. Call this rate α_{dil}—the *dil* is for *dilation*.

The net or total *loss of protein* is called α and hence

$$\alpha = \alpha_{des} + \alpha_{dil}$$

The net rate of change of the protein concentration is then our familiar model

$$\frac{dY^*}{dt} = \underbrace{\beta}_{\text{constant growth}} - \underbrace{\alpha\, Y^*}_{\text{loss term}}$$

We usually do not make a distinction between the gene Y and its transcribed protein Y^*. We usually treat the letters Y and Y^* as the same even though it is not completely correct. Hence, we just write as our model

$$Y' = \beta - \alpha\, Y$$
$$Y(0) = Y_0$$

and then solve it using the integrating factor method even though, strictly speaking, Y is the gene!, p. 149.

R

Response Time In a protein synthesis model

$$y'(t) = -\alpha\, y(t) + \beta$$
$$y(0) = y_0$$

the time it takes the solution to go from its initial concentration y_0 to a value halfway between the initial amount and the steady state value is called the **response time**. It is denoted by t_r and $t_r = \ln(2)/\alpha$ so it is functionally the same as the **half life** in an exponential decay model, p. 149.

Riemann Integral If a function on the finite interval $[a, b]$ is bounded, we can define a special limit which, if it exists, is called the Riemann Integral of the function on the interval $[a, b]$. Select a finite number of points from the interval $[a, b]$, $\{t_0, t_1, \ldots, t_{n-1}, t_n\}$. We don't know how many points there are, so a different selection from the interval would possibly gives us more or less points. But for convenience, we will just call the last point t_n and the first point t_0. These points are not arbitrary—t_0 is always a, t_n is always b and they are ordered like this:

$$t_0 = a < t_1 < t_2 < \cdots < t_{n-1} < t_n = b$$

The collection of points from the interval $[a, b]$ is called a Partition of $[a, b]$ and is denoted by some letter—here we will use the letter **P**. So if we say P is a partition of $[a, b]$, we know it will have $n + 1$ points in it, they will be labeled from t_0 to t_n and they will be ordered left to right with strict inequalities. But, we will not know what value the positive integer n actually is. The simplest Partition P is the two point partition $\{a, b\}$. Note these things also:

1. Each partition of $n + 1$ points determines n subintervals of $[a, b]$
2. The lengths of these subintervals always adds up to the length of $[a, b]$ itself, $b - a$.
3. These subintervals can be represented as

$$\{[t_0, t_1], [t_1, t_2], \ldots, [t_{n-1}, t_n]\}$$

   or more abstractly as $[t_i, t_{i+1}]$ where the index i ranges from 0 to $n - 1$.
4. The length of each subinterval is $t_{i+1} - t_i$ for the indices i in the range 0 to $n - 1$.

Now from each subinterval $[t_i, t_{i+1}]$ determined by the Partition P, select any point you want and call it s_i. This will give us the points s_0 from $[t_0, t_1]$, s_1 from $[t_1, t_2]$ and so on up to the last point, s_{n-1} from $[t_{n-1}, t_n]$. At each of these points, we can evaluate the function f to get the value $f(s_j)$. Call these points an **Evaluation Set** for the partition P. Let's denote such an evaluation set by the letter E. If the function f was nice enough to be positive always and continuous, then the product $f(s_i) \times (t_{i+1} - t_i)$ can be interpreted as the area of a rectangle; in general, though, these products are not areas. Then, if we add up all these products, we get a sum which is useful enough to be given a special name: the Riemann sum for the function f associated with the Partition P and our choice of evaluation set $E = \{s_0, \ldots, s_{n-1}\}$. This sum is represented by the symbol $S(f, P, E)$ where the things inside the parenthesis are there to remind us that this sum depends on our choice of the function f, the partition P and the evaluation set E. The Riemann sum is normally written as

$$S(f, P, E) = \sum_{i \in P} f(s_i)\,(t_{i+1} - t_i)$$

and we just remember that the choice of P will determine the size of n. Each partition P has a maximum subinterval length—let's use the symbol $\| P \|$ to denote this length. We read the symbol $\| P \|$ as the **norm** of P. Each partition P and evaluation set E determines the number $S(f, P, E)$ by a simple calculation. So if we took a collection of partitions P_1, P_2 and so on with associated evaluation sets E_1, E_2 etc., we would construct a sequence of real numbers $\{S(f, P_1, E_1), S(f, P_2, E_2), \ldots, S(f, P_n, E_n), \ldots\}$. Let's assume the norm of the partition P_n gets smaller all the time; i.e. $\lim_{n \to \infty} \| P_n \| = 0$. We could then ask if this sequence of numbers converges to something. What if the sequence of Riemann sums we construct above converged to the same number I no matter what sequence of partitions whose norm goes to zero and associated evaluation sets we chose? Then, we would have that the value of this limit is *independent* of the choices above. This is what we mean by the **Riemann Integral** of f on the interval $[a, b]$. If there is a number I so that

$$\lim_{n \to \infty} S(f, P_n, E_n) = I$$

no matter what sequence of partitions $\{P_n\}$ with associated sequence of evaluation sets $\{E_n\}$ we choose as long as $\lim_{n \to \infty} \| P_n \| = 0$, we say that the Riemann Integral of f on $[a, b]$ exists and equals the value I. The value I is dependent on the choice of f and interval $[a, b]$. So we often denote this value by $I(f, [a, b])$ or more simply as, $I(f, a, b)$. Historically, the idea of the Riemann integral was developed using area approximation as an application, so the summing nature of the Riemann Sum was denoted by the 16th century *letter S* which resembled an elongated or stretched letter S which looked like what we call the integral sign $\int$. Hence, the common notation for the Riemann Integral of f on $[a, b]$, when this value exists, is $\int_a^b f$. We usually want to remember what the independent variable of f is also and we want to remind ourselves that this value is obtained as we let the norm of the partitions go to zero. The symbol dt for the independent variable t is used as a reminder that $t_{i+1} - t_i$ is going to zero as the norm of the partitions goes to zero. So it has been very convenient to add to the symbol $\int_a^b f$ this information and use the augmented symbol $\int_a^b f(t)\, dt$ instead. Hence, if the independent variable was x instead of t, we would use $\int_a^b f(x)\, dx$. Since for a function f, the name we give to the independent variable is a matter of personal choice, we see that the choice of variable name we use in the symbol $\int_a^b f(t)\, dt$ is very arbitrary. Hence, it is common to refer to the independent variable we use in the symbol $\int_a^b f(t)\, dt$ as the dummy variable of integration, p. 129.

S

Separation of Variables Method A common PDE model is the general cable model which is given below is fairly abstract form.

$$\beta^2 \frac{\partial^2 \Phi}{\partial x^2} - \Phi - \alpha \frac{\partial \Phi}{\partial t} = 0, \text{ for } 0 \leq x \leq L, \ t \geq 0,$$
$$\frac{\partial \Phi}{\partial x}(0, t) = 0,$$
$$\frac{\partial \Phi}{\partial x}(L, t) = 0,$$
$$\Phi(x, 0) = f(x).$$

The domain is the usual half infinite $[0, L] \times [0, \infty)$ where the spatial part of the domain corresponds to the length of the dendritic cable in an excitable nerve cell. We won't worry too much about the details of where this model comes from as we will discuss that in another volume. The boundary conditions $u_x(0, t) = 0$ and $u_x(L, t) = 0$ are called *Neumann Boundary conditions*. The conditions $u(0, t) = 0$ and $u(L, t) = 0$ are known as *Dirichlet Boundary conditions*. One way to find the solution is to assume we can separate the variables so that we can write $\Phi(x, t) = u(x)w(t)$. We assume a solution of the form $\Phi(x, t) = u(x)\,w(t)$ and compute the needed partials. This leads to a the new equation

$$\beta^2 \frac{d^2 u}{dx^2}\, w(t) - u(x)w(t) - \alpha u(x) \frac{dw}{dt} = 0.$$

Rewriting, we find for all x and t, we must have

$$w(t)\left(\beta^2 \frac{d^2 u}{dx^2} - u(x)\right) = \alpha u(x) \frac{dw}{dt}.$$

This tells us

$$\frac{\beta^2 \frac{d^2 u}{dx^2} - u(x)}{u(x)} = \frac{\alpha \frac{dw}{dt}}{w(t)}, \quad 0 \leq x \leq L, \ t > 0.$$

The only way this can be true is if both the left and right hand side are equal to a constant that is usually called the **separation constant** Θ. This leads to the decoupled equations

$$\alpha \frac{dw}{dt} = \Theta\, w(t), \ t > 0,$$
$$\beta^2 \frac{d^2 u}{dx^2} = (1 + \Theta)\, u(x), \ 0 \leq x \leq L,$$

We also have boundary conditions.

$$\frac{du}{dx}(0) = 0$$
$$\frac{du}{dx}(L) = 0.$$

This gives us a second order ODE to solve in x and a first order ODE to solve in t. We have a lot of discussion about this in the text which you should study. In general, we find there is an infinite family of solutions that solve these coupled ODE models which we can label $u_n(x)$ and $w_n(t)$. Thus, any finite combination $\Phi_n(x, t) = \sum_{n=0}^{N} a_n u_n(x) w_n(t)$ will solve these ODE models, but we are still left with satisfying the last condition that $\Phi(x, 0) = f(x)$. We do this by finding a series solution. We can show that the data function f can be written as a series $f(x) = \sum_{n=0}^{\infty} b_n u_n(x)$ for a set of constants $\{b_0, b_1, \ldots\}$ and we can also show that the series $\Phi(x, t) = \sum_{n=0}^{\infty} a_n u_n(x) w_n(t)$ solves the last boundary condition $\Phi(x, 0) = \sum_{n=0}^{\infty} a_n u_n(x) w_n(0) = f(x)$ as long as we choose $a_n = b_n$ for all n. The idea of a series and the mathematical machinery associated with that takes a while to explain, so Chap. 16 is devoted to that, p. 495.

T

Tangent plane error We can characterize the error made when a function of two variables is replaced by its tangent plane at a point better if we have access to the second order partial derivatives of f. The value of f at the point $(x_0 + \Delta x, y_0 + \Delta y)$ can be expresses as follows:

$$\begin{aligned} f(x_0 + \Delta x, y_0 + \Delta y) &= f(x_0, y_0) + f_x(x_0, y_0)\Delta x + f_y(x_0, y_0)\Delta y \\ &\quad + \frac{1}{2} \begin{bmatrix} \Delta x \\ \Delta y \end{bmatrix}^T \boldsymbol{H}(x_0 + c\Delta x, y_0 + c\Delta y) \begin{bmatrix} \Delta x \\ \Delta y \end{bmatrix} \end{aligned}$$

where c between 0 and 1 so that the tangent plane error is given by, p. 435

$$E(x_0, y_0, \Delta x, \Delta y) = \frac{1}{2} \begin{bmatrix} \Delta x \\ \Delta y \end{bmatrix}^T \boldsymbol{H}(x_0 + c\Delta x, y_0 + c\Delta y) \begin{bmatrix} \Delta x \\ \Delta y \end{bmatrix},$$

Index

J.K. Peterson, *Calculus for Cognitive Scientists*,
Cognitive Science and Technology, DOI 10.1007/978-981-287-877-9

Printed in the United States
By Bookmasters